First-Order Partial Differential Equations

Volume 1: Theory and Application of Single Equations

Hyun-Ku Rhee
Seoul National University
Seoul, Korea

Rutherford Aris
University of Minnesota
Minneapolis, Minnesota

Neal R. Amundson
University of Houston
Houston, Texas

DOVER PUBLICATIONS, INC.
Mineola, New York

Published in Canada by General Publishing Company, Ltd., 895 Don Mills Road, 400-2 Park Centre, Toronto, Ontario M3C 1W3.

Published in the United Kingdom by David & Charles, Brunel House, Forde Close, Newton Abbot, Devon TQ12 4PU.

Bibliographical Note

This Dover edition, first published in 2001, is an unabridged reprint of the work originally published by Prentice-Hall, Inc., Englewood Cliffs, N.J., in 1986.

Library of Congress Cataloging-in-Publication Data

Rhee, Hyun-Ku, 1939–

Theory and application of single equations / Hyun-Ku Rhee, Rutherford Aris, Neal R. Amundson.

p. cm. — (First-order partial differential equations ; v. 1)

Originally published: Englewood Cliffs, N.J. : Prentice-Hall, c1986, in series: Prentice-Hall international series in the physical and chemical engineering sciences.

Includes index.

ISBN 0-486-41993-2 (pbk.)

1. Differential equations, Partial. I. Aris, Rutherford. II. Amundson, Neal Russell, 1916– III. Title.

QA374 .R47 2001 vol. 1
515'.353 s—dc21
[515'.353]

2001041182

Manufactured in the United States of America
Dover Publications, Inc., 31 East 2nd Street, Mineola, N.Y. 11501

Dedication

This book is dedicated to the enormous number of excellent students from Seoul National University who have come to the United States for graduate study in chemical engineering. They have set a standard that will not be surpassed.

Contents

Preface

In the preface to the earlier version of this work we said:

> Very frequently the subject of first-order partial differential equations receives rather short shrift in the courses an engineer is likely to attend, as well as in books he may use. It is disposed of in a chapter or two as an hors d'oeuvres before the three-course meal of the second order equations of mathematical physics. There is some justification for this, as, indeed, there are some exceptions to the custom, but it is difficult to find an extended treatment of first-order equations and, particularly, a treatment that includes the wide variety of applications and is accessible to those who are not mathematicians. This is what we have attempted to provide in the present volume, for though it bears the rubric of "Mathematical Methods in Chemical Engineering," its examples will be found to have been drawn from a considerable range of scientific and engineering disciplines. The first-order equation arises naturally in the model of any situation that is dominated by convective rather than dissipative or dispersive phenomena, and we have discussed rather fully the craft of constructing such models. The method of characteristics then becomes the natural tool by which the structure of the solution and the way in which it is generated can be seen and felt, and the differences between linear, quasi-linear, and nonlinear equations can be appreciated. Besides giving the traditional methods of the Laplace transform and the connection with the calculus of variations, we have tried to introduce some of the ideas of weak solutions and give the engineer an entree into the work of Lax and others, which at first sight he may find a little forbidding. . . .

That was written in 1972 and, in truth, the situation is not substantially different in 1985, although some small changes have occurred. The references at the end of each chapter will attest to the fact that some progress has been made. Pure mathematicians have discovered the fertile ground of the field, and we hope we have included at least a sample of their work so that the interested reader may dig further.

This is to be the first volume of two, and some remarks are in order. This volume will include only the theory and application of single first-order equations, while the second will be concerned with coupled systems of first-order equations with primary application to the field of chromatographic separation. The engineering scientist is well aware that the equations of fixed- and countercurrent moving-bed adsorption systems are the same in form, if not in detail, as a variety of other fixed- and moving-bed problems.

It is appropriate here to mention where we hope our book will be used. Engineering students, particularly those with a mathematical inclination, attend courses on partial differential equations, some at the senior level, but most while first-year graduate students. Those who do so would profit by spending some time on first-order partial differential equations. Those whose primary interest is in transport of energy and mass should probably indulge somewhat in the first-order equation since it is at the root of convective transport, because for many systems, as indicated previously, the dispersive effects are small, and the convective model is the simplest nontrivial one, the solutions for which are much easier to obtain, at the same time, elucidating the basic structure of the solution space. Finally, we hope that mathematicians might be interested in it as a source of applications. Those parts of the book involving weak solutions, Chapter 7, are mathematically sophisticated and could be omitted in an engineering course, while Chapter 8 may also be trying for that audience. Aside from these, however, the understanding of the solution methods for first-order single nonlinear partial differential equations is not technically difficult and demands little more than a sound understanding of the calculus and elementary ordinary differential equations. Many universities have applied mathematics offerings for first-year graduate students in engineering, and it is to these that we address our efforts.

A substantial portion of the book has been used at various times as a basis for lectures at Seoul National University, the University of Minnesota, and at the University of Houston in courses in applied mathematics in the chemical engineering departments. Thesis students in all three departments have used the material in their researches, largely as a starting point for more advanced endeavors, although the forthcoming second volume is more pertinent in this regard.

Some chapters from the earlier version have been completely rewritten, some remain almost untouched, while some have been omitted and will occur in the second volume. One will discover immediately that the authorship has been expanded, and these volumes would never have been completed without the generosity of Seoul National University and the University of Houston, who granted leaves and stipends for one of us (H.-K. R.) to spend almost two years on the writing. Funds were also generously supplied by the University of Houston to cover the preparation of the manuscript. All of the art work from the previous version has been redone.

Through the years we have received the encouragement of numerous people who expressed wonderment that we had not reintroduced the former book into circulation in some form. We hope they are not disappointed with the result.

Hyun-Ku Rhee
Rutherford Aris
Neal R. Amundson

First-Order
Partial Differential
Equations

Mathematical Preliminaries

0

0.1 Functions and Their Derivatives

In analytic geometry it is convenient to think of the function $y = f(x)$ as a curve in the (x, y)-plane. The curve is said to be continuous at a point (x_0, y_0) if given an $\varepsilon > 0$ there exists a $\delta > 0$ such that

$$|f(x) - f(x_0)| < \varepsilon$$

whenever

$$|x - x_0| < \delta$$

The curve is said to be uniformly continuous in the interval (a, b) if δ does not depend on x. The function $f(x)$ is said to be single valued at x_0 if only one value of y_0 is defined by $y_0 = f(x_0)$. The function $y = f(x)$ has a derivative at x_0 if the limit

$$\lim_{h \to 0} \frac{f(x_0 + h) - f(x_0)}{h}$$

exists. It is denoted by $f'(x_0)$, $(df/dx)_{x=x_0}$, dy/dx, or y', each evaluated at x_0. The geometric interpretation of the derivative is the slope of the curve $y = f(x)$ or the slope of the tangent to the curve at the point x_0.

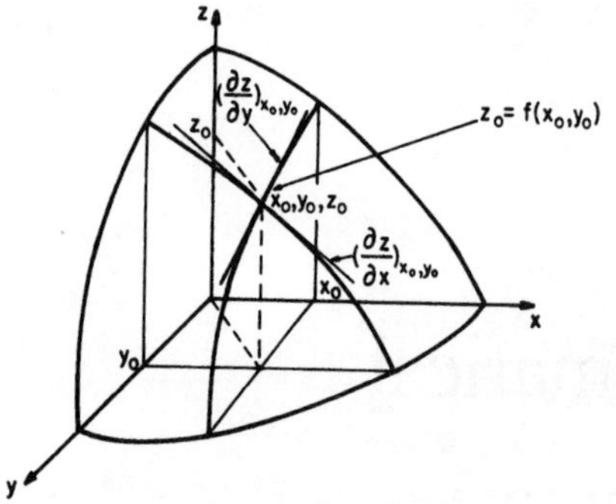

Figure 0.1

Similarly, for functions of several variables, say of two in particular, $z = f(x, y)$, the function $f(x, y)$ is said to be single valued at (x_0, y_0) if only one value of z_0 is defined at that point by $z_0 = f(x_0, y_0)$. A function of two variables is said to be continuous at a point (x_0, y_0) if given an $\varepsilon > 0$ there exists a $\delta > 0$ such that

$$|f(x, y) - f(x_0, y_0)| < \varepsilon$$

whenever

$$|x - x_0| < \delta, \qquad |y - y_0| < \delta$$

We think of $z = f(x, y)$ as defining a surface in the space of x, y, and z. For example, $ax + by + c = z$ is a plane, whereas $z = 2x^2 + 3y^2$ is a paraboloid defined for all x and y and for $z > 0$. Any plane parallel to the (x, y)-plane cuts the paraboloid in an ellipse; planes passing through the z-axis give parabolas.

The partial derivatives of $f(x, y) = z$ at (x_0, y_0) are defined as

$$\left(\frac{\partial f}{\partial x}\right)_{x_0,y_0} = \lim_{h\to 0}\frac{f(x_0 + h, y_0) - f(x_0, y_0)}{h} = (f_x)_{\substack{x=x_0\\ y=y_0}} \tag{0.1.1}$$

$$\left(\frac{\partial f}{\partial y}\right)_{x_0,y_0} = \lim_{k\to 0}\frac{f(x_0, y_0 + k) - f(x_0, y_0)}{k} = (f_y)_{\substack{x=x_0\\ y=y_0}} \tag{0.1.2}$$

These are the rates of change of $f(x, y)$ in planes parallel to the (x, z)-plane and (y, z)-plane, respectively, and are therefore the slopes of the curves of intersection of these planes with the surface at (x_0, y_0), as shown in Fig. 0.1.

In general, if $u = f(x_1, \ldots, x_n)$ is a function of n variables, the partial derivative $\partial f/\partial x_i$ is defined by

$$\frac{\partial f}{\partial x_i} = \lim_{h_i\to 0}\frac{1}{h_i}[f(x_1, x_2, \ldots, x_i + h_i, \ldots, x_n) - f(x_1, \ldots, x_i, \ldots, x_n)] \tag{0.1.3}$$

As before, a change is made in the ith coordinate only, and the limiting value of the quotient, when it exists, defines the partial derivative.

0.2 Functions of Functions and Their Derivatives

Given a function $u = f(x_1, x_2, x_3, \ldots, x_n)$ with the independent variables $x_1, x_2, \ldots, x_n$, the total differential of u is defined as

$$du = \frac{\partial f}{\partial x_1} dx_1 + \frac{\partial f}{\partial x_2} dx_2 + \cdots + \frac{\partial f}{\partial x_n} dx_n \tag{0.2.1}$$

If we suppose that each x_i is a function of m variables t_j, $j = 1, \ldots, m$,

$$x_i = x_i(t_1, t_2, \ldots, t_m) \tag{0.2.2}$$

then

$$dx_i = \frac{\partial x_i}{\partial t_1} dt_1 + \frac{\partial x_i}{\partial t_2} dt_2 + \cdots + \frac{\partial x_i}{\partial t_m} dt_m \tag{0.2.3}$$

Thus

$$\begin{aligned} du &= \frac{\partial f}{\partial x_1} \sum_{j=1}^{m} \frac{\partial x_1}{\partial t_j} dt_j + \frac{\partial f}{\partial x_2} \sum_{j=1}^{m} \frac{\partial x_2}{\partial t_j} dt_j + \cdots + \frac{\partial f}{\partial x_n} \sum_{j=1}^{m} \frac{\partial x_n}{\partial t_j} dt_j \\ &= \left(\frac{\partial f}{\partial x_1}\frac{\partial x_1}{\partial t_1} + \frac{\partial f}{\partial x_2}\frac{\partial x_2}{\partial t_1} + \cdots + \frac{\partial f}{\partial x_n}\frac{\partial x_n}{\partial t_1}\right) dt_1 \\ &\quad + \left(\frac{\partial f}{\partial x_1}\frac{\partial x_1}{\partial t_2} + \frac{\partial f}{\partial x_2}\frac{\partial x_2}{\partial t_2} + \cdots + \frac{\partial f}{\partial x_n}\frac{\partial x_n}{\partial t_2}\right) dt_2 \\ &\quad \vdots \\ &\quad + \left(\frac{\partial f}{\partial x_1}\frac{\partial x_1}{\partial t_m} + \frac{\partial f}{\partial x_2}\frac{\partial x_2}{\partial t_m} + \cdots + \frac{\partial f}{\partial x_n}\frac{\partial x_n}{\partial t_m}\right) dt_m \\ &= \sum_{j=1}^{m} \left(\sum_{i=1}^{n} \frac{\partial f}{\partial x_i}\frac{\partial x_i}{\partial t_j}\right) dt_j \end{aligned} \tag{0.2.4}$$

is the total differential of u considered as a function of m variables t_j.

If the partial derivative of u with respect to t_j is desired, then, since u can be regarded as a function of the t_j, its total derivative is

$$du = \sum_{j=1}^{m} \frac{\partial u}{\partial t_j} dt_j$$

and we have, by comparison with Eq. (0.2.4),

$$\frac{\partial u}{\partial t_j} = \frac{\partial f}{\partial x_1}\frac{\partial x_1}{\partial t_j} + \frac{\partial f}{\partial x_2}\frac{\partial x_2}{\partial t_j} + \cdots + \frac{\partial f}{\partial x_n}\frac{\partial x_n}{\partial t_j} \tag{0.2.5}$$

This is the familiar chain rule of the differential calculus. If all the x_i are functions of a single variable t, then

$$\frac{du}{dt} = \frac{\partial f}{\partial x_1}\frac{dx_1}{dt} + \frac{\partial f}{\partial x_2}\frac{dx_2}{dt} + \cdots + \frac{\partial f}{\partial x_n}\frac{dx_n}{dt} \tag{0.2.6}$$

As an example, let us consider the differential equation

$$V\frac{\partial u}{\partial z} + \frac{\partial u}{\partial t} = 0 \tag{0.2.7}$$

where V is a constant, and let us make the transformation to new independent variables

$$y = t - \frac{z}{V}, \qquad x = z \tag{0.2.8}$$

Then

$$\begin{aligned}\frac{\partial u}{\partial z} &= \frac{\partial u}{\partial y}\frac{\partial y}{\partial z} + \frac{\partial u}{\partial x}\frac{\partial x}{\partial z}\\ &= \frac{\partial u}{\partial y}\left(-\frac{1}{V}\right) + \frac{\partial u}{\partial x}(1)\end{aligned}$$

and

$$\begin{aligned}\frac{\partial u}{\partial t} &= \frac{\partial u}{\partial y}\frac{\partial y}{\partial t} + \frac{\partial u}{\partial x}\frac{\partial x}{\partial t}\\ &= \frac{\partial u}{\partial y}(1) + \frac{\partial u}{\partial x}(0)\end{aligned}$$

Hence the equation becomes

$$V\frac{\partial u}{\partial x} = 0 \tag{0.2.9}$$

Again, if we consider the equation

$$\frac{\partial^2 v}{\partial x^2} = \alpha^2\frac{\partial^2 v}{\partial \theta^2} \tag{0.2.10}$$

we may ask whether there are linear relations

$$x_1 = a_1 x + b_1\theta$$

$$x_2 = a_2 x + b_2\theta$$

that will simplify the partial differential equation to $\partial^2 v/\partial x_1 \partial x_2 = 0$. But

$$\frac{\partial v}{\partial x} = \frac{\partial v}{\partial x_1}a_1 + \frac{\partial v}{\partial x_2}a_2$$

$$\frac{\partial^2 v}{\partial x^2} = \frac{\partial^2 v}{\partial x_1^2}a_1^2 + 2\frac{\partial^2 v}{\partial x_1\,\partial x_2}a_1 a_2 + \frac{\partial^2 v}{\partial x_2^2}a_2^2$$

and $$\frac{\partial^2 v}{\partial \theta^2} = \frac{\partial^2 v}{\partial x_1^2} b_1^2 + 2 \frac{\partial^2 v}{\partial x_1 \, \partial x_2} b_1 b_2 + \frac{\partial^2 v}{\partial x_2^2} b_2^2$$

Substituting in Eq. (0.2.10),

$$\frac{\partial^2 v}{\partial x_1^2} a_1^2 + 2 \frac{\partial^2 v}{\partial x_1 \, \partial x_2} a_1 a_2 + \frac{\partial^2 v}{\partial x_2^2} a_2^2 = \alpha^2 \left(\frac{\partial^2 v}{\partial x_1^2} b_1^2 + 2b_1 b_2 \frac{\partial^2 v}{\partial x_1 \, \partial x_2} + \frac{\partial^2 v}{\partial x_2^2} b_2^2 \right)$$

We can remove both the pure second derivatives, $\partial^2 v/\partial x_1^2$ and $\partial^2 v/\partial x_2^2$, by setting

$$a_1^2 - \alpha^2 b_1^2 = 0 \quad \text{and} \quad a_2^2 - \alpha^2 b_2^2 = 0 \tag{0.2.11}$$

so that the equation becomes

$$2(a_1 a_2 - \alpha^2 b_1 b_2) \frac{\partial^2 v}{\partial x_1 \, \partial x_2} = 0 \tag{0.2.12}$$

Now Eq. (0.2.11) can be satisfied by taking

$$a_1 = \pm \alpha b_1, \qquad a_2 = \pm \alpha b_2 \tag{0.2.13}$$

But if we choose the same sign in each case, the coefficient $(a_1 a_2 - \alpha^2 b_1 b_2)$ in Eq. (0.2.12) will also vanish. However, if we take opposite signs, giving

$$x_1 = b_1(\theta + \alpha x) \quad \text{and} \quad x_2 = b_2(\theta - \alpha x)$$
or
$$x_1 = b_1(\theta - \alpha x) \quad \text{and} \quad x_2 = b_2(\theta + \alpha x) \tag{0.2.14}$$

then $a_1 a_2 - \alpha^2 b_1 b_2 = \pm 2\alpha^2 b_1 b_2 \neq 0$, and Eq. (0.2.12) reduces to

$$\frac{\partial^2 v}{\partial x_1 \, \partial x_2} = 0 \tag{0.2.15}$$

which is the form for which we were aiming.

Exercises

0.2.1. Consider Eq. (0.2.7) under the change of variables $\sigma = z - Vt$, $\tau = t$. Give a physical interpretation to this change of variables and to that of Eq. (0.2.8).

0.2.2. Consider the pair of equations in which x and t are independent variables and V, ν, k_2, γ, and K are all positive constants,

$$V \frac{\partial c}{\partial x} + \frac{\partial c}{\partial t} + \nu \frac{\partial n}{\partial t} = 0, \qquad \frac{\partial n}{\partial t} = k_2[(\gamma - Kn)c - n]$$

and make the transformation

$$y = \frac{1}{\nu}\left(t - \frac{x}{V}\right), \qquad z = \frac{x}{V}, \qquad k = \nu k_2$$

Could you make a further transformation to remove explicit mention of the constant k? What advantages or disadvantages might this have?

0.3 Implicit Functions

We will often want to change the form of an equation defining the relationship between two or more variables. Here, for example, are three of the types of questions we may want to ask.

1. If $y = f(x)$, can we define the inverse relationship $x = g(y)$ uniquely?
2. If $F(x, y, z) = 0$, can we solve the equation and express z in the form $z = f(x, y)$?
3. Does the transformation $u = u(x, y)$, $v = v(x, y)$ map a region of the (x, y)-plane onto a region of the (u, v)-plane in a one-to-one fashion? (This question, although it is related to the first two, will be answered in Sec. 0.4.)

The first question can be illustrated by Figs. 0.2 and 0.3. The relation $y = f(x)$ is plotted in Fig. 0.2 as a curve in the (x, y)-plane, and x_0 and $y_0 = f(x_0)$ are the coordinates of a point on the curve. It seems clear that if the curve is monotonically increasing, as it is to the left of $\bar{x}_0$ in the figure, then for any y in the immediate neighborhood of y_0 we can find the corresponding x uniquely. We write this inverse relation as $x = g(y)$ and recognize that it is single valued. However, we should be in difficulty in the neighborhood of a point like $(\bar{x}_0, \bar{y}_0)$, where $f'(\bar{x}_0) = 0$. The figure shows that, for $y < \bar{y}_0$, two values of x correspond to y, whereas for $y > \bar{y}_0$ there is no value of x at all. Thus in an interval $a \leqslant x \leqslant b$ for which $f'(x)$ does not become zero (it could be negative just as well as positive), we can find $x = g(y)$, and the derivative $g'(y)$ is defined. In Fig. 0.3 there is an inflection point at $x = x_0$ with $f'(x_0) = 0$. In this case $x = g(y)$ is uniquely defined, but $g'(y)$ does not exist (i.e., is infinite) at $y = y_0$. Evidently, monotonicity is a necessary and sufficient condition for the inversion of the relation $y = f(x)$. What we are asking for in the third question is a generalization of this idea to a pair of functions of two variables.

In the case of the second question we have raised, the example $F(x, y, z) = x^2 + y^2 + z^2 + 1 = 0$ shows that it may be impossible to find real solutions of the form $z = f(x, y)$. However, if a solution of this form can be found,

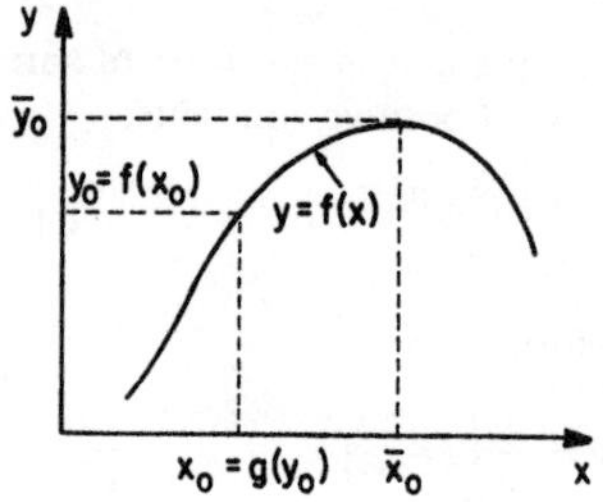

Figure 0.2

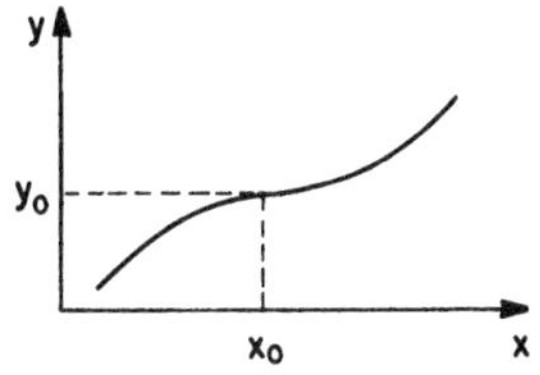

Figure 0.3

then $F(x, y, f(x, y)) = 0$ is an identity in x and y. The possibility of the solution in such a case is embodied in the following more general theorem. (See, for example, Courant, Vol. II, p. 117.)

> THEOREM: Let $F(x_1, x_2, \ldots, x_n, z)$ be a continuous function of the independent variables $x_1, x_2, \ldots, x_n$, and z with continuous partial derivatives, $F_{x_1}, F_{x_2}, \ldots, F_z$ in some region R'. Let $(x_1^0, x_2^0, \ldots, x_n^0, z^0)$ be an interior point of this region, and let $F(x_1^0, x_2^0 \ldots, x_n^0, z^0) = 0$, but $F_z(x_1^0, x_2^0, \ldots, x_n^0, z^0) \neq 0$. Then there is an interval $z_1 < z < z_2$ containing z^0 and a region R containing $(x_1^0, x_2^0, \ldots, x_n^0, z^0)$ within the region R' such that for every $(x_1, x_2, \ldots, x_n, z)$ in R the equation $F(x_1, x_2, \ldots, x_n, z) = 0$ is satisfied by exactly one value of z in the interval $z_1 < z < z_2$. For this value of z, which is denoted by $z = f(x_1, x_2, \ldots, x_n)$, the equation
>
> $$F(x_1, x_2, \ldots, x_n, f(x_1, x_2, \ldots, x_n)) = 0 \qquad (0.3.1)$$
>
> holds identically in R and $z^0 = f(x_1^0, x_2^0, \ldots, x_n^0)$. The function f is a continuous function and has continuous partial derivatives in R.

For a function $F(x, y, z)$, the differential dF is defined as

$$dF = \frac{\partial F}{\partial x}\,dx + \frac{\partial F}{\partial y}\,dy + \frac{\partial F}{\partial z}\,dz \qquad (0.3.2)$$

where x, y, and z are independent variables and dx, dy, and dz are the differentials of these independent variables. If $F = K$, a constant, then

$$dF = \frac{\partial F}{\partial x}\,dx + \frac{\partial F}{\partial y}\,dy + \frac{\partial F}{\partial z}\,dz = 0$$

and, since the partial derivatives are computed at some definite point, dx, dy, and dz cannot be independent. However, from the theorem on implicit functions, assuming the hypotheses are satisfied, $z = f(x, y)$, and therefore

$$dz = \frac{\partial f}{\partial x}\,dx + \frac{\partial f}{\partial y}\,dy \qquad (0.3.3)$$

Substituting this in the preceding equation, one obtains

$$\frac{\partial F}{\partial x}\,dx + \frac{\partial F}{\partial y}\,dy + \frac{\partial F}{\partial z}\left(\frac{\partial f}{\partial x}\,dx + \frac{\partial f}{\partial y}\,dy\right) = 0$$

or

$$\left(\frac{\partial F}{\partial x} + \frac{\partial F}{\partial z}\frac{\partial f}{\partial x}\right)dx + \left(\frac{\partial F}{\partial y} + \frac{\partial F}{\partial z}\frac{\partial f}{\partial y}\right)dy = 0$$

which must hold for all dx and dy. Thus

$$\frac{\partial F}{\partial x} + \frac{\partial F}{\partial z}\frac{\partial f}{\partial x} = 0$$

$$\frac{\partial F}{\partial y} + \frac{\partial F}{\partial z}\frac{\partial f}{\partial y} = 0$$

or

$$\frac{\partial f}{\partial x} = \frac{\partial z}{\partial x} = -\frac{\partial F/\partial x}{\partial F/\partial z} \tag{0.3.4}$$

$$\frac{\partial f}{\partial y} = \frac{\partial z}{\partial y} = -\frac{\partial F/\partial y}{\partial F/\partial z} \tag{0.3.5}$$

and this gives the differentiation formulas for the function f in terms of the partial derivatives of F.

A word should be inserted here about notation. Given a function $f(x, y, z)$ with x, y, and z as independent variables, the meaning of $\partial f/\partial x$, $\partial f/\partial y$, and $\partial f/\partial z$ is clear. In applications, and particularly in thermodynamics, it is customary to write

$$\left(\frac{\partial f}{\partial x}\right)_{y,z}, \quad \left(\frac{\partial f}{\partial y}\right)_{x,z}, \quad \left(\frac{\partial f}{\partial z}\right)_{x,y}$$

in which the variables under consideration are explicitly specified. This is necessary, for a quantity like the free energy may be considered as being a function of different sets of variables, some being more convenient in one situation than another. For example, $(\partial F/\partial T)_p$ is not the same as $(\partial F/\partial T)_V$, and this is obvious when written in this form. In purely mathematical analyses, the variables that are taken to be independent are usually apparent, whereas in physics there is frequently a choice.

The function $F(x, y, z) = K$ represents a one-parameter family of surfaces since, in principle, provided that $\partial F/\partial z$ is not zero, one can solve for z as $z = f(x, y, K)$.

0.4 Sets of Functions

In mathematical applications we are frequently led to the problem of inverting a set of functions, such as

$$
\begin{aligned}
u &= u(x, y) \\
v &= v(x, y)
\end{aligned}
\tag{0.4.1}
$$

That is, we are asking whether it is possible to find an inverse pair of functions

$$
\begin{aligned}
x &= x(u, v) \\
y &= y(u, v)
\end{aligned}
\tag{0.4.2}
$$

The answer to this question is embodied in the following theorem. (See Courant, Vol. II, p. 152.)

> THEOREM: If in the neighborhood of a point (x_0, y_0) the functions $u(x, y)$ and $v(x, y)$ are continuously differentiable with $u_0 = u(x_0, y_0)$, $v_0 = v(x_0, y_0)$, and if in addition the Jacobian determinant
>
> $$J = \begin{vmatrix} \dfrac{\partial u}{\partial x} & \dfrac{\partial u}{\partial y} \\ \\ \dfrac{\partial v}{\partial x} & \dfrac{\partial v}{\partial y} \end{vmatrix} \tag{0.4.3}$$

is not zero at (x_0, y_0), then in the neighborhood of the point (x_0, y_0) the system of equations $u = u(x, y)$, $v = v(x, y)$ has a unique inverse

$$x = x(u, v)$$

$$y = y(u, v)$$

with $x_0 = x(u_0, v_0)$, $y_0 = y(u_0, v_0)$ and such that

$$u \equiv u(x(u, v), y(u, v)) \tag{0.4.4}$$

$$v \equiv v(x(u, v), y(u, v)) \tag{0.4.5}$$

in some neighborhood of (u_0, v_0). Also in that neighborhood

$$\frac{\partial x}{\partial u} = \frac{1}{J}\frac{\partial v}{\partial y}, \qquad \frac{\partial x}{\partial v} = -\frac{1}{J}\frac{\partial u}{\partial y} \tag{0.4.6}$$

$$\frac{\partial y}{\partial u} = -\frac{1}{J}\frac{\partial v}{\partial x}, \qquad \frac{\partial y}{\partial v} = \frac{1}{J}\frac{\partial u}{\partial x} \tag{0.4.7}$$

These last formulas may be easily obtained by differentiation of Eqs.

(0.4.4) and (0.4.5), since

$$1 = \frac{\partial u}{\partial x}\frac{\partial x}{\partial u} + \frac{\partial u}{\partial y}\frac{\partial y}{\partial u}$$

$$0 = \frac{\partial u}{\partial x}\frac{\partial x}{\partial v} + \frac{\partial u}{\partial y}\frac{\partial y}{\partial v}$$

$$1 = \frac{\partial v}{\partial x}\frac{\partial x}{\partial v} + \frac{\partial v}{\partial y}\frac{\partial y}{\partial v}$$

$$0 = \frac{\partial v}{\partial x}\frac{\partial x}{\partial u} + \frac{\partial v}{\partial y}\frac{\partial y}{\partial u}$$

where the first and fourth are obtained by differentiating with respect to u and the second and third by differentiating with respect to v; Eqs. (0.4.6) and (0.4.7) are then obtained by solving the first and fourth and then the second and third as simultaneous algebraic equations. These formulas have an obvious generalization to a higher number of dimensions.

If the Jacobian J vanishes at a point (x_0, y_0), the possibility exists that the two functions $u = u(x, y)$, $v = v(x, y)$ cannot be inverted in the neighborhood of the point, and, even if the inverse does exist, the derivatives of the inverse do not, as may be seen from the preceding formulas for the partial derivatives. For example,

$$u = a \cos x \cos y, \qquad v = a \cos x \sin y$$

has a Jacobian

$$\begin{vmatrix} -a \sin x \cos y & -a \cos x \sin y \\ -a \sin x \sin y & a \cos x \cos y \end{vmatrix} = -a^2 \cos x \sin x$$

This Jacobian will vanish if $x = 0, \pi/2$, and so on. Now if $x = 0$, $u = a \cos y$, $v = a \sin y$, and (u, v) traverses the boundary of the circle of radius a as y varies from 0 to 2π. If $0 < x < \pi/2$, the point (u, v) goes around a circle of smaller radius $a \cos x$. But if $x = \pi/2$, that radius has shrunk to zero and the whole line segment $x = \pi/2$, $0 \leqslant y \leqslant 2\pi$, is mapped onto a point. Since the locus of $x = c$ coincides with that of $x = -c$, it is evident that the line $x = 0$, on which $J = 0$, is the boundary of a unique relationship between (x, y) and (u, v). This is also true for $x = \pi/2$, since $x = \pi/2 - c$ and $x = \pi/2 + c$ would have the same image in the (u, v)-plane. In addition, we see a geometric degeneracy associated with $x = \pi/2$, for the line $x = \pi/2$, $0 \leqslant y \leqslant 2\pi$, which is a one-dimensional structure, is mapped into a point, $u = v = 0$, that is of zero dimension.

The general theorem (see, for example, Goursat, Vol. I, p. 45) on implicit functions is as follows.

THEOREM: Consider a system of n equations

$$F_j(x_1, x_2, \ldots, x_p; u_1, u_2, \ldots, u_n) = 0, \qquad j = 1, \ldots, n$$

Suppose that these equations are satisfied for a set of values $x_i = x_i^0$, i

$= 1, \ldots, p$, $u_j = u_j^0$, $j = 1, \ldots, n$ and that the functions F_j are continuous along with all their first partial derivatives in the neighborhood of this point. Suppose the Jacobian

$$J = \begin{vmatrix} \frac{\partial F_1}{\partial u_1} & \frac{\partial F_1}{\partial u_2} & \cdots & \frac{\partial F_1}{\partial u_n} \\ \frac{\partial F_2}{\partial u_1} & \frac{\partial F_2}{\partial u_2} & \cdots & \frac{\partial F_2}{\partial u_n} \\ \vdots & & & \\ \frac{\partial F_n}{\partial u_1} & \frac{\partial F_n}{\partial u_2} & \cdots & \frac{\partial F_n}{\partial u_n} \end{vmatrix}$$

does not vanish at the point. Then there exists one and only one system of continuous functions

$$u_j = \phi_j(x_1, x_2, \ldots, x_p)$$

that satisfies the equations $F_j = 0$; that is,

$$F_j(x_1, x_2, \ldots, x_p; \phi_1, \phi_2, \ldots, \phi_n) = 0$$

and which reduces to u_j^0 at $x_1 = x_1^0$, $x_2 = x_2^0, \ldots, x_p = x_p^0$.

An interesting situation, however, arises when the Jacobian vanishes *identically* in a region. We will show in this case that the two functions u and v are dependent; that is, there exists a function $F(u, v) = 0$ in that region. For cxample, if $u = x + y$, $v = x^2 + 2xy + y^2$, $J \equiv 0$, and obviously $u^2 - v = 0$.

Suppose first that $u_x = 0$ and $u_y = 0$ (then $J \equiv 0$); it follows that u is a constant. As x and y range over a region, then the image u and v will range over $u =$ constant, and hence a region of the (x, y)-plane is mapped into a single line $u =$ constant in the (u, v)-plane, and there is no possibility of a one-to-one mapping from one plane to the other. On the other hand, if $u_x = 0$ and $v_x = 0$, then u and v are each only functions of y, and therefore u is some function of v. If $u_x = 0$ and the Jacobian vanishes, then either $v_x = 0$ or $u_y = 0$. Thus we need consider further only the case where $J = 0$, but the derivatives themselves do not vanish.

Since $\partial v/\partial y \neq 0$, we can solve $v = v(x, y)$ for y to obtain $y = y(x, v)$, and therefore $u = u(x, y(x, v)) = F(x, v)$. We will now show that if $J = 0$ then F cannot depend on x. Since

$$\begin{aligned} du &= \frac{\partial F}{\partial x} dx + \frac{\partial F}{\partial v} dv \\ &= \frac{\partial F}{\partial x} dx + \left(\frac{\partial v}{\partial x} dx + \frac{\partial v}{\partial y} dy\right) \frac{\partial F}{\partial v} \end{aligned}$$

and

$$du = \frac{\partial u}{\partial x}\,dx + \frac{\partial u}{\partial y}\,dy$$

it follows by subtraction that

$$\left(\frac{\partial u}{\partial x} - \frac{\partial F}{\partial x} - \frac{\partial F}{\partial v}\frac{\partial v}{\partial x}\right)dx + \left(\frac{\partial u}{\partial y} - \frac{\partial F}{\partial v}\frac{\partial v}{\partial y}\right)dy = 0$$

Because x and y are independent variables, dx and dy are independent, and hence

$$\frac{\partial u}{\partial x} - \frac{\partial F}{\partial x} - \frac{\partial F}{\partial v}\frac{\partial v}{\partial x} = 0$$

and

$$\frac{\partial u}{\partial y} - \frac{\partial F}{\partial v}\frac{\partial v}{\partial y} = 0$$

Therefore,

$$\begin{aligned}\frac{\partial F}{\partial x} &= \frac{\partial u}{\partial x} - \frac{\partial v}{\partial x}\left(\frac{\partial u/\partial y}{\partial v/\partial y}\right)\\ &= \left(\frac{\partial v}{\partial y}\right)^{-1} J\end{aligned}$$

But since $J = 0$, the function F does not depend on x, and thus we have obtained a sufficient condition for dependence. The condition $J = 0$ is obviously also a necessary one. For, if there exists a functional relation $\pi(u, v) = 0$ between $u = u(x, y)$ and $v = v(x, y)$, then

$$\frac{\partial \pi}{\partial u}\frac{\partial u}{\partial x} + \frac{\partial \pi}{\partial v}\frac{\partial v}{\partial x} = 0$$

$$\frac{\partial \pi}{\partial u}\frac{\partial u}{\partial y} + \frac{\partial \pi}{\partial v}\frac{\partial v}{\partial y} = 0$$

which, considered as a set of linear simultaneous algebraic equations in π_u, π_v, demands that $J = 0$ if a nontrivial solution is to exist.

The general theorem may be stated. (See, for example, Goursat, Vol. I, p. 52.)

THEOREM: Let $u_1, u_2, \ldots, u_n$ be n functions of the n independent variables $x_1, x_2, \ldots, x_n$. In order that there exists a function $\pi(u_1, u_2, \ldots, u_n) = 0$ that does not involve the x_i explicitly, it is necessary and sufficient that the Jacobian

$$J = \begin{vmatrix} \frac{\partial u_1}{\partial x_1} & \frac{\partial u_2}{\partial x_1} & \cdots & \frac{\partial u_n}{\partial x_1} \\ \frac{\partial u_2}{\partial x_2} & & \cdots & \frac{\partial u_n}{\partial x_2} \\ \cdot & & & \\ \cdot & & & \\ \cdot & & & \\ \frac{\partial u_1}{\partial x_n} & \frac{\partial u_2}{\partial x_n} & \cdots & \frac{\partial u_n}{\partial x_n} \end{vmatrix} = \frac{\partial(u_1, \ldots, u_n)}{\partial(x_1, \ldots, x_m)}$$

should vanish identically.

Remark: If the functions $u_1, u_2, \ldots, u_n$ involve certain other variables $y_1, y_2, \ldots, y_m$ besides the x_i, then the vanishing of the same Jacobian is necessary and sufficient for the existence of a functional relation among the u_i independent of the x_i but perhaps still containing the y_i.

A simple computation will show that if $u_i = u_i(x_1, x_2, \ldots, x_n)$ and $x_j = x_j(y_1, y_2, \ldots, y_n)$, $i = 1$ to n, $j = 1$ to n, then the Jacobian of the transformation from the y_j's directly to the u_i's is the product of the two individual Jacobians; that is,

$$\begin{vmatrix} \frac{\partial u_1}{\partial x_1} & \frac{\partial u_1}{\partial x_2} & \cdots & \frac{\partial u_1}{\partial x_n} \\ \frac{\partial u_2}{\partial x_1} & \frac{\partial u_2}{\partial x_2} & \cdots & \frac{\partial u_2}{\partial x_n} \\ \cdot & & & \\ \cdot & & & \\ \cdot & & & \\ \frac{\partial u_n}{\partial x_1} & \frac{\partial u_n}{\partial x_2} & \cdots & \frac{\partial u_n}{\partial x_n} \end{vmatrix} \begin{vmatrix} \frac{\partial x_1}{\partial y_1} & \frac{\partial x_1}{\partial y_2} & \cdots & \frac{\partial x_1}{\partial y_n} \\ \frac{\partial x_2}{\partial y_1} & \frac{\partial x_2}{\partial y_2} & \cdots & \frac{\partial x_2}{\partial y_n} \\ \cdot & & & \\ \cdot & & & \\ \cdot & & & \\ \frac{\partial x_n}{\partial y_1} & \frac{\partial x_n}{\partial y_2} & \cdots & \frac{\partial x_n}{\partial y_n} \end{vmatrix} = \begin{vmatrix} \frac{\partial u_1}{\partial y_1} & \frac{\partial u_1}{\partial y_2} & \cdots & \frac{\partial u_1}{\partial y_n} \\ \frac{\partial u_2}{\partial y_1} & \frac{\partial u_2}{\partial y_2} & \cdots & \frac{\partial u_2}{\partial y_n} \\ \cdot & & & \\ \cdot & & & \\ \cdot & & & \\ \frac{\partial u_n}{\partial y_1} & \frac{\partial u_n}{\partial y_2} & \cdots & \frac{\partial u_n}{\partial y_n} \end{vmatrix}$$

From this formula it is apparent that $\partial(u_1, u_2, \ldots, u_n)/\partial(u_1, u_2, \ldots, u_n) = 1$ or that $\partial(u_1, u_2, \ldots, u_n)/\partial(x_1, x_2, \ldots, x_n)$ and $\partial(x_1, x_2, \ldots, x_n)/\partial(u_1, u_2, \ldots, u_n)$ are reciprocals.

0.5 Differentiation of Implicit Functions

Given a function $F(x, y, z) = 0$, the partial derivatives may be computed from the formula

$$0 = \frac{\partial F}{\partial x} dx + \frac{\partial F}{\partial y} dy + \frac{\partial F}{\partial z} dz$$

For example, if this is written as

$$dy = -\left(\frac{\partial F/\partial x}{\partial F/\partial y}\right) dx - \left(\frac{\partial F/\partial z}{\partial F/\partial y}\right) dz$$

and y is regarded as a function of x and z, so that

$$dy = \frac{\partial y}{\partial x} dx + \frac{\partial y}{\partial z} dz$$

we see that

$$\frac{\partial y}{\partial x} = -\frac{\partial F/\partial x}{\partial F/\partial y}, \qquad \frac{\partial y}{\partial z} = -\frac{\partial F/\partial z}{\partial F/\partial y} \tag{0.5.1}$$

With a pair of functions, say

$$F(x, y, z, w) = 0$$
$$G(x, y, z, w) = 0$$

we have

$$0 = \frac{\partial F}{\partial x} dx + \frac{\partial F}{\partial y} dy + \frac{\partial F}{\partial z} dz + \frac{\partial F}{\partial w} dw$$

$$0 = \frac{\partial G}{\partial x} dx + \frac{\partial G}{\partial y} dy + \frac{\partial G}{\partial z} dz + \frac{\partial G}{\partial w} dw$$

The second of these may be solved for

$$dw = -\left(\frac{\partial G}{\partial w}\right)^{-1} \left(\frac{\partial G}{\partial x} dx + \frac{\partial G}{\partial y} dy + \frac{\partial G}{\partial z} dz\right)$$

and substituted into the first to give

$$0 = \left(\frac{\partial F}{\partial x} - \frac{\partial F/\partial w}{\partial G/\partial w}\frac{\partial G}{\partial x}\right)dx + \left(\frac{\partial F}{\partial y} - \frac{\partial F/\partial w}{\partial G/\partial w}\frac{\partial G}{\partial y}\right)dy + \left(\frac{\partial F}{\partial z} - \frac{\partial F/\partial w}{\partial G/\partial w}\frac{\partial G}{\partial z}\right)dz$$

Since dx, dy, and dz are not independent, it follows that

$$\left(\frac{\partial y}{\partial x}\right)_z = -\frac{\dfrac{\partial F}{\partial x}\dfrac{\partial G}{\partial w} - \dfrac{\partial F}{\partial w}\dfrac{\partial G}{\partial x}}{\dfrac{\partial F}{\partial y}\dfrac{\partial G}{\partial w} - \dfrac{\partial F}{\partial w}\dfrac{\partial G}{\partial y}} = -\frac{\dfrac{\partial(F, G)}{\partial(x, w)}}{\dfrac{\partial(F, G)}{\partial(y, w)}}$$

with similar formulas for other partial derivatives. They may be easily generalized to other situations with a larger number of functions and variables. For example, suppose we are given n functions

$$F_1(x_1, x_2, \ldots, x_p; u_1, u_2, \ldots, u_n) = 0$$

$$F_2(x_1, x_2, \ldots, x_p; u_1, u_2, \ldots, u_n) = 0$$

$$\cdot$$

$$\cdot$$

$$\cdot$$

$$F_n(x_1, x_2, \ldots, x_p; u_1, u_2, \ldots, u_n) = 0$$

where we have assumed that the independent variables are $x_1, x_2, \ldots, x_p$. (Note that we must always decide which variables are to be taken as independent.) Then

$$\frac{\partial F_j}{\partial x_i} + \frac{\partial F_j}{\partial u_1}\frac{\partial u_1}{\partial x_i} + \frac{\partial F_j}{\partial u_2}\frac{\partial u_2}{\partial x_i} + \cdots + \frac{\partial F_j}{\partial u_n}\frac{\partial u_n}{\partial x_i} = 0, \qquad i = 1, \ldots, p,\ j = 1, \ldots, n$$

and we can compute

$$\frac{\partial u_k}{\partial x_i} = -\frac{\dfrac{\partial(F_1, F_2, \ldots, F_n)}{\partial(u_1, \ldots, u_{k-1}, x_i, u_{k+1}, \ldots, u_n)}}{\dfrac{\partial(F_1, F_2, \ldots, F_n)}{\partial(u_1, u_2, \ldots, u_n)}},$$

$$i = 1, \ldots, p,\ k = 1, \ldots, n \qquad (0.5.2)$$

where the Jacobian in the denominator is assumed to be nonzero.

0.6 Surfaces

Previously, we mentioned that the relation $F(x, y, z) = K$ represents a surface for a fixed value of K and a one-parameter family of surfaces if the parameter K is allowed to vary. Clearly, $G(x, y, z, K) = 0$ is a one-parameter family of surfaces where the parameter is implicit in the function G. Consider now the three functions

$$x = \phi(u, v), \qquad y = \psi(u, v), \qquad z = \chi(u, v) \qquad (0.6.1)$$

where the functions ϕ, ψ, and χ are continuous with continuous partial derivatives. We will assume also that the three Jacobians

$$\begin{vmatrix} \phi_u & \phi_v \\ \psi_u & \psi_v \end{vmatrix}, \qquad \begin{vmatrix} \psi_u & \psi_v \\ \chi_u & \chi_v \end{vmatrix}, \qquad \begin{vmatrix} \chi_u & \chi_v \\ \phi_u & \phi_v \end{vmatrix}$$

do not vanish; then the functions ϕ, ψ, and χ are independent, and the first two may be solved for u and v and substituted into the third to give

$$z = \chi(u(x, y), v(x, y)) = f(x, y) \qquad (0.6.2)$$

to give the equations of a surface. (*Note:* Actually only one of the Jacobians need be nonzero for the substitution to be valid.) Thus the representation given by Eq. (0.6.1) is a two-parameter representation of a surface. As the point (u, v) ranges over a plane, the point (x, y, z) sweeps out a surface.

Consider next the parametric representation

$$x = \xi(t), \qquad y = \eta(t), \qquad z = \zeta(t) \tag{0.6.3}$$

If t is eliminated between the first two, one obtains a function $f(x, y) = 0$, and if t is eliminated between the second and third, a function $g(y, z) = 0$. These two functions then define cylinders, the first with generators parallel to the z-axis and the second parallel to the x-axis. The intersection of these cylinders will generally define a curve in space. Thus Eqs. (0.6.3) are the parametric representation of a space curve.

In Fig. 0.4 we have a surface defined by the two-parameter representation given by Eq. (0.6.1) and a set of curves defined on the surface given by the same set of equations, but with v fixed in one direction and u fixed in the other. Thus the numbers (u, v) may be treated as coordinates on the surface.

Now, if we have a curve $u = u(t)$, $v = v(t)$ in the (u, v)-plane, it will be mapped by $x = \phi(u, v)$, $y = \psi(u, v)$, $z = \chi(u, v)$ into a curve on the surface. In the (x, y, z)-space,

$$\left(\frac{ds}{dt}\right)^2 = \left(\frac{dx}{dt}\right)^2 + \left(\frac{dy}{dt}\right)^2 + \left(\frac{dz}{dt}\right)^2$$

where s is the length of arc along the curve on the surface. But

$$\frac{dx}{dt} = \frac{\partial x}{\partial u}\frac{du}{dt} + \frac{\partial x}{\partial v}\frac{dv}{dt}$$

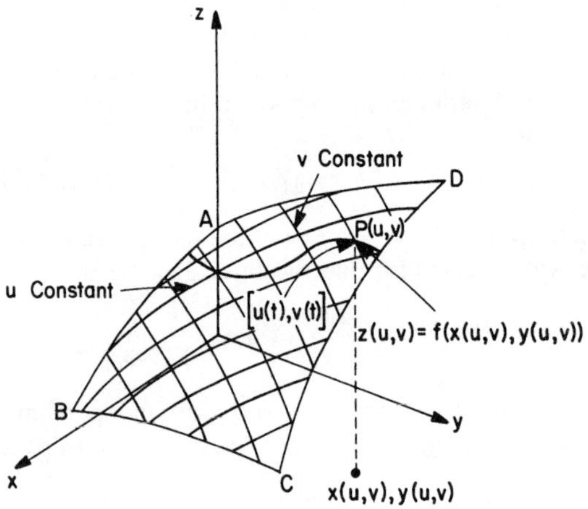

Figure 0.4

and so on, so

$$\left(\frac{ds}{dt}\right)^2 = E\left(\frac{du}{dt}\right)^2 + 2F\frac{du}{dt}\frac{dv}{dt} + G\left(\frac{dv}{dt}\right)^2 \tag{0.6.4}$$

where

$$\begin{aligned} E &= \left(\frac{\partial x}{\partial u}\right)^2 + \left(\frac{\partial y}{\partial u}\right)^2 + \left(\frac{\partial z}{\partial u}\right)^2 \\ F &= \frac{\partial x}{\partial u}\frac{\partial x}{\partial v} + \frac{\partial y}{\partial u}\frac{\partial y}{\partial v} + \frac{\partial z}{\partial u}\frac{\partial z}{\partial v} \\ G &= \left(\frac{\partial x}{\partial v}\right)^2 + \left(\frac{\partial y}{\partial v}\right)^2 + \left(\frac{\partial z}{\partial v}\right)^2 \end{aligned} \tag{0.6.5}$$

We note that ds, the element of arc length on the surface, is independent of the parameterization of the curve in the (u, v)-plane and depends only on the surface itself. In fact, we may write

$$(ds)^2 = E(du)^2 + 2F\,du\,dv + G(dv)^2 \tag{0.6.6}$$

0.7 Tangents and Normals

If we are given the equation of a surface in explicit form $z = f(x, y)$, the equation of the tangent plane at the point (x_0, y_0, z_0), $z_0 = f(x_0, y_0)$ is

$$z - z_0 = \left(\frac{\partial f}{\partial x}\right)_0 (x - x_0) + \left(\frac{\partial f}{\partial y}\right)_0 (y - y_0) \tag{0.7.1}$$

while the equation of the normal to the plane at that point is

$$\frac{x - x_0}{(\partial f/\partial x)_0} = \frac{y - y_0}{(\partial f/\partial y)_0} = \frac{z - z_0}{-1} \tag{0.7.2}$$

If the equation of the surface is given in implicit form $F(x, y, z) = 0$, then the tangent plane is

$$\left(\frac{\partial F}{\partial x}\right)_0 (x - x_0) + \left(\frac{\partial F}{\partial y}\right)_0 (y - y_0) + \left(\frac{\partial F}{\partial z}\right)_0 (z - z_0) = 0 \tag{0.7.3}$$

and the normal line is

$$\frac{x - x_0}{(\partial F/\partial x)_0} = \frac{y - y_0}{(\partial F/\partial y)_0} = \frac{z - z_0}{(\partial F/\partial z)_0} \tag{0.7.4}$$

The direction cosines of the normal are therefore proportional to

$$\left(\frac{\partial F}{\partial x}\right)_0, \quad \left(\frac{\partial F}{\partial y}\right)_0, \quad \left(\frac{\partial F}{\partial z}\right)_0$$

or

$$\left(\frac{\partial z}{\partial x}\right)_0, \quad \left(\frac{\partial z}{\partial y}\right)_0, \quad -1$$

If the equation of the surface is given in parametric form,

$$x = \phi(u, v), \qquad y = \psi(u, v), \qquad z = \chi(u, v) \tag{0.7.5}$$

we can compute $\partial z/\partial x$ and $\partial z/\partial y$ as follows. For $\partial z/\partial x$, we have

$$\frac{\partial z}{\partial x} = \frac{\partial z}{\partial u}\frac{\partial u}{\partial x} + \frac{\partial z}{\partial v}\frac{\partial v}{\partial x}$$

but from Eqs. (0.4.7),

$$\frac{\partial u}{\partial x} = J\frac{\partial y}{\partial v}, \qquad \frac{\partial v}{\partial x} = -J\frac{\partial y}{\partial u}$$

where

$$J = \frac{\partial(u, v)}{\partial(x, y)}$$

Hence

$$\frac{\partial z}{\partial x} = J\frac{\partial z}{\partial u}\frac{\partial y}{\partial v} - J\frac{\partial z}{\partial v}\frac{\partial y}{\partial u} = \frac{\partial(u, v)}{\partial(x, y)}\frac{\partial(z, y)}{\partial(u, v)}$$

and likewise

$$\frac{\partial z}{\partial y} = \frac{\partial(u, v)}{\partial(x, y)}\frac{\partial(x, z)}{\partial(u, v)}$$

Therefore, the equation of the plane is

$$z - z_0 = \frac{\partial(u, v)}{\partial(x, y)_0}\frac{\partial(z, y)}{\partial(u, v)_0}(x - x_0) + \frac{\partial(u, v)}{\partial(x, y)_0}\frac{\partial(x, z)}{\partial(u, v)_0}(y - y_0)$$

or

$$\frac{\partial(x, y)}{\partial(u, v)_0}(z - z_0) = \frac{\partial(z, y)}{\partial(u, v)_0}(x - x_0) + \frac{\partial(x, z)}{\partial(u, v)_0}(y - y_0) \tag{0.7.6}$$

and hence the equation of the normal to the plane is

$$\frac{x - x_0}{\dfrac{\partial(y, z)}{\partial(u, v)_0}} = \frac{y - y_0}{\dfrac{\partial(z, x)}{\partial(u, v)_0}} = \frac{z - z_0}{\dfrac{\partial(x, y)}{\partial(u, v)_0}}$$

where the direction cosines are proportional to the appropriate Jacobians in the denominator.

Exercise

0.7.1. Two surfaces are given by $f(x, y, z) = a$ and $g(x, y, z) = b$. They have a common point at (x_0, y_0, z_0). Show that they also have a common tangent plane if $f_x g_y - f_y g_x = 0$ and $f_x g_z - f_z g_x = 0$. What is the value of $f_y g_z - f_z g_y$ at this point?

0.8 Direction Cosines and Space Curves

If we suppose that s is arc length along a curve, the direction cosines of the tangent line to a space curve are given by

$$\frac{dx}{ds} = \cos\alpha, \qquad \frac{dy}{ds} = \cos\beta, \qquad \frac{dz}{ds} = \cos\gamma \tag{0.8.1}$$

If the curve is given parametrically as $x = \xi(t)$, $y = \eta(t)$, $z = \zeta(t)$, then the cosines are given by

$$\begin{aligned}
\cos\alpha &= \frac{\xi'(t)}{\sqrt{\xi'(t)^2 + \eta'(t)^2 + \zeta'(t)^2}} \\
\cos\beta &= \frac{\eta'(t)}{\sqrt{\xi'(t)^2 + \eta'(t)^2 + \zeta'(t)^2}} \\
\cos\gamma &= \frac{\zeta'(t)}{\sqrt{\xi'(t)^2 + \eta'(t)^2 + \zeta'(t)^2}}
\end{aligned} \tag{0.8.2}$$

If the space curve is given as the intersection of two cylinders

$$f(x, y) = 0, \qquad g(x, z) = 0 \tag{0.8.3}$$

the direction cosines are given by

$$\begin{aligned}
\cos\alpha &= \frac{1}{\sqrt{1 + (dy/dx)^2 + (dz/dx)^2}} \\
\cos\beta &= \frac{dy/dx}{\sqrt{1 + (dy/dx)^2 + (dz/dx)^2}}, \qquad \frac{dy}{dx} = -\frac{\partial f/\partial x}{\partial f/\partial y}, \quad \frac{dz}{dx} = -\frac{\partial g/\partial x}{\partial g/\partial z} \\
\cos\alpha &= \frac{dz/dx}{\sqrt{1 + (dy/dx)^2 + (dz/dx)^2}}
\end{aligned} \tag{0.8.4}$$

If the space curve is given by the intersection of two surfaces

$$F(x, y, z) = 0, \qquad G(x, y, z) = 0 \tag{0.8.5}$$

then the direction cosines are proportional to

$$\begin{vmatrix} F_y & F_z \\ G_y & G_z \end{vmatrix}, \quad \begin{vmatrix} F_z & F_x \\ G_z & G_x \end{vmatrix}, \quad \begin{vmatrix} F_x & F_y \\ G_x & G_y \end{vmatrix} \tag{0.8.6}$$

Exercise

0.8.1. Prove the assertions of this section.

0.9 Directional Derivatives

Suppose we are given a function of three variables $F(x, y, z)$ and a space curve given parametrically as $x = \xi(t)$, $y = \eta(t)$, $z = \zeta(t)$. Then, as we move along the curve, the function values of F will in general change continuously. If s is arc length, then it is clear that dF/ds is the rate of change of F along the curve in the direction of the tangent. The arc length s could be chosen as a parameter along the curve, and although the actual parametric representation of a curve in terms of arc length is often not easy to obtain, in general, its use in theoretical considerations greatly facilitates the discussion. Now

$$\frac{dF}{dt} = \frac{\partial F}{\partial x}\frac{dx}{ds}\frac{ds}{dt} + \frac{\partial F}{\partial y}\frac{dy}{ds}\frac{ds}{dt} + \frac{\partial F}{\partial z}\frac{dz}{ds}\frac{ds}{dt}$$

or

$$\frac{dF}{dt}\cdot\frac{dt}{ds} = \frac{\partial F}{\partial x}\cos\alpha + \frac{\partial F}{\partial y}\cos\beta + \frac{\partial F}{\partial z}\cos\gamma \tag{0.9.1}$$

The left side is dF/ds, and it is clear that the directional derivative is a generalization of the partial derivatives $\partial F/\partial x$, $\partial F/\partial y$, and $\partial F/\partial z$, since by appropriate choice of the space curve these can be obtained as special cases. The directional derivative at a point is not dependent on a particular space curve, as is apparent, but its value at a given point does depend on the direction at that point. We can then reasonably ask the question: Of all possible directions at a fixed point, in what direction is the function increasing the most rapidly? To see this most clearly, we note that dF/ds is the scalar product of a vector

$$\left[\frac{\partial F}{\partial x}, \frac{\partial F}{\partial y}, \frac{\partial F}{\partial z}\right]$$

with a unit vector $[\cos\alpha, \cos\beta, \cos\gamma]$. However, the first vector is a vector normal to the surface $F(x, y, z) = K$, and dF/ds may be written

$$\frac{dF}{ds} = \frac{dF}{dn}\cdot\frac{dn}{ds} \tag{0.9.2}$$

where dn/ds is the rate of change of the distance along the normal with respect to distance in the direction given by the vector $[\cos\alpha, \cos\beta, \cos\gamma]$. Now (Fig. 0.5 shows the two-dimensional version)

$$\frac{dn}{ds} = \cos\theta \tag{0.9.3}$$

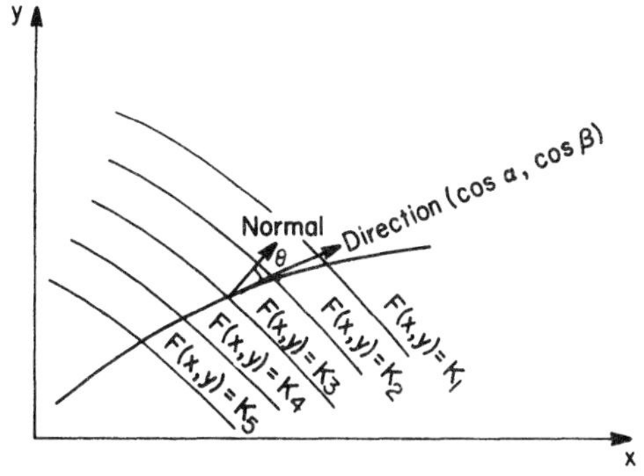

Figure 0.5

where θ is the angle between the normal direction and the vector [cos α, cos β, cos γ]. dF/ds is thus a maximum when $\theta = 0$, and the value of the directional derivative normal to the surface $F(x, y, z) = K$ is

$$\frac{dF}{dn} = \sqrt{\left(\frac{\partial F}{\partial x}\right)^2 + \left(\frac{\partial F}{\partial y}\right)^2 + \left(\frac{\partial F}{\partial z}\right)^2} \tag{0.9.4}$$

and is called the *gradient*.

In the sequel, much use will be made of the directional derivative, for the theory of first-order partial differential equations leads to the idea of characteristic curves, and solution of the equations is obtained by solving ordinary differential equations in the direction of these characteristic curves.

0.10 Envelopes

In this section we will discuss envelopes of curves and surfaces since these will play an important role in finding solutions of first-order partial differential equations.

Suppose we are given a one-parameter family of curves in the (x, y)-plane expressed analytically as

$$f(x, y, \alpha) = 0 \tag{0.10.1}$$

Then as α is varied, different members of the family will be generated. Now it may happen when one considers the totality of such curves that each is tangent to a single curve. If this is the case, then the tangent curve is called an *envelope*. For example, consider the family of straight lines

$$2\alpha y = 2x + \alpha^2$$

As shown in Fig. 0.6, as α varies through positive and negative values, it appears that all straight lines will be tangent to a curve having the appearance of a parabola.

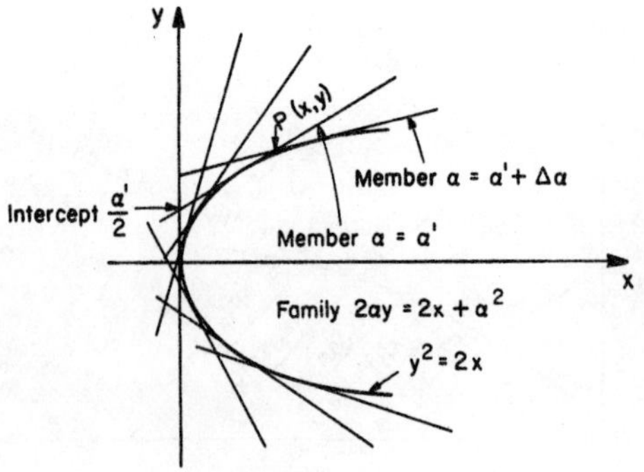

Figure 0.6

The method of computation of the envelope is based on the idea of finding the point of intersection of *adjacent* members of the family. Let α be fixed at α'; then

$$f(x, y, \alpha') = 0 \tag{0.10.2}$$

and

$$f(x, y, \alpha' + \Delta\alpha) = 0$$

is also a member of the family. If these two curves intersect at a point $P(x, y)$, then any linear combination of the equations gives a curve that passes through the point of intersection P, and therefore the curve

$$f(x, y, \alpha' + \Delta\alpha) - f(x, y, \alpha') = 0$$

also passes through P. By the law of the mean,

$$f_\alpha(x, y, \alpha' + \theta\,\Delta\alpha) = 0, \qquad 0 < \theta < 1$$

and as $\Delta\alpha \to 0$, we have

$$f_\alpha(x, y, \alpha') = 0 \tag{0.10.3}$$

Thus we have two equations, each of which is a curve passing through the point $P(x, y)$ at the point of tangency of two adjacent members of the family, and therefore the result of eliminating α' between these should produce the locus of the tangencies and therefore the envelope. In the case of our example, these equations are

$$2\alpha y = 2x + \alpha^2$$

$$2y - 2\alpha = 0$$

or the envelope is $y^2 = 2x$, a parabola.

There are examples in which the preceding method leads to loci that are not true envelopes. For instance, $(y - \alpha)^2 = x^3$ is a family of curves lying in the right half-plane of which the member α has a cusp on the y-axis at $y = \alpha$. Differentiating with respect to α gives $y - \alpha = 0$, and substituting back gives $x = 0$. This is indeed a singular locus, but it is a locus of cusps rather than an envelope, since the tangent to the member of the family does not coincide with the tangent to the locus. A sufficient condition for the locus defined by Eqs. (0.10.2) and (0.10.3) to be an envelope that is valid for most situations is given by the following:

THEOREM: Let $f(x, y, \alpha)$ be continuous and have continuous partial derivatives ($f_x, f_y, f_\alpha, f_{x\alpha}, f_{y\alpha}, f_{\alpha\alpha}$) in the neighborhood of a point (x_0, y_0, α_0) so that

$$f(x_0, y_0, \alpha_0) = 0, \qquad f_\alpha(x_0, y_0, \alpha_0) = 0$$

$$\begin{vmatrix} f_x & f_y \\ f_{x\alpha} & f_{y\alpha} \end{vmatrix} \neq 0, \qquad f_{\alpha\alpha} \neq 0$$

at (x_0, y_0, α_0). Then the equations

$$f(x, y, \alpha) = 0 \quad \text{and} \quad f_\alpha(x, y, \alpha) = 0$$

define a curve that is the envelope of the family $f(x, y, \alpha) = 0$ in the neighborhood of (x_0, y_0, α_0).

Let us consider now the corresponding situation for surfaces. Let S be a one-parameter family of surfaces

$$f(x, y, z, \alpha) = 0 \tag{0.10.4}$$

If there exists a surface T that is tangent to each of the members of S along a curve C, then T is said to be the envelope of S. The tangency curve C is called a *characteristic curve*. If an envelope to S is to exist, then there must be a curve on each member of S, and the locus of these curves must be tangent to each surface of S along the corresponding curves C on S. As before, then,

$$f(x, y, z, \alpha + \Delta\alpha) - f(x, y, z, \alpha) = 0$$

is a surface passing through the intersection of the two surfaces and hence through the point of tangency in the limit of the envelope and the surface. In the limit

$$f_\alpha(x, y, z, \alpha) = 0 \tag{0.10.5}$$

and hence the elimination of α between this equation and $f(x, y, z, \alpha) = 0$ should give the equation for the envelope.

We can also consider a two-parameter family of surfaces

$$f(x, y, z, \alpha, \beta) = 0 \tag{0.10.6}$$

and ask whether there is an envelope. If again we consider a nearby surface

$$f(x, y, z, \alpha + \Delta\alpha, \beta + \Delta\beta) = 0$$

then

$$f(x, y, z, \alpha + \Delta\alpha, \beta + \Delta\beta) - f(x, y, z, \alpha, \beta) = 0$$

is a surface passing through the intersection of the two surfaces, and in the limit

$$\frac{\partial f}{\partial \alpha} d\alpha + \frac{\partial f}{\partial \beta} d\beta = 0$$

Since α and β are independent, so also are $d\alpha$ and $d\beta$, and hence

$$\begin{aligned} f_\alpha(x, y, z, \alpha, \beta) &= 0 \\ f_\beta(x, y, z, \alpha, \beta) &= 0 \end{aligned} \tag{0.10.7}$$

and the locus of the point $P(x, y, z)$, the result of the elimination of α and β from these two equations and $f(x, y, z, \alpha, \beta) = 0$, is the envelope (if it exists).

Now the envelope in this case is somewhat different than in the two-dimensional case or the single-parameter case. For if we consider the equations in pairs,

$$\begin{cases} f(x, y, z, \alpha, \beta) = 0, \\ f_\alpha(x, y, z, \alpha, \beta) = 0, \end{cases} \qquad \begin{cases} f(x, y, z, \alpha, \beta) = 0 \\ f_\beta(x, y, z, \alpha, \beta) = 0 \end{cases}$$

it is apparent that in general no curve of contact with the envelope can exist here, since for fixed β the first pair will define one curve, while for fixed α the second pair will define a second curve. Since these two lines must lie on a surface, they must intersect in general, and hence the envelope is the locus of points of contact (tangencies) rather than being a locus of characteristic curves.

If one supposes that $\beta = \phi(\alpha)$, then

$$f(x, y, z, \alpha, \phi(\alpha)) = 0$$

is a one-parameter family of surfaces, and the envelope is given by the simultaneous elimination of α from

$$f(x, y, z, \alpha, \phi(\alpha)) = 0$$

and

$$\frac{\partial f}{\partial \alpha} + \frac{\partial f}{\partial \beta} \phi'(\alpha) = 0$$

The envelope then contacts the surface in a characteristic curve, but the characteristic depends on the particular function $\phi(\alpha)$. The totality of all characteristics for different $\phi(\alpha)$, in general, will not sweep out a surface envelope.

As an example, consider the family of unit spheres with centers in $z = 0$:

$$f = (x - \alpha)^2 + (y - \beta)^2 + z^2 - 1 = 0$$

Then

$$-\tfrac{1}{2} f_\alpha = x - \alpha = 0$$

$$-\tfrac{1}{2} f_\beta = y - \beta = 0$$

and the envelope is given by $z = \pm 1$, that is, the pair of horizontal planes with point contacts with each of the spheres with centers on the (x, y)-plane and unit radius.

Suppose now that we pick out a one-parameter family whose centers lie on $y = x^2$, $z = 0$ by setting $\beta = \alpha^2$; then

$$f = (x - \alpha)^2 + (y - \alpha^2)^2 + z^2 - 1 = 0$$

$$f_\alpha = -2(x - \alpha) - 4\alpha(y - \alpha^2) = 0$$

The characteristic curves are circles, of which a typical one is the intersection of the plane $f_\alpha = 0$ with the sphere $f = 0$. The equation of the envelope may be obtained by elimination of α between the two equations, giving a rather complicated expression representing the parabolic tube shown in Fig. 0.7.

A one-parameter family of straight lines generates a *ruled surface*. If the family is given by the two planes

$$x = f_1(\alpha)z + f_2(\alpha)$$

$$y = g_1(\alpha)z + g_2(\alpha)$$

where α is a parameter, then a surface will be generated that depends on the four functions $f_1(\alpha)$, $f_2(\alpha)$, $g_1(\alpha)$, and $g_2(\alpha)$. Consider now a one-parameter family of planes

$$z = \alpha x + f(\alpha)y + \phi(\alpha)$$

The envelope of this family is obtained by eliminating α between this equation and

$$x + yf'(\alpha) + \phi'(\alpha) = 0$$

These two equations represent a straight line, and therefore the characteristic curves are straight lines and the envelope is a ruled surface. This envelope is called a *developable surface*.

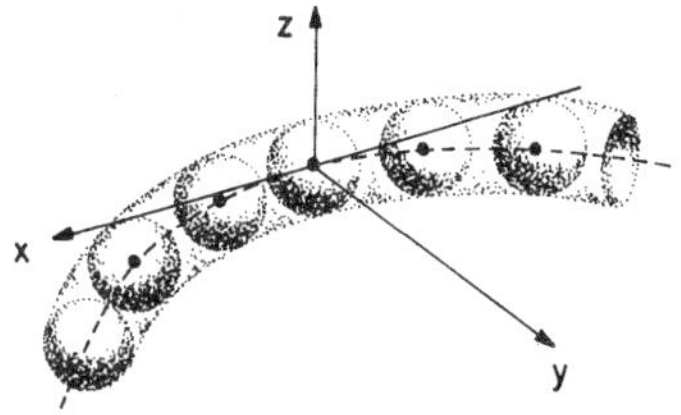

Figure 0.7

Exercises

0.10.1. Draw the family of lines

$$y = \alpha x - \frac{\alpha}{1 + \alpha}$$

for various values of α, and observe that an envelope is formed touching the x- and y-axes at $(1, 0)$ and $(0, -1)$, respectively.

0.10.2. Show that the family of straight lines

$$y = \alpha x - (1 - \kappa)\alpha(\kappa + \alpha)^{-1}$$

where κ is a constant $0 < \kappa < 1$, envelopes a parabola lying in the fourth quadrant and touching the x- and y-axes at $((1 - \kappa)/\kappa, 0)$ and $(0, -(1 - \kappa))$, respectively.

0.10.3. For the family of lines

$$y = \left[1 + \frac{k}{(1 + \alpha)^2}\right](x - 1 + \alpha)$$

where k is a positive constant, determine the envelope in terms of the parameter α; that is, find $x(\alpha)$ and $y(\alpha)$ for the envelope. When $0 \leqslant \alpha \leqslant 1$, sketch the lines and envelope for (a) $k = 3$, (b) $k = 6.75$, and (c) $k = 12$.

0.10.4. Find the envelope of the planes

$$x \cos \alpha + y \sin \alpha + z \cot \theta = 0, \qquad 0 < \theta < \frac{\pi}{2}$$

0.11 Differential Equations

For the system of differential equations

$$\frac{dx}{dt} = f(x, y, z, t)$$

$$\frac{dy}{dt} = g(x, y, z, t) \tag{0.11.1}$$

$$\frac{dz}{dt} = h(x, y, z, t)$$

with the initial conditions

$$x = x_0, \qquad y = y_0, \qquad z = z_0 \qquad \text{at} \qquad t = 0$$

we have the fundamental existence and uniqueness theorem for solutions.

THEOREM: Let the functions $f(x, y, z, t)$, $g(x, y, z, t)$, and $h(x, y, z, t)$ be defined and continuous in a domain D and satisfy a Lipschitz condition in the same domain,

$$| f(x, y, z, t) - f(x', y', z', t') | \leq K(| x - x' | + | y - y' | + | z - z' |)$$

with similar conditions for g and h. Then for each point $\bar{t}$, $\bar{x}$, $\bar{y}$, and $\bar{z}$ in D there exists a unique solution

$$x = x(t)$$
$$y = y(t)$$
$$z = z(t)$$

which is continuous and has continuous derivatives such that

$$\bar{x} = x(\bar{t})$$
$$\bar{y} = y(\bar{t})$$
$$\bar{z} = z(\bar{t})$$

A solution defined in an interval $a < t < b$ is said to be complete if there is no solution defined in a larger interval and agreeing with the given solution in $a < t < b$. If $b < \infty$ and the solution is continuous from the left at b but remains in D, then it is not complete and the solution can be extended beyond b. If $b < \infty$ and the solution is complete, then as $t \to b$ from the left a boundary point of D must be approached. A similar argument holds for $t = a$. Under the hypothesis of the theorem it may be shown that a unique complete solution exists.

Now suppose we are given a curve in (x, y, z)-space in parametric form:

$$x = x_0(\xi)$$
$$y = y_0(\xi) \qquad (0.11.2)$$
$$z = z_0(\xi)$$

And suppose the initial conditions for the system of differential equations are given at a particular value of ξ, say ξ_0. Then the solution of Eqs. (0.11.1) will be of the form

$$x = x(t, x_0(\xi_0))$$
$$y = y(t, y_0(\xi_0))$$
$$z = z(t, z_0(\xi_0))$$

But ξ_0 can be any point on the initial curve; so, dropping the subscript zero, the family of solutions that pass through the curve (0.11.2) is given by

$$x = f_1(t, \xi)$$
$$y = f_2(t, \xi)$$

$$z = f_3(t, \xi)$$

This is a two-parameter family and therefore defines a surface that is generated by the solution curves of the differential equations. It may be shown that the solution depends continuously on the initial data; that is, two solution curves that are close together initially stay close together. Therefore, if the initial curve is continuous, the surface defined by the solution curves will be continuous. (See Fig. 0.8.)

0.12 Strips

We have seen that the three differential equations

$$\frac{dx}{dt} = f(x, y, z), \qquad \frac{dy}{dt} = g(x, y, z), \qquad \frac{dz}{dt} = h(x, y, z)$$

subject to $x = x_0$, $y = y_0$, $z = z_0$ at $t = 0$, define a space curve through the point. We have also seen that the direction normal to the surface $z = z(x, y)$ is defined by the ratio $p:q:-1$, where $p = z_x$ and $q = z_y$ are the partial derivatives of z with respect to x and y. The set of five quantities (x, y, z, p, q) may be thought of as defining a surface element located at (x, y, z) with normal in the direction given by the ratio $p:q:-1$, as shown in Fig. 0.9.

Now suppose that we have a set of five equations with prescribed values at $t = 0$:

$$\begin{aligned}
\frac{dx}{dt} &= f(x, y, z, p, q), & x &= x_0 \\
\frac{dy}{dt} &= g(x, y, z, p, q), & y &= y_0 \\
\frac{dz}{dt} &= h(x, y, z, p, q), & z &= z_0 \\
\frac{dp}{dt} &= k(x, y, z, p, q), & p &= p_0 \\
\frac{dq}{dt} &= l(x, y, z, p, q), & q &= q_0
\end{aligned} \tag{0.12.1}$$

These can be integrated (subject to the usual conditions) and will give the quintuple $x(t)$, $y(t)$, $z(t)$, $p(t)$, $q(t)$. Thus we will have a space curve $x(t)$, $y(t)$, $z(t)$ together with a normal direction $p(t):q(t):-1$ defined at each point. We may think of this as a ribbon or strip, as shown in Fig. 0.9.

However, the differential equations and initial conditions cannot be quite arbitrary if the strip is to be continuous and not have a break of the sort shown in the insert to Fig. 0.9. In the first place, the functions f, g, h, k, and l must be continuous and satisfy a Lipschitz condition as in the previous

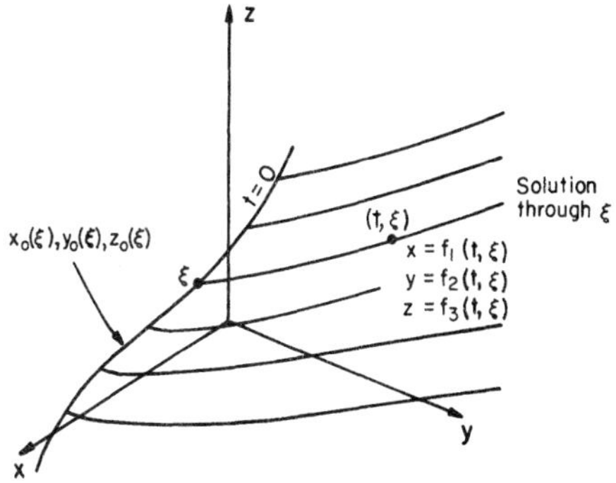

Figure 0.8

section. But we must also have

$$dz = p\,dx + q\,dy$$

for any increments; so

$$pf(x, y, z, p, q) + qg(x, y, z, p, q) - h(x, y, z, p, q) \equiv 0 \qquad (0.12.2)$$

This is the strip condition. It must also be satisfied by the initial values.

Exercise

0.12.1. F and G are functions of the n independent variables $x_i (i = 1, \ldots, n)$ and the n derivatives $p_i = \partial u/\partial x_i$ $(i = 1, \ldots, n)$ of $u(x_1, \ldots, x_n)$. Show that the Poisson bracket

$$(F, G) = \sum_{i=1}^{n} \left(\frac{\partial F}{\partial x_i}\frac{\partial G}{\partial p_i} - \frac{\partial F}{\partial p_i}\frac{\partial G}{\partial x_i} \right)$$

is invariant under change of independent variables.

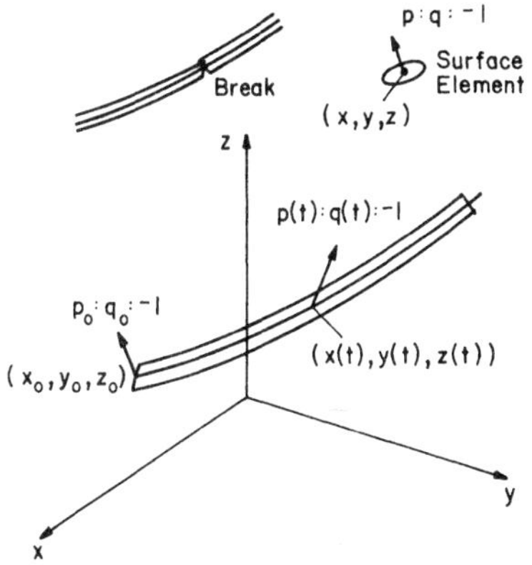

Figure 0.9

REFERENCES

Most of this chapter contains material that is usually presented in courses in advanced calculus or elementary analysis, for which there is a wide variety of textbooks and reference works. Those found the most useful are as follows:

Richard Courant, *Differential and Integral Calculus*:
Volume I, *Single Variable*, 1936
Volume II, *Multiple Variables*, 1937
Nordemann Publishing Co., New York

A. E. Taylor, *Advanced Calculus*, Xerox College Publishing, Greenwich, Conn., 1955.

E. B. Wilson, *Advanced Calculus*, Ginn and Company, Boston, 1912.

E. Goursat, *A Course in Mathematical Analysis*:
Volume I (translated by E. R. Hedrick), 1904
Volume II-1 (translated by E. R. Hedrick and O. Dunkel), 1916
Volume II-2 (translated by E. R. Hedrick and O. Dunkel), Ginn and Company, Boston, 1917
(Volume III, *Partial Differential Equations*, has not been translated.)

Mathematical Models That Give First-order Partial Differential Equations

1

1.1 Introduction

First-order partial differential equations arise most naturally from the simplest models of exchange processes and other physical conservation laws. These will be our primary concern, but we will also see how they can arise from geometrical or purely analytical considerations, for this turns out to be an excellent way of illustrating the structure of the solution. We will also obtain first-order partial differential equations from variational problems and by the introduction of generating functions or limiting assumptions in kinetic equations.

By the phrase "simplest models" of physical processes we mean those that neglect the dispersive effects of heat conduction, diffusion, or viscosity and take cognizance only of convective transport. If a quantity $u(z, t)$ is a function of position and time in a flow in which the velocity is V, then a particle at position z at time t will be found at $z + V\,\delta t$ at time $t + \delta t$. The difference quotient $[u(z + V\,\delta t, t + \delta t) - u(z, t)]/\delta t$ has the limit of

$$V\frac{\partial u}{\partial z} + \frac{\partial u}{\partial t}$$

as $\delta t \to 0$, and we see both first derivatives with respect to distance and with respect to time in this *convective derivative*, as it is called. When diffusion, or conduction, is brought into the picture, we have fluxes that are proportional to the first derivatives with respect to z, and these become second derivatives when built into the balance equations (cf. Exercise 1.11.7). In these simple models, the time derivative appears as the rate of change of a capacity term; there are no acceleration terms to introduce second derivatives with respect to time.

In most physical problems a single partial differential equation can only be obtained by gross idealization or by taking a limiting case. This is because physical phenomena are rarely isolated and are more likely to be coupled together. Thus, in the adsorption of a solute on a packed bed, the process of adsorption will generate heat, which will change the adsorption characteristics of the system. Consequently, one should consider both a mass–balance equation for the concentration and a coupled heat–balance equation for the temperature and its effect. If the heat of adsorption is small or the feed very dilute, it may be possible to consider the system as isothermal, but this is the kind of idealization that is needed to reach a single equation. Even when the assumption of isothermality is made, we should take into account the finite rate of adsorption, and this would give a second equation. However, in the limiting case of extremely rapid adsorption, we can use an equilibrium relationship for the adsorbed concentration and recover a single equation.

It is clear from general mathematical experience that single equations are easier to solve than systems, and linear equations are simpler than nonlinear. In building up an understanding of solutions and learning the techniques for finding them, it is obviously necessary to begin with the simplest kinds. But these should not be overlooked from the physical point of view either, for the kind of mathematical model that is constructed for a physical system depends not only on the nature of the system but also on the kinds of answers that the model is to give. The mark of a good scientist or engineer is to be able to write down a model that suffices for the job—neither more nor less—and this is one of the creative aspects of science and engineering. The creative part of mathematics in this context is to discover the solution, either quantitatively, by giving formulas and numerical results, or qualitatively, by discerning the general behavior or topological features of the solution. At one time these functions were considered to be different tasks to be performed by different people. Now their essential interdependence is much more clearly recognized, and they are more likely to be performed by the same individual.

1.2 Chromatography of a Single Solute

Chromatography is a process by which a separation of chemical species is obtained by selective adsorption on a solid medium. The general problem is extremely complicated. We will consider in this volume only a few of the simple problems; the more complex ones are to be treated in a second volume.

For the moment we will suppose that we have a single chemical species that is adsorbed from a dilute solution in an inert solvent moving through the interstices of a finely divided solid bed of particles, much like water filtering through fine sand. The solute is adsorbed on the solid surface and there are no distributed parameters in the system. Let

c = concentration of solute in moles per unit volume in the fluid phase
n = concentration of solute on the solid in moles per unit volume of solid
ε = fractional void volume in the bed
A = cross-sectional area of the bed
V = interstitial velocity of fluid through the bed
z = space variable along the bed
t = time
$c_0(z)$ = initial distribution of solute in the interstitial fluid
$n_0(z)$ = initial distribution of adsorbed solute
$c_f(t)$ = influent concentration of solute as a function of time

If we consider Fig. 1.1 and make a simple shell balance, we obtain

$$(Vc\varepsilon A)_{z,t} - (Vc\varepsilon A)_{z+\Delta z,t} = \left\{\frac{\partial}{\partial t}[c\varepsilon A\,\Delta z + n(1-\varepsilon)A\,\Delta z]\right\}_{\bar{z},t}, \qquad z < \bar{z} < z + \Delta z$$

where the left side is the rate of flow of solute into the shell minus the rate of flow from the shell, and the right side is the rate of accumulation of solute

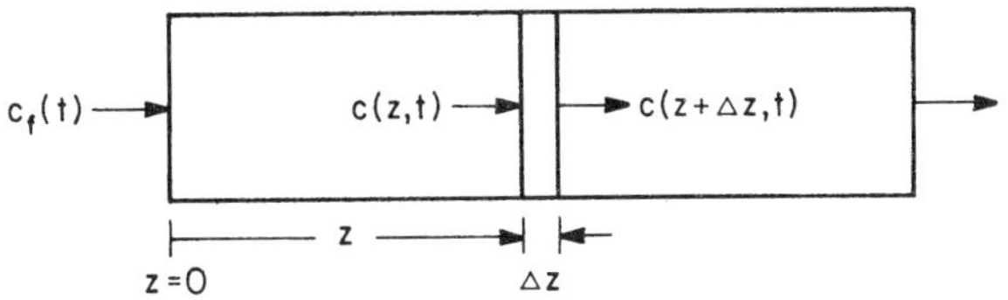

Figure 1.1

in the shell. If we assume that V, ε, and A are constant and take the limit, we obtain

$$\varepsilon V \frac{\partial c}{\partial z} + \varepsilon \frac{\partial c}{\partial t} + (1 - \varepsilon) \frac{\partial n}{\partial t} = 0 \tag{1.2.1}$$

This is the basic conservation equation and, as stated earlier, contains no second-order derivatives, since all transport mechanisms except forced convection has been neglected. This equation involves no assumption about the mechanism of adsorption, and before any solution can be attempted something must be said about this. If there is a finite rate of transfer of solute from solution to adsorbent, then we must assume that this rate depends on the pertinent parameters of the problem; that is,

$$(1 - \varepsilon) \frac{\partial n}{\partial t} = F(c, n, V, \varepsilon, \text{solute, inert carrier, etc.})$$

A wide variety of choices is available here, and we shall investigate only a few. We might assume, as chemical engineers frequently do, that the rate of transfer of solute is determined by the rate of transfer of mass through the *stagnant* film about the particle, other processes occurring at equilibrium. This produces an expression

$$(1 - \varepsilon) \frac{\partial n}{\partial t} = a_v k(c - c^*) \tag{1.2.2}$$

where c^* would be the concentration in the fluid phase if it were in equilibrium with the adsorbed phase, a_v is the superficial surface area for mass transfer per unit bed volume, and k is a lumped constant diffusional conductance. The equilibrium relation is of the form

$$n = f(c) \tag{1.2.3}$$

where the function f is, in general, temperature dependent. Here k depends on the velocity V and the physical properties of the interstitial fluid, as well as the geometry of the solid particles, and is a well-correlated function available in standard works. Equations (1.2.1) to (1.2.3) may then be combined to give as a mathematical model

$$\begin{aligned} \varepsilon V \frac{\partial c}{\partial z} + \varepsilon \frac{\partial c}{\partial t} + (1 - \varepsilon) \frac{\partial n}{\partial t} &= 0 \\ (1 - \varepsilon) \frac{\partial n}{\partial t} &= a_v k(c - c^*) \\ n &= f(c^*) \end{aligned} \tag{1.2.4}$$

and to these equations we should append the initial and boundary values

$$\text{at } x = 0, \qquad c = c_f(t)$$
$$\text{at } t = 0, \qquad c = c_0(z) \tag{1.2.5}$$
$$\text{at } t = 0, \qquad n = n_0(z)$$

Another form the rate of adsorption might assume is

$$\frac{1 - \varepsilon}{a_v} \frac{\partial n}{\partial t} = k_1 c(N - n) - k_2 n \tag{1.2.6}$$

where k_1 and k_2 are temperature-dependent rate constants and N is the saturation concentration of the solute. This is a rate law, which under equilibrium conditions corresponds to the Langmuir isotherm

$$n = N \frac{Kc}{1 + Kc}, \qquad K = \frac{k_1}{k_2} \tag{1.2.7}$$

Alternatively, one might assume that the rate of adsorption is a linear function,

$$\frac{1 - \varepsilon}{a_v} \frac{\partial n}{\partial t} = k_1 c - k_2 n \tag{1.2.8}$$

This enables simple mathematical solutions, which, however, are not physically realistic for anything but very dilute solutions. In all these models we have ignored the effect of changes in temperature, which may be justified for dilute solutions because of the great heat capacity of the liquid and solid. We have also assumed that the problem is one dimensional in the space variable.

If one makes the assumption of local equilibrium (i.e., at each point in the column there is equilibrium between solution and adsorbed species), then Eq. (1.2.3) may be substituted into Eq. (1.2.1) to give

$$\varepsilon V \frac{\partial c}{\partial z} + [\varepsilon + (1 - \varepsilon) f'(c)] \frac{\partial c}{\partial t} = 0 \tag{1.2.9}$$

Here we have arrived at a single equation, and the appropriate auxiliary conditions would be

$$c(0, t) = c_f(t), \qquad c(z, 0) = c_0(z) \tag{1.2.10}$$

The specification of the initial adsorbed concentration drops out, since consistency with the assumption of equilibrium requires that $n_0(z) = f(c_0(z))$. The function $f(c)$ is often called the *adsorption isotherm*, since it is fixed when the temperature is fixed. We see in Eq. (1.2.9) that its derivative occurs in the combination $\varepsilon + (1 - \varepsilon) f'(c)$. In fact, $\varepsilon c + (1 - \varepsilon) f(c) = \varepsilon g(c)$ is the total amount of solute per unit volume of the column held both as adsorbate and as solution in the interstices. The function $g(c)$ is sometimes called the *column isotherm*, and Eq. (1.2.9) can be written

$$V\frac{\partial c}{\partial z} + g'(c)\frac{\partial c}{\partial t} = 0 \tag{1.2.11}$$

Another modification of the equation was discussed in Sec. 0.2.

These forms of the chromatographic equations will often be used to illustrate the principles and methods of solution.

Exercises

1.2.1. Obtain an equation for $\partial n/\partial t$ when there is a mass-transfer resistance given by Eq. (1.2.2) and a finite rate of adsorption at the surface given by an equation similar to Eq. (1.2.6). Discuss the limiting cases.

1.2.2. Consider a chromatographic disc of depth d in which the solvent flows radially from an inner surface of radius a to an outer surface of radius b. If ε is the fractional free space and Q the volume flow rate of solvent, derive an equation connecting $c(r, t)$ and $n(r, t)$, the concentrations in the two phases. What are the boundary and initial conditions? Transform r and t so that the equation may be reduced to Eq. (1.2.1).

1.2.3. In the preparation of a spherical catalyst particle, the dry pellet is sometimes immersed in a large volume of liquid containing various precursors. The liquid is sucked into the porous pellet by capillary imbibition and, if the motion is completely symmetrical, will trap and compress the air initially in the pores into a central spherical region. The various components, present in concentrations c_{fi} in the bulk solution, are adsorbed on the surface as the fluid moves in: $c(r, t)$ and $n(r, t)$ being the concentrations within the pore and on the adjacent solid surface at radius r and time t. Set up the equations for c and n under various assumptions.

1.2.4. Using the adsorption equilibrium case in the previous problem, take $x = (a^3 - r^3)/a^3$ and $y = (3/a^3)\int_0^t r^2 v(r, t)\,dt$ = fractional volume imbibed at time t and show that the equations of column chromatography are regained. Here a denotes the radius of the catalyst pellet and $v(r, t)$ the speed of liquid imbibition.

1.3 Chromatography of Several Solutes

For each solute species present, we can write a conservation equation of exactly the same form as before. Letting a suffix i distinguish the species, we therefore have

$$V\frac{\partial c_i}{\partial z} + \frac{\partial c_i}{\partial t} + \frac{1-\varepsilon}{\varepsilon}\frac{\partial n_i}{\partial t} = 0 \tag{1.3.1}$$

We could make assumptions about the finite rate of adsorption similar to those made before. However, there is little enough information on multi-component isotherms and virtually nothing on simultaneous rates of adsorption, so there is little point in going beyond the equilibrium case.

The multicomponent Langmuir isotherm is given by

$$n_i = \frac{N_i K_i c_i}{1 + \Sigma K_j c_j}, \qquad i = 1, 2, \ldots, m \tag{1.3.2}$$

where the summation is taken over all solutes present and the K_i are functions of temperature, having units of a reciprocal concentration. The parameter N_i denotes the limiting value of n_i, which is defined as the maximum number of moles of solute i that can be adsorbed per unit volume of adsorbent. According to Kemball, Rideal, and Guggenheim [*Trans. Faraday Soc.*, **44** 948 (1944)], the value of N_i must be the same for all solutes; otherwise, there is a contradiction with the Gibb's adsorption isotherm. In practice, however, the experimental value of N_i varies from one solute to the other.

The derivatives of n_i are worth noting, for we see that

$$\frac{\partial n_i}{\partial c_i} = \frac{N_i K_i (1 + \Sigma' K_j c_j)}{(1 + \Sigma K_j c_j)^2} \tag{1.3.3}$$

where in Σ' the term $K_i c_i$ is omitted, and for $k \neq i$,

$$\frac{\partial n_i}{\partial c_k} = -\frac{N_i K_i K_k c_i}{(1 + \Sigma K_j c_j)^2} \tag{1.3.4}$$

Clearly, $\partial n_i/\partial c_i$ is positive, showing that an increase in concentration of any one species, with other concentrations kept constant, increases the adsorbed concentration of that species. On the other hand, $\partial n_i/\partial c_k$ is negative, corresponding to the fact that an increase in concentration of solute k will decrease the adsorbed concentrations of the other solutes. For the general isotherm

$$n_i = f_i(c_1, c_2, \ldots, c_m) \tag{1.3.5}$$

we might well lay down the restrictions

$$\frac{\partial f_i}{\partial c_j} \begin{cases} > 0, & \text{if } i = j \\ < 0, & \text{if } i \neq j \end{cases} \tag{1.3.6}$$

To apply the rate equation for the mass-transfer limitation case, we would need to express c_i^*, the equilibrium concentration c_i, as a function of the n_j. Now, from Eq. (1.3.2),

$$\left(1 + \sum K_j c_j\right) \frac{n_i}{N_i} = K_i c_i$$

so

$$\left(1 + \sum K_j c_j\right) \sum \frac{n_i}{N_i} = \sum K_i c_i$$

or changing the dummy suffix i to j,

$$1 + \sum K_j c_j = \left(1 - \sum \frac{n_j}{N_j}\right)^{-1}$$

Thus

$$c_i^* = \frac{(1/K_i)\,(n_i/N_i)}{1 - \sum (n_j/N_j)} \tag{1.3.7}$$

and then

$$\frac{1-\varepsilon}{a_v}\frac{\partial n_i}{\partial t} = k_i(c_i - c_i^*) \tag{1.3.8}$$

The adsorption itself will be rate determining if we write

$$\frac{1-\varepsilon}{a_v N_i}\frac{\partial n_i}{\partial t} = k_{1i}\, c_i\left(1 - \sum \frac{n_i}{N_i}\right) - k_{2i}\frac{n_i}{N_i} \tag{1.3.9}$$

If k_{1i} and k_{2i} all become large, with $k_{1i}/k_{2i} = K_i$, then we recover the equilibrium relations of Eq. (1.3.2).

Exercises

1.3.1. Show that the multicomponent Langmuir isotherm is convex in the following sense. If $c_i^{(1)}$ and $c_i^{(2)}$ are two sets of concentrations with $c_i^{(1)} \geqslant c_i^{(2)}$, and $n_i = f_i(c)$ denotes Eq. (1.3.2), then

$$f_i(\lambda c^{(1)} + (1-\lambda)c^{(2)}) \geqslant \lambda f_i(c^{(1)}) + (1-\lambda) f_i(c^{(2)})$$

for $0 \leqslant \lambda \leqslant 1$.

1.3.2. The Langmuir isotherm, Eq. (1.3.2), carries an underlying assumption that the solvent or the carrier behaves as an inert component. If the solvent is the least adsorbable species and the total pressure remains fixed, show that the isotherm may be given by Eq. (1.3.2) with the summation taken over solutes only. How would the parameters N_i and K_i be modified?

1.3.3. Show that the constant-separation-factor equilibrium relations employed for ion-exchange systems [see, for example, Tondeur and Klein, *Ind. Eng. Chem. Fundam.* **6**, 351 (1967)] can be rewritten in the form of Eq. (1.3.2) in which the summation excludes the species with the smallest separation factor.

1.3.4. In Wilhelm's parapump, the fluid is passed forth and back through a fixed-bed adsorber with an oscillating velocity. In his notation [*Ind. Eng. Chem. Fundam.* **7**, 337 (1968)], the basic equations of the process are

$$\alpha f(t)\frac{\partial \phi_{fi}}{\partial z} + \frac{\partial \phi_{fi}}{\partial t} + \kappa\frac{\partial \phi_{si}}{\partial t} - \eta\frac{\partial^2 \phi_{fi}}{\partial z^2} = 0$$

$$\frac{\partial \phi_{si}}{\partial t} + \lambda(\phi_{fi}^* - \phi_{fi}) = 0$$

$$\phi_{fi}^* = \phi_{fi}^*(T(t), \phi_{sj})$$

Identify each of the terms, explaining its physical significance and relating it to other physical quantities where possible. Under total reflux, the exit stream is collected for half a cycle, is mixed, and returns through the bed during the next half-cycle. Ignore the last term in the first equation so that it becomes first order, and write down the boundary conditions, assuming $f(t) \geqslant 0$, $n\theta \leqslant t \leqslant (n + \frac{1}{2})\theta$; $f(t) \leqslant 0$, $(n + \frac{1}{2})\theta \leqslant t \leqslant (n + 1)\theta$; and that the length of the column is 1.

1.4 Chromatography with Heat Effects

The mass balance equations we have derived are valid also in the nonisothermal case, provided that the velocity V is constant. If V is not constant, it must be kept within the differentiation with respect to z. We will proceed to the general case in which the heat effects must be considered. The equations for the conservation of the several species are†

$$\varepsilon \frac{\partial(Vc_i)}{\partial z} + \varepsilon \frac{\partial c_i}{\partial t} + (1 - \varepsilon) \frac{\partial n_i}{\partial t} = 0, \qquad i = 1, 2, \ldots, m \tag{1.4.1}$$

There are m equations, and these are independent if the inert carrier is not included among the species. If the inert carrier is included among these equations, they may be multiplied by the appropriate molecular weight of each species M_i and summed to give

$$\varepsilon \frac{\partial\left(V \sum c_i M_i\right)}{\partial z} + \varepsilon \frac{\partial \sum c_i M_i}{\partial t} + (1 - \varepsilon) \frac{\partial}{\partial t} \sum n_i M_i = 0$$

or (1.4.2)

$$\varepsilon \frac{\partial(V\rho)}{\partial z} + \varepsilon \frac{\partial \rho}{\partial t} + (1 - \varepsilon) \frac{\partial}{\partial t}\left(\sum n_i M_i\right) = 0$$

where $\rho = \sum c_i M_i$ is the density, and the sums both here and in all that follows run from 1 to $m + 1$. This is the continuity equation for mass and hence is not independent. Thus in either case there are $m + 1$ equations (i.e., equations for either m components plus the inert or m components plus the overall continuity equation).

We will now write the energy balance over a shell of thickness Δz as before, making use of the energy relation and assuming an adiabatic system

$$\left[\frac{\partial(\rho U)}{\partial t}\right]_{\bar{z}} \Delta z = (V\rho U)_z - (V\rho U)_{z+\Delta z} + (Vp)_z - (Vp)_{z+\Delta z}, \qquad z < \bar{z} < z + \Delta z$$

where U is the internal energy per unit mass and p represents the pressure.

†Note that the ε appearing here and in the following assumes that the adsorbent is not porous. If the adsorbent is porous, then the first ε is different from the second and third.

Using the partial molar internal energy $\tilde{U}_i$ in our balances, the right side becomes

$$(\varepsilon \sum Vc_i\tilde{U}_i)_z - (\varepsilon \sum Vc_i\tilde{U}_i)_{z+\Delta z} + \varepsilon\,(Vp)_z - \varepsilon(Vp)_{z+\Delta z}$$

The left side is

$$\left\{\frac{\partial}{\partial t}\left[\Delta z\varepsilon \sum \tilde{U}_i c_i + \Delta z(1-\varepsilon)\sum \tilde{U}_{is} n_i + \Delta z(1-\varepsilon)\rho_s\hat{U}_s\right]\right\}, \qquad z < \bar{z} < z + \Delta z$$

where $\tilde{U}_{is}$ is the partial molar internal energy of the adsorbed species and $\rho_s\hat{U}_s$ the internal energy of the adsorbent. Equating these two quantities and passing to the limit, one obtains

$$-\varepsilon\frac{\partial}{\partial z}\sum Vc_i\tilde{U}_i - \varepsilon\frac{\partial(Vp)}{\partial z} = \frac{\partial}{\partial t}\left[\varepsilon\sum c_i\tilde{U}_i + (1-\varepsilon)\sum \tilde{U}_{is}n_i + (1-\varepsilon)\rho_s\hat{U}_s\right]$$

This may be written

$$-\varepsilon\frac{\partial}{\partial z}\left[V\left(\sum c_i\tilde{H}_i - p\sum c_i\tilde{V}_i + p\right)\right] - \varepsilon\sum\tilde{U}_i\frac{\partial c_i}{\partial t} - (1-\varepsilon)\sum\tilde{U}_{is}\frac{\partial n_i}{\partial t} = \varepsilon\sum c_i\frac{\partial\tilde{U}_i}{\partial t} + (1-\varepsilon)\sum n_i\frac{\partial\tilde{U}_{is}}{\partial t} + (1-\varepsilon)\rho_s\frac{\partial\hat{U}_s}{\partial t},$$

where $\tilde{H}_i$ and $\tilde{V}_i$ are the partial molar enthalpy and volume, respectively, and

$$\tilde{H}_i = \tilde{U}_i + p\tilde{V}_i$$

If we make use of the conservation equation for each species and assume that heat capacities are constant, this equation may be written

$$-\varepsilon\sum(\tilde{H}_i - \tilde{U}_i)\frac{\partial(Vc_i)}{\partial z} - \varepsilon V\sum c_i\tilde{C}_{pi}\frac{\partial T}{\partial z} - (1-\varepsilon)\sum(\tilde{U}_{is} - \tilde{U}_i)\frac{\partial n_i}{\partial t} = \left[\varepsilon\sum c_i\tilde{c}_{vi} + (1-\varepsilon)\sum n_i\tilde{c}_{vis} + (1-\varepsilon)\rho_s\hat{C}_s\right]\frac{\partial T}{\partial t}$$

Within the accuracy of this model, it is probably reasonable to neglect $\tilde{H}_i - \tilde{U}_i$ and to assume that $\tilde{U}_{is} - \tilde{U}_i = \Delta H_i$, which is the molar heat of adsorption of the ith species. Thus we can write

$$-\varepsilon V\frac{\partial T}{\partial z}\sum c_i\tilde{C}_{pi} + (1-\varepsilon)\sum(-\Delta H_i)\frac{\partial n_i}{\partial t} = \left[\varepsilon\sum c_i\tilde{C}_{vi} + (1-\varepsilon)\sum n_i\tilde{C}_{vis} + (1-\varepsilon)\rho_s\hat{C}_s\right]\frac{\partial T}{\partial t} \qquad (1.4.3)$$

which is the heat balance under the assumption of complete thermal and physical equilibrium. The term that dominates in the brackets on the right side is the heat capacity of the solid adsorbent.

The momentum balance on the interstitial fluid may be written as

$$\frac{\partial p}{\partial z} + \frac{\partial(\rho V^2)}{\partial z} + \frac{f}{A}\rho V^2 = \frac{\partial(\rho V)}{\partial t}$$

where f, the friction factor, is a function of the Reynolds number and A is the gross cross-sectional area of the bed. This may be written

$$\frac{\partial p}{\partial z} + 2\rho V\frac{\partial V}{\partial z} + V^2\frac{\partial \rho}{\partial z} + \frac{f}{A}\rho V^2 = \frac{\partial(\rho V)}{\partial t} \tag{1.4.4}$$

If we assume the ideal gas law is valid, then

$$p = RT\sum c_i \tag{1.4.5a}$$

$$pM = \rho RT, \qquad \rho = \sum c_i M_i \tag{1.4.5b}$$

where M is the average molecular weight of the mixture, and the pressure terms could be eliminated from all the equations.

Now the dependent variables in the system are $c_1, c_2, \ldots, c_{m+1}, T, \rho, V$, and p. The adsorbed concentrations n_i are also unknown, but these are assumed to be related by the equilibrium expressions

$$n_i = n_i\,(c_1, c_2, \ldots, c_{m+1}, T, p)$$

to the quantities c_i. If we assume that the inert species in the gas phase is one of the c_j, then there are m unknown c_i's and thus there are $m + 4$ unknown dependent variables. There are also $m + 4$ equations connecting these variables [Eqs. (1.4.1), (1.4.3), (1.4.4), (1.4.5a), and (1.4.5b)], and hence the problem seems to be properly posed. We would specify the initial distribution of the solute concentrations c_i and T, and this fixes the initial pressure and the density, as well as the initial adsorbed concentrations. The initial velocity would also have to be given at $z = 0$; that is at the inlet, one would specify the concentrations and the temperature, as well as the pressure and velocity. The method of procedure to obtain the solution is more or less straightforward, but to the author's knowledge a problem of this generality has never been solved in complete detail.

If it is assumed that the interstitial velocity is constant, we are to solve Eq. (1.4.3) simultaneously with the set of m equations in Eq. (1.4.1). Let us put

$$C_f = \sum c_i\tilde{C}_{pi} \cong \sum c_i\tilde{C}_{vi} = \text{heat capacity per unit volume of the fluid phase} \tag{1.4.6a}$$

$$C_s = \sum n_i\tilde{C}_{vis} + \rho_s\hat{C}_s = \text{heat capacity per unit volume of the solid phase} \tag{1.4.6b}$$

Note that in Eq. (1.4.6a) the summation covers all solutes and the inert carrier.

Then Eq. (1.4.3) becomes

$$-\varepsilon V C_f \frac{\partial T}{\partial z} + (1 - \varepsilon) \sum (-\Delta H_i) \frac{\partial n_i}{\partial t} = \varepsilon C_f \frac{\partial T}{\partial t} + (1 - \varepsilon) C_s \frac{\partial T}{\partial t} \qquad (1.4.7)$$

If we let

$$c_{m+1} = C_f(T - T_0) = \text{energy concentration per unit volume of the fluid phase} \qquad (1.4.8a)$$

$$n_{m+1} = C_s(T - T_0) - \sum (-\Delta H_i) n_i = \text{energy concentration per unit volume of the solid phase} \qquad (1.4.8b)$$

where T_0 denotes the reference temperature, Eq. (1.4.7) can be reduced to the form

$$V \frac{\partial c_{m+1}}{\partial z} + \frac{\partial c_{m+1}}{\partial t} + \frac{1 - \varepsilon}{\varepsilon} \frac{\partial n_{m+1}}{\partial t} = 0 \qquad (1.4.9)$$

and the function $n_{m+1} = f_{m+1}(c_1, c_2, \ldots, c_m, c_{m+1})$ satisfies Eq. (1.3.6). Consequently, the problem with heat effects may be regarded as the same problem with one more solute.

While Eq. (1.3.2) can be used to describe the adsorption equilibrium, the parameter K_i is no longer a constant but shows a strong temperature dependence as

$$K_i = K_{0i} T^{1/2} e^{-\Delta H_i / RT}, \qquad i = 1, 2, \ldots, m \qquad (1.4.10)$$

in which k_{0i} is a constant, ΔH_i the heat of adsorption per mole of the ith species, and R the gas law constant. Since adsorption is an exothermic process, $\Delta H_i < 0$ for all i, and this guarantees that the function $n_{m+1} = f_{m+1}$ defined by Eq. (1.4.8b) will satisfy the inequality conditions in Eq. (1.3.6).

There are many levels of model building for this problem, and it seems appropriate to consider the model in which there are heat and mass-transfer resistances at the particle surface but in which the adsorption process itself occurs under equilibrium conditions. In this case the concentrations and temperature are different in the interstitial fluid and in the adsorbent particle volume. Since this problem is somewhat complicated, we will summarize the notation:

c_i = concentration of ith species in the interstitial volume
c_{is} = concentration of the ith species in the intraparticle voids
T = temperature of interstitial fluid
T_s = solid temperature
n_i = adsorbed species concentration

k_i = mass-transfer coefficient for ith species
h = heat-transfer coefficient
a_v = superficial surface area of adsorbent per unit volume of bed
M_i = molecular weight of the ith species
p = pressure
V = interstitial velocity
ε = fractional void volume in bed
α = fractional void volume in particle
ρ = density
$\tilde{U}_i$ = partial molar internal energy of the ith species in the intersticial fluid
$\tilde{U}'_{is}$ = partial molar internal energy of the ith species in the intraparticle fluid
$\tilde{U}_{is}$ = partial molar internal energy of ith species in the adsorbed state
$\hat{U}_s$ = internal energy of adsorbent per unit mass
$\tilde{H}_{is}$ = partial molar enthalpy of the ith species in the adsorbed state
$\tilde{H}_i$ = partial molar enthalpy of the ith species in the interstitial fluid
$\tilde{C}_{vi}, \tilde{C}'_{vis}, C_{vis}$ = molar heat capacities corresponding to the preceding internal energies
$\tilde{C}_{pi}$ = molar heat capacity corresponding to the enthalpy, $\tilde{H}_i$
$\hat{C}_s$ = specific heat of adsorbent particles
ρ_s = density of the adsorbent particles

The conservation equation for each of the species in the interstitial fluid is

$$\varepsilon \frac{\partial(Vc_i)}{\partial z} + \varepsilon \frac{\partial c_i}{\partial t} + a_v k_i(c_i - c_{is}) = 0, \qquad i = 1, 2, \ldots, m \tag{1.4.11}$$

For the solid phase we have

$$\alpha(1 - \varepsilon) \frac{\partial c_{is}}{\partial t} + (1 - \varepsilon) \frac{\partial n_i}{\partial t} = a_v k_i(c_i - c_{is}) \tag{1.4.12}$$

The continuity equation for the interstitial fluid is

$$\varepsilon \frac{\partial(\rho V)}{\partial z} + \varepsilon \frac{\partial \rho}{\partial t} + a_v \sum k_i M_i(c_i - c_{is}) = 0 \tag{1.4.13}$$

As before, this equation is not independent of Eq. (1.4.11) if all the species as well as the inert are included in that equation.

For the energy equation, we have some choices, since we can write conservation equations on the fluid, on the solid, and an overall balance. We choose to write the first two, although it is clear that the third may be obtained by summing the first two. The conservation equation on the interstitial fluid is

$$\varepsilon \frac{\partial}{\partial t}\left(\sum c_i \tilde{U}_i\right) = -\varepsilon \frac{\partial}{\partial z}\left(\sum c_i \tilde{U}_i V\right) - \varepsilon \frac{\partial (pV)}{\partial z} + a_v h(T_s - T)$$

$$-a_v \sum k_i(c_i - c_{is})\left[\tilde{H}_{is} - (\tilde{H}_{is} - \tilde{H}_i)\frac{1}{2}\left(1 + \frac{c_i - c_{is}}{|c_i - c_{is}|}\right)\right] \tag{1.4.14}$$

The complicated nature of the last term accounts for the fact that one must know which species are leaving the interstitial fluid and which species are entering it from the particle. The conservation equation for the solid phase is

$$(1 - \varepsilon)\frac{\partial}{\partial t}[\alpha \sum c_{is}\tilde{U}'_{is} + \sum n_i \tilde{U}_{is} + \rho_s \hat{U}_s] - ha_v(\mathrm{T} - \mathrm{T}_s)$$

$$= a_v \sum k_i(c_i - c_{is})\left[\tilde{H}_{is} - (\tilde{H}_{is} - \tilde{H}_i)\frac{1}{2}\left(1 + \frac{c_i - c_{is}}{|c_i - c_{is}|}\right)\right] \tag{1.4.15}$$

In these equations we have assumed that when mass is interchanged between the fluid and solid it carries with it its partial molar enthalpy. Equations (1.4.11) and (1.4.14) may be combined to give

$$\varepsilon \sum (\tilde{U}_i - \tilde{H}_i)\frac{\partial c_i}{\partial t} + \varepsilon \sum c_i \tilde{C}_{vi}\frac{\partial T}{\partial t} + \varepsilon V \frac{\partial T}{\partial z}\sum c_i \tilde{C}_{pi}$$

$$= a_v h(T_s - T) - a_v \sum k_i(c_i - c_{is})(\tilde{H}_{is} - \tilde{H}_i)\frac{1}{2}\left(1 - \frac{c_i - c_{is}}{|c_i - c_{is}|}\right) \tag{1.4.16}$$

As before, we can probably neglect the term with $\tilde{\mathrm{U}}_i - \tilde{\mathrm{H}}_i$. The heat carried with the diffusing molecules may be small and this might be neglected, so the heat balance could be written as

$$\varepsilon \frac{\partial T}{\partial t}\left(\sum c_i \tilde{C}_{vi}\right) + \varepsilon V \frac{\partial T}{\partial z}\left(\sum c_i \tilde{C}_{pi}\right) = a_v h(T_s - T) \tag{1.4.17}$$

the summations being the static and dynamic heat capacities of the interstitial fluid. Combination of Eqs. (1.4.12) and (1.4.15) gives

$$(1 - \varepsilon)\left[\alpha \sum (\tilde{U}'_{is} - \tilde{H}_i)\frac{\partial c_{is}}{\partial t} + \sum (\tilde{U}_{is} - \tilde{H}_i)\frac{\partial n_i}{\partial t}\right]$$

$$+ (1 - \varepsilon)\left(\alpha \sum c_{is}\tilde{C}'_{vis} + \sum n_i \tilde{C}_{vis} + \hat{\mathrm{C}}_s \rho_s\right)\frac{\partial \mathrm{T}_s}{\partial \mathrm{t}} \tag{1.4.18}$$

$$= a_v \sum k_i(c_i - c_{is})(\tilde{H}_{is} - \tilde{H}_i)\frac{1}{2}\left(1 - \frac{c_i - c_{is}}{|c_i - c_{is}|}\right) + ha_v(T - T_s)$$

Note that the second term on the left side is related to the heat of adsorption, while the last term is the capacity term for the adsorbent. Quite possibly all the other terms on the left side are small compared to these. The heat-transfer term on the right side will dominate the other term, generally, so we can write

$$(1 - \varepsilon) \sum \Delta H_i \frac{\partial n_i}{\partial t} + (1 - \varepsilon) \hat{C}_s \rho_s \frac{\partial T_s}{\partial t} = h a_v (T - T_s) \tag{1.4.19}$$

To solve this problem, then, one would have to consider Eqs. (1.4.11), (1.4.12), (1.4.17), (1.4.19), (1.4.4), and (1.4.5) and appropriate initial and influent specifications. The dependent variables are c_i, T, T_s, n_i, V, and p. The system of equations is made up of five first-order partial differential equations (actually $m + 4$) and one algebraic equation, Eq. (1.4.5a).

Exercise

1.4.1. Derive a pair of first-order partial differential equations to describe the drying of a fixed bed using the following notation.

M = mass flow rate of drying air
l = distance from inlet to bed
L = total length of bed
x = mass fraction of water vapor in air at (l, t)
y = mass fraction of water on solid
W = mass of solid per unit length of bed
t = time
T = temperature at (l, t)
m = mass of air in interstices of a unit length of bed ($m << W$)
C_s = specific heat of solid phase
C_f = specific heat of air
ΔH = latent heat of evaporation

1.5 Countercurrent Adsorber

From an engineering point of view, it is desirable to have a continuous operation for an adsorber, and to do this we make the adsorbent move countercurrently against the fluid phase. In this case we have to consider the convective transport by the solid phase in Eq. (1.3.1). Using the notation as denoted in Fig. 1.2, we have the equation

$$\varepsilon V_f \frac{\partial c_i}{\partial z} - (1 - \varepsilon) V_s \frac{\partial n_i}{\partial z} + \varepsilon \frac{\partial c_i}{\partial t} + (1 - \varepsilon) \frac{\partial n_i}{\partial t} = 0 \tag{1.5.1}$$

from the material balance for species i. Note that V_f and V_s are the interstitial velocity of the fluid phase and the speed of the solid phase, respectively. If we have equilibrium established everywhere, then the isotherm relates n_i as

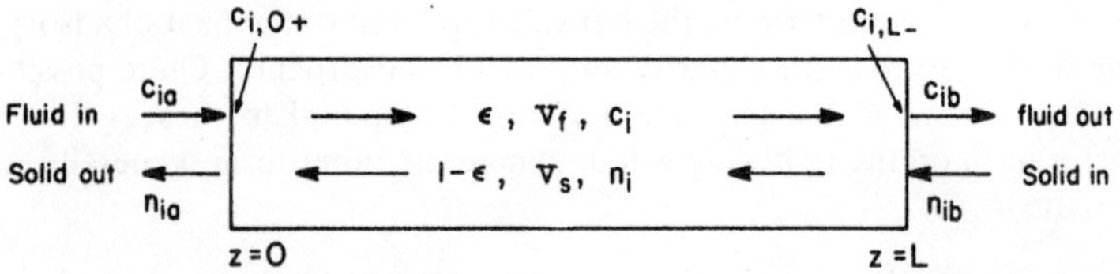

Figure 1.2

a function of c_i's. Thus we have

$$n_i = f_i(c_1, c_2, \ldots, c_m) \tag{1.5.2}$$

in general, or we can use the Langmuir isotherm, Eq. (1.3.2), as a specific example. Otherwise, we will have to introduce appropriately one of the rate expressions discussed in Sec. 1.2 for single solute cases and in Sec. 1.3 for the case of several solutes. For instance, if the adsorption is rate limiting in a multicomponent system, we can write

$$\frac{1-\varepsilon}{a_v N_i}\left\{-V_s\frac{\partial n_i}{\partial z} + \frac{\partial n_i}{\partial t}\right\} = k_{1i}c_i\left(1 - \sum \frac{n_j}{N_j}\right) - k_{2i}\frac{n_i}{N_i} \tag{1.5.3}$$

The initial distribution of solutes will be prescribed separately in each phase

$$\text{at } t = 0, \qquad c_i = c_{i0}(z) \quad \text{and} \quad n_i = n_{i0}(z) \tag{1.5.4}$$

except for the case of equilibrium, in which the two are related by Eq. (1.5.2). At boundaries, we would specify only the concentrations of incoming streams, $c_{ia}(t)$ and $n_{ib}(t)$. For the outgoing streams, there is no reason why they should be in equilibrium with the incoming streams even if we have equilibrium inside the column. These concentrations are really dependent on the nature of contact between the two phases at boundaries. Within the framework of the present model, it is reasonable to assume that there is no accumulation of material at boundaries, and so a simple material balance gives

$$\text{at } z = 0, \qquad \varepsilon V_f(c_{ia} - c_{i,0+}) = (1-\varepsilon)V_s(n_{i,0+} - n_{ia}) \tag{1.5.5}$$

$$\text{at } z = L, \qquad \varepsilon V_f(c_{i,L-} - c_{ib}) = (1-\varepsilon)V_s(n_{ib} - n_{i,L-}) \tag{1.5.6}$$

where the subscripts $0+$ and $L-$ are used to denote the states just inside the respective boundaries. When the states just inside the boundaries are known, these equations will directly determine the concentrations of outgoing streams, n_{ia} and c_{ib}.

Another interesting system to look into is that of the adsorption column involved with chemical reactions in the solid phase. For simplicity we will consider the case of a single solute and assume that the solute, when adsorbed on the solid, reacts to give a product that desorbs immediately. If the reaction rate per unit volume of the column is $k_r n$, Eq. (1.5.1) is modified to give

$$\varepsilon V_f\frac{\partial c}{\partial z} - (1-\varepsilon)\,V_s\frac{\partial n}{\partial z} + \varepsilon\frac{\partial c}{\partial t} + (1-\varepsilon)\frac{\partial n}{\partial t} = -k_r n \tag{1.5.7}$$

In the equilibrium case, we have the isotherm

$$n = f(c) \tag{1.5.8}$$

so that the equation to be solved is a single partial differential equation of first order. To this we should append the initial and boundary conditions, the latter being given by Eqs. (1.5.5) and (1.5.6) since there is no reaction at both boundaries. If the solid phase is fixed, the second term in Eq. (1.5.7) drops out, and the initial and boundary conditions are simply

$$\begin{aligned} \text{at } t = 0, \quad & c = c_0(z) \\ \text{at } z = 0, \quad & c = c_f(t) \end{aligned} \tag{1.5.9}$$

Such a system is usually referred to as a *chromatographic reactor*.

In the case of limitation by the rate of adsorption, we would include the reaction term in Eq. (1.5.3) and combine with Eq. (1.5.7) to have

$$\begin{aligned} \varepsilon\left\{V_f \frac{\partial c}{\partial z} + \frac{\partial c}{\partial t}\right\} &= -k_1'c(N - n) + k_2'n \\ (1 - \varepsilon)\left\{-V_s \frac{\partial n}{\partial z} + \frac{\partial n}{\partial t}\right\} &= k_1'c(N - n) - (k_2' + k_r)n \end{aligned} \tag{1.5.10}$$

where $k_1' = a_v k_1$ and $k_2' = a_v k_2$.

Exercises

1.5.1. Put Eq. (1.5.1) for a single solute into the form of Eq. (1.2.11). What is the function $g'(c)$ in this case. Is this function convex if $f(c)$ is convex?

1.5.2. Suppose that the reaction on the solid is reversible and first order in both directions and that adsorption of both the reactant and the product is rate limiting. Using appropriate symbols, write the material balance equations for both species.

1.6 Heat Exchanger

The simplest form of heat exchanger consists of a tube immersed in a bath of temperature T_a. If the temperature of the fluid flowing through the tube is $T(z, t)$ at a point z from the inlet and at time t, and the heat transfer through the wall is described by the overall heat transfer coefficient U, then a balance over the section between z and $z + \Delta z$ over the time interval $(t, t + \Delta t)$ gives

$$\begin{aligned} A\,\Delta z C_p\,[T(z, t + \Delta t) - T(z, t)] &= VA\,\Delta t C_p[T(z, t) - T(z + \Delta z, t)] \\ &\quad + UP\,\Delta z(T_a - T)\,\Delta t \end{aligned}$$

where C_p is the heat capacity per unit volume, A the area of the tube, and P the perimeter of the section. Dividing by $\Delta z\,\Delta t$ and passing to the limit,

we have

$$\frac{\partial T}{\partial t} + V\frac{\partial T}{\partial z} = H(T_a - T) \tag{1.6.1}$$

where

$$H = \frac{UP}{AC_p} \tag{1.6.2}$$

If V varies with position or temperature, we must write

$$\frac{\partial T}{\partial t} + \frac{\partial}{\partial z}VT = H(T_a - T) \tag{1.6.3}$$

The inlet temperature $T(0, t) = T_i(t)$ and the initial temperature distribution $T(z, 0) = T_0(z)$ would normally be specified.

The countercurrent heat exchanger is shown in Fig. 1.3. The same notation may be used with suffixes 1 and 2, but since the second stream is going in the negative z-direction we will denote its velocity by $-V_2$, $V_2 > 0$. Then a similar balance on the first side gives

$$A_1C_{p1}\left(\frac{\partial T_1}{\partial t} + V_1\frac{\partial T_1}{\partial z}\right) = UP(T_2 - T_1)$$

while on the second we have

$$A_2C_{p2}\left(\frac{\partial T_2}{\partial t} - V_2\frac{\partial T_2}{\partial z}\right) = UP(T_1 - T_2)$$

These may be written

$$\frac{\partial T_1}{\partial t} + V_1\frac{\partial T_1}{\partial z} = H_1(T_2 - T_1) \tag{1.6.4}$$

$$\frac{\partial T_2}{\partial t} - V_2\frac{\partial T_2}{\partial z} = H_2(T_1 - T_2) \tag{1.6.5}$$

using an obvious extension of Eq. (1.6.2). Here the initial temperature distributions $T_1(z, 0) = T_{10}(z)$ and $T_2(z, 0) = T_{20}(z)$ can be given, but the inlet temperatures are specified at opposite ends:

$$T_1(0, t) = T_{1i}(t), \qquad T_2(L, t) = T_{2i}(t) \tag{1.6.6}$$

Heat transfer in a packed bed is another example of interest here. We will assume that thermal equilibrium is established everywhere at any time. If we let C_f and C_s denote the heat capacities per unit volume of the fluid and solid phases, respectively, and take z in the direction of flow from the inlet, an energy balance yields

$$\varepsilon VC_f\frac{\partial T}{\partial z} + \frac{\partial}{\partial t}[\varepsilon C_fT + (1 - \varepsilon)C_sT] = \frac{UP}{A}(T_a - T)$$

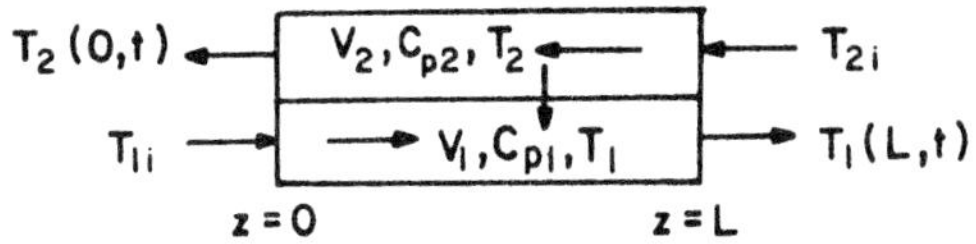

Figure 1.3

where V is the interstitial velocity, ε the fractional void volume of the bed, and T_a the ambient temperature. This may be rewritten

$$(1 + \beta)\frac{\partial T}{\partial t} + V\frac{\partial T}{\partial z} = H(T_a - T) \tag{1.6.7}$$

where

$$\beta = \frac{(1 - \varepsilon)C_s}{\varepsilon C_f} \tag{1.6.8}$$

and

$$H = \frac{UP}{\varepsilon A C_f} \tag{1.6.9}$$

With gas–solid systems, the parameter β usually assumes an order of magnitude of 10^3. If the packed bed is adiabatic, $H = 0$ and so the equation becomes homogeneous. The initial and boundary conditions may be specified as

$$\begin{aligned} \text{at } t = 0, &\quad T = T_0(z) \\ \text{at } z = 0, &\quad T = T_i(t) \end{aligned} \tag{1.6.10}$$

Exercises

1.6.1. Find the steady-state solution $T_s(z)$ of Eq. (1.6.1) when $T(0, t) = T_i$.

1.6.2. Find the steady-state solutions of the pair of equations (1.6.4) and (1.6.5) when the inlet temperatures specified in Eq. (1.6.6) are constant.

1.6.3. If the solid phase is allowed to move countercurrently against the fluid phase, derive an equation similar to Eq. (1.6.7).

1.7 Polymerization in a Batch Reactor

We will consider here an addition polymerization carried out in a batch reactor with a first-order initiation step, second-order propagation, and a second-order monomer termination step. If M_1 is the monomer, P_j is active $j$mer, and M_j is dead $j$mer, then the chemical mechanism will be assumed to be

$$M_1 \xrightarrow{k_i} P_1 \qquad \text{(initiation)}$$

$$P_j + M_1 \xrightarrow{k_p} P_{j+1} \qquad \text{(propagation)}$$

$$P_j + M_1 \xrightarrow{k_t} M_{j+1} \qquad \text{(termination)}$$

Note that M_1, P_j, and M_j stand not only for chemical species but also for their concentrations. With these assumptions on reaction order, the differential equations describing the system are

$$\frac{dM_1}{dt} = -k_i M_1 - \sum_{j=1}^{\infty} k_p P_j M_1 - \sum_{j=1}^{\infty} k_t P_j M_1 \tag{1.7.1}$$

$$\frac{dP_j}{dt} = k_p P_{j-1} M_1 - k_p P_j M_1 - k_t P_j M_1, \qquad j = 2, 3, 4, \ldots \tag{1.7.2}$$

$$\frac{dM_j}{dt} = k_t P_{j-1} M_1, \qquad j = 2, 3, 4, \ldots \tag{1.7.3}$$

$$\frac{dP_1}{dt} = k_i M_1 - k_p P_1 M_1 - k_t P_1 M_1 \tag{1.7.4}$$

We have indicated that j may become arbitrarily large, but of course this is impossible since there can be no finite nonzero concentration of species of infinite molecular weight. These equations have been solved as a large system of differential equations, but it is also instructive to proceed as follows.

Define $P_T = \sum_{j=1}^{\infty} P_j$; then, by adding Eqs. (1.7.2) and (1.7.4), we have

$$\frac{dP_T}{dt} = k_i M_1 - k_t P_T M_1 \tag{1.7.5}$$

But

$$\frac{dM_1}{dt} = -k_i M_1 - k_p M_1 P_T - \mathrm{k}_t M_1 P_T$$

$$\frac{dP_1}{dt} = k_i M_1 - k_p P_1 M_1 - \mathrm{k}_t P_1 M_1$$

These three equations may be written as

$$\frac{dM_1}{-k_i - (k_p + k_t)P_T} = \frac{dP_1}{k_i - (k_p + k_t)P_1} = \frac{dP_T}{k_i - k_t P_T} = M_1 dt \tag{1.7.6}$$

These may be integrated directly to give P_T as a function of M_1 and P_1 also as a function of M_1. Each of M_1, P_1, and P_T may then be found as functions of time. In general, one is interested in obtaining the various statistics of the system, such as the mean average polymer length, and for these one needs to know the individual concentrations P_j and M_j. While these may be found successively from the differential equations, we will proceed on a somewhat different tack.

We will suppose that we are interested in systems in which j can become very large and further that the concentration P_{j+1} does not differ much from the concentration P_j. In fact, we can write $P_j(t)$ as $P(j, t)$, a notation that lends itself to thinking of j as a continuous rather than discrete variable. We may then approximate the difference between P_j and P_{j+1} by

$$\frac{P_{j+1}(t) - P_j(t)}{(j + 1) - j} = \frac{P(j + 1, t) - P(j, t)}{(j + 1) - j} \cong \frac{\partial P(j, t)}{\partial j}$$

or, alternatively, truncate the Taylor series for $P(j + 1, t)$ after the linear term so that

$$P(j + 1, t) = P(j, t) + \frac{\partial P(j, t)}{\partial j}$$

The equation

$$\frac{dP_j}{dt} = k_p P_{j-1} M_1 - k_p P_j M_1 - k_t P_j M_1, \qquad j = 2, 3, 4, \ldots$$

then becomes

$$\frac{1}{M_1}\frac{\partial P}{\partial t} = -k_p \frac{\partial P}{\partial j} - k_t P \tag{1.7.7}$$

where $P = P_j(t) = P(j, t)$.

If we make the substitution

$$\int_0^t M_1(t)dt = \tau \tag{1.7.8}$$

Eq. (1.7.7.) becomes

$$\frac{\partial P}{\partial \tau} + k_p \frac{\partial P}{\partial j} = -k_t P \tag{1.7.9}$$

Now P is being considered as a function of j and τ, and we can specify $P(1, \tau)$ because it can be computed from Eq. (1.7.4). Also, $P(j, 0)$ denotes the initial distribution of active polymer, which is known. For a batch reactor loaded with pure M_1, $P(j, 0)$ will be zero.

Exercises

1.7.1. If $P_j(0) = 0, j = 1, 2, \ldots$, show that the solution of Eq. (1.7.9) is given by

$$P(j, \tau) = \frac{k_i}{k_p + k_t}[e^{-(k_t/k_p)(j-1)} - e^{(j-1)-(k_p+k_t)\tau}]$$

This is valid for $j \leqslant 1 + k_p\tau$; for $j > 1 + k_p\tau$, $P \equiv 0$. Find also $P_T(\tau)$ and $M_1(\tau)$.

1.7.2. A very long sequence of first-order reactions $A_{r-1} \to A_r$ (rate constant k_r) is idealized to a continuum of reactions in a continuous mixture. In the continuous mixture, r becomes a continuous variable and $c(r, t)\, dr$ is the concentration of material of index in the range $(r, r + dr)$ at time t. Show that the analogous first-order process would be described by the partial differential equation

$$\frac{\partial}{\partial t} c(r, t) + \frac{\partial}{\partial r} [k(r)c(r, t)] = 0$$

1.8 Other Problems in Chemical Kinetics

Mikovsky and Wei have considered the exchange of deuterium with hydrides and obtained an explicit solution. It is instructive to approach their problem using the concept of a generating function to obtain a partial differential equation. A hydride with n hydrogen atoms can exist in $(n + 1)$ isomeric states $P_0, \ldots, P_n$, where P_i denotes that i of the hydrogen atoms have been replaced by deuterium. The reaction

$$P_i + D \rightleftharpoons P_{i+1} + H$$

is reversible, and it is reasonable to assume that the forward rate is $k(n - i)$, since any of the remaining $(n - i)$ hydrogen atoms may be exchanged, while the back reaction rate would be $k(i + 1)$. Let $p_0, \ldots, p_n$ denote the concentrations of $P_0, \ldots, P_n$ in a closed system, and let p and q be the atomic concentrations of hydrogen and deuterium, respectively. Then there are three invariant combinations of concentration:

1. Total hydride, $\sum p_i = \pi$
2. Total free hydrogen + deuterium, $p + q = 2\psi$ (1.8.1)
3. Total deuterium, $q + \sum ip_i = \delta$

All summations are from 0 to n unless otherwise stated.

For p and q, we have the equations

$$\frac{dp}{dt} = -\frac{dq}{dt} = -kp \sum ip_i + kq \sum (n - i)p_i \tag{1.8.2}$$

whereas for p_i

$$\frac{dp_i}{dt} = kq(n - i + 1)p_{i-1} - k[q(n - i) + pi]p_i + kp(i + 1)p_{i+1} \tag{1.8.3}$$

The last equation is valid for $i = 0$ and $i = n$ if we interpret p_{-1} and p_{n+1} as zero.

Now let us set

$$P(\mu, t) = \sum_{i=0}^{n} \mu^i p_i(t) \tag{1.8.4}$$

where μ is an artificial variable, $\mu > 0$, introduced to form this generating function. Notice that the equation

$$p_i(t) = \frac{1}{i!}\left[\left(\frac{\partial}{\partial\mu}\right)^i P(\mu, t)\right]_{\mu=0} \tag{1.8.5}$$

will allow us to recover any of the concentrations from the generating function. The first and third invariants can be expressed in terms of the generating function:

$$P(1, t) = \pi$$

$$q + P_\mu(1, t) = \delta$$

where P_μ denotes $\partial P/\partial\mu$. The equation for q, Eq. (1.8.2), can be written

$$\begin{aligned}\frac{dq}{dt} &= k(p + q)\sum ip_i - nkq\sum p_i \\ &= 2\psi kP_\mu(1, t) - nk\pi q \qquad (1.8.6)\\ &= 2\psi k\delta - k(2\psi + n\pi)q\end{aligned}$$

If $q = q_0$ at $t = 0$, this can be immediately integrated to give

$$q = q_0e^{-(2\psi+n\pi)kt} + \frac{2\psi\delta}{2\psi + n\pi}[1 - e^{-(2\psi+n\pi)kt}] \tag{1.8.7}$$

To obtain an equation for the generating function, we multiply each equation in (1.8.3) by the corresponding μ^i and sum from $i = 0$ to $i = n$. Then

$$\begin{aligned}\frac{1}{k}\frac{\partial P}{\partial t} &= q\mu\left[n\sum \mu^{i-1}p_{i-1} - \mu\sum (i - 1)\mu^{i-2}p_{i-1}\right] \\ &\quad - \left[nq\sum \mu^i p_i + (p - q)\mu\sum i\mu^{i-1}p_i\right] \\ &\quad + p\left[\sum (i + 1)\mu^i p_{i+1}\right]\end{aligned}$$

To the first term on the right side we add and subtract $n\mu^n p_n$, making the first sum nP and the second μP_μ. Then, combining the various terms, we have

$$\frac{1}{k}\frac{\partial P}{\partial t} - (\mu - 1)[(p + \mu q)\frac{\partial P}{\partial\mu} - nqP] = 0 \tag{1.8.8}$$

Here again is a first-order differential equation to be solved subject to the boundary conditions

$$P(\mu, 0) = \sum \mu^i p_i(0) \tag{1.8.9}$$

and

$$P(1, t) = \pi \tag{1.8.10}$$

On the left side of Eq. (1.8.8) we have the variables p and q, but these are already determined as functions of time.

A more complicated case is that of an N-place enzyme E, which can form complexes $ES, ES_2 \dots, ES_N$ with a substrate S, any one of which can break down into the product and the previous complex. If we denote ES_n by C_n, $n = 0, \dots, N$, we have $2N$ reactions:

$$C_{n-1} + S \underset{k_{-n}}{\overset{k_n}{\rightleftharpoons}} C_n \xrightarrow{k'_n} C_{n-1} + P$$

Now since C_n is formed by

(i) complexing of C_{n-1}
(ii) product formation from C_{n+1}
(iii) dissociation of C_{n+1}

and destroyed by

(iv) dissociation of C_n
(v) product formation from C_n
(vi) complexing of C_n

we have the following balance:

$$\frac{dC_n}{dt} = \underset{\text{(i)}}{k_n C_{n-1} S} - (\underset{\text{(iv)}}{k_{-n}} + \underset{\text{(v)}}{k'_n} + \underset{\text{(vi)}}{k_{n+1}S})C_n + (\underset{\text{(ii)}}{k'_{n+1}} + \underset{\text{(iii)}}{k_{-(n+1)}})C_{n+1} \tag{1.8.11}$$

Also, if the sites on the enzyme are equivalent, it is reasonable to suppose that k_n is proportional to the number of vacant sites on C_{n-1}; that is,

$$k_n = k(N + 1 - n) \tag{1.8.12}$$

Likewise, k'_n and k_{-n} should be proportional to the number of occupied sites

$$k_{-n} = k_{-}n, \qquad k'_n = k'n \tag{1.8.13}$$

Let

$$\tau = kS_0 t, \qquad \sigma = \frac{S}{S_0}, \qquad \gamma_n = \frac{C_n}{\sum C_n}$$
$$K = \frac{k_-}{kS_0}, \qquad M = \frac{k + k'}{kS_0} \tag{1.8.14}$$

where S_0 is the initial concentration of S. Then

$$\frac{d\gamma_n}{d\tau} = (N + 1 - n)\sigma\gamma_{n-1} - [Mn + (N - n)\sigma]\gamma_n + M(n + 1)\gamma_{n+1} \tag{1.8.15}$$

and this equation applies for $n = 0$ and $n = N$, with the understanding that $\gamma_{-1} = \gamma_{N+1} = 0$.

For the concentration of the substrate itself, we have the equation

$$\frac{dS}{dt} = \sum k_{-n}C_n - S \sum k_n C_{n-1}$$

or, in dimensionless form,

$$\frac{1}{\gamma}\frac{d\sigma}{d\tau} = K \sum n\gamma_n - \sigma \sum (N - n)\gamma_n \tag{1.8.16}$$

where $\gamma = \Sigma\, C_n/S_0$.

Let us introduce a generating function

$$\Gamma(\mu, \tau) = \sum_{n=0}^{N} (1 + \mu)^n \gamma_n(\tau) \tag{1.8.17}$$

Since the total amount of enzyme present, $\Sigma\, C_n$, must be constant,

$$\Gamma(0, \tau) = 1 \tag{1.8.18}$$

The value of Γ when $\tau = 0$ is given by the initial distribution among the different complexes, but, assuming that the enzyme is originally present only as $E = C_0$,

$$\Gamma(\mu, 0) = 1 \tag{1.8.19}$$

Multiplying each of the equations (1.8.15) by the corresponding $(1 + \mu)^n$ and adding, we have

$$\frac{\partial \Gamma}{\partial \tau} + \mu[M + \sigma(1 + \mu)]\frac{\partial \Gamma}{\partial \mu} = N\mu\sigma\Gamma \tag{1.8.20}$$

The dimensionless substrate concentration σ is a function of time, and Eq. (1.8.16) becomes

$$\frac{1}{\gamma}\frac{d\sigma}{d\tau} = (K + \sigma)\Gamma_\mu(0, \tau) - N\sigma \tag{1.8.21}$$

This is an interesting pair of equations, for the second contains the derivative of the solution of the first evaluated along a particular line, $\mu = 0$.

Other kinetic problems can be treated by their translation into continuous variables, as in the previous section, or by using a generating function, as in this section. The translation to a continuous analogue involves a degree of approximation; the generating function is accurate, but it requires a certain regularity in the problem (e.g., $k'_n = k'n$, etc.) before it can be applied.

Exercises

1.8.1. Show that a slight simplification of Eqs. (1.8.8) and (1.8.20) results from the substitutions $n\pi(\mu, t) = \log P(\mu, t)$ and $N\gamma(\mu, \tau) = \log \Gamma(\mu, \tau)$, respectively. What does Eq. (1.8.21) becomes when γ is used in place of Γ?

1.8.2. Show that an ordinary differential equation for $\Pi(\mu, \tau) = \sum_{j=1}^{\infty} \mu^j P_j(\tau)$ is obtained from Eqs. (1.7.2), (1.7.4), and (1.7.8). Solve it when $P_j(0) = 0$, $j = 1, 2, \ldots,$ and compare the solution to that given in Exercise 1.7.1.

1.9 Tubular Reactor

We will consider the lumped constant tubular reactor in which m coupled reactions are taking place. These may be written

$$\sum_{i=1}^{n} a_{ij} A_i = 0, \qquad j = 1, 2, \ldots, m \tag{1.9.1}$$

where there are n chemical species A_i and the stoichiometric coefficients a_{ij} are subject only to the convention that $a_{ij} > 0$ if A_i is regarded as a product of the jth reaction. The rate of formation of species A_i in moles per unit time per unit volume by the jth reaction is denoted by f_{ij}. By g_i we will designate the concentration of A_i in moles per unit mass of reaction mixture. W is the mass flux and A the reactor cross-sectional area. Then a simple shell balance gives

$$(g_i WA)_z - (g_i WA)_{z+\Delta z} + \sum_{j=1}^{m} f_{ij} A\, \Delta z$$

$$= \left[\frac{\partial}{\partial t}(g_i \rho A\, \Delta z)\right]_{\bar{z}}, \qquad z \leqslant \bar{z} \leqslant z + \Delta z$$

In the limit as $\Delta z \to 0$, this reduces to

$$-\frac{\partial}{\partial z}(g_i W) + \sum_{j=1}^{m} f_{ij} = \frac{\partial}{\partial t}(\rho g_i) \tag{1.9.2}$$

An overall mass balance or continuity equation can be obtained by multiplying each equation by the molecular weight of A_i and adding:

$$-\frac{\partial W}{\partial z} = \frac{\partial \rho}{\partial t} \tag{1.9.3}$$

Also the rates of formation of each species by any one reaction are proportional to their stoichiometric coefficients in that reaction, so

$$\frac{f_{ij}}{a_{ij}} = \frac{f_{kj}}{a_{kj}} = f_j, \qquad i, k = 1, 2, \ldots, n$$

Thus

$$-W\frac{\partial g_i}{\partial z} + \sum_{j=1}^{m} a_{ij} f_j = \rho\, \frac{\partial g_i}{\partial t} \tag{1.9.4}$$

A momentum balance may be written in the same way, giving

$$(\rho V^2 A)_z - (\rho V^2 A)_{z+\Delta z} + (pA)_z - (pA)_{z+\Delta z} - \tau P\,\Delta z = \left[\frac{\partial}{\partial t}(\rho V A\,\Delta z)\right]_{\bar{\bar{z}}}, \qquad z \leqslant \bar{\bar{z}} \leqslant z + \Delta z$$

and in the limit

$$\frac{\partial(\rho V^2)}{\partial z} - \frac{\partial p}{\partial z} - \tau\frac{P}{A} = \frac{\partial(\rho V)}{\partial t} \tag{1.9.5}$$

where τ is the shear stress between the fluid and tube-wall surface having perimeter, P, $V = W/\rho$ is the linear velocity of flow, and p is the pressure. For highly turbulent flow,

$$\tau = \rho V^2 f(\mathrm{Re}) \tag{1.9.6}$$

where Re is the Reynolds number. Hence we may write

$$-W\frac{\partial V}{\partial z} - \frac{\partial p}{\partial z} - \frac{P}{A}\rho V^2 f = \rho\frac{\partial V}{\partial t} \tag{1.9.7}$$

The basic form of the energy balance follows in much the same way as in the derivation for the nonisothermal chromatographic column and reduces to

$$\frac{\partial}{\partial t}\left(\sum_{i=1}^{n} \rho g_i \tilde{U}_i\right) = -\frac{\partial}{\partial z}\left(\sum_{i=1}^{n} W g_i \tilde{U}_i\right) - \frac{\partial}{\partial z} pV + \frac{UP}{A}(T_a - T)$$

where U is the overall heat-transfer coefficient at the wall and T_a is the ambient temperature. We now use the fact that

$$\sum_{i=1}^{n} \rho g_i \tilde{U}_i = \sum_{i=1}^{n} \rho g_i \tilde{H}_i - p$$

where $\tilde{H}_i$ is the partial molar enthalpy. Then

$$\frac{\partial}{\partial t}\left(\sum \rho g_i \tilde{H}_i - p\right) + \frac{\partial}{\partial z}\left(V \sum \rho g_i \tilde{H}_i\right) = \frac{UP}{A}(T_a - T)$$

The enthalpy depends principally on the temperature T and

$$\frac{d}{dT}\left(\sum \rho g_i \tilde{H}_i\right) = \sum \rho g_i \tilde{C}_{pi} = C_p$$

the heat capacity per unit volume of the reaction mixture. Hence

$$\frac{\partial}{\partial t}\left(\sum \rho g_i \tilde{H}_i\right) = \sum \tilde{H}_i \frac{\partial}{\partial t}\rho g_i + \left(\sum \rho g_i C_{pi}\right)\frac{\partial T}{\partial t}$$

and when these are inserted and Eq. (1.9.4) is used, we obtain

$$C_p\left(\frac{\partial T}{\partial t} + V\frac{\partial T}{\partial z}\right) = \frac{\partial p}{\partial t} + \sum_{j=1}^{m}(-\Delta H_j)f_j + \frac{UP}{A}(T_a - T) \qquad (1.9.8)$$

where $\Delta H_j = \Sigma_{i=1}^{n} a_{ij}\tilde{H}_i$ is the enthalpy change of the jth reaction.

Thus the pertinent equations for the simplest model of the tubular reactor are all first-order partial differential equations with two independent variables.

Exercises

1.9.1. Under what conditions can the n equations (1.9.4) be replaced by a set of m equations for ξ_j by means of the substitution $g_i = g_{i0} + \Sigma_{j=1}^{m} a_{ij}\xi_j$?

1.9.2. A fixed bed consists of spherical particles of Fe_3O_4 that are being oxidized isothermally to Fe_2O_3 by a stream of oxygen. In each spherical particle of radius a the oxygen must diffuse through a layer of Fe_2O_3 before reacting at a sharp interface of radius r, which is of course a function of t and the position of the pellet in the bed. It is assumed that this reaction front moves slowly through the pellet so that at any instant the profile of concentration within the pellet is given by the steady-stage diffusion equation. If the reaction is first order with respect to oxygen and c is the concentration of oxygen outside the pellet, show that $\dot{n}$, the flux of oxygen into the pellet, is

$$\dot{n} = 4\pi c\left[\frac{1}{D}\left(\frac{1}{r} - \frac{1}{a}\right) + \frac{1}{kr^2}\right]^{-1}$$

where D is the diffusion coefficient of oxygen through the oxidized layer and k the rate constant. The rate of consumption of Fe_3O_4 is proportional to n. Recognizing that c and r will be functions of position in the bed z and time t, derive the simultaneous equations

$$\frac{\partial x}{\partial t} + V\frac{\partial x}{\partial z} + A\frac{xy^2}{y(1-y) + h} = 0$$

$$\frac{\partial y}{\partial t} + B\frac{x}{y(1-y) + h} = 0$$

for $x = c/c_f$, $y = r/a$, where c_f is the concentration of oxygen fed to the bed at $z = 0$. Explicate the constants A, B, and h and state the boundary conditions.

1.10 Enhanced Oil Recovery

Water flooding and polymer flooding are just two examples among a variety of processes by which recovery of oil from a reservoir can be enhanced. The general problem is very complicated, so we will consider here only the simplest models.

A waterflood is carried out by injecting water into the entrance face of the reservoir to displace the oil and eventually induce a breakthrough of oil from the opposite side. If we neglect the effects of dispersion, viscous fingering, and capillarity, a shell balance for water at time t over a section between planes at distances z and $z + \Delta z$ from the entrance gives

$$(q_T s_w)_{z,t} - (q_T s_w)_{z+\Delta z,t} = \left\{ \frac{\partial}{\partial t} (\phi A \; \Delta z s_w) \right\}_{\bar{z},t}, \qquad z < \bar{z} < z + \Delta z$$

where s_w = volume fraction of water (or water saturation) in the pore space
f_w = fractional flow rate of water
q_T = total flow rate
ϕ = fractional void volume in the porous medium
A = cross-sectional area of the porous medium.

If we assume q_T, ϕ, and A are constant and take the limit as Δz approaches zero, we obtain

$$q_T \frac{\partial f_w}{\partial z} + \phi A \frac{\partial s_w}{\partial t} = 0 \qquad (1.10.1)$$

This is the basic conservation equation, which was first derived by Buckley and Leverett in 1942. In the derivation of Eq. (1.10.1), it was assumed that the two phases are completely immiscible. If the displacement occurs in a horizontal system, the fractional flow f_w can be expressed in the following form, the basis of which is Darcy's law:

$$f_w = \frac{1}{1 + \dfrac{\mu_w}{\mu_o} \dfrac{K_{ro}}{K_{rw}}} \qquad (1.10.2)$$

in which μ and K_r are the viscosity and the relative permeability, respectively, and the subscripts o and w denote the oil and water phases, respectively. This is usually called the *fractional flow equation*.

The quantity f_w is a function of water saturation s_w inasmuch as both K_{ro} and K_{rw} are themselves functions of s_w. As the water saturation increases, the value of K_{ro} declines, whereas that of K_{rw} rises, with the result that the value of f_w increases. Empirical equations, known as *Corey type*, are widely applied for K_{ro} and K_{rw}:

$$K_{ro} = (K_{ro})_0 \, (1 - \psi)^p$$
$$K_{rw} = (K_{rw})_0 \, \psi^n \qquad (1.10.3)$$

where $(\quad)_0$ denotes the value of the quantity inside when the flow consists of its own phase only, p and n are constants, and ψ is defined as

$$\psi = \frac{s_w - s_{wc}}{1 - s_{or} - s_{wc}} \qquad (1.10.4)$$

Here s_{wc} is the connate water saturation and s_{or} the residual oil saturation.

Thus ψ represents the water saturation normalized between two limits $s_w = s_{wc}$ and $s_w = 1 - s_{or}$.

Now we can rewrite Eq. (1.10.1) in the form

$$F'(s_w)\frac{\partial s_w}{\partial z} + \frac{\phi A}{q_T}\frac{\partial s_w}{\partial t} = 0 \tag{1.10.5}$$

in which $f_w = F(s_w)$ represents the fractional flow equation, and to this equation we should append the initial and boundary conditions

$$\begin{aligned} &\text{at } t = 0, \qquad s_w = s_{wc} \\ &\text{at } z = 0, \qquad s_w = s_{wf} \end{aligned} \tag{1.10.6}$$

In polymer flooding the displacing fluid contains a polymer species and, hence, in addition to Eq. (1.10.1), we have to consider the conservation equation for the polymer. If we assume that the polymer does not dissolve in the oil phase but adsorbs on the surface of the immobile rock phase, a simple material balance gives

$$q_T\frac{\partial}{\partial z}(f_w c) + \phi A\frac{\partial}{\partial t}\{(s_w c) + c_s\} = 0 \tag{1.10.7}$$

where c and c_s denote the polymer concentrations in the mobile and immobile phases, respectively, the former being expressed in mass per unit volume of the mobile phase and the latter in mass per unit volume of the pore space.

Now we find in Eq. (1.10.2) that K_{ro} and K_{rw} are functions of s_w and c, while μ_w is dependent on c and thus f_w is a function of s_w and c. Adsorption of polymer is usually assumed to be in equilibrium, so c_s may be expressed as a function of c only. If we make use of the conservation equation for water, Eq. (1.10.1), Eq. (1.10.7) can be reduced to a simpler form. This equation may then be combined with Eqs. (1.10.1) and (1.10.2) to give as a mathematical model

$$\begin{gathered} \left(\frac{\partial f_w}{\partial s_w}\right)\frac{\partial s_w}{\partial z} + \left(\frac{\partial f_w}{\partial c}\right)\frac{\partial c}{\partial z} + \frac{\phi A}{q_T}\frac{\partial s_w}{\partial t} = 0 \\ f_w\frac{\partial c}{\partial z} + \frac{\phi A}{q_T}\{s_w + c_s'(c)\}\frac{\partial c}{\partial t} = 0 \\ f_w = f_w(s_w, c) \\ c_s = c_s(c) \end{gathered} \tag{1.10.8}$$

for which the initial and boundary conditions are specified as

$$\begin{aligned} &\text{at } t = 0, \qquad s_w = s_{wc} \quad \text{and} \quad c = 0 \\ &\text{at } z = 0, \qquad s_w = s_{wf} \quad \text{and} \quad c = c_f \end{aligned} \tag{1.10.9}$$

Exercises

1.10.1. Derive the second differential equation in Eq. (1.10.8) by combining Eqs. (1.10.1) and (1.10.7).

1.10.2. Consider the displacement of a system with m components and p phases through a porous medium. The adsorbed phase may be counted as a separate phase. Let s_j denote the volume fraction (or saturation) of phase j in the pore space (including the volume occupied by the adsorbed phase), f_j the fractional flow of phase j, and c_{ij} the volume fraction of component i in phase j. Using the overall concentration C_i and the overall fractional flow F_i of component i defined by

$$\left.\begin{aligned} C_i &= \sum_{j=1}^{p} s_j c_{ij} \\ F_i &= \sum_{j=1}^{p} f_j c_{ij} \end{aligned}\right\}, \qquad i = 1, 2, \ldots, n$$

derive the material balance equation for component i and show that this equation reduces to Eq. (1.10.5) or (1.10.7) under the pertinent assumptions.

1.11 Kinematic Waves in General

The type of equation we have been deriving from mass and energy balances has come to be known as a *conservation equation* or *kinematic wave equation*. The latter term is due to Lighthill and Whitham, who contrasted the purely convective or kinematic terms that go into the equations with the dynamical terms that we have called diffusive or dispersive, and that give rise to higher-order derivatives. The applications that they give and some others will be mentioned.

If, in considering any physical quantity in a one-dimensional flow situation, we can define (1) the flux, $q(z, t)$, that is, the rate at which the quantity passes the point z at time t, and (2) the concentration $k(z, t)$, that is, the amount of this quantity per unit length at the point z and time t, then a balance over the section between z and $z + \Delta z$ and time interval t to $t + \Delta t$ gives

$$[k(z, t + \Delta t) - k(z, t)]\,\Delta z = [q(z, t) - q(z + \Delta z, t)]\,\Delta t$$

Dividing by $\Delta t\,\Delta z$ and passing to the limit gives

$$\frac{\partial k}{\partial t} + \frac{\partial q}{\partial z} = 0 \tag{1.11.1}$$

This is a single partial differential equation with two independent variables. If there is an equilibrium type of relationship

$$q = q(k, z) \tag{1.11.2}$$

then

$$\frac{\partial q}{\partial t} + c\frac{\partial q}{\partial z} = 0 \tag{1.11.3}$$

where

$$c = \frac{\partial q}{\partial k} = c(q, z) \tag{1.11.4}$$

provides us with a single equation in a single unknown, the flux. Clearly, c has the dimensions of a velocity and is amenable to the following physical interpretation. If we consider the solution $q = q(z, t)$ at a point z and time t and ask where the flux has the same value at time $t + \delta t$, the answer is that it is at $z + \delta z$, where

$$q(z + \delta z, t + \delta t) = q(z, t)$$

But to the first order in differential quantities, this implies that

$$\delta t \frac{\partial q}{\partial t} + \delta z \frac{\partial q}{\partial z} = 0$$

and, comparing this with Eq. (1.11.3), we see that $\delta z = c\,\delta t$. Thus c is the speed with which a point of constant flux is moving; it is called the *wave velocity*. We will see later that these lines of constant flux are the characteristics around which the whole solution is built. It is not always convenient to obtain the equation in this form; but if we substitute for k, then we have an additional term in the equation:

$$\frac{\partial k}{\partial t} + c\frac{\partial k}{\partial z} + \left(\frac{\partial q}{\partial z}\right)_k = 0 \tag{1.11.5}$$

Only if q is independent of z will it also be true that c is the speed with which a point of constant concentration moves. The *mean velocity* at z and t might be defined as

$$v(z, t) = \frac{q(z, t)}{k(z, t)} \tag{1.11.6}$$

and we see that this is less or greater than the wave velocity according to whether q/k is less or greater than $\partial q/\partial k$.

The movement of floods on long rivers gives an illustration of the case where q increases more than linearly with k. We can define k as the volume of water per unit length, that is, as the cross-sectional area of the river at z and t, and q as the volume flowing past a given point per unit time. Then a balance of the frictional force $f\rho v^2 L$ (where f is a friction factor, ρ the density, and L the wetted perimeter) against the gravitational pull $\rho g S k$ (where

S is the slope of the river) gives

$$v = \left(\frac{gSk}{fL}\right)^{1/2}$$

If S, f, and L were independent of k, we would have

$$q = vk = \alpha k^{3/2}$$

where α is a constant of proportionality, and

$$\frac{\partial q}{\partial t} + \frac{3}{2}\alpha k^{1/2}\frac{\partial q}{\partial z} = 0$$

or

$$\frac{\partial q}{\partial t} + \frac{3}{2}(\alpha^2 q)^{1/3}\frac{\partial q}{\partial z} = 0$$

However, L will certainly depend on k and the shape of the river bed; for example, a triangular cross section would make $k \propto L^2$, while a rectangular section of breadth b gives $k = \frac{1}{2}b(L - b)$. If f remains constant, this gives

$$q = \alpha' k^{5/4}$$

for the triangular and

$$q = \alpha'' k^{3/2}\left(1 + \frac{2k}{b^2}\right)^{-1/2}$$

for the rectangular. But the friction factor will also depend on L, and in fact a known correlation approximates it by $f \propto (k/L)^{-1/3}$. In this case, $v \propto (k/L)^{2/3}$, giving $q \propto k^{4/3}$ for a triangular section and $k^{4/3}\ (1 + 2kb/b^2)^{-1/3}$ for a rectangular section. Thus the relation between q and k might be represented approximately by

$$q = (\alpha k)^{\beta} \tag{1.11.7}$$

where α and β are constants and β lies between 1 and 1.5. This would give the equation

$$\frac{\partial q}{\partial t} + \alpha\beta q^{\frac{\beta - 1}{\beta}}\frac{\partial q}{\partial z} = 0 \tag{1.11.8}$$

An illustration of the case in which $v > c$ is provided by Lighthill and Whitham's study of traffic flow on roads. Here q, the flux of vehicles per hour, is again the product of their concentration k (vehicles per mile) and average speed v (miles per hour), but the speed will usually decrease as the concentration increases. Indeed, we may expect that there is a jamming concentration k_j at which the flow of traffic ceases. The simplest relation between v and k is given by supposing that

$$v = v_m\left(1 - \frac{k}{k_j}\right)$$

where v_m is the maximum speed on the uncluttered highway. In spite of its obvious simplicity, this has some experimental support, and one curve based on data of B. D. Greenshields [*Proc. Highway Research Board* **14**, 448–474 (1935)] gives $k_j = 200$, $v_m = 40$, approximately. Here

$$q = vk = v_m k_j \frac{k}{k_j}\left(1 - \frac{k}{k_j}\right) \tag{1.11.9}$$

and

$$\frac{\partial k}{\partial t} + v_m\left(1 - 2\frac{k}{k_j}\right)\frac{\partial k}{\partial z} = 0 \tag{1.11.10}$$

Thus we see that the wave velocity actually becomes negative when $k > \frac{1}{2} k_j$.

The theory of sedimentation, the settling of slurries, or the subsidence of slimes can be set up in much the same term (see Exercise 1.11.5), and some insight can be gained into the movement of glaciers (see Exercise 1.11.6). Thus the theory of kinematic waves or conservation equations plays varied roles in natural philosophy and is fecund with applications.

Exercises

1.11.1. Show that

$$c = v + k\frac{\partial v}{\partial k}$$

and hence that $c = \beta v$ for a relation of the type given in Eq. (1.11.7).

1.11.2. A flood is often described by its *stage*, or the height of the water above an arbitrary reference point. If the breadth of the river at the point z and stage h is $b(h, z)$ and the observed flow relationship is $q(h, z)$, show that the stage satisfies the equation

$$\frac{\partial h}{\partial t} + c\frac{\partial h}{\partial z} + d = 0$$

where $c = (1/b)(\partial q/\partial h)_z$ and $d = (1/b)(\partial q/\partial z)_h$. Examine this further for a river of triangular cross section.

1.11.3. Examine the floodwater equations when there is a term representing the runoff water and when a tributary enters the main river.

1.11.4. Obtain the equilibrium chromatographic equations as kinematic waves by identifying the flux and concentration. What is the wave velocity?

1.11.5. A uniform column of slurry is set up with $z = 0$ at the free surface and the positive z-axis going downward. If $n(z, t)$ is the number density of particles at z and t and their settling velocity is $V(n)$, obtain an equation for $n(z, t)$. Make this dimensionless by taking suitable natural coordinates, letting H be the height of the column, $V(0) = V_m$, $V(n_m) = 0$. What modifications must be made to the equations if the cross-sectional area is a function $A(z)$?

1.11.6. The flux of ice in a glacier may be represented by $q = ak^{n+2} + v_b k$, where v_b is the velocity at the bottom and n is an empirical power-law constant thought to be about 3 or 4. One theory makes $v_b = bk^{(n+1)/2}$. Derive the equation for k in this case. Observations on several glaciers in the Yakutat Bay area of Alaska showed that bulges moved with speeds three to four times faster than the mean speed of the glacier. Is this reasonable?

1.11.7. Show that, if the flux includes a Fickian diffusion term, then a second-order derivative is introduced into the equation.

1.11.8. A well-mixed crystallizer of liquor volume $V(t)$ is fed by a stream of flow rate $Q_i(t)$ in which there are $n_i(s, t)\,ds$ crystals in the size range $(s, s + ds)$ per unit volume. In the crystallizer itself the size increases by growth at a rate $r(s, t)$, and a crystal of a given size may "die" by breakage with probability $\beta(s)dt$ in the time interval $(t, t + dt)$. If it breaks into two fragments of sizes σs and $(1 - \sigma)s$ with probability density $p(\sigma)$, show that $n(s, t)\,ds$, the number of crystals in the size range $(s, s + ds)$ at time t, is governed by the equation

$$\frac{\partial n}{\partial t} + \frac{\partial}{\partial s} rn = \frac{Q_i}{V}(n_i - n) - \beta(s)n + 2\int_0^1 p(\sigma)n\left(\frac{s}{\sigma}, t\right)\beta\left(\frac{s}{\sigma}\right) d\sigma$$

1.12 Equations of Compressible Fluid Flow

The equations of compressible fluid flow are derived in so many engineering textbooks that we will not belabor them here. If $\rho(x, y, z, t)$ is the density at a point (x, y, z) at time t, and u, v, and w are the velocities in the x, y, and z-directions, a mass balance gives

$$\frac{\partial \rho}{\partial t} + \frac{\partial}{\partial x}\rho u + \frac{\partial}{\partial y}\rho v + \frac{\partial}{\partial z}\rho w = 0 \tag{1.12.1}$$

Balance of momentum, when viscosity is ignored, gives three equations:

$$\frac{\partial u}{\partial t} + u\frac{\partial u}{\partial x} + v\frac{\partial u}{\partial y} + w\frac{\partial u}{\partial z} + \frac{1}{\rho}\frac{\partial p}{\partial x} = 0 \tag{1.12.2}$$

$$\frac{\partial v}{\partial t} + u\frac{\partial v}{\partial x} + v\frac{\partial v}{\partial y} + w\frac{\partial v}{\partial z} + \frac{1}{\rho}\frac{\partial p}{\partial y} = 0 \tag{1.12.3}$$

$$\frac{\partial w}{\partial t} + u\frac{\partial w}{\partial x} + v\frac{\partial w}{\partial y} + w\frac{\partial w}{\partial z} + \frac{1}{\rho}\frac{\partial p}{\partial z} = 0 \tag{1.12.4}$$

where p is the pressure. To these must be added an equation of state

$$p = f(\rho, S) \tag{1.12.5}$$

expressing the pressure as a function of density and specific entropy, S. If changes of state are adiabatic,

$$\frac{\partial S}{\partial t} + u\frac{\partial S}{\partial x} + v\frac{\partial S}{\partial y} + w\frac{\partial S}{\partial z} = 0 \tag{1.12.6}$$

These equations are more compactly written if we use vector notation, writing $\mathbf{v}$ for the vector of velocity with components (u, v, w) and $\boldsymbol{\nabla}$ for the vector operator $(\partial/\partial x, \partial/\partial y, \partial/\partial z)$. Thus

$$\frac{\partial \rho}{\partial t} + \boldsymbol{\nabla} \cdot (\rho \mathbf{v}) = 0 \tag{1.12.7}$$

$$\frac{\partial \mathbf{v}}{\partial t} + (\mathbf{v} \cdot \boldsymbol{\nabla})\mathbf{v} + \frac{1}{\rho}\nabla p = 0 \tag{1.12.8}$$

$$\frac{\partial S}{\partial t} + (\mathbf{v} \cdot \boldsymbol{\nabla})S = 0 \tag{1.12.9}$$

For unsteady, isentropic flow in one dimension, the equations can be reduced to two in number. For

$$p = k\rho^{\gamma} \tag{1.12.10}$$

and then

$$\frac{\partial \rho}{\partial t} + u\frac{\partial \rho}{\partial x} + \rho\frac{\partial u}{\partial x} = 0 \tag{1.12.11}$$

$$\rho\frac{\partial u}{\partial t} + \rho u\frac{\partial u}{\partial x} + c^2\frac{\partial \rho}{\partial x} = 0 \tag{1.12.12}$$

where

$$c^2 = \frac{\partial p}{\partial \rho} = \gamma\frac{p}{\rho} = \gamma k \rho^{\gamma - 1} \tag{1.12.13}$$

is the square of the velocity of sound.

Exercises

1.12.1. Put Eqs. (1.12.8) into the form

$$\frac{\partial}{\partial t}\rho \mathbf{v} + \boldsymbol{\nabla} \cdot (\rho \mathbf{v}\mathbf{v}) + \nabla p = 0$$

1.13 Flow of Electricity and Heat and Propagation of Light

When electricity flows down a long cable, the potential v (volts) and current i (amperes) are functions of position and time. The potential drop is due to the resistance R (ohms per unit length) and inductance L (henries per unit length) of the wire, and over an element $(z, z + dz)$ we have

$$v(z + dz, t) - v(z, t) = -R\, i\, dz - L \frac{di}{dt} dz$$

or in the limit

$$\frac{\partial v}{\partial z} + L \frac{\partial i}{\partial t} + Ri = 0 \tag{1.13.1}$$

Similarly, the loss of current is by leakage, G (siemens per unit length), and by the capacitance, C (farads per unit length), and leads to an equation

$$\frac{\partial i}{\partial z} + C \frac{\partial v}{\partial t} + Gv = 0 \tag{1.13.2}$$

Here we have a pair of equations for $v(z, t)$ and $i(z, t)$ with constant coefficients.

In deriving the basic equations of heat conduction from Fourier's law, the objection is often raised that the velocity of propagation of heat is assumed to be infinitely large. It is a valid objection, although in most practical cases the differences would not be appreciable and the usual parabolic heat conduction equation is found to give good agreement with observation. However, for high rate processes in rarefied media, it is desirable to take into account the finiteness of the velocity of heat propagation. There is a relaxation time, τ_r (of the order of 10^{-9} to 10^{-11} second, and the flux of heat, $\mathbf{q}$, is no longer simply proportional to the temperature gradient, ∇T, as in Fourier's law, but is given by

$$\mathbf{q} = -\lambda \nabla T - \tau_r \frac{\partial \mathbf{q}}{\partial t} \tag{1.13.3}$$

where λ is the conductivity and $\partial \mathbf{q}/\partial t$ the rate of change of heat flux. Such an equation is similar to the constitutive laws used for viscoelastic fluids.

In one space dimension where the temperature is $T(z, t)$ and the heat flux $q(z, t)$, a heat balance over an element between z and $z + dz$ gives the usual kinematic wave equation

$$\frac{\partial q}{\partial z} + c\rho \frac{\partial T}{\partial t} = 0 \tag{1.13.4}$$

where c is the specific heat and ρ the density. Then Eq. (1.13.3) may be written

$$\tau_r \frac{\partial q}{\partial t} + \lambda \frac{\partial T}{\partial z} + q = 0 \tag{1.13.5}$$

and the last two equations constitute a pair of first-order partial differential equations to be solved for q and T. Clearly, they are basically the same as the equations for the flow of electricity.

Geometrical optics can be regarded as the limiting case of wave optics. In the latter, the propagation of light is governed by the second-order wave equation

$$\frac{\partial^2 w}{\partial x^2} + \frac{\partial^2 w}{\partial y^2} + \frac{\partial^2 w}{\partial z^2} + k^2 w = 0 \tag{1.13.6}$$

where x, y, and z are the spatial coordinates and $k = 2\pi/\lambda$ is the wave number. This equation degenerates as $k \to \infty$ or $\lambda \to 0$, and this is the limiting case of geometrical optics. Let us assume a special form for w,

$$w = Ae^{ik_0 u}, \qquad k_0 = \frac{2\pi}{\lambda} \tag{1.13.7}$$

where A, the amplitude factor, and u, the *eikonal*, are slowly varying functions of position. Then as $k_0 \to \infty$, w will become a very rapidly varying function of position. Now

$$w_x = ik_0 w u_x + w(\ln A)_x$$

$$w_{xx} = -k_0^2 w u_x^2 + ik_0 w u_{xx} + 2ik_0 w(\ln A)_x u_x + w(\ln A)_x^2 + w(\ln A)_{xx},$$

so

$$w_{xx} + w_{yy} + w_{zz} + k^2 w = -wk_0^2\left(u_x^2 + u_y^2 + u_z^2 - \frac{k^2}{k_0^2}\right)$$
$$+ ik_0 w[u_{xx} + u_{yy} + u_{zz} + 2u_x(\ln A)_x + 2u_y(\ln A)_y + 2u_z(\ln A)_z] + \cdots$$

where the omitted terms are bounded as $k_0 \to \infty$. Evidently, we will be able to retain some meaning to the equation as $k_0 \to \infty$ if the coefficients of k_0 and k_0^2 vanish. Thus u satisfies

$$u_x^2 + u_y^2 + u_z^2 = n^2 = \lim_{k_0 \to \infty} \frac{k^2}{k_0^2} \tag{1.13.8}$$

where n is the index of refraction of the medium. The amplitude factor is obtained from the equation

$$\nabla u \cdot \nabla(\ln A) = -\frac{1}{2}\nabla^2 u \tag{1.13.9}$$

If the medium through which light is propagated is inhomogeneous, Eq.

(1.13.8) becomes

$$(\nabla u)^2 = n^2(x, y, z) \tag{1.13.10}$$

This is the *eikonal equation.*

Exercises

1.13.1. If $u(x, y) = ct$ represents the position of a wave front at time t, show that it is always moving in a direction normal to itself with velocity c if $u_x^2 + u_y^2 = 1$.

1.13.2. Show that Eqs. (1.13.1) and (1.13.2) lead to a second-order equation

$$\frac{\partial^2 u}{\partial z^2} = a\frac{\partial^2 u}{\partial t^2} + b\frac{\partial u}{\partial t} + cu$$

which is satisfied by both i and v. Relate a, b, and c to L, R, C, and G. This is the telegrapher's equation.

1.13.3. Eliminate q between Eqs. (1.13.4) and (1.13.5) to give a form of the telegrapher's equation. Show that, in the notation of the previous exercise, $a = b\tau_r$ and $1/b$ is the thermal diffusivity, and deduce that the usual heat-conduction equation is obtained when $\tau_r \to 0$.

1.14 Two Problems in Optimization

Using the simplest model of the tubular reactor, we may set up the following optimization problem. For a single reaction $\Sigma\, a_i A_i = 0$ and steady state of operation, we may write Eq. (1.9.4) in the form

$$W\frac{dg_i}{dz} = a_i f \tag{1.14.1}$$

The reaction rate f is a function of the g_i and temperature T. But, by the stoichiometry of the reaction $\Sigma\, a_i A_i = 0$, the g_i are constrained in such a way that the change in the number of moles of A_i is proportional to a_i; that is,

$$g_i = g_{i0} + a_i \xi \tag{1.14.2}$$

In this expression, g_{i0} may be taken to be the feed composition so that $\xi = 0$ at $z = 0$, and then, on substituting in Eq. (1.14.1), we have one equation:

$$W\frac{d\xi}{dz} = f(\xi, T), \qquad \xi(0) = 0 \tag{1.14.3}$$

If p_i is the price of 1 mole of A_i, then $\Sigma\, p_i g_i$ is the price per pound of a mixture of composition g_i, and we will take this as an index of the value of the operation. Thus the rate at which profit is being made by a reactor of length

y is

$$W \sum p_i(g_i - g_{i0}) = W\left(\sum a_i p_i\right) [\xi(y) - \xi(0)]$$
$$= \left(\sum a_i p_i\right) \int_0^y f(\xi, T)dz \qquad (1.14.4)$$

Presumably, $\Sigma a_i p_i > 0$ or we would not be trying to make the reaction go in this direction, and it will therefore be sensible to try and maximize the integral in Eq. (1.14.4). Many reaction rates have the property that for each composition, ξ, there is a temperature $T_m(\xi)$ at which $f(\xi, T)$ is greatest; that is,

$$F(\xi) = f(\xi, T_m(\xi)) > f(\xi, T), \qquad T \neq T_m \qquad (1.14.5)$$

Now it would seem intuitively obvious that, if the integral is to be made as large as possible, then the integrand should always be as large as possible. Intuition happens to be correct in this case, but even here it must be proved; in more complicated cases it may be misleading. We will outline the dynamic programming approach to this particular problem since it leads quite naturally to a first-order partial differential equation.

The problem to be solved is this: For the system governed by Eq. (1.14.3), we wish to maximize the integral in Eq. (1.14.4) by choosing $T(z)$ in the interval $0 \leqslant z \leqslant y$. This can best be done by embedding the problem in a slightly larger one in which for the moment we will not insist that $\xi(0)$ be zero, but let it have a certain value, x. Then, when we have discovered how to maximize the integral by the correct choice of $T(z)$, the maximum it attains will depend only on x and y. Let

$$u(x, y) = \max \int_0^y f(\xi, T)dz \qquad (1.14.6)$$

where

$$W\frac{d\xi}{dz} = f(\xi, T), \qquad \xi(0) = x \qquad (1.14.7)$$

To derive an equation for $u(x, y)$, we first observe that the integral to be maximized can be written as the sum of two parts,

$$\int_0^\delta f(\xi, T)dz + \int_\delta^y f(\xi, T)dz$$

The first of these two integrals is affected by the choice of $T(z)$ for $0 \leqslant z \leqslant \delta$ and the second by the choice of $T(z)$ on the remainder of the interval $\delta \leqslant z \leqslant y$. Suppose that for the moment we have made a choice over the first interval, and we ask how the choice for the remaining interval should be made. The length of the reactor remaining is $y - \delta$, and the value of the extent ξ at $z = \delta$, the beginning of the second section, is $\xi(\delta)$. Now, if we make any choice other than the optimal one for the interval $\delta \leqslant z \leqslant y$, the second integral will be smaller than it need be, and so the sum of the two integrals will not achieve its greatest value. On the other hand, if we do make the optimal choice of $T(z)$ in the second integral, its value will be $u(\xi(\delta), y - \delta)$, for its inlet extent is $\xi(\delta)$ and length is $y - \delta$. Thus, if we are to make the best choice of $T(z)$ over the whole interval $(0, y)$, we must not throw anything away by making anything less than the best choice over any subinterval (δ, y). Now we are left with the problem of making the choice over $(0, \delta)$, and we can write

$$u(x, y) = \max_{\substack{T(z) \\ 0 \leqslant z \leqslant \delta}} \left[\int_0^\delta f(\xi, T)dz + u(\xi(\delta), y - \delta) \right] \tag{1.14.8}$$

We are going to let δ become small and make a choice T_0 over $0 \leqslant z \leqslant \delta$. Then the following approximations can be made:

$$\int_0^\delta f(\xi, T)dz = f(x, T_0)\delta + 0(\delta^2)$$

$$\xi(\delta) = x + \frac{1}{W} f(x, T_0)\delta + 0(\delta^2)$$

and

$$u(\xi(\delta), y - \delta) = u(x, y) + u_x(x, y)\,[W^{-1}f(x, T_0)\delta] - u_y(x, y)\delta + 0(\delta^2)$$

But T_0 only affects the terms with $f(x, T_0)$, and so the other terms can be taken outside the maximization. Thus Eq. (1.14.8) can be written

$$u(x, y) = u(x, y) - u_y\,\delta + \max_{T_0}\,[f(x, T_0)(1 + W^{-1}u_x)\delta] + 0(\delta^2)$$

We can subtract $u(x, y)$ from both sides, divide through by δ, and let $\delta \rightarrow 0$; then

$$u_y = \max_{T_0} f(x, T_0)(1 + W^{-1}u_x) \tag{1.14.9}$$

But referring back to the property of $f(x, T)$ given by Eq. (1.14.5), we see that this maximum is given by

$$T_0 = T_m(x) \tag{1.14.10}$$

and

$$u_y = F(x)(1 + W^{-1}u_x) \tag{1.14.11}$$

Two comments should be made on this equation. First, in replacing $f(x, T_0)$ by its maximum, we have assumed that $1 + W^{-1}u_x$ is positive. But u_x is the rate of increase of the maximum of $[\xi(L) - x]$ with x. Now $\xi(L)$ certainly increases with x, although it may do but slowly, so $W^{-1}u_x > -1$.

Second, if the choice of temperature is subject to restrictions, this merely means that the maximum of $f(x, T_0)$ must be found within these restrictions. No greater complication is introduced into the equations, and in some cases a numerical search for the maximum may be made easier.

Finally, we note that the boundary conditions are given by the platitude that if the reactor has zero length it can make only zero profit; that is,

$$u(x, 0) = 0 \tag{1.14.12}$$

In this simple case it appears that we really have no need for a partial differential equation, for, having found that we should always choose $T = T_m(\xi)$, we can go back to the original equation and write

$$W\frac{d\xi}{dz} = F(\xi)$$

solving it with initial condition $\xi = 0$. However, this is a peculiar feature of this particularly simple example. Let us turn to a more complicated situation in which it is no longer true that the choice maximizes the desirable reaction rate at each point.

A batch reactor is to be used for the simultaneous reactions $A \to B \to C$, and the temperature is to be so chosen as to maximize the yield of B. For definiteness, let us suppose that the activation energy of the second reaction $B \to C$ is greater than that of the first. Then if the reactor is initially charged with A, we will expect to start at a high temperature and decrease it as the reaction proceeds. For if we do not start at a high temperature, the first reaction will never get going, and if we do not reduce the temperature, the side reaction $B \to C$ will detract from the yield of B at the later stages. If $a(t)$ and $b(t)$ denote the concentrations of A and B at time t, we have the following problem: Given

$$\frac{da}{dt} = -k_1 a, \qquad a(0) = a_0 \tag{1.14.13}$$

$$\frac{db}{dt} = k_1 a - k_2 b, \qquad b(0), = 0 \tag{1.14.14}$$

maximize

$$b(z) - b(0) = \int_0^z (k_1 a - k_2 b)\, dt$$

by choice of $T(t)$ during the course of the reaction $0 \leqslant t \leqslant z$. Since

$$k_1 = A_1 e^{-E_1/RT} \quad \text{and} \quad k_2 = A_2 e^{-E_2/RT}$$

we can set

$$k_2 = A_2 A_1{}^{-E_2/E_1}(k_1)^{E_2/E_1} = \rho k_1^r \tag{1.14.15}$$

and for convenience drop the suffix 1 on k_1. Since $k(T)$ is monotonic, it is just as good to choose the rate constant $k(T)$ as it is to choose the temperature itself.

As before, we embed the problem in a slightly more general one, taking $a(0) = x$ and $b(0) = y$ as variables and setting

$$u(x, y, z) = \max[b(z) - y] = \max \int_0^z (ka - \rho k^r b)\, dt \qquad (1.14.16)$$

Then we can solve our original problem by finding $u(a_0, 0, z)$. Again we write

$$u(x, y, z) = \max \left[\int_0^\delta (ka - \rho k^r b)\, dt + \int_\delta^z (ka - \rho k^r b)\, dt \right] \qquad (1.14.17)$$

and argue that, unless the optimal choice of k is made for $\delta \leqslant t \leqslant z$, we will never attain the maximum for the whole interval. If the optimal choice is made for the latter integral, its value is $u(a(\delta), b(\delta), z - \delta)$. But, as before,

$$\int_0^\delta (ka - \rho k^r b)\, dt = (kx - \rho k^r y)\delta + 0(\delta^2)$$

$$a(\delta) = x - kx\delta + 0(\delta^2)$$

$$b(\delta) = y + (kx - \rho k^r y)\delta + 0(\delta_2)$$

and

$$u(a(\delta), b(\delta), z - \delta)$$

$$= u(x, y, z) - u_x kx\delta + u_y(kx - \rho k^r y)\delta - u_z\delta + 0(\delta^2)$$

Hence, observing that $u(x, y, z)$ can be canceled on both sides and that we can then divide through by δ and let it tend to zero, we obtain

$$u_z = \max[kx(1 + u_y - u_x) - \rho k^r y(1 + u_y)] \qquad (1.14.18)$$

where the choice of $k = k(T(0))$ is to be made optimally. This maximization is done rather simply by differentiating and eliminating k (see Exercise 1.14.1), giving

$$k = \left[\frac{x(1 + u_y - u_x)}{r\rho y(1 + u_y)} \right]^{1/(r-1)} \qquad (1.14.19)$$

and

$$u_z = \left\{ \frac{(r - 1)^{r-1}}{r^r} \frac{[x(1 + u_y - u_x)]^r}{\rho y(1 + u_y)} \right\}^{1/(r-1)} \qquad (1.14.20)$$

This clearly is a highly nonlinear partial differential equation of the first order. As before, it would have to be solved subject to the boundary condition

$$u(x, y, 0) = 0 \qquad (1.14.21)$$

These are just two of many optimization problems that might be posed, and we will have more to say about the solution of the second one in Sec. 3.6. One feature that is characteristic of these problems is worth noting, however. We started out by wanting to find $u(a_0, 0, z)$, but the boundary condition on the equation is on $z = 0$, and we have no way of knowing at what point $(x, y, 0)$ we should start in order to hit $(a_0, 0, z)$. Some form of this two-point boundary-value problem always arises.

Exercises

1.14.1. Establish Eqs. (1.14.19) and (1.14.20). If T is constrained to lie between T_* and T^*, what modifications need to be made?

1.14.2. A batch reactor for the reaction $A \rightarrow B \rightarrow C$ is to be operated, not for a specified length of time z, but so as to maximize the "profit" $(b(z) - b(0) - \lambda z)$. Show that a partial differential equation in two variables $x = a(0)$, $y = b(0)$ can be obtained.

1.14.3. Show that $u(x, y, z)$ in Eq. (1.14.20) can be written as $u(x, y, z) = xv(y/x, z) = yw(x/y, z)$ and derive equations for v and w.

1.14.4. Consider the general control problem of a system governed by the equations

$$\dot{\mathbf{x}} = \mathbf{f}(\mathbf{x}, \mathbf{u}), \quad \mathbf{x}(0) = \boldsymbol{\xi}$$

for which a control $\mathbf{u}(t)$, $0 \leqslant t \leqslant \theta$, must be found that minimizes a loss function by the integral

$$\int_0^\theta f^0(\mathbf{x}, \mathbf{u})dt$$

of a function f^0 of the state $\mathbf{x}(t)$ and the controls $\mathbf{u}(t)$. Without paying too much attention to rigor, obtain a partial differential equation for $g(\boldsymbol{\xi}, \theta)$, the minimum value of the loss.

1.15 An Estimation Problem

In the context of modern control theory, the word *filtering* refers to the problem of deducing the best estimate of a quantity when the measurements of it are indirect and corrupted by noise. There are several approaches to this problem, and the one we give here has considerable affinity with the dynamic programming approach to optimization problems.

Suppose that we have a physical process whose state at time t is given by a single variable $x(t)$ that is governed by an ordinary differential equation

$$\frac{dx}{dt} = g(x, t) \qquad (1.15.1)$$

This process is observed, but instead of being able to measure x we can only get an approximate value,

$$y(t) = x(t) + \xi(t) \tag{1.15.2}$$

where $\xi(t)$ is the observational error. If $y(t)$ is known over the interval $0 \leqslant t \leqslant \theta$, how can we make the best estimate of the value of $x(\theta)$? This of course requires a definition of "best," and a simple choice is as follows. Suppose that $x(\theta)$ is chosen to be equal to c; then the equation for x could be integrated back from $t = \theta$ to $t = 0$ and the total squared error,

$$E = \int_0^\theta [y(t) - x(t)]^2 \, dt \tag{1.15.3}$$

could be evaluated. A reasonable choice of c would then be one that minimizes E.

Let us recognize that the value of E will be a function of c and θ by writing

$$f(c, \theta) = \int_0^\theta [y(t) - x(t)]^2 \, dt \tag{1.15.4}$$

so that the best estimate of x in the sense we have defined is the value of c for which

$$\frac{\partial f}{\partial c} = 0 \tag{1.15.5}$$

However, we can obtain a partial differential equation for $f(c, \theta)$ as follows. Suppose we made the choice c at time $\theta + \delta\theta$ instead of at θ. This means that on looking back we can say that the value of $x(\theta)$ would have had to have been $c - g(c, \theta)\delta\theta$. But the integral can be written

$$f(c, \theta + \delta\theta) = \int_0^\theta [y(t) - x(t)]^2 \, dt + \int_\theta^{\theta+\delta\theta} [y(t) - x(t)]^2 \, dt$$

The first integral is just $f(c - g(c, \theta)\delta\theta, \theta)$ since it is taken over $(0, \theta)$ and $x(\theta) = c - g(c, \theta)\delta\theta$. The second integral can be approximated by giving $x(t)$ its value at $t = \theta + \delta\theta$, that is, c. Thus

$$f(c, \theta + \delta\theta) = f(c - g(c, \theta)\delta\theta, \theta) + [y(\theta) - c]^2 \, \delta\theta + O(\delta\theta)$$

where $O(\delta\theta)$ denotes terms that go to zero more rapidly than $\delta\theta$. Expanding the two f's by Taylor's theorem, subtracting $f(c, \theta)$ from both sides, and dividing through by $\delta\theta$, we have, as $\delta\theta \to 0$,

$$\frac{\partial f}{\partial \theta} + g(c, \theta) \frac{\partial f}{\partial c} = [y(\theta) - c]^2 \tag{1.15.6}$$

As a boundary condition for this equation, we could embody our a priori estimate of the state. For example, if c_0 were the best available estimate, we would certainly want $f(c, \theta)$ to be minimum at $c = c_0$. Then by finding the surface $f(c, \theta)$ and tracing out the valley along which $f_c = 0$, we will find the best estimate in the sense we have defined.

Exercises

1.15.1. If the right side of Eq. (1.15.1) is linear, $g(x, t) = ax + h(t)$, and the initial condition is quadratic, $f(c, 0) = \alpha(c - c_0)^2$, show that $f(c, \theta)$ is quadratic in c. Find equations for the coefficients and the best estimate.

1.15.2. Let $\mathbf{x}$ be an n-vector and $\mathbf{H}$ a matrix of order $p \times n$. The system is governed by

$$\dot{\mathbf{x}} = g(\mathbf{x}, t)$$

and the observations are $\mathbf{y} = \mathbf{Hx} + \boldsymbol{\xi}$, where $\boldsymbol{\xi}$ is error. The current value of $\mathbf{x}$ is to be estimated so as to minimize

$$\int_0^\theta [\mathbf{y}(t) - \mathbf{Hx}(t)]^T Q(t) [\mathbf{y}(t) - \mathbf{Hx}(t)] \, dt$$

Find an equation for $f(\mathbf{c}, \theta)$, the value of this integral when $\mathbf{x}(\theta) = \mathbf{c}$.

1.16 Geometrical Origins

From an engineering point of view, the geometrical and analytical origins of first-order partial differential equations are less important than the physical situations that give rise to them. However, it is valuable to see how they do arise from the elimination of arbitrary constants or functions, for it will be possible to tie back the solutions to this construction in a very clear way.

If we are given a family of surfaces depending on two parameters

$$f(x, y, z, \alpha, \beta) = 0 \tag{1.16.1}$$

we can eliminate α and β by the following procedure. Differentiating partially with respect to x and y and treating z as a function of these variables gives two equations,

$$\begin{aligned} f_x + f_z z_x &= 0 \\ f_y + f_z z_y &= 0 \end{aligned} \tag{1.16.2}$$

Now f_x, f_y, and f_z will generally be functions of x, y, z, α and β just as f itself is. But we have three equations and can eliminate α and β from them; this will then give us an equation of the form

$$F(x, y, z, z_x, z_y) = 0 \tag{1.16.3}$$

As is traditional in discussing the equation in two independent variables, we will use the abbreviations

$$p = z_x, \qquad q = z_y \tag{1.16.4}$$

It is evident that when there are n independent variables, $x_1, \ldots, x_n$, and n parameters, $\alpha_1, \ldots, \alpha_n$, we can proceed in the same manner. Thus the equation

$$f(x_1, \ldots, x_n, z, \alpha_1, \ldots, \alpha_n) = 0 \tag{1.16.5}$$

can be differentiated with respect to x_i to give

$$f_{x_i} + f_z p_i = 0 \tag{1.16.6}$$

where

$$p_i = \frac{\partial z}{\partial x_i}$$

From the $(n + 1)$ equations (1.16.5) and (1.16.6), we can eliminate the n constants to give

$$F(x_1, \ldots, x_n, z, p_1, \ldots, p_n) = 0 \tag{1.16.7}$$

To illustrate this, consider the family of unit spheres with centers on the plane $z = 0$,

$$f \equiv (x - \alpha)^2 + (y - \beta)^2 + z^2 - 1 = 0 \tag{1.16.8}$$

From this we have

$$\begin{aligned} 2(x - \alpha) + 2zp &= 0 \\ 2(y - \beta) + 2zq &= 0 \end{aligned} \tag{1.16.9}$$

so that

$$z^2(p^2 + q^2 + 1) = 1 \tag{1.16.10}$$

This may be put into a number of equivalent forms; one that we will use later in Sec. 9.1 is

$$F(x, y, z, p, q) \equiv z(p^2 + q^2)^{1/2} - (1 - z^2)^{1/2} = 0 \tag{1.16.11}$$

First-order differential equations also arise when an arbitrary function is eliminated. For example, if

$$z = f(x, y)\phi(g(x, y))$$

where ϕ is arbitrary, we may again differentiate with respect to x and y, giving

$$p = f_x\phi + f\phi' g_x$$

$$q = f_y\phi + f\phi' g_y$$

Then both ϕ and ϕ' can be eliminated to give

$$\begin{vmatrix} f & 0 & z \\ f_x & fg_x & p \\ f_y & fg_y & q \end{vmatrix} = z\frac{\partial(f, g)}{\partial(x, y)}f - pf^2 g_y + qf^2 g_x = 0$$

One way in which an arbitrary function may be introduced is to select an arbitrary one-parameter family from a two-parameter family by putting $\beta = \phi(\alpha)$,

$$f(x, y, z, \alpha, \phi(\alpha)) = 0$$

Now if $f_\phi \neq 0$, this equation can be solved for

$$\phi(\alpha) = g(x, y, z, \alpha)$$

Differentiating with respect to x and y gives the two equations

$$g_x + g_z p = 0 \quad \text{and} \quad g_y + g_z q = 0$$

from which we can eliminate the parameter α. Evidently, we will recover the same equation as from the elimination of α and β from the two-parameter family. For example, if $\beta = \phi(\alpha)$ in

$$(x - \alpha)^2 + (y - \beta)^2 + z^2 = 1$$

then

$$\phi(\alpha) = y - [1 - z^2 - (x - \alpha)^2]^{1/2}$$

But by differentiation,

$$0 = [pz + (x - \alpha)][1 - z^2 - (x - \alpha)^2]^{-1/2}$$

and

$$0 = 1 + qz[1 - z^2 - (x - \alpha)^2]^{-1/2}$$

and substituting from the first into the second, we again have

$$z^2(p^2 + q^2 + 1) = 1$$

Exercises

1.16.1. Show that the family of planes

$$z = \alpha x + \beta y + f(\alpha, \beta)$$

gives rise to the equation of Clairaut:

$$z - px - qy = f(p, q)$$

1.16.2. Find the differential equation of the family of planes tangent to a unit sphere. Show that it is satisfied by any cone of half-angle θ provided its vertex lies on a concentric sphere of radius cosec θ and its axis goes through the center of the spheres.

1.16.3. If $\phi(u,v)$ is an arbitrary function and u and v are known functions of x, y, and z, show that the solution $z = z(x, y)$ of $\phi(u, v) = 0$ must satisfy an equation of the form $Pp + Qq = R$, where P, Q, and R are certain functions of x, y, and z. Show also that $Pu_x + Qu_y + Ru_z = 0$ and $Pv_x + Qv_y + Rv_z = 0$.

1.16.4. Show that $(n - 1)$ arbitrary functions can be eliminated when there are n independent variables. Choose a form such as $z = f_1(x)\phi_1(f_2(x)) \ldots \phi_{n-1}(f_n(x))$, where $\phi_1, \ldots, \phi_{n-1}$ are arbitrary and $f_1, \ldots, f_n$ are given functions to illustrate this.

1.16.5. If $f(x_1, \ldots, x_n)$ is a homogeneous function of degree p [i.e., $f(\lambda x_1, \ldots, \lambda x_n) = \lambda^p f(x_1, \ldots, x_n)$], show that it satisfies Euler's relation,

$$\sum x_i \frac{\partial f}{\partial x_i} = pf$$

1.17 Cauchy–Riemann Equations

If $z = x + iy$ is a complex number and $f(z)$ a function of z, it is important to know when $f(z)$ is differentiable. In textbooks on the theory of functions of complex variables, the following theorem is proved. Let

$$f(z) = u(x, y) + iv(x, y) \tag{1.17.1}$$

be a function of a complex variable defined in a domain D of the complex plane. A necessary and sufficient condition for $f(z)$ to be differentiable at $z = z_0 = x_0 + iy_0$, a point of D, is that $u(x, y)$ and $v(x, y)$ should be differentiable functions of the real variables x and y and that the first derivatives should satisfy the Cauchy–Riemann conditions

$$\frac{\partial u}{\partial x} = \frac{\partial v}{\partial y}, \qquad \frac{\partial v}{\partial x} = -\frac{\partial u}{\partial y} \tag{1.17.2}$$

It is evident that the curves of constant $u(x, y)$ that may be drawn in the (x, y)-plane always intersect the curves of constant $v(x, y)$ at right angles, for

$$\left(\frac{dy}{dx}\right)_u = -\frac{u_x}{u_y} = -\frac{v_y}{-v_x} = \frac{1}{v_x/v_y} = -\frac{1}{(dy/dx)_v}$$

Also $u(x, y)$ and $v(x, y)$ satisfy Laplace's equation, since $u_{xx} = v_{xy} = v_{yx} = -u_{yy}$, and likewise for $v(x, y)$.

If we are given $u(x, y)$, it is interesting to know if we can find the conjugate potential function $v(x, y)$. The answer is that we can do so to within an arbitrary constant, and we may proceed as follows. Since

$$v_y(x, y) = u_x(x, y)$$

we have, on integrating with respect to y,

$$v(x, y) = \int_{y_0}^{y} u_x(x, y')dy' - \phi(x)$$

where $\phi(x)$ is an arbitrary function of x. But then

$$v_x(x, y) = \int_{y_0}^{y} u_{xx}(x, y')dy' - \phi'(x) = -u_y(x, y)$$

so

$$\phi'(x) = u_y(x, y) + \int_{y_0}^{y} u_{xx}(x, y')dy'$$

Integrating now with respect to x,

$$\begin{aligned}\phi(x) &= \int_{x_0}^{x} u_y(x', y)\,dx' + \int_{x_0}^{x} dx' \int_{y_0}^{y} u_{xx}(x', y')dy' + C \\ &= \int_{x_0}^{x} u_y(x', y)\,dx' - \int_{x_0}^{x} dx' \int_{y_0}^{y} u_{yy}(x', y')dy' + C \\ &= \int_{x_0}^{x} [u_y(x', y) - u_y(x', y) + u_y(x', y_0)]\,dx' + C \\ &= \int_{x_0}^{x} u_y(x', y_0)dx' + C\end{aligned}$$

Then

$$v(x, y) = \int_{y_0}^{y} u_x(x, y')dy' - \int_{x_0}^{x} u_y(x', y_0)\,dx' + \text{constant}$$

Exercises

1.17.1. Discuss the differentiability of the following functions: (a) $f(z) = z^n$, n integral; (b) $f(z) = \bar{z} = x - iy$; (c) $f(z) = \cos z$.

1.17.2. Show that the following can be real parts of functions of a complex variable and find the imaginary part: (a) $u(x, y) = x^2 - y^2$; (b) $u(x, y) = e^x \sin y$.

1.17.3. The magnetostatic potential around a wire passing perpendicularly through the origin of the (x, y)-plane and carrying a current I is $\phi(x, y) = 2I \tan^{-1}\frac{y}{x}$. Find the force field $\psi(x, y)$, the potential field orthogonal to $\phi(x, y)$.

REFERENCES

1.2 and 1.3. The basic equations of chromatography are found in the literature of the 1940s, if not before. See, for example:

J. N. Wilson, "A Theory of Chromatography," *J. Amer. Chem. Soc.* **62**, 1583–1591 (1940).

D. deVault, "The Theory of Chromatography," *J. Amer. Chem. Soc.* **65**, 532–540 (1943).

H. C. Thomas, "Heterogeneous Ion Exchange in a Flowing System," *J. Amer. Chem. Soc.* **66**, 1664–1666 (1944).

J. E. Walter, "Multiple Adsorption from Solutions," *J. Chem. Phys.* **13**, 229–234 (1945).

An independent approach using the plate model is treated in

A. J. P. Martin and R. L. M. Synge, "A New Form of Chromatogram Employing Two Liquid Phases," *Biochem. J.* **35**, 1358–1368 (1941).

Parametric pumping, an invention of R. H. Wilhelm, is discussed in

R. H. Wilhelm, A. W. Rice, and A. R. Bendelius, "Parametric Pumping: A Dynamic Principle for Separating Fluid Mixtures," *Ind. Eng. Chem. Fundam.* **5**, 141–144 (1966).

R. H. Wilhelm, A. W. Rice, R. W. Rolke, and N. H. Sweed, "Parametric Pumping," *Ind. Eng. Chem. Fundam.* **7**, 337–349 (1968).

1.4. Earlier works associated with heat effects in fixed-bed adsorption are found in

F. W. Leavitt, "Non-isothermal Adsorption in Large Fixed Beds," *Chem. Eng. Prog.* **58**, 54–59 (1962).

N. R. Amundson, R. Aris, and R. Swanson, "On Simple Exchange Waves in Fixed Beds," *Proc. Roy. Soc. Lond.* **A286**, 129–139 (1965).

More general treatment is to be found in

H.-K. Rhee, E. D. Heerdt, and N. R. Amundson, "An Analysis of an Adiabatic Adsorption Column: Parts I, II, III, and IV," *Chem. Eng. J.* **1**, 241–254 and 279–290 (1970), and *Chem. Eng. J.* **3**, 22–34 and 121–135 (1972).

Temperature dependence of the Langmuir isotherm is discussed in

J. H. deBoer, *The Dynamical Character of Adsorption*, 2nd ed., Oxford University Press, New York, 1968.

1.5. Basic equations for the countercurrent adsorber are found in

H.-K. Rhee, R. Aris, and N. R. Amundson, "Multicomponent Adsorption in Continuous Countercurrent Exchangers, *Phil. Trans. Roy. Soc. Lond.* **A269**, 187–215 (1971).

The countercurrent adsorber with reaction was the subject of

S. Viswanathan, "Chromatographic Reactors," Ph.D. thesis, University of Minnesota, 1973.

B. K. Cho, "Studies of Continuous Chromatographic Reactors," Ph.D. thesis, University of Minnesota, 1979.

The case with eq. (1.5.10) is well discussed in

R. Aris, *Mathematical Modelling Techniques*, Pitman, London, 1978, p. 86–103.

1.7. For a survey of the differential equations of polymerization, see

S. L. Liu and N. R. Amundson, "Analysis of Polymerization Kinetics and the Use of a Digital Computer," *Rubber Chem. Tech.* **34**, 995–1133 (1961).

The use of partial differential equations is given in

R. J. Zeman and N. R. Amundson, "Continuous Models for Polymerization," *AIChE J.* **9**, 297–302 (1963).

R. J. Zeman and N. R. Amundson, "Continuous Polymerization Models, Parts I and II," *Chem. Eng. Sci.* **20**, 331–361 and 637–664 (1965).

The notions of continuous mixtures are treated in

R. Aris and G. R. Gavalas, "On the Theory of Reactions in Continuous Mixtures," *Phil. Trans. Roy. Soc. Lond.* **A260**, 351–393 (1966).

1.8. Exchange with hydrides is discussed in

R. J. Mikovsky and J. Wei, "A Kinetic Analysis of the Exchange of Deuterium with Hydrides," *Chem. Eng. Sci.* **18**, 253–258 (1962).

1.9. Reduction to the minimum number of equations by use of reaction extents is discussed in

R. Aris, *Introduction to the Analysis of Chemical Reactors*, Prentice-Hall, Englewood Cliffs, N.J., 1965.

1.10. The basic equations of water flooding are found in

M. C. Leverett, "Flow of Oil–Water Mixtures through Unconsolidated Sands," *Trans. AIME* **132**, 149–171 (1939).

M. C. Leverett, "Capillary Behavior in Porous Solids," *Trans. AIME* **142**, 152–169 (1941).

S. E. Buckley and M. C. Leverett, "Mechanism of Fluid Displacement in Sands," *Trans. AIME* **146**, 107–116 (1942).

This model for polymer flooding is discussed in

J. T. Patton, K. H. Coats, and G. T. Colegrove, "Prediction of Polymer Flood Performance," *Soc. Pet. Eng. J.* **11**, 72–84 (March 1971).

Formulation of multicomponent, multiphase displacement is treated in

F. G. Helfferich, "Theory of Multicomponent, Multiphase Displacement in Porous Media," *Soc. Pet. Eng. J.* **21**, 51–62 (Feb. 1981).

1.11. The work of Lighthill and Whitham is to be found in

M. J. Lighthill and G. B. Whitham, "On Kinematic Waves: I. Flood Movements in Long Rivers," *Proc. Roy. Soc. Lond.* **A229**, 281–316 (1955); "II. A Theory of Traffic Flow on Long Crowded Roads," ibid., 317–345.

References to the work on floods and traffic flow are given there. For the latter topic see also

W. D. Ashton, *The Theory of Road Traffic Flow*, Methuen, London, 1966.

Sedimentation was first treated on these lines by

G. J. Kynch; for reference see those given for Sec. 6.7.

The application to the movement of glaciers is from

J. F. Nye, "Surges in Glaciers," *Nature* **181**, 1450–1451 (May 1958).

The conservation equations of crystal growth and breakage are discussed in

A. D. Randolph, "Effect of Crystal Breakage on Crystal Size Distribution in a Mixed Suspension Crystalizer," *Ind. Eng. Chem. Fundam.* **8**, 58–63 (1969).

1.12. For the development of the equations of fluid mechanics, see, for example,

L. M. Milne-Thompson, *Theoretical Hydrodynamics*, Macmillan, New York, 1938.

R. Aris, *Vectors, Tensors and the Basic Equations of Fluid Mechanics*, Prentice-Hall, Englewood Cliffs, N.J., 1962.

The form given here follows closely the brief treatment given at the beginning of

R. Courant and K. O. Friedrichs, *Supersonic Flow and Shock Waves*, Wiley-Interscience, New York, 1948.

Applications to wave motion in incompressible fluids, gas dynamics, plasticity, and the flow of granular materials are given in

M. B. Abbott, *An Introduction to the Method of Characteristics*, Thames and Hudson, London, 1966.

1.13. A detailed discussion of the basic equation of heat conduction, with references to some of the earlier work is given in

J. Meixner, "On the Linear Theory of Heat Conduction," *Arch. Rat. Mech. Anal.* **39**, 108–130 (1970).

A chemical engineering situation that demands a finite relaxation time for heat conduction is considered in

S. H. Chan, M. J. D. Low, and W. K. Mueller, "Hyperbolic Heat Conduction on Catalytic Supported Crystallites," *AIChE J.* **17**, 1499–1501 (1971).

E. Ruckenstein and C. A. Petty, "On the Aging of Supported Metal Catalyst Due to Hot Spots," *Chem. Eng. Sci.* **27**, 937–946 (1972).

For a detailed development of the eikonal equation and further references to the foundations of geometrical optics, see chapter III of

M. Born and E. Wolf, *Principles of Optics*, 2nd ed., Pergamon Press, New York, 1964.

The two-dimensional eikonal equation is treated in detail in

N. Bleistein, *Mathematical Methods for Wave Phenomena*, Academic Press, Orlando, Florida, 1984.

1.14. Much more on the dynamic programming approach to optimization problems, which we have sketched in this section, is to be found in

R. E. Bellman, *Dynamic Programming*, Princeton University Press, Princeton, N.J., 1957.

See also

R. E. Bellman and S. Dreyfus, *Applied Dynamic Programming*, Princeton University Press, Princeton, N.J., 1962.

There are numerous introductory texts:

S. E. Dreyfus, *Dynamic Programming and the Calculus of Variations*, Academic Press, New York, 1965.

G. L. Nemhauser, *Introduction to Dynamic Programming*, Wiley, New York, 1966.

D. S. Wilde, *Optimum Seeking Methods*, Prentice-Hall, Englewood Cliffs, N.J., 1964.

Problems on chemical reactors are treated in

R. Aris, *The Optimal Design of Chemical Reactors*, Academic Press, New York, 1961.

1.15. This treatment of the filtering problem is taken from

R. E. Bellman, H. H. Kaginada, R. E. Kalaba, and R. Sridhar, "Invariant Imbedding and Nonlinear Filtering Theory," RM-4374-PR, Rand Corporation, Dec. 1964, Santa Monica, Calif.

Motivations, Classifications, and Some Methods of Solution

2

2.1 Comparisons between Ordinary and Partial Differential Equations

Some illuminating analogies may be drawn between ordinary and partial differential equations that will trade on the reader's knowledge of the former and motivate some of the development that follows.

The ordinary differential equation

$$\frac{dy}{dx} = f(x, y) \tag{2.1.1}$$

means that if the curve $y = \phi(x)$ is a solution, then at the point (x_0, y_0), $\phi'(x_0) = f(x_0, y_0)$. In other words, if we attach to each point of the (x, y)-plane a short line segment of slope $f(x, y)$, we will have a diagram in which the solution curves can be roughly discerned, much as lines of magnetic force can be demonstrated by iron filings on a piece of paper held over a magnet. The method of isoclines consists of drawing the loci $f(x, y) = s$, where s is a constant, in the (x, y)-plane, marking them with short lines of slope s and fairing a solution curve by eye. The complete solution of the ordinary differential equation is a one-parameter family of curves covering a certain region of the (x, y)-plane. We only get a particular solution of the ordinary differential equation (2.1.1) when we specify a particular point; for example, $y(x_0) = y_0$. Thus the complete solution is a family of one-dimensional objects (solution curves) in two-dimensional space, and to get a particular solution the solution must be required to pass through a zero-dimensional object (point).

The partial differential equation in two independent variables

$$F(x, y, z, p, q) = 0 \tag{2.1.2}$$

has solution surfaces that can be written

$$z = \phi(x, y) \tag{2.1.3}$$

When $p = \phi_x(x, y)$ and $q = \phi_y(x, y)$ are calculated at any point of the solution surface and substituted back into Eq. (2.1.2), this equation is satisfied identically. But, in contrast to the ordinary differential equation, the direction of the tangent plane is not in general uniquely specified by the partial differential equation. For even if the ordinary differential equation (2.1.1) were written by analogy with (2.1.2) as $F(x, y, p) = 0$, where $p = dy/dx$, we could always recover the original form by solving for p. However, the single equation (2.1.2) could not be solved for both p and q, and it therefore represents only a restriction on possible pairs of values of p and q, rather than a prescription for their actual values. Nor is this surprising in the light of knowing that the partial differential equation can arise from a two-parameter family of surfaces when only a one-parameter family of surfaces is needed to fill three-dimensional space. The complete solution of the partial differential equation can be represented as a family of two-dimensional objects (solution surfaces) in three-dimensional space. But just as the ordinary differential equation required the specification of an object (a point) of one less dimension than its solution (a curve), so here we will expect to get a unique solution only as we require it to pass through a curve (i.e., an object of one less dimension).

In the general case of a first-order equation with n independent variables,

$$F(x_1, x_2, \ldots, x_n, z, p_1, p_2, \ldots, p_n) = 0 \tag{2.1.4}$$

we need $(n + 1)$ dimensions to show the solutions that are a family of n-dimensional manifolds. We will pick a particular solution out of this family if we require it to pass through a given $(n - 1)$-dimensional manifold.

In the case of the ordinary differential equation, the envelope of the family of solution curves, if there be one, is also a solution of the ordinary differential equation. For example, the ordinary differential equation

$$2x\left(\frac{dy}{dx}\right)^2 - 2y\frac{dy}{dx} + 1 = 0$$

has the family of lines

$$2\alpha y = 2x + \alpha^2$$

as its complete solution. But as we saw in Sec. 0.10, this family has the parabola $y^2 = 2x$ for an envelope; and it is easy to see that the envelope also satisfies the differential equation, being what is known as a singular solution. Similarly, we saw in Sec. 1.16 that the family of unit spheres

$$(x - \alpha)^2 + (y - \beta)^2 + z^2 = 1$$

gave rise to the partial differential equation

$$z^2(p^2 + q^2 + 1) = 1$$

But in Sec. 0.10 we proved the geometrically obvious fact that the two planes $z = \pm 1$ envelop this family. Since $p = q = 0$ on these planes, they clearly satisfy the partial differential equation. However, we can also select a one-parameter family and find its envelope and obtain a surface that satisfies the equation. For example, if $\alpha = r \cos \theta$, $\beta = r \sin \theta$, $r > 1$, the envelope will clearly be a torus, which by differentiating with respect to θ and eliminating θ we can show to have the equation

$$[(x^2 + y^2)^{1/2} - r]^2 + z^2 = 1$$

It is a simple enough matter to confirm that this satisfies the partial differential equation. It would, in fact, be the unique solution surface passing through the initial curve $x^2 + y^2 = r^2$, $z = 1$. Since any enveloping surface is touched at every point by one of the family of surfaces it envelops, the values of x, y, z, p, and q must be the same for the envelope and tangent member, and the envelope therefore satisfies the same differential equation as the family.

The general solution to the partial differential equation is built up as the envelope of an arbitrary one-parameter family selected from the two-parameter family that forms the complete integral. A particular solution satisfying certain boundary conditions or containing a given initial curve can be selected from the general solution by specifying the arbitrary function that pulls out the one-parameter family. By comparison with the ordinary differential equation, we see that in each case the envelope of the complete solution, if there be one, provides a singular solution. Since the complete solution of the ordinary differential equation is only a one-parameter family, we cannot select from it a zero-parameter family except by taking just one member. From the two-parameter family, however, we have the option of selecting a one-parameter family to provide a particular solution. We will find these geometrical ideas constantly turning up in one form or another.

Exercise

2.1.1. Show that any right cylinder of unit radius whose axis lies in the (x, y)-plane is a solution of $z^2(p^2 + q^2 + 1) = 1$.

2.2 Classification of Equations

Given two independent variables x and y, a function z, and its partial derivatives, $p = \partial z/\partial x$ and $q = \partial z/\partial y$, it is possible to conceive of a function

$$F(x, y, z, p, q) = 0 \tag{2.2.1}$$

of these five quantities. Such a function, in general, will not be linear. In many applications the function F takes on special forms and these special forms will occupy us mostly in this work. Later, however, the general theory of finding solutions for more general equations will be discussed. Special forms of the function F are as follows.

Linear: $P(x, y)p + Q(x, y)q = R(x, y, z)$. (2.2.2)

Strictly linear if R is independent of z.

Linear if R is a linear function of z.

Semilinear if R depends on z in a nonlinear fashion.

Quasi-linear: $P(x, y, z)p + Q(x, y, z)q = R(x, y, z)$. (2.2.3)

Any equation that does not fit into one of these forms may be properly called nonlinear.

If there are n independent varibles $x_1, x_2, \ldots, x_n$, the general equation

$$F(x_1, x_2, \ldots, x_n, z, p_1, p_2, \ldots, p_n) = 0 \tag{2.2.4}$$

where $p_i = \partial z/\partial x_i$, is similarly classified:

Linear: $F = \Sigma P_i(\mathbf{x})p_i - R(\mathbf{x}, z)$. (2.2.5)

Strictly linear if R is independent of z.

Linear if R is a linear function of z.

Semilinear if R depends on z in a nonlinear fashion.

Quasi-linear: $F = \Sigma P_i(\mathbf{x}, z)p_i - R(\mathbf{x}, z)$. (2.2.6)

Since we will consider in this book mainly quasi-linear systems of first-order partial differential equations, it is important to see that a general nonlinear equation can be reduced to a system of quasi-linear partial differential equations of first order. This is embodied in the following theorem. (See, for example, Jeffrey and Taniuti (1964), p. 24.)

THEOREM: The initial-value problem for a system of general nonlinear partial differential equations with noncharacteristic initial data may be reduced to a noncharacteristic initial-value problem for a system of quasi-linear partial differential equations of first order.

First, we carry out the reduction for the first-order differential equation

$$F(x, y, z, p, q) = 0 \tag{2.2.7}$$

where

$$p = z_x, \qquad q = z_y \tag{2.2.8}$$

Let us suppose that Eq. (2.2.7) can be solved for p; that is, $\partial F/\partial p \neq 0$, assuming the form

$$p = f(x, y, z, q) \tag{2.2.9}$$

The initial-value problem now is to find a solution $z(x, y)$ of Eq. (2.2.9), which becomes a prescribed function

$$z(0, y) = \phi(y) \qquad \text{for } x = 0 \tag{2.2.10}$$

From Eq. (2.2.8), we have

$$p_y = \frac{\partial}{\partial y}\left(\frac{\partial z}{\partial x}\right) = \frac{\partial^2 z}{\partial y\, \partial x} = \frac{\partial}{\partial x}\left(\frac{\partial z}{\partial y}\right) = q_x$$

while differentiation of Eq. (2.2.9) with respect to x gives

$$p_x = f_x + f_z p + f_q q_x$$

where f_q, for example, denotes the partial derivative of f with respect to q, holding x, y, and z constant. Consequently, we obtain for the three quantities z, p, and q the system of first-order partial differential equations

$$\begin{aligned} z_x &= p \\ p_x &= f_x + f_z p + f_q p_y \\ q_x &= p_y \end{aligned} \tag{2.2.11}$$

and this is a quasi-linear system because f_z and f_q are functions of x, y, z, and q. On the other hand, we note that with the function $z(0, y) = \phi(y)$ prescribed for $x = 0$ the initial values $q(0, y) = \phi'(y)$ are automatically prescribed also, and Eq. (2.2.9) further yields the initial values for p. Therefore, the initial conditions are specified as

$$\begin{aligned} z(0, y) &= \phi(y) \\ p(0, y) &= f(0, y, \phi(y)\ \phi'(y)) \\ q(0, y) &= \phi'(y) \end{aligned} \tag{2.2.12}$$

Next we prove that this initial-value problem is equivalent to the original one by showing that, with a solution z, p, and q of the system of equations (2.2.11) and (2.2.12), the equations

$$z_x = p = f(x, y, z, q) \quad \text{and} \quad z_y = q$$

are satisfied. From the first equation of Eq. (2.2.11), we have

$$z_{xx} = p_x = \frac{\partial f}{\partial x}$$

which, upon integration with respect to x, gives

$$z_x = p = f(x, y, z, q) + a(y)$$

But since $p = f$ holds for $x = 0$, it follows that $a(y) = 0$, and thus $z_x = p = f(z, y, z, q)$. The last equation of Eq. (2.2.11) implies

$$z_{xy} = q_x$$

which may be integrated with respect to x to give

$$z_y = q + b(y)$$

By using again the initial conditions, we see that $z_y = q$ for all x and y. Therefore, $z(x, y)$ is the solution to the original problem.

Here the argument has been illustrated for a general first-order nonlinear partial differential equation with the dependent variable z and the two independent variables x and y. But a similar procedure can be applied to replace a higher-order differential equation† (see Exercise 2.2.1) or a system of differential equations by a first-order quasi-linear system of partial differential equations. The argument can also be extended to equations with n independent variables.

The system of equations (2.2.11) is nonhomogeneous in the derivatives, and the term $(f_x + f_z p)$, as well as the coefficient f_q, contains the independent variables x and y. It is often convenient to pass to another equivalent quasi-linear system of first-order partial differential equations that is homogeneous and whose coefficients are not dependent on the independent variables x and y. For this purpose, we introduce two additional dependent variables ξ and η defined as

$$\xi = x, \qquad \eta = y \tag{2.2.13}$$

and replace x and y in f_x, f_z, and f_q by ξ and η, respectively. Since $\xi_x = \eta_y$

†The converse is by no means true. Not every system of two first-order partial differential equations, let alone one of three first-order differential equations, is equivalent to a second-order differential equation (see Courant and Hilbert, 1962).

$= 1$, $\xi(0, y) = 0$, and $\eta(0, y) = y$, we may rewrite Eqs. (2.2.11) and (2.2.12) into the equivalent system of differential equations for z, p, q, ξ, and η:

$$\begin{aligned} z_x &= p\eta_y \\ p_x &= f_q p_y + (f_x + pf_z)\eta_y \\ q_x &= p_y \\ \xi_x &= \eta_y \\ \eta_x &= 0 \end{aligned} \tag{2.2.14}$$

with the initial conditions

$$\begin{aligned} z(0, y) &= \phi(y) \\ p(0, y) &= f(0, y, \phi(y), \phi'(y)) \\ q(0, y) &= \phi'(y) \\ \xi(0, y) &= 0 \\ \eta(0, y) &= y \end{aligned} \tag{2.2.15}$$

The system of equations (2.2.14) is homogeneous in the derivatives, and the independent variables x and y no longer appear explicitly in the coefficients. The initial-value problem given by Eqs. (2.2.14) and (2.2.15) is obviously equivalent to that for Eqs. (2.2.7) and (2.2.10). The same argument can be applied to the initial-value problem of second order to obtain an equivalent initial-value problem for a system of homogeneous, first-order differential equations for eight dependent variables.

All the initial-value problems formulated in this manner will have the form of a quasi-linear system of first-order equations

$$\frac{\partial z_i}{\partial x} = \sum_{j=1}^{m} G_{ij}(z_1, z_2, \ldots, z_m) \frac{\partial z_j}{\partial y}, \quad i = 1, 2, \ldots, m \tag{2.2.16}$$

with initial conditions of the form

$$z_i(0, y) = \phi_i(y) \tag{2.2.17}$$

In case of n variables $y_1, y_2, \ldots, y_n$, Eq. (2.2.16) will be written as

$$\frac{\partial z_i}{\partial x} = \sum_{k=1}^{n} \sum_{j=1}^{m} G_{ijk}(z_1, z_2, \ldots, z_m) \frac{\partial z_j}{\partial y_k}, \qquad i = 1, 2, \ldots, m \tag{2.2.18}$$

The preceding arguments may be summarized in the following theorem:

> THEOREM: The initial-value problem for a system of partial differential equations of any order with noncharacteristic initial data can be

reduced to an equivalent noncharacteristic initial-value problem for a system of quasi-linear, homogeneous, first-order partial differential equations.

Exercise

2.2.1. Consider the initial-value problem for the second-order differential equation $F(x, y, z, p, q, r, s, t) = 0$, with the abbreviation $p = z_x$, $q = z_y$, $r = z_{xx} = p_x$, $s = z_{xy} = p_y = q_x$, and $t = z_{yy} = q_y$. The initial values of z and z_x are prescribed at $x = 0$; that is, $z(0, y) = \phi(y)$ and $z_x(0, y) = \psi(y)$. Find an equivalent problem for a system of quasi-linear differential equations of first order and justify that the new problem is equivalent to the original one. Can you formulate another equivalent initial-value problem for a system of homogeneous differential equations of first order in which the independent variables x and y do not appear explicitly in the coefficients?

2.3 When Has an Equation Been Solved?

As one learns rapidly enough with ordinary differential equations, it is rare for the solution to be obtainable in a simple explicit formula. This, however, does not prevent us from writing a solution in the form

$$z = f(x, y) \tag{2.3.1}$$

and asserting that it is such that, when substituted back into the equation as

$$F(x, y, f(x, y), f_x(x, y), f_y(x, y)) = 0 \tag{2.3.2}$$

it yields an identity in x and y. Although we may not often be able to obtain a simple formula for the solution, we will generally be proceeding constructively and will show the steps by which the values of x, y, and z at each point of the solution surface may be obtained. We will obtain the solution in parametric form, as in Secs. 0.6 and 0.7.

$$x = \phi(s, \xi), \qquad y = \psi(s, \xi), \qquad z = \chi(s, \xi) \tag{2.3.3}$$

In this case, since p and q can be calculated as functions of s and ξ by

$$p\frac{\partial(\phi, \psi)}{\partial(s, \xi)} = \frac{\partial(\chi, \psi)}{\partial(s, \xi)}, \qquad q\frac{\partial(\phi, \psi)}{\partial(s, \xi)} = \frac{\partial(\phi, \chi)}{\partial(s, \xi)} \tag{2.3.4}$$

we have a solution surface when

$$F(\phi(s, \xi), \psi(s, \xi), \chi(s, \xi), p(s, \xi), q(s, \xi)) = 0$$

identically in s and ξ.

A solution of the form

$$z = f(x, y; \alpha, \beta) \quad \text{or} \quad f(x, y, z; \alpha, \beta) = 0 \tag{2.3.5}$$

with two arbitrary constants is called the *complete solution*. A solution of the form

$$z = u(x, y)g(v(x, y)) \tag{2.3.6}$$

or

$$h(u(x, y, z), v(x, y, z)) = 0 \tag{2.3.7}$$

where g or h is an arbitrary function, is called a *general solution*. A solution with no arbitrary elements in it and that does not derive its uniqueness from the imposition of a boundary condition is a *singular solution*. Such is the envelope of the complete solution. A solution satisfying an imposed condition is the *particular solution* of a given problem, and it is this that is generally needed in a physical situation.

We often expect solutions to be represented by smooth surfaces, and a solution that is not everywhere differentiable is called *weak*. The commonest form of weak solution is one that has discontinuities in first derivatives across a line so that the solution surface appears to have a ridge. We will also meet *discontinuous solutions*, where z itself and not merely p and q are discontinuous. This kind of discontinuity is often referred to as a *shock*. An important feature of quasi-linear and nonlinear equations is that their solutions may only be strong in part of the region of interest and may develop discontinuities as they move away from the initial data.

In physical situations, we have seen that the two independent variables that most naturally arise are one of position, z, and one of time, t. The natural boundary conditions are then most often that the dependent variable is specified for all time at some fixed position, for example, the inlet, and for all positions at a fixed time, for example, initially. In physical problems we will regard the problem as solved when the particular solution that satisfies the equation and the boundary conditions has been constructed.

All that we have said here for the case of two independent variables carries over, mutatis mutandis, for n independent variables.

Exercise

2.3.1. Show that

$$c(z, t) = c_0\left[z - \frac{Vt}{g'(c(z, t))}\right]$$

is a solution (in implicit form) of the equation

$$V\frac{\partial c}{\partial z} + g'(c)\frac{\partial c}{\partial t} = 0$$

for which $c(z, 0) = c_0(z)$. What sort of limitations do you think would be imposed on this by physical common sense?

2.4 Special Methods for Certain Equations

We have already seen (Exercise 1.16.1) that the family of planes

$$z = \alpha x + \beta y + f(\alpha, \beta) \tag{2.4.1}$$

gives rise to the equation, known as *Clariaut's*,

$$z = px + qy + f(p, q) \tag{2.4.2}$$

It follows that the family of planes is the complete solution of Eq. (2.4.2). A general solution may be found by setting $\beta = \phi(\alpha)$, an arbitrary function, and eliminating α between

$$z = \alpha x + \phi(\alpha)y + f(\alpha, \phi(\alpha)) \tag{2.4.3}$$

and

$$0 = x + \phi'(\alpha)y + f_\alpha + f_\beta\phi'(\alpha) \tag{2.4.4}$$

As we remarked in Sec. 0.10, these are equations of two planes, whose intersection is a line and hence the typical generator of a ruled surface. A singular solution is obtained by eliminating both α and β from Eq. (2.4.1) and

$$x + f_\alpha(\alpha, \beta) = 0, \qquad y + f_\beta(\alpha, \beta) = 0 \tag{2.4.5}$$

Singular solutions should always be checked to see that they satisfy the differential equation, for the process of elimination can also produce loci of cusps and nodes that do not satisfy the equation.

There are some other special cases of equations that lend themselves to particular methods. For example, if only p and q are present so that the equation is

$$F(p, q) = 0 \quad \text{or} \quad q = f(p) \tag{2.4.6}$$

we may take p and q to be constants constrained by the equation; that is,

$$p = \alpha, \qquad q = f(\alpha)$$

where α is arbitrary. Then since $dz = p\,dx + q\,dy$ and p and q are constant, we have

$$z = \alpha x + f(\alpha)y + \beta \tag{2.4.7}$$

a complete solution with two arbitrary constants. Again this is a two-parameter family of planes.

An important example of this is the equation for light propagation in two dimensions (Sec. 1.13),

$$u_x^2 + u_y^2 = 1 \tag{2.4.8}$$

for which we immediately have

$$u(x, y) = \alpha x + \sqrt{1 - \alpha^2}\, y + \beta \tag{2.4.9}$$

or

$$u(x, y) = x \cos \gamma + y \sin \gamma + \beta \tag{2.4.10}$$

Thus the locus u = constant is a line in the plane whose normal makes an angle γ with the x-axis. If we make β an arbitrary function of γ, we have a general solution given by eliminating γ between the two equations

$$\begin{aligned} u &= x \cos \gamma + y \sin \gamma + \phi(\gamma) \\ 0 &= -x \sin \gamma + y \cos \gamma + \phi'(\gamma) \end{aligned} \tag{2.4.11}$$

By writing these equations as

$$\begin{aligned} x &= u \cos \gamma + \phi'(\gamma) \sin \gamma - \phi(\gamma) \cos \gamma \\ y &= u \sin \gamma - \phi'(\gamma) \cos \gamma - \phi(\gamma) \sin \gamma \end{aligned} \tag{2.4.12}$$

we see that the characteristics (i.e., the lines of constant γ along which the envelope touches each member of the family with parameter γ) are straight lines. These are the light rays, which in the geometrical optics of a homogeneous medium follow a straight path.

A second special form is that in which only z, p, and q are present,

$$F(z, p, q) = 0 \tag{2.4.13}$$

Here we assume that z is a function of the combination $x + \alpha y = \xi$, say. Thus

$$p = \frac{\partial z}{\partial x} = \frac{dz}{d\xi}, \qquad q = \frac{\partial z}{\partial y} = \frac{dz}{d\xi}\alpha$$

and

$$F\left(z, \frac{dz}{d\xi}, \alpha\frac{dz}{d\xi}\right) = 0$$

is an ordinary differential equation. Its solution will involve a new arbitrary constant β, so we again have a complete solution.

As an example of this, we may treat the equation

$$z^2(p^2 + q^2 + 1) = 1$$

which we obtained in Sec. 1.16. Here

$$(1 + \alpha^2)\left(\frac{dz}{d\xi}\right)^2 = \frac{1 - z^2}{z^2}$$

or

$$\frac{dz}{d\xi} = \frac{\pm 1}{(1 + \alpha^2)^{1/2}} \frac{(1 - z^2)^{1/2}}{z}$$

Thus

$$d(1 - z^2)^{1/2} = \pm d\frac{x + \alpha y}{(1 + \alpha^2)^{1/2}} = \pm d(x \cos \gamma + y \sin \gamma)$$

where $\alpha = \tan\gamma$, and so

$$z^2 + (x\cos\gamma + y\sin\gamma + \beta)^2 = 1 \tag{2.4.14}$$

It is interesting to note that the equation was derived from the family of unit spheres with centers on the (x, y)-plane and that we have recovered the family of unit cylinders with axes lying in the (x, y)-plane (see Exercise 2.1.1). Clearly, these families are equivalent, for the sphere of center (a, b) can be obtained by the envelope of the one-parameter family for which $\beta = -a\cos\gamma - b\sin\gamma$ in Eq. (2.4.14). Similarly, the cylinder given by Eq. (2.4.14) can be obtained as the envelope of all unit spheres of center (a, b) for which $a\cos\gamma + b\sin\gamma + \beta = 0$.

A third special form arises when z is absent and the independent variables separate, and the equation can be written

$$F(x, p) = G(y, q) \tag{2.4.15}$$

Then we put each side equal to an arbitrary constant, α, and solve for p and q, respectively:

$$p = f(x; \alpha), \qquad q = g(y; \alpha) \tag{2.4.16}$$

The complete solution is then obtained by integrating $dz = p\,dx + q\,dy$ to give

$$z = \int_{x_0}^{x} f(x'; \alpha)dx' + \int_{y_0}^{y} g(y'; \alpha)dy' + \beta \tag{2.4.17}$$

An example would be given by light propagation in an inhomogeneous medium with $n(x, y) = \nu(x^2 + y^2) + \mu$ [cf. Eq. (1.13.10)]. Then

$$p^2 + q^2 = \nu x^2 + \nu y^2 + \mu$$

or

$$p = (\alpha + \nu x^2)^{1/2}, \qquad q = (-\alpha + \nu y^2 + \mu)^{1/2}$$

so

$$z = \int (\alpha + \nu x^2)^{1/2}\,dx + \int (\mu - \alpha + \nu y^2)^{1/2}\,dy + \beta$$

It now generally becomes very much more difficult to form an envelope, but in principle it is still possible to set $\beta = \phi(\alpha)$ in Eq. (2.4.17) and eliminate α between it and

$$0 = \int_{x_0}^{x} f_\alpha(x'; \alpha)dx' + \int_{y_0}^{y} g_\alpha(y'; \alpha)dy' + \phi'(\alpha) \tag{2.4.18}$$

Exercises

2.4.1. By considering a plane wave satisfying $u_x^2 + u_y^2 = n$ incident at angle γ_1 to the y-axis in a medium for which $n = n_1$, $x < 0$, $n = n_2$, $x > 0$, prove Snell's law of refraction.

2.4.2. Show that the ordinary differential equation of Clairaut's form can be solved in essentially the same way as the partial differential equation. Generalize to n independent variables.

2.4.3. Show that the general solution of $xp + yq = 0$ is $z = f(x/y)$, where f is an arbitrary function. How would you extend this to n independent variables?

2.5 Method of Characteristics for Quasi-linear Equations

The subject of this section has been discussed the most comprehensively by Courant and Hilbert in their second volume of *Methods of Mathematical Physics*. We will try to cover here the backbone of their material, for this will be frequently referred to in subsequent chapters. Although the treatment here is directed toward quasi-linear equations, linear equations are also included.

We consider the quasi-linear, first-order partial differential equation, Eq. (2.2.3):

$$P(x, y, z)p + Q(x, y, z)q = R(x, y, z) \tag{2.5.1}$$

where P, Q, and R are given functions of x, y, and z that are continuously differentiable and satisfy the condition $P^2 + Q^2 \neq 0$. Suppose that the solution surface $z = z(x, y)$ has been determined, and consider the two vectors, $[P, Q, R]$ and $[p, q, -1]$, at a point (x, y, z) on the surface. Since Eq. (2.5.1) indicates that their scalar product vanishes, the two vectors are perpendicular to each other at the point. On the other hand, we have the relation

$$dz = z_x\,dx + z_y\,dy \tag{2.5.2}$$

which implies that the vector $[z_x, z_y, -1] = [p, q, -1]$ has the direction of the normal to the solution surface. Thus the partial differential equation (2.5.1) may be interpreted geometrically as a requirement that any solution surface $z = z(x, y)$ through a point $B(x, y, z)$ must be tangent there to a prescribed vector $[P, Q, R]$. This is depicted in Fig. 2.1.

It then follows that the tangent planes of all integral surfaces of Eq. (2.5.1) through the point $B(x, y, z)$ belong to a single pencil of planes whose axis is given by the relations

$$\frac{dx}{P} = \frac{dy}{Q} = \frac{dz}{R} \tag{2.5.3}$$

at the point B. Such a pencil and its axis are called the *Monge pencil* and the *Monge axis*, respectively. Equation (2.5.3) defines a direction field in the (x, y, z)-space, and its integral curves are called the *characteristic curves* of the partial differential equation (2.5.1). The projections of the charac-

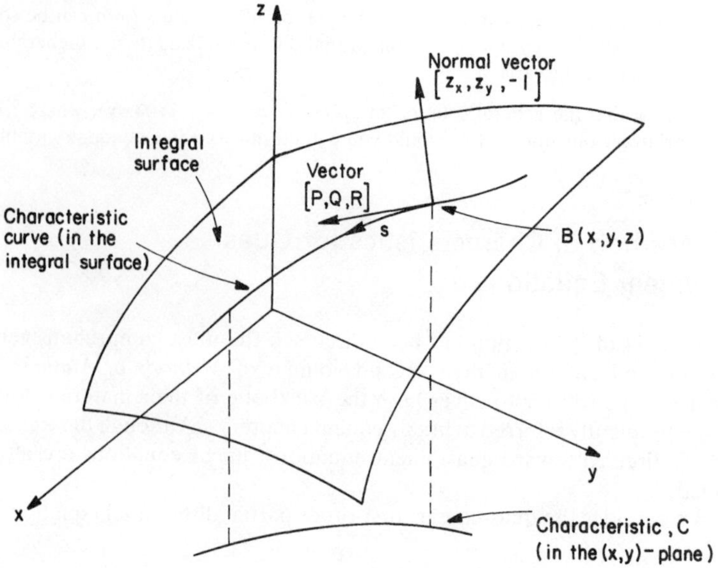

Figure 2.1

teristic curves onto the (x, y)-plane are referred to as the *characteristic base (or ground) curves* or simply as the *characteristics* and are usually denoted by the symbol C (see Fig. 2.1). If we introduce a parameter s running along the characteristic curves, Eq. (2.5.3) can be written as

$$\frac{dx}{ds} = P(x, y, z)$$

$$\frac{dy}{ds} = Q(x, y, z) \tag{2.5.4}$$

$$\frac{dz}{ds} = R(x, y, z)$$

which will be called the *characteristic differential equations* of Eq. (2.5.1).

It is now clear that integrating Eq. (2.5.1) is equivalent to finding surfaces that are at every point tangent to the Monge axis. Such surfaces are determined in the form of a one-parameter family of characteristic curves by integrating Eq. (2.5.4). Thus we see that every surface $z = z(x, y)$ generated by a one-parameter family of characteristic curves is an integral surface of Eq. (2.5.1), and conversely. The converse can be verified without difficulty by using Eq. (2.5.2) (see Exercise 2.5.1).

Since the parameter s does not appear explicitly in the differential equations, an additive constant in the parameter s is inessential, and we may regard the initial values of x, y, and z as specified for $s = 0$. With these initial conditions, the solution to the system of equations, Eq. (2.5.4), is uniquely determined (see Sec. 0.11), and this leads us to the following theorem.

THEOREM: Every characteristic curve that has one point in common with an integral surface lies entirely on the integral surface. Furthermore, every integral surface is generated by a one-parameter family of characteristic curves.

Let us now consider an initial-value problem for Eq. (2.5.1). We define a space curve I, to be called the *initial curve*, by prescribing x, y, and z as functions of a parameter ξ:

$$I: \quad x = x_0(\xi), \qquad y = y_0(\xi), \qquad z = z_0(\xi) \tag{2.5.5}$$

where $x_\xi^2 + y_\xi^2 \neq 0$ and the curve I has a simple projection I_0 on the (x, y)-plane. We let $s = 0$ along the initial curve I. The solution surface is then determined by drawing through each point of the curve I a characteristic curve, which is a solution curve of Eq. (2.5.4). This can be done in a unique way within a certain neighborhood of the curve I. The situation is shown in Fig. 2.2, while Fig. 2.3 represents the projection on the (x, y)-plane. The family of characteristic curves will be given by the form

$$\begin{aligned} x &= x(s, \xi) \\ y &= y(s, \xi) \\ z &= z(s, \xi) \end{aligned} \tag{2.5.6}$$

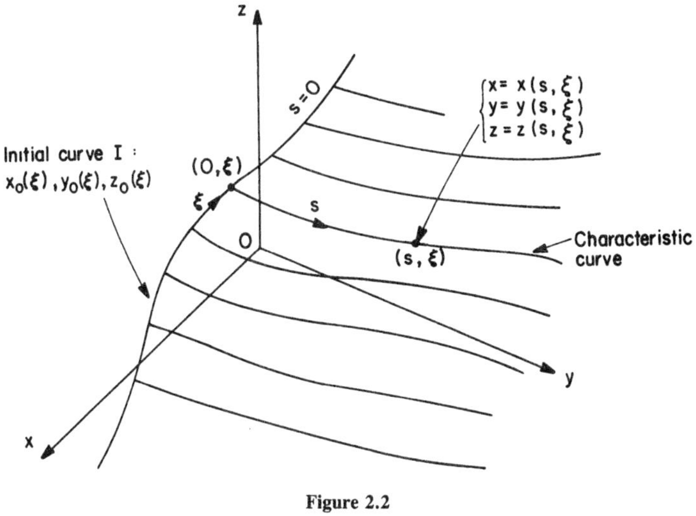

Figure 2.2

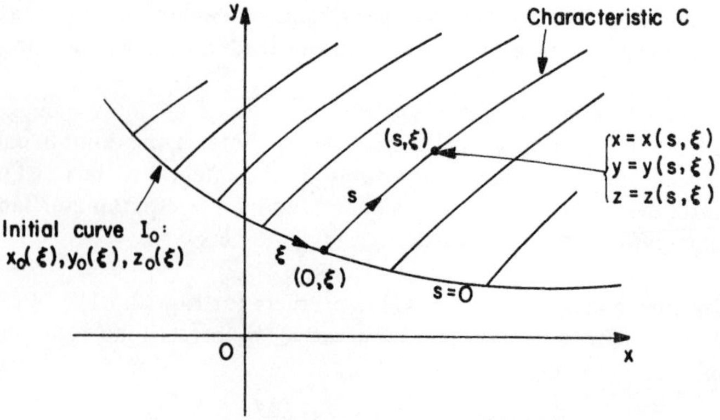

Figure 2.3

in which ξ appears as a parameter. These curves generate a surface $z = z(x, y)$ if we can express s and ξ as functions of x and y. A sufficient condition for this is that the Jacobian

$$J = \frac{\partial(x, y)}{\partial(s, \xi)} = \frac{\partial x}{\partial s}\frac{\partial y}{\partial \xi} - \frac{\partial y}{\partial s}\frac{\partial x}{\partial \xi} = Py_\xi - Qx_\xi \tag{2.5.7}$$

does not vanish along the initial curve I.

If $J \neq 0$ everywhere along the curve I, z can be expressed in terms of x and y. The last equation in Eq. (2.5.4) is then equivalent to the original partial differential equation, Eq. (2.5.1), because

$$\frac{dz}{ds} = \frac{\partial z}{\partial x}\frac{dx}{ds} + \frac{\partial z}{\partial y}\frac{dy}{ds} = Pz_x + Qz_y = R$$

Thus the surface $z = z(x, y)$ is the solution of the initial-value problem and its uniqueness follows from the theorem established previously.

If the condition $J = 0$ prevails everywhere along the curve I, we find that

$$\frac{x_\xi}{y_\xi} = \frac{P}{Q}$$

from Eq. (2.5.7). Hence if there is to exist a solution surface that passes through the curve I, we must have

$$\frac{x_\xi}{P} = \frac{y_\xi}{Q} = \frac{z_x x_\xi + z_y y_\xi}{z_x P + z_y Q} = \frac{z_\xi}{R} \tag{2.5.8}$$

This is identical to Eq. (2.5.4), and we see that the initial-value problem can be solved only when the initial curve I is itself a characteristic curve. Clearly, the solution is not unique here since such an initial curve can belong to a variety of integral surfaces determined by Eq. (2.5.4). Consequently, in the case of vanishing Jacobian, the initial-value problem does not have a solution unless the initial curve I is a characteristic curve, and then there are infinitely many solutions of the problem.

For the purpose of illustration we consider a simple example, the equation

$$z_x + zz_y = 1 \tag{2.5.9}$$

with the initial condition $z(0, y) = \alpha y$, where α is a constant. The characteristic differential equations are immediately written as

$$\frac{dx}{ds} = 1, \qquad \frac{dy}{ds} = z, \qquad \frac{dz}{ds} = 1 \tag{2.5.10}$$

We define the initial curve I by prescribing $x = 0$, $y = \xi$, and $z = \alpha\xi$ for $s = 0$ and integrate Eq. (2.5.10) to obtain the family of characteristic curves emanating from the curve I:

$$\begin{aligned} x &= s \\ y &= \frac{s^2}{2} + \alpha\xi s + \xi \\ z &= s + \alpha\xi \end{aligned} \tag{2.5.11}$$

By using the first two equations, s and ξ can be expressed in terms of x and y; that is,

$$s = x, \qquad \xi = \frac{y - x^2/2}{1 + \alpha x}$$

Therefore, the solution is given by

$$z = \alpha \frac{y - x^2/2}{1 + \alpha x} + x = \frac{\alpha y + x + \alpha x^2/2}{1 + \alpha x} \tag{2.5.12}$$

This is a unique solution since $J = Py_\xi - Qx_\xi = 1$ along the curve I. However, we expect that there would be trouble if $\alpha < 0$ because z diverges for $x = -1/\alpha$. This is typical of the behavior of quasi-linear equations, which will be discussed in Chapter 5.

If the initial curve I is chosen to be a characteristic curve, that is, $x = \xi$, $y = \xi^2/2$, and $z = \xi$ for $s = 0$, then Eq. (2.5.10) gives

$$\begin{aligned} x &= s + \xi \\ y &= \frac{s^2}{2} + \xi s + \frac{\xi^2}{2} = \frac{1}{2}(s + \xi)^2 \\ z &= s + \xi \end{aligned} \tag{2.5.13}$$

which is an expression for the same curve I, since an additive constant in the parameter s is immaterial. Although $J = 0$ along the curve I, the solution to Eq. (2.5.9) may be given implicitly by the equation

$$y = \frac{z^2}{2} + g(z - x) \tag{2.5.14}$$

where g is an arbitrary function with $g(0) = 0$. If Eq. (2.5.14) determines z uniquely as a function of x and y, then the integral surface $z = z(x, y)$ passes through the initial curve I. But there are infinitely many possibilities, because the function g may be chosen arbitrarily, and hence there exists an infinity of solutions.

Exercises

2.5.1. Prove that every integral surface $z = z(x, y)$ of Eq. (2.5.1) is generated by a one-parameter family of characteristic curves.

2.5.2. Consider Eq. (2.5.9) with the initial curve I defined by the prescribed data: $x = 2\xi$, $y = \xi^2$, and $z = \xi$ for $s = 0$. Here the curve I is noncharacteristic. Can you establish an argument that this initial-value problem cannot be solved? (*Note*: Here the projection I_0 of the curve I is a characteristic, while the curve I itself is not a characteristic curve.)

2.6 Alternative Treatment of the Quasi-linear Equations

We consider the quasi-linear equation

$$P(x, y, z)p + Q(x, y, z)q = R(x, y, z) \tag{2.6.1}$$

where P, Q, and R are given functions of x, y, and z. Indeed, P and Q can be independent of z so that linear equations are not excluded. We suppose that $g(x, y, z) = k$ is a solution of this equation. Now if x and y are considered as independent variables, then

$$\frac{\partial g}{\partial x} + \frac{\partial g}{\partial z}\frac{\partial z}{\partial x} = 0$$

$$\frac{\partial g}{\partial y} + \frac{\partial g}{\partial z}\frac{\partial z}{\partial y} = 0$$

so that we have

$$p = \frac{\partial z}{\partial x} = -\frac{\partial g/\partial x}{\partial g/\partial z}$$

$$q = \frac{\partial z}{\partial y} = -\frac{\partial g/\partial y}{\partial g/\partial z}$$

By substitution in Eq. (2.6.1), we obtain the equation

$$P\frac{\partial g}{\partial x} + Q\frac{\partial g}{\partial y} + R\frac{\partial g}{\partial z} = 0 \tag{2.6.2}$$

which may be written in the form

$$[P, Q, R]\begin{bmatrix} \dfrac{\partial g}{\partial x} \\ \\ \dfrac{\partial g}{\partial y} \\ \\ \dfrac{\partial g}{\partial z} \end{bmatrix} = 0$$

where

$$\nabla g = \begin{bmatrix} \dfrac{\partial g}{\partial x} \\ \\ \dfrac{\partial g}{\partial y} \\ \\ \dfrac{\partial g}{\partial z} \end{bmatrix}$$

is the gradient of the solution surface $g(x, y, z) = k$ in the (x, y, z)-space and is normal to that surface. Hence the vector

$$\mathbf{v} = \begin{bmatrix} P \\ Q \\ R \end{bmatrix}$$

is perpendicular to ∇g and therefore is tangent to the surface $g(x, y, z) = k$ and to a curve lying in that surface and passing through the point of tangency. If we consider a point in the surface $g(x, y, z) = k$, which moves in such a way that it is always tangent to the vector $\mathbf{v}$, we see that it traces out a curve on the surface. Consider a vector $\mathbf{r}$ from the origin to this curve and a parameter s running along the curve; then

$$\mathbf{r} = \begin{bmatrix} x \\ y \\ z \end{bmatrix} \quad \text{and} \quad \frac{d\mathbf{r}}{ds} = \begin{bmatrix} \dfrac{dx}{ds} \\ \\ \dfrac{dy}{ds} \\ \\ \dfrac{dz}{ds} \end{bmatrix}$$

If the curve traced out by $\mathbf{r}$ is to have the direction of the vector $d\mathbf{r}/ds$ at each point, then

$$\frac{dx/ds}{P} = \frac{dy/ds}{Q} = \frac{dz/ds}{R}$$

or

$$\frac{dx}{P} = \frac{dy}{Q} = \frac{dz}{R} \tag{2.6.3}$$

We see that these are two simultaneous ordinary differential equations, for they might be written in several alternative ways; for example,

$$\frac{dy}{dx} = \frac{Q(x, y, z)}{P(x, y, z)}, \qquad \frac{dz}{dx} = \frac{R(x, y, z)}{P(x, y, z)} \tag{2.6.4}$$

In this form we might expect solutions $y = y(x, k_1)$, $z = z(x, k_2)$, where k_1 and k_2 are arbitrary constants. But let us be more general and notice that, however we arrange the three equations into two, solutions can always be represented by two functions:

$$g_1(x, y, z) = k_1, \qquad g_2(x, y, z) = k_2 \tag{2.6.5}$$

Since Eq. (2.6.3) is equivalent to the characteristic differential equations of Eq. (2.6.1), we see that the intersections of these two surfaces are indeed the characteristic curves. As k_1 varies with k_2 fixed, the family of g_1 surfaces will intersect the g_2 surfaces in a family of characteristic curves, and conversely. Thus Eq. (2.6.5) represents a two-parameter family of curves that generates integral surfaces satisfying the partial differential equation (2.6.1).

Suppose now we have an initial-value problem so that the integral surface is required to pass through a given initial curve I that is not a characteristic curve. It then follows that all the characteristic curves in the integral surface must pass through the curve I, one characteristic curve passing through each point on the curve I. Thus a relation between the two constants, k_1 and k_2, must be found such that the intersections of the g_1 surfaces and the g_2 surfaces all pass through the initial curve I. This implies that k_1 and k_2 are functionally related by, for example, $k_2 = f(k_1)$ or $F(k_1, k_2) = 0$,† and hence a one-parameter family of characteristic curves is sufficient to determine an integral surface. Such a surface can be represented in the form

$$F(g_1(x, y, z), g_2(x, y, z)) = 0 \tag{2.6.6}$$

We will show now that this is in fact a solution and is the general solution since $F(k_1, k_2)$ was arbitrary.

†As an example, consider the initial data $z(x, 0) = \phi(x)$, where the x-axis is nowhere characteristic. We then have to require that the two equations, $g_1(x, 0, \phi(x)) = k_1$ and $g_2(x, 0, \phi(x)) = k_2$ be satisfied. These may be written in the form $f_1(x) = k_1$ and $f_2(x) = k_2$, respectively. From the first equation we have $x = f_3(k_1)$, and this is substituted in the second to give $k_2 = f(k_1)$.

We consider $g_1 = g_1(x, y, z)$ and $g_2 = g_2(x, y, z)$, where for the moment x, y, and z are considered to be independent. Then

$$\frac{\partial F}{\partial x} = \frac{\partial F}{\partial g_1}\frac{\partial g_1}{\partial x} + \frac{\partial F}{\partial g_2}\frac{\partial g_2}{\partial x}$$

$$\frac{\partial F}{\partial y} = \frac{\partial F}{\partial g_1}\frac{\partial g_1}{\partial y} + \frac{\partial F}{\partial g_2}\frac{\partial g_2}{\partial y}$$

$$\frac{\partial F}{\partial z} = \frac{\partial F}{\partial g_1}\frac{\partial g_1}{\partial z} + \frac{\partial F}{\partial g_2}\frac{\partial g_2}{\partial z}$$

and thus we find by using Eq. (2.6.2)

$$\begin{aligned} P\frac{\partial F}{\partial x} + Q\frac{\partial F}{\partial y} &= \frac{\partial F}{\partial g_1}\left(P\frac{\partial g_1}{\partial x} + Q\frac{\partial g_1}{\partial y}\right) + \frac{\partial F}{\partial g_2}\left(P\frac{\partial g_2}{\partial x} + Q\frac{\partial g_2}{\partial y}\right) \\ &= -R\left(\frac{\partial F}{\partial g_1}\frac{\partial g_1}{\partial z} + \frac{\partial F}{\partial g_2}\frac{\partial g_2}{\partial z}\right) \\ &= -R\frac{\partial F}{\partial z} \end{aligned}$$

which was to be shown.

In order to gain insight into the behavior of solutions, we consider the linear differential equation

$$z_x + \sigma z_y = 1 \tag{2.6.7}$$

in which σ is a positive constant. From Eq. (2.6.3), we have

$$\frac{dx}{1} = \frac{dy}{\sigma} = \frac{dz}{1}$$

or

$$\frac{dy}{dx} = \sigma, \qquad \frac{dz}{dx} = 1$$

Direct integration gives the equations

$$y - \sigma x = k_1, \qquad z - x = k_2$$

which represents a two-parameter family. These equations give the projections of the characteristic curves on the (x, y)- and (x, z)-planes, respectively. The former being the characteristics C, we might say that $(z - x)$ remains invariant along each characteristic. Since there is a functional relationship $k_2 = f(k_1)$, we can write

$$z - x = f(y - \sigma x) \tag{2.6.8}$$

where f is an arbitrary function of one variable. This is the general solution of Eq. (2.6.7).

If we have the initial data $z(0, y) = \phi(y)$ prescribed along the line $x = 0$, which is not a characteristic, then we see that $y = k_1$, and $\phi(y) = k_2$ along the initial curve I. Thus we have $k_2 = \phi(k_1)$, and the solution is given by

$$z - x = \phi(y - \sigma x) \tag{2.6.9}$$

Suppose now that the initial data are prescribed along a characteristic C so that $z(x, y) = \psi(x, y)$ on $y - \sigma x = \alpha$, where α is a constant. From Eq. (2.6.8), we have the condition $\psi(x, y) - x = f(\alpha) =$ constant. This cannot be satisfied in general. (Note that this is the case with $J = 0$ along the curve I in the previous section.) But if we further make the initial curve I become a characteristic curve by specifying $z(x, y) = x + \alpha_0$ on $y - \sigma x = \alpha$, then Eq. (2.6.8) satisfies the initial data with the requirement $f(\alpha) = \alpha_0$. We see, however, that there is an infinite number of solutions, because any function $f(y - \sigma x)$ with $f(\alpha) = \alpha_0$ will give a solution in the form of Eq. (2.6.8).

As an example for the quasi-linear equation, let us find the general solution of

$$y^2 z \frac{\partial z}{\partial x} + z^2 x \frac{\partial z}{\partial y} = -xy^2 \tag{2.6.10}$$

The characteristic differential equations are

$$\frac{dx}{y^2 z} = \frac{dy}{z^2 x} = \frac{dz}{-xy^2}$$

The first and third of these give $z^2 + x^2 = k_1$, and the second and third give $y^3 + z^3 = k_2$, so the general solution is

$$F(z^2 + x^2, y^3 + z^3) = 0 \tag{2.6.11}$$

If a curve is given in (x, y, z)-space as the intersection of two surfaces, say†

$$f_1(x, y, z) = 0, \qquad f_2(x, y, z) = 0$$

and two independent solutions to Eq. (2.6.3) are also known,

$$g_1(x, y, z) = k_1, \qquad g_2(x, y, z) = k_2$$

then the possibility exists of eliminating x, y, and z among the four functions f_1, f_2, g_1, and g_2 to obtain a relation between k_1 and k_2, say

$$F(k_1, k_2) = 0$$

Then

$$F(g_1(x, y, z), g_2(x, y, z)) = 0$$

is a solution to the partial differential equation that passes through the given

†Note that these equations represent one way of prescribing the initial data.

curve. For example, we might be given the differential equation

$$(ax + b)\frac{\partial z}{\partial x} + (cy + e)\frac{\partial z}{\partial y} = \alpha z + \beta, \qquad x > 0, y > 0 \tag{2.6.12}$$

where all the constants are positive and be asked to find the solution that passes through the intersection of†

$$z = \psi(x, y) \quad \text{and} \quad lx + my = n \tag{2.6.13}$$

where l, m, and n are positive constants and $\psi(x, y)$ is a prescribed function. In this case the characteristic differential equations are

$$\frac{dx}{ax + b} = \frac{dy}{cy + e} = \frac{dz}{\alpha z + \beta}$$

and two solutions may be written as

$$\frac{(ax + b)^{1/a}}{(cy + e)^{1/c}} = k_1$$

$$\frac{(\alpha z + \beta)^{1/\alpha}}{(ax + b)^{1/a}} = k_2$$

The general solution may then be written as

$$F\left(\frac{(ax + b)^{1/a}}{(cy + e)^{1/c}}, \frac{(\alpha z + \beta)^{1/\alpha}}{(ax + b)^{1/a}}\right) = 0$$

or

$$\frac{(\alpha z + \beta)^{1/\alpha}}{(ax + b)^{1/a}} = G\left(\frac{(ax + b)^{1/a}}{(cy + e)^{1/c}}\right)$$

This could be solved for z in the form

$$z = \frac{1}{\alpha}(ax + b)^{\alpha/a}\left[G\left(\frac{(ax + b)^{1/a}}{(cy + e)^{1/c}}\right)\right]^{\alpha} - \frac{\beta}{\alpha} \tag{2.6.14}$$

where G is an arbitrary function.

Now from $z = \psi(x, y)$, where $lx + my = n$, it follows that $z = \psi(x, -lx/m + n/m) = \phi(x)$. Therefore,

$$\phi(x) = \frac{1}{\alpha}(ax + b)^{\alpha/a}\left[G\left(\frac{(ax + b)^{1/a}m^{1/c}}{(-clx + nc + me)^{1/c}}\right)\right]^{\alpha} - \frac{\beta}{\alpha}$$

or

$$G\left(\frac{(ax + b)^{1/a}m^{1/c}}{(-clx + nc + me)^{1/c}}\right) = [\alpha\phi(x) + \beta]^{1/\alpha}(ax + b)^{-1/a}$$

Let

$$\frac{(ax + b)^{1/a} m^{1/c}}{(-clx + nc + me)^{1/c}} = u = f(x)$$

and $x = g(u)$ its inverse function; then

$$G(u) = [\alpha\phi(g(u)) + \beta]^{1/\alpha} (ag(u) + b)^{-1/a}$$

Hence

$$z = \frac{1}{\alpha}(ax + b)^{\alpha/a} \left[\alpha\phi\left(g\left(\frac{(ax + b)^{1/a}}{(cy + e)^{1/c}}\right)\right) + \beta\right]\left[ag\left(\frac{(ax + b)^{1/a}}{(cy + e)^{1/c}}\right) + b\right]^{-\alpha/a} - \frac{\beta}{\alpha} \tag{2.6.15}$$

From these examples one thing is obvious. In finding the solution of a partial differential equation that passes through a given curve, one probably would not proceed in this way, since it is apparent that only the simplest of problems could be fully solved. Normally, the characteristic differential equations would be integrated from a starting point on the initial curve as discussed in the previous section. Nevertheless, it is useful to see how the different notions of the complete, general, and particular solutions are related to one another.

Exercises

2.6.1. Find the general solution of

$$(y - xz)p + (x + yz)q = x^2 + y^2$$

2.6.2. Show that the general solution of

$$\frac{\partial}{\partial x}\left(\frac{z}{a(x)}\right) + \frac{\partial}{\partial y}\left(\frac{z}{b(y)}\right) = 0$$

can be written

$$z(x, y) = a(x)b(y)f\left(\int_0^x a(x')dx' - \int_0^y b(y')dy'\right)$$

What restrictions should be placed on the functions $a(x)$ and $b(y)$?

REFERENCES

Most of the materials in this chapter are usually discussed in one way or another in reference books on partial differential equations. The most extensive discussion is obviously found in

R. Courant and D. Hilbert, *Methods of Mathematical Physics: Volume II. Partial Differential Equations*, Interscience, New York, 1962.

The following may also be found useful:

P. R. Garabedian, *Partial Differential Equations*, Wiley, New York, 1964.

A. Jeffrey and T. Taniuti, *Non-linear Wave Propagation*, Academic Press, New York, 1964.

R. L. Streeter, *The Analysis and Solution of Partial Differential Equations*, Brooks/Cole Publishing Co., Monterey, Calif., 1973.

A. Jeffrey, *Quasilinear Hyperbolic Systems and Waves*, Pitman, London, 1976.

N. Bleistein, *Mathematical Methods for Wave Phenomena*, Academic Press, Orlando, Florida, 1984.

Linear and Semilinear Equations

3

The classification of equations that we gave in Sec. 2.2 sets apart those that are linear in the first derivatives. These are somewhat simpler to solve, and it will be worthwhile to present the method of characteristics in this context first. In fact, we will spend some time on the simplest case of all, the case of constant coefficients. Because it is so easy to present the geometrical aspect of the method of characteristics in three dimensions, we will develop each stage of the argument for the case of two independent variables first, using the abbreviations p and q for $\partial z/\partial x$ and $\partial z/\partial y$. The extensions to n independent variables are immediate.

3.1 Linear and Semilinear Equations with Constant Coefficients

The equation we wish to consider is

$$Pp + Qq = P\frac{\partial z}{\partial x} + Q\frac{\partial z}{\partial y} = R(x, y) \qquad (3.1.1)$$

where P and Q are constants and $P^2 + Q^2 \neq 0$, but R may be a function of position in the (x, y)-plane. This is the strictly linear case with constant coefficients. Later, in the linear or semilinear case, we will make R a function of z also.

We will apply the method of characteristics discussed in Sec. 2.5 to the solution of Eq. (3.1.1). The characteristic curves are determined by integrating the characteristic differential equations

$$\frac{dx}{ds} = P$$
$$\frac{dy}{ds} = Q \tag{3.1.2}$$
$$\frac{dz}{ds} = R(x, y)$$

where s is the characteristic parameter that runs along the characteristic curve (see Sec. 2.5). Since $dy/dx = Q/P =$ constant, all the characteristic ground curves, C, in the (x, y)-plane are straight lines having the same slope.

For simplicity, let us suppose that the initial data are given on the whole real axis so that

$$z(x, 0) = Z(x), \qquad -\infty < x < \infty \tag{3.1.3}$$

If we then set

$$x = \xi, \qquad y = 0, \quad \text{and} \quad z = Z(\xi) \qquad \text{for } s = 0 \tag{3.1.4}$$

the first two equations of Eq. (3.1.2) can be integrated to give

$$\begin{aligned} x &= Ps + \xi \\ y &= Qs \end{aligned} \tag{3.1.5}$$

and hence we have specified each point (x, y) in terms of two coordinates, s and ξ. On the line $s = 0$, we are given the initial or boundary data. This is shown in Fig. 3.1.

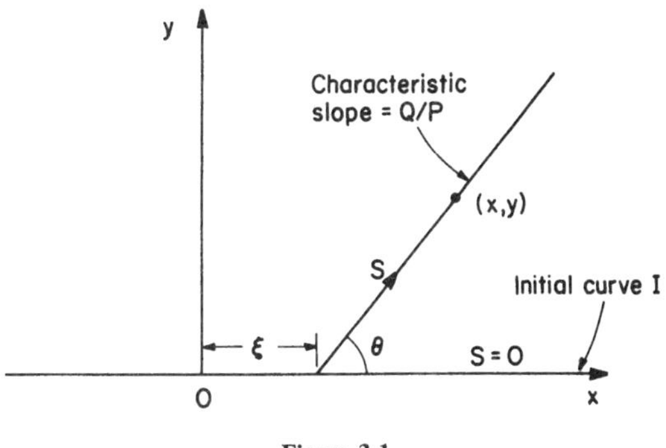

Figure 3.1

By using Eq. (3.1.5), we can express R as a function of s and ξ. Since we are regarding ξ as constant along a characteristic, it appears in the function R parametrically and, holding it constant, we can integrate the last equation of Eq. (3.1.2) from $s = 0$, where z is known, to the general point s. Thus

$$z(s, \xi) = Z(\xi) + \int_0^s R(Ps' + \xi, Qs')\, ds' \tag{3.1.6}$$

Finally, we can express s and ξ in terms of x and y by inverting Eq. (3.1.5),

$$\begin{aligned} s &= \frac{y}{Q} \\ \xi &= x - \frac{Py}{Q} \end{aligned} \tag{3.1.7}$$

so that

$$z(x, y) = Z\left(x - \frac{Py}{Q}\right) + \int_0^{y/Q} R\left(Ps' + x - \frac{Py}{Q}, Qs'\right) ds' \tag{3.1.8}$$

This procedure has been successful because the ratio Q/P is nonvanishing. It is clear that we could have been given data on a much more general curve than the real axis, but that we would run into trouble had that curve ever been tangent to the characteristic direction. This is illustrated by the three cases shown in Fig. 3.2.

In case (a), the curve I along which z is given can be parameterized by ξ and given by the equations

$$x_0 = X(\xi) \quad \text{and} \quad y_0 = Y(\xi), \qquad \text{for } s = 0 \tag{3.1.9}$$

and the general point in the (x, y) plane would be

$$\begin{aligned} x &= Ps + X(\xi) \\ y &= Qs + Y(\xi) \end{aligned} \tag{3.1.10}$$

By direct analogy with Eq. (3.1.6), we have

$$z(s, \xi) = Z(\xi) + \int_0^s R(Ps' + X(\xi), Qs' + Y(\xi))\, ds' \tag{3.1.11}$$

If the functions X and Y were actually given, we could substitute for s and ξ and obtain a form of solution analogous to Eq. (3.1.8). This can be done as long as the Jacobian

$$J = \frac{\partial(x, y)}{\partial(s, \xi)} = PY' - QX' \tag{3.1.12}$$

does not vanish anywhere along the curve I, or since $J = 0$ implies $Y'/X' = (dy/dx)_\xi = Q/P$, if the curve I becomes nowhere tangent to the characteristic direction.

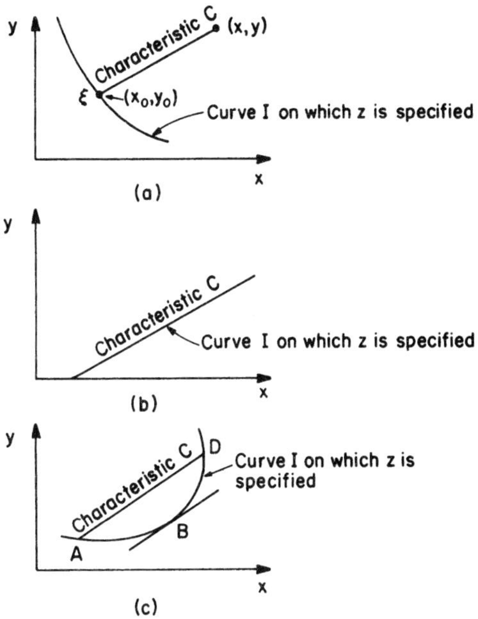

Figure 3.2

If z were specified along a characteristic, as in Fig. 3.2(b), it could only be arbitrarily specified at one point; on the rest of the "curve" it would have to satisfy Eq. (3.1.11) in order to be consistent with the differential equation. We also see that the solution could not leave the initial curve. It would represent an uninformative situation and would suggest that the problem had been improperly formulated. Trouble also arises if the initial curve I is anywhere tangent to a characteristic, as at point B in Fig. 3.2(c). The value of z could be arbitrarily specified along AB, but the value of z at a point beyond the point of tangency, say D, would have to be related to the value on the same characteristic (here A) by Eq. (3.1.11) in order to be consistent with the differential equation. We also notice that the transformations from (s, ξ) to (x, y), and vice versa, are not unique if the initial curve is anywhere characteristic.

In fact, Eq. (3.1.1) could have been solved without knowing the method of characteristics. The left side of Eq. (3.1.1) is proportional to the derivative of z with respect to arc length r in a direction inclined at an angle

$$\theta = \tan^{-1}\frac{Q}{P} \tag{3.1.13}$$

to the x-axis (see Sec. 0.9 for directional derivatives). Suppose that the initial data are given in the form of Eq. (3.1.3). If we then set

$$x = r\cos\theta + \xi \qquad y = r\sin\theta \tag{3.1.14}$$

we have specified each point (x, y) in terms of two coordinates, ξ, the intersection of the line of slope $\tan\theta$ with the real axis, and r, the distance of (x, y) from $(0, \xi)$ along such a line. On the line $r = 0$, we are given the initial data. In Fig. 3.1 the parameter s is now replaced by the arc length r. The advantage of this is that we can write†

$$\left(\frac{\partial z}{\partial r}\right)_\xi = \frac{\partial z}{\partial x}\frac{\partial x}{\partial r} + \frac{\partial z}{\partial y}\frac{\partial y}{\partial r} = p\cos\theta + q\sin\theta$$
$$= \frac{P}{\sqrt{P^2 + Q^2}}\,P + \frac{Q}{\sqrt{P^2 + Q^2}}\,q = \frac{R(x, y)}{\sqrt{P^2 + Q^2}} \tag{3.1.15}$$

The left side is the derivative of z with respect to r, keeping ξ constant, and the right side can be expressed as a function of r and ξ by using Eq. (3.1.14). Regarding ξ as constant, we can integrate both sides of Eq. (3.1.15) from $r = 0$ to the general point r to obtain

$$z(r, \xi) = Z(\xi) + \int_0^r R(r'\cos\theta + \xi, r'\sin\theta)\frac{dr'}{\sqrt{p^2 + Q^2}} \tag{3.1.16}$$

By the inversion of Eq. (3.1.14), we have

$$r = y\,\text{cosec}\,\theta \qquad \xi = x - y\cot\theta \tag{3.1.17}$$

and thus

$$z(x, y) = Z(x - y\cot\theta) + \int_0^{y\,\text{cosec}\,\theta} R(r'\cos\theta + x - y\cot\theta, r'\sin\theta)\frac{dr'}{\sqrt{p^2 + Q^2}} \tag{3.1.18}$$

It can be seen immediately that Eq. (3.1.18) is identical to Eq. (3.1.8).

†Note that

$$\left(\frac{dx}{dr}\right)^2 + \left(\frac{dy}{dr}\right)^2 = 1$$

whereas

$$\left(\frac{dx}{ds}\right)^2 + \left(\frac{dy}{ds}\right)^2 = P^2 + Q^2$$

or

$$\left[\left(\frac{dx}{dr}\right)^2 + \left(\frac{dy}{dr}\right)^2\right]\left(\frac{dr}{ds}\right)^2 = \left(\frac{dr}{ds}\right)^2 = P^2 + Q^2$$

Thus the parameters r and s are proportional.

Let us note two other features about a characteristic before turning to practical examples. First, a discontinuity in the derivative of the initial data (or a *disturbance*) is propagated along a characteristic. For suppose $Z(\xi)$ were continuous at $\xi = \xi_0$ but that it had a corner there so that

$$Z'(\xi_0 - 0) \neq Z'(\xi_0 + 0) \tag{3.1.19}$$

Then since, from Eq. (3.1.6),

$$\left(\frac{\partial z}{\partial \xi}\right)_s = Z'(\xi) + \int_0^s R_x(Ps' + \xi, Qs')\, ds' \tag{3.1.20}$$

this discontinuity will be propagated as a discontinuity in $(\partial z/\partial \varepsilon)_s$ and hence in the derivative in any direction that crosses the characteristic. Notice that

$$\left(\frac{\partial z}{\partial s}\right)_\xi = R(Ps + \xi, Qs)$$

does not suffer any discontinuity. Second, in this case, a discontinuity in the value of z itself will also be propagated along a characteristic. For if $Z(\xi)$ is undefined at $\xi = \xi_0$ and $Z(\xi_0 - 0) \neq Z(\xi_0 + 0)$, then, from Eq. (3.1.6),

$$z(s, \xi_0 + 0) - z(s, \xi_0 - 0) = Z(\xi_0 + 0) - Z(\xi_0 - 0) \neq 0 \tag{3.1.21}$$

This discontinuity of value is often referred to as a *contact discontinuity* [see P. D. Lax, *Comm. Pure Appl. Math.* **10**, 537 (1957)], and it should be distinguished from a "shock," which we will encounter in Chapter 5 in relation to quasi-linear equations. A contact discontinuity arises only when there is one in the initial data, as seen from Eq. (3.1.21), and propagates along a characteristic. This is one of the characteristic features of linear problems.

The term R in Eq. (3.1.1) may depend on the variable z in a linear or nonlinear way; in the latter case we call the equation *semilinear*. These cases with constant P and Q are not essentially more difficult. Here we have

$$P\frac{\partial z}{\partial x} + Q\frac{\partial z}{\partial y} = R(x, y, z) \tag{3.1.22}$$

and, if the initial data are given in the form of Eq. (3.1.3), we can use Eq. (3.1.5) to obtain

$$\frac{dz}{ds} = R(Ps + \xi, Qs, z) \tag{3.1.23}$$

in the direction of a characteristic.

This is now an ordinary differential equation for z as a function of s with ξ held constant. Discontinuities in derivative and value would be propagated along characteristics, but the size of the discontinuity would no longer remain constant.

Since the information prescribed on the initial curve I is propagated along characteristics, it is clear that the data specified at a given point A of the

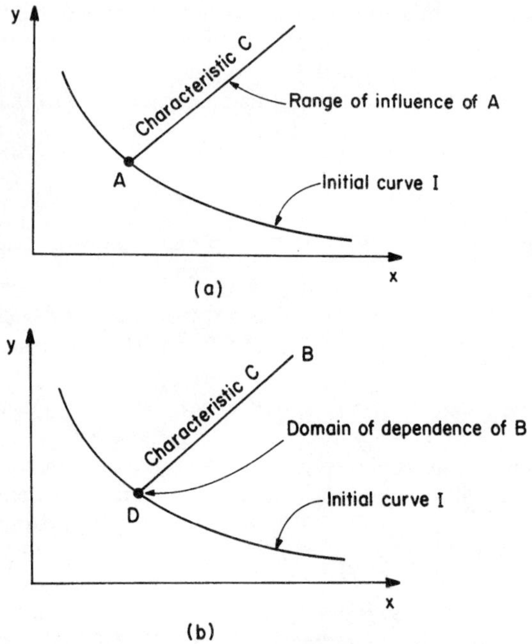

Figure 3.3

curve I can affect the solution only at those points that lie on the characteristic C through the point. In this sense, we call the characteristic C the *range of influence* of the data at point A. Starting from a specific point B of the solution, we can draw a characteristic passing through the point to find the intersection D with the curve I. The point D is called the *domain of dependence* of the solution at point B because the solution there depends only on the data specified at point D. These are shown in Fig. 3.3. We also visualize how the propagation of discontinuities along characteristics will divide the (x, y)-plane into regions of analytically different solutions.

Exercises

3.1.1. Show that Eq. (3.1.18) is identical to Eq. (3.1.8). Also, express the solution (3.1.8) or (3.1.18) in the forms

$$z(x, y) = Z(x - y \cot \theta) + \int_0^y R(x - (y - y') \cot \theta, y') \frac{dy'}{Q}$$

and

$$z(x, y) = Z(x - y \cot \theta) + \int_{x - y\cot\theta}^{x} R(x', (x' - x) \tan \theta + y) \frac{dx'}{P}$$

where $\tan \theta = Q/P$.

3.1.2. Show that in the strictly linear case the angle of a "corner" in the initial data is propagated without change along a characteristic.

3.1.3. Discuss the case of n independent variables when the coefficients of the derivatives are constants. How should the initial data be specified in this case?

3.1.4. Show how the semilinear equation can be solved by quadratures in case

$$R(x, y, z) = S(x, y)T(z)$$

3.2 Examples of Linear and Semilinear Equations

In this section we turn to practical cases and consider two examples. First, let us consider the plug flow reactor in which a first-order, irreversible reaction, $A \rightarrow$ products, takes place. If we assume that the density and temperature of the reaction mixture remain uniform and constant, the material balance for the reactant at a distance x from the inlet at time t yields the equation

$$\frac{\partial c}{\partial t} + V\frac{\partial c}{\partial x} = -kc \tag{3.2.1}$$

in which c is the molar concentration of the reactant, V the linear velocity of the reaction mixture, and k the reaction rate constant. [Compare Eq. (3.2.1) with Eq. (1.9.4)]. To this linear equation, we would append the arbitrary initial and boundary conditions

$$\begin{aligned} &\text{at } t = 0, \quad c = f(x), \quad \text{for } x \geqslant 0 \\ &\text{at } x = 0, \quad c = g(t), \quad \text{for } t \geqslant 0 \end{aligned} \tag{3.2.2}$$

The characteristic differential equations are

$$\begin{aligned} \frac{dt}{ds} &= 1 \\ \frac{dx}{ds} &= V \\ \frac{dc}{ds} &= -kc \end{aligned} \tag{3.2.3}$$

and so the characteristics C in the (x, t)-plane are straight lines of slope $1/V$. Since the initial curve I has a corner at the origin, the characteristic OA through the origin divides the (x, t)-plane into two regions as shown in Fig. 3.4. Below OA the solution is influenced by the initial data $f(x)$, while above OA it depends on the feed data $g(t)$ only. The solution in each region can

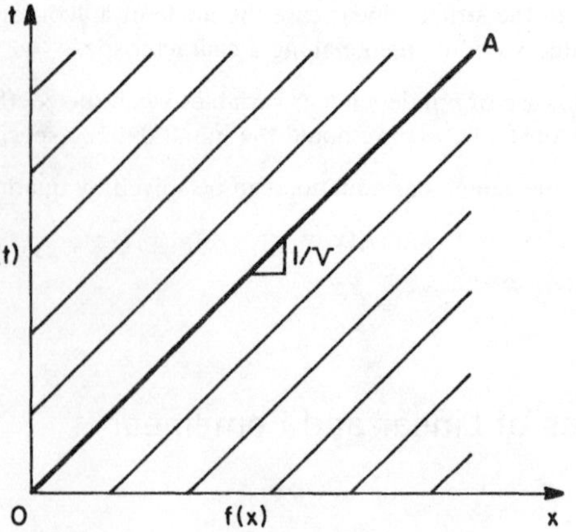

Figure 3.4

be determined in a straightforward manner because the last equation in Eq. (3.2.3) does not contain x and t explicitly.

The initial data along the x-axis may be expressed in terms of a parameter ξ; that is,

$$t = 0, \quad x = \xi, \quad \text{and} \quad c = f(\xi) \qquad \text{for } s = 0 \tag{3.2.4}$$

so Eq. (3.2.3) can be integrated from $s = 0$ to an arbitrary s. Thus

$$t = s, \quad x = Vs + \xi, \quad c(s, \xi) = f(\xi)\exp(-ks) \tag{3.2.5}$$

From the first two equations, $s = t$ and $\xi = x - Vt$, and substituting these into the last equation, we find

$$c(x, t) = f(x - Vt)\exp(-kt), \qquad \text{for } x > Vt \tag{3.2.6}$$

This is the solution for the region below OA or for $x > Vt$. It is always determined uniquely because the x-axis never becomes tangent to a characteristic.

For the solution in the region above OA or for $x < Vt$, we may use another parameter η to rewrite the feed data along the t-axis:

$$t = \eta, \quad x = 0, \quad \text{and} \quad c = g(\eta) \qquad \text{for } s = 0 \tag{3.2.7}$$

Integrating Eq. (3.2.3) with Eq. (3.2.7), we obtain

$$t = s + \eta, \quad x = Vs, \quad c(s, \eta) = g(\eta)\exp(-ks) \tag{3.2.8}$$

From the first two equations, $s = x/V$ and $\eta = t - x/V$, and so

$$c(x, t) = g\left(t - \frac{x}{V}\right)\exp\left(\frac{-kx}{V}\right) \qquad \text{for } x > Vt \tag{3.2.9}$$

In case the feed concentration (at $x = 0$) is constant, that is, $g(t) = c_F =$ constant, this equation reduces to

$$c(x, t) = c_F \exp\left(\frac{-kx}{V}\right) \tag{3.2.10}$$

which is independent of t. Hence, the reactor immediately attains the steady state as the entering feed stream passes through it and pushes out the initial reaction mixture. This shows the typical nature of the plug flow model.

If the reaction rate in the preceding example is given by a nonlinear function of c, Eq. (3.2.1) becomes semilinear, but this incurs no further difficulty in the solution scheme because the term R still does not contain x and t (see Exercise 3.1.4). For an nth order irreversible reaction, we have

$$\frac{\partial c}{\partial t} + V\frac{\partial c}{\partial x} = -kc^n, \qquad n \neq 1 \tag{3.2.11}$$

and so the last equation of Eq. (3.2.3) becomes

$$\frac{dc}{ds} = -kc^n \tag{3.2.12}$$

which is readily integrated along a characteristic. With the data in Eq. (3.2.2), therefore, we find

$$\begin{aligned} c(x,t) &= f(x - Vt)\{1 + (n-1)kt[f(x - Vt)]^{n-1}\}^{-\nu}, \qquad \text{for } x > Vt \\ c(x,t) &= g\left(t - \frac{x}{V}\right)\left\{1 + (n-1)k\left(\frac{x}{V}\right)\left[g\left(t - \frac{x}{V}\right)\right]^{n-1}\right\}^{-\nu} \qquad \text{for } x < Vt \end{aligned} \tag{3.2.13}$$

where $\nu = 1/(n - 1)$.

The second example is concerned with the heat exchanger immersed in a bath of varying temperature. If $T(x, t)$ is the temperature within the cooling tube at a distance x from the inlet at time t and $T_a(t)$ is the ambient temperature around the tube,

$$\frac{\partial T}{\partial t} + V\frac{\partial T}{\partial x} = H(T_a - T) \tag{3.2.14}$$

(see Sec. 1.6). The natural boundary conditions to specify are

$$\begin{aligned} T(x, 0) &= T_0(x), \qquad 0 \leqslant x \leqslant L \\ T(0, t) &= T_F(t), \qquad t \geqslant 0 \end{aligned} \tag{3.2.15}$$

The characteristics are again straight lines in the (x, t)-plane of slope $1/V$, as shown in Fig. 3.5. We see that the situation beneath the characteristic OA, where the effect of the initial temperature distribution is still felt, is rather different from that above OA, where everything depends on the inlet temperature. Although we can apply the same procedure as in the previous case, we will use the second approach presented in Sec. 3.1.

Consider first a point (x, t) lying below OA, that is, $x > Vt$. Now the inclination of the characteristic is

$$\theta = \tan^{-1}\left(\frac{1}{V}\right) \tag{3.2.16}$$

so that if $r = \sqrt{1 + V^2}t$ and $\xi = x - Vt$, as indicated in Fig. 3.5,

$$\frac{dT}{dr} = \frac{H}{\sqrt{1 + V^2}}\left[T_a\left(\frac{r}{\sqrt{1 + V^2}}\right) - T\right] \tag{3.2.17}$$

[Compare with Eq. (3.1.15).] This is a linear differential equation for T with data specified as $T = T_0(\xi)$ at $r = 0$. Thus

$$T(r, \xi) = T_0(\xi) \exp\left(\frac{-Hr}{\sqrt{1 + V^2}}\right)$$

$$+ H\int_0^r T_a\left(\frac{r'}{\sqrt{1 + V^2}}\right) \exp\left[\frac{-H(r - r')}{\sqrt{1 + V^2}}\right] \frac{dr'}{\sqrt{1 + V^2}} \tag{3.2.18}$$

and substituting back for r and ξ (see Exercise 3.1.1),

$$T(x, t) = T_0(x - Vt) \exp(-Ht)$$

$$+ H\int_0^t T_a(t') \exp[-H(t - t')]\, dt' \tag{3.2.19}$$

Above the characteristic OA, we may take $\eta = t - x/V > 0$ to be the

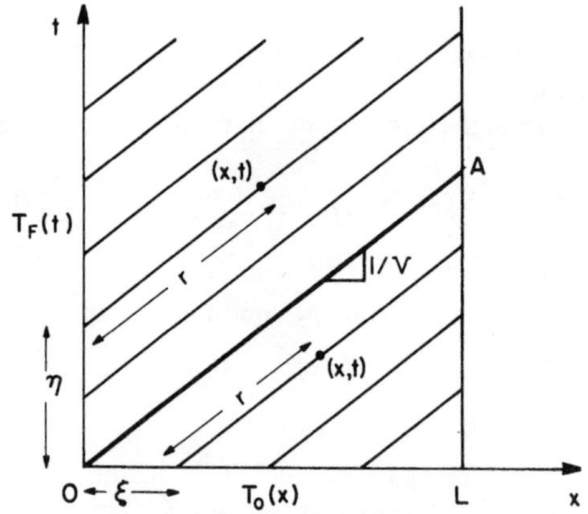

Figure 3.5

values of t at which the characteristics through (x, t) meets the t-axis. Then the equation for T is again given by Eq. (3.2.17) and the solution by Eq. (3.2.18), but with $T_F(\eta)$ in place of $T_0(\xi)$. Now the value of t when $r = 0$ is $t - x/V$, so

$$T(x, t) = T_F(t - x/V) \exp\left(\frac{-Hx}{V}\right) + H\int_{t-x/V}^{t} T_a(t') \exp\left[-H(t - t')\right] dt' \qquad (3.2.20)$$

In case T_a and T_F are constant, we can see how the transient solution approaches the steady state. For taking $t > L/V$ with constant T_a and T_F in Eq. (3.2.20) gives

$$\begin{aligned} T(x, t) &= T_F \exp\left(\frac{-Hx}{V}\right) + HT_a \int_{t-x/V}^{t} \exp[-H(t - t')]\, dt' \\ &= T_F \exp\left(\frac{-Hx}{V}\right) + T_a\left[1 - \exp\left(\frac{-Hx}{V}\right)\right] \end{aligned}$$

or

$$\frac{T(x, t) - T_F}{T_a - T_F} = 1 - \exp\left(-\frac{Hx}{V}\right) \qquad (3.2.21)$$

Since the right side of the equation does not contain the time t, we see that the steady state is established immediately after the initial contents of the tube have been washed through. This is a feature of the simplicity of the model, in particular, of not taking into account the heat capacity of the tube.

If the overall heat transfer coefficient H varies as a function of temperature T, we find an example of a semilinear equation in Eq. (3.2.14). Now Eq. (3.2.17) becomes a nonlinear ordinary differential equation, and the prospect for an analytic solution seems remote except for some special forms of $H(T)$. However, if we assume T_a remains constant, the solution subject to the data in Eq. (3.2.15) can be expressed in the form

$$\begin{aligned} t &= \frac{r}{\sqrt{1 + V^2}} = \int_{T_0(x-Vt)}^{T} \frac{dT}{H(T)(T_a - T)} \qquad \text{for } x > Vt \\ \frac{x}{V} &= \frac{r}{\sqrt{1 + V^2}} = \int_{T_F(t-x/V)}^{T} \frac{dT}{H(T)(T_a - T)} \qquad \text{for } x < Vt \end{aligned} \qquad (3.2.22)$$

Exercises

3.2.1. Apply the Laplace transform to solve Eq. (3.2.1) subject to the conditions in Eq. (3.2.2), and compare the solution with Eqs. (3.2.6) and (3.2.9).

3.2.2. Find the equations for the characteristic curves for Eq. (3.2.14) and follow the same procedure as for the plug flow reactor. Can you express the solution in the form given by Eqs. (3.2.19) and (3.2.20)?

3.2.3. If T_a is constant and T_F fluctuates about a constant mean value $\overline{T}_F$ according to the formula

$$T_F(t) = \overline{T}_F + \alpha(T_a - \overline{T}_F) \cos \omega t$$

show that the rate of heat removal $Q(t)$ fluctuates about its average value $\overline{Q}$ according to the formula

$$\frac{Q(t)}{\overline{Q}} = 1 + \alpha \left\{ H^2 \frac{[1 - 2\exp(-HL/V)]\cos^2(\omega L/V) + \exp(-2HL/V)}{(H^2 + \omega^2)[1 - \exp(-HL/V)]^2} \right\}^{1/2} \cos(\omega t - \phi)$$

$$\tan\phi = \frac{\omega[1 - \exp(-HL/V)]\cos(\omega L/V) - H[\exp(-HL/V)]\sin(\omega L/V)}{H[1 - \exp(-HL/V)]\cos(\omega L/V) + \omega\exp(-HL/V)\sin(\omega L/V)}$$

3.2.4. Consider an enzyme reaction in a plug flow reactor for which the reaction rate is given by the Michaelis–Menten equation

$$\text{Rate} = \frac{kc}{K_M + c}$$

where k and K_M are constants. Using the data given by Eq. (3.2.2), determine the concentration field $c(x, t)$.

3.2.5. The term *pulsed reactor* has been given to a short isothermal bed to which a pulse of reactant is fed [Kokes and others, *J. Amer. Chem. Soc.* **17**, 5860 (1955)]. Ignoring diffusion and other dispersive effects, taking the rate of disappearance of the reactant to be $r(c)$ when its concentration is c, and taking a Gaussian concentration input, set up and solve the equations for the reactor in terms of certain integrals. In particular, for an nth-order reaction show that the fraction unconverted is

$$\frac{1}{2\pi} \int_{-\infty}^{\infty} e^{-\lambda^2/2}[1 + \kappa(2\pi)^{-(n-1)/2} e^{-(n-1)\lambda^2/2}]^{-\nu} \, d\lambda$$

where $\kappa = k(n - 1)\theta(Q/\sigma)^{n-1}$, $\nu = 4/(n - 1)$, θ is the residence time, Q the amount of reactant in the pulse, and σ its variance.

3.2.6. If $\kappa^2 \leq (2\pi)^{n-1}$, show that the fraction unconverted is

$$X = \sum_{p=0}^{\infty} \frac{\nu(\nu + 1) \ldots (\nu + p - 1)}{p!} \kappa^p (2\pi)^{-(n-1)p/2}[1 + (n - 1)p]^{-1/2}$$

and that though the emerging pulse is not Gaussian its variance is

$$\frac{\sigma^2}{X} \left\{ \sum_{p=0}^{\infty} \frac{\nu(\nu + 1) \ldots (\nu + p - 1)}{p!} \kappa^p (2\pi)^{-(n-1)p/2}[1 + (n - 1)p]^{-3/2} \right\}$$

3.3 Homogeneous Equations

Although it may be only a special case of linear systems, the problem with vanishing R in Eq. (3.1.1) represents a variety of physical systems, as we have seen in Chapter 1, and shows some interesting features. It is also of importance as the first step to the treatment of quasilinear equations in Chapter 5, because the characteristic features here have their counterparts in homogeneous quasi-linear equations.

The example we would like to consider is the chromatography of a single solute with a linear equilibrium relation (see Sec. 1.2). Thus, if $c(z, t)$ and $n(z, t)$ are the concentrations in the fluid and solid phases, respectively, at a distance z from the inlet at time t, we have the conservation equation

$$\varepsilon V\frac{\partial c}{\partial z} + \varepsilon\frac{\partial c}{\partial t} + (1-\varepsilon)\frac{\partial n}{\partial t} = 0 \tag{3.3.1}$$

and the equilibrium relation

$$n = Kc \tag{3.3.2}$$

where K is the adsorption equilibrium constant. The two equations may be combined to give

$$V\frac{\partial c}{\partial z} + (1+\nu K)\frac{\partial c}{\partial t} = 0 \tag{3.3.3}$$

in which the parameter ν defined by

$$\nu = \frac{1-\varepsilon}{\varepsilon} \tag{3.3.4}$$

represents the volume ratio of the solid phase to the fluid phase.

The equations for the characteristic curves are

$$\begin{aligned} \frac{dz}{ds} &= V \\ \frac{dt}{ds} &= 1+\nu K \\ \frac{dc}{ds} &= 0 \end{aligned} \tag{3.3.5}$$

The first two equations, if put together, give the characteristic direction σ in the (z, t)-plane, that is,

$$\frac{dt}{dz} = \frac{1+\nu K}{V} \equiv \sigma \tag{3.3.6}$$

which is a constant in this case, whereas the last equation in Eq. (3.3.5) requires that the concentration c remain invariant along a characteristic. Consequently, not only the characteristics C in the (z, t)-plane are straight lines (of slope σ), but also the characteristic curves in the (z, t, c)-space are straight and parallel to the (z, t)-plane. This implies that a given state of concentration would be simply propagated along a characteristic direction without changing its value. In this regard, the reciprocal of σ is often called the *propagation speed* of the state of concentration.

If the initial state of the bed is $c(z, 0) = f(z)$, we use the parameter ξ along the z-axis so that $z = \xi$ and $c = f(\xi)$ for $s = 0$ [cf. Eq. (3.2.4)]. Thus, from Eqs. (3.3.5) and (3.3.6),

$$\begin{cases} t = \sigma(z - \xi) \\ c = f(\xi) \end{cases}$$

or, after elimination of ξ,

$$c(z, t) = f\left(z - \frac{t}{\sigma}\right) \qquad \text{for } z > \frac{t}{\sigma} \tag{3.3.7}$$

Likewise, with the feed concentration $c(0, t) = g(t)$, we let $t = \eta$ and $c = g(\eta)$ along the t-axis on which $s = 0$, and so

$$\begin{cases} t = \sigma z + \eta \\ c = g(\eta) \end{cases}$$

or

$$c(z, t) = g(t - \sigma z) \qquad \text{for } z < \frac{t}{\sigma} \tag{3.3.8}$$

Equations (3.3.7) and (3.3.8) imply that the data f and g are propagated in the characteristic direction σ without change of shape. In particular, if the solute is fed in the form of an impulse to a clean bed, we have $f(z) = 0$ and $g(t) = c_0\tilde{\delta}(t)$, where $\tilde{\delta}(t) = 0$ for $t \geq 0$ and $\int_0^\infty \tilde{\delta}(t)\, dt = 1$. Then the solution is

$$c(z, t) = c_0\tilde{\delta}(t - \sigma z) \tag{3.3.9}$$

which says that the impulse is propagated in the characteristic direction σ from the origin, and it will reach the outlet of the bed at

$$t_L = \sigma L = \frac{L}{V}(1 + \nu K) = \frac{L}{V}\left(1 + \frac{1 - \varepsilon}{\varepsilon} K\right) \tag{3.3.10}$$

and this is what we usually call the *retention time*. Since, however, the term L/V represents the time required to pass through the void space, the solute is actually retained by the solid phase for the time period of $(1 - \varepsilon)KL/\varepsilon V$. This is depicted in Fig. 3.6. In Fig. 3.7 we show the propagation of a parabolic pulse in invariant form.

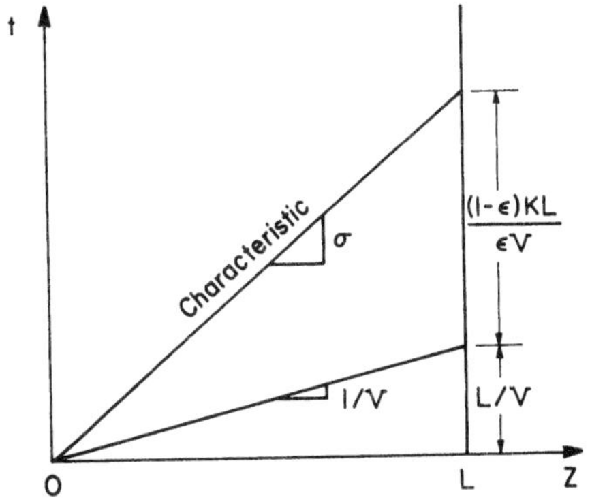

Figure 3.6

Suppose now the bed is initially clean and the feed concentration takes a step change at $t = 0$ so that $f(z) = 0$ and $g(t) = c_0 H(t)$, where $H(t)$ is the Heaviside unit function. Then, from Eq. (3.3.8),

$$c(z, t) = c_0 H(t - \sigma z) \tag{3.3.11}$$

and this solution is sketched in Fig. 3.8. Indeed, the discontinuity in the data propagates along the characteristic through the origin, and so the solution surface contains a vertical face (designated by C_0–0–6–7 in the figure). The line segments 1–2–3–4 represent a typical profile of concentration, and we can clearly see how the jump $\overline{23}$ is propagated. The breakthrough curve is observed at the exit ($z = L$) and is given by the line segments 5–6–7–8.

If the feed concentration increases linearly with time over a time interval

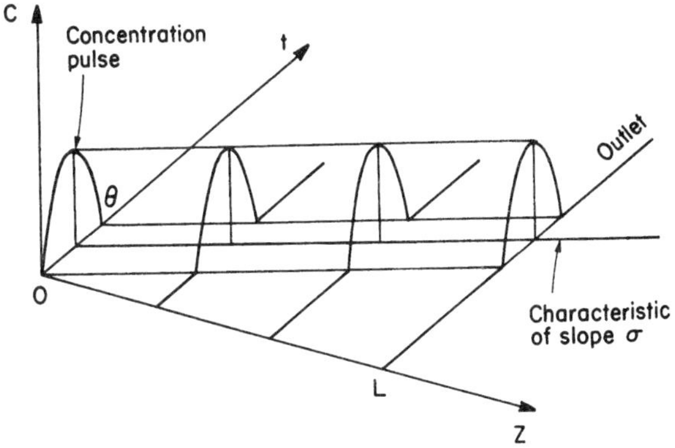

Figure 3.7

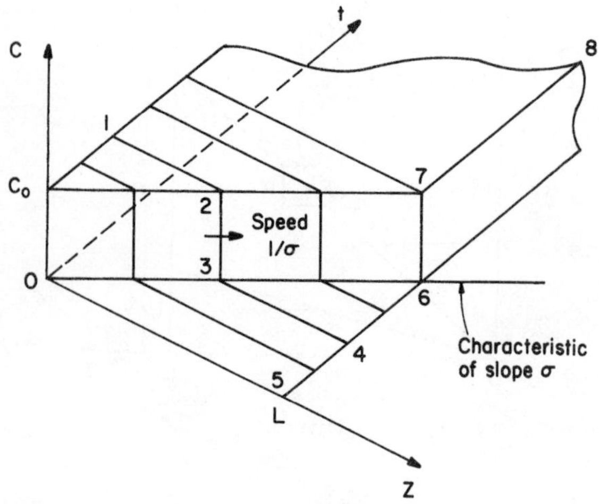

Figure 3.8

θ; that is,

$$c(0, t) = g(t) = \begin{cases} c_0 \dfrac{t}{\theta}, & \text{for } 0 \le t \le \theta \\ c_0, & \text{for } t \ge \theta \end{cases} \tag{3.3.12}$$

the solution can be expressed in the form

$$c(z, t) = \begin{cases} 0, & \text{for } t - \sigma z \le 0 \\ \dfrac{c_0}{\theta}(t - \sigma z), & \text{for } 0 \le t - \sigma z \le \theta \\ c_0 & \text{for } t - \sigma z > \theta \end{cases} \tag{3.3.13}$$

which is sketched in Fig. 3.9. Clearly, the jumps in the derivative of the data propagate along characteristics. The breakthrough curve is given by the line segments 5–6–7–8, where the time at point 6 is $t = \sigma L$ and, at point 7, $t = \theta + \sigma L$. Since the value of σ is greater than unity, the profile of concentration becomes more slanted than the data, for example, the line segment $\overline{23}$ in the profile versus $\overline{67}$ in the breakthrough curve.

Figures 3.8 and 3.9 illustrate the fact that not only jumps in the derivative of the solution but that in the solution itself may occur along characteristics. But these jumps all originated from the corresponding jumps in the initial and feed data. The propagation of a particular form of concentration profile may be considered as a wave phenomenon, and it is evident that a concentration wave propagates with a constant speed without changing its form. This is typical of *linear waves*.

Another example is found in heat transfer in a packed bed under adiabatic

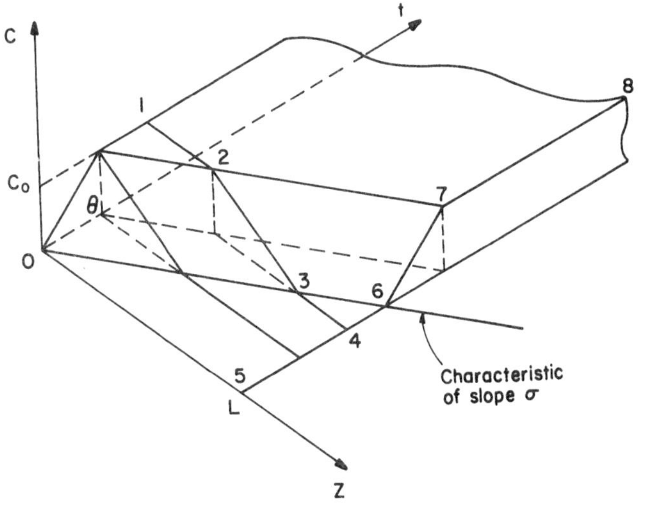

Figure 3.9

conditions. From Eq. (1.6.7) we have the equation for the temperature $T(z, t)$:

$$(1 + \beta)\frac{\partial T}{\partial t} + V\frac{\partial T}{\partial z} = 0 \tag{3.3.14}$$

Hence, the characteristic direction σ in the (z, t)-plane is given by

$$\frac{dt}{dz} = \frac{1 + \beta}{V} \equiv \sigma \tag{3.3.15}$$

in which $\beta = (1 - \varepsilon)C_s/\varepsilon C_f$ is the heat capacity ratio of the solid phase to the fluid phase and usually assumes a very large value. Not only disturbances but discontinuities in the temperature will propagate along characteristics in the (z, t)-plane, which are now very steep. Since the propagation speed is the reciprocal of the σ value, a temperature signal fed to the bed will arrive at the outlet $(z = L)$ of the bed after a time interval of

$$t_L = \sigma L = \frac{L}{V}(1 + \beta) = \frac{L}{V}\left[1 + \frac{(1 - \varepsilon)C_s}{\varepsilon C_f}\right] \tag{3.3.16}$$

In particular, if both the initial and feed temperatures are constant, that is, $T(z, 0) = T$ and $T(0, t) = T_F$, then the discontinuity $| T_F - T_0 |$ in temperature simply propagates through the bed with constant speed $1/\sigma$ and makes a breakthrough at $t = t_L$. This is shown in Fig. 3.10. The discontinuity $|T_F - T_0|$ is often referred to as a *thermal wave* and is evidently a contact discontinuity.

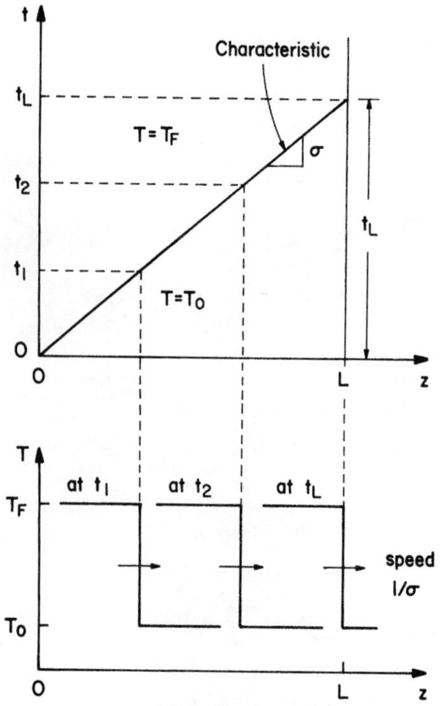

Figure 3.10

Exercises

3.3.1. If a first-order, irreversible reaction, $A \rightarrow$ products, takes place only in the solid phase, how would Eq. (3.3.3) be modified? Suppose a mixture of constant concentration c_0 is fed to a clean bed. Determine the concentration field $c(z, t)$ inside the bed and sketch the solution surface in the (z, t, c)-space.

3.3.2. Consider the adsorption of a single solute in a countercurrently moving bed system (see Exercise 1.6.3). For a linear equilibrium relation, derive the mass balance equation for the solute and determine the characteristic direction. Discuss whether there is a possibility that the characteristic direction takes a negative sign and, if so, how you would proceed to find the solution. In particular, if the initial and feed concentrations are constant in a semi-infinitely long column, what is the concentration in the solid phase leaving the column?

3.3.3. In case both the initial and feed temperatures are constant, apply the Laplace transformation to find the solution of Eq. (3.3.14) and confirm that the solution is of the form shown in Fig. 3.10. If the solid phase can be put in motion with an acceleration, would it be possible to make the thermal wave stationary inside the column?

†3.4 Equilibrium Theory of the Parametric Pump

Pigford and others (1969) have given an equilibrium theory of the parametric pump (see Exercise 1.3.4), which can be treated within the framework of what we have done so far. In the "parapump" the fluid in a fixed bed is passed back and forth in a cyclic manner. A solute can be adsorbed from the fluid onto the fixed solid, and in the equilibrium theory it is assumed that the equilibrium composition is instantaneously attained. The pumping effect is attained by varying the temperature synchronously with the flow so that the adsorption is stronger when the flow is in one direction. If a square wave is used both for the flow and the temperature, then in each half-cycle the coefficients will be constants.

Consider a single solute of concentration c in the fluid phase and an equilibrium concentration on the solid phase of

$$n = K(T)c \tag{3.4.1}$$

This rather restrictive linear relation is also necessary to attain constant coefficients in the differential equation, but K is a function of temperature, which is also to be made a function of time. Then Eq. (1.2.1) becomes‡

$$V(t)\frac{\partial c}{\partial z} + [1 + \nu K(t)]\frac{\partial c}{\partial t} = -\nu\frac{dK}{dt}c \tag{3.4.2}$$

where ν is defined by Eq. (3.3.4). We now suppose that square waves are imposed on the velocity and temperature so that during

$$n\theta < t < \left(n + \frac{1}{2}\right)\theta, \qquad V(t) = V, \qquad 1 + \nu K(t) = \frac{1}{\mu_+} \tag{3.4.3}$$

while during

$$\left(n + \frac{1}{2}\right)\theta < t < (n+1)\theta, V(t) = -V, \qquad 1 + \nu K(t) = \frac{1}{\mu_-} \tag{3.4.4}$$

The signs + and − will be used to indicate the parts of the cycle during which the velocity is positive and negative, respectively. The characteristics in the two half-cycles are

$$\frac{dz}{ds} = V, \qquad \frac{dt}{ds} = \frac{1}{\mu_+} \quad \text{or} \quad \frac{dz}{dt} = \mu_+ V \tag{3.4.5}$$

†Since this section is devoted to a rather particular example, it may be omitted without loss of continuity in the development.

‡Note that this treatment is not very realistic since it neglects the effect of heat capacity and the resulting temperature waves in the bed. This would be a considerably more difficult problem. In spite of these obvious shortcomings, however, the equilibrium theory compares quite well with certain of the experiments.

and

$$\frac{dz}{ds} = -V, \qquad \frac{dt}{ds} = \frac{1}{\mu_-} \quad \text{or} \quad \frac{dz}{dt} = -\mu_- V \tag{3.4.6}$$

respectively, and along them c remains constant, since dK/dt is zero within each half-cycle.

We must examine what happens when the temperature is suddenly changed from T_+ to T_-, causing a change in K from K_+ to K_-. In a segment of the bed of length dz, we have in the one case an amount of solute

$$dz[\varepsilon c_+ + (1-\varepsilon)n_+] = dz[\varepsilon c_+ + (1-\varepsilon)K_+ c_+] = \varepsilon \frac{c_+}{\mu_+}\, dz$$

and in the other an amount

$$dz[\varepsilon c_- + (1-\varepsilon)n_-] = dz[\varepsilon c_- + (1-\varepsilon)K_- c_-] = \varepsilon \frac{c_-}{\mu_-}\, dz$$

Since these must be the same, the effect of a temperature change is to change the concentration in the ratio

$$\frac{c_-}{c_+} = \frac{\mu_-}{\mu_+} \equiv \lambda < 1 \tag{3.4.7}$$

which is to be made less than 1 as denoted. In agreement with the assumption of equilibrium, we suppose that these changes take place instantaneously at intervals of $\theta/2$.

The boundary and initial conditions can best be seen by reference to Fig. 3.11. The t-axis is divided into half-cycles, and the length of the column is L. Initially, the column is at constant concentration c_0, and during the first half-cycle this is the concentration in the feed. The emerging concentration at the top during the first half of the nth cycle $(+)$ will be denoted by $\Gamma_n(t)$. It is mixed to a uniform concentration $\overline{\Gamma}_n$ and becomes the feed to the top of the column, $z = L$, during the next half-cycle. The concentration emerging from the bottom of the column during the second half of the nth cycle $(-)$ will be denoted by $\gamma_n(t)$, and its mean value $\overline{\gamma}_n$ is the feed at $z = 0$ during the first half of the $(n+1)$th cycle.

For the sake of simplicity, we will assume that

$$L = \frac{V\mu_+\theta}{2} \tag{3.4.8}$$

so that at the end of the first half-cycle the characteristic from the origin (O) has just reached the top of the bed (B). Since characteristics bear constant concentrations and all initial concentrations (on OA) are c_0, we must have all concentrations on (AB) also equal to c_0; that is,

$$\Gamma_1(t) = c_0 = \overline{\Gamma}_1 \tag{3.4.9}$$

Similarly, all points on the $+$ side of BC (since they lie on characteristics that emanate from OC) must correspond to a concentration of c_0; we can

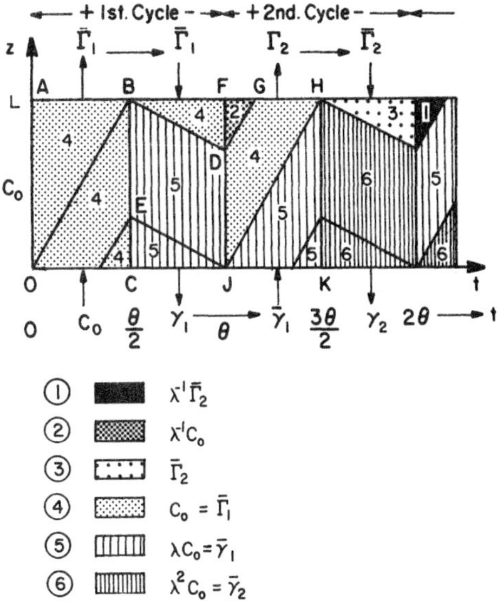

Figure 3.11

write this

$$c(z, \frac{\theta}{2} - 0) = c_0$$

But at $t = \theta/2$ there is a temperature change, and according to Eq. (3.4.7) a c_+ changes to a $c_- = \lambda c_+$. Thus

$$c(z, \frac{\theta}{2} + 0) = \lambda c_0 \tag{3.4.10}$$

We can now move to the second half of the first cycle, taking the boundary conditions from Eqs. (3.4.9) and (3.4.10). The characteristics in this rectangle all have slope $-V\mu_-$, so the point D where the characteristic through B meets the vertical $t = \theta$ is

$$z = L - \frac{V\mu_-\theta}{2} = L\left(1 - \frac{\mu_-}{\mu_+}\right) = L(1 - \lambda) \tag{3.4.11}$$

Since $c = c_0$ on BF, the characteristics will bear this concentration to any point on the minus side of DF; by the change of temperature, it will become $\lambda^{-1}c_0$ on the plus side. On the other hand, the characteristics arriving at a point on DJ will have come from a point on BE bearing the concentration λc_0, and after the temperature jump the concentration on the plus side of DJ will return to $c_0 = \lambda^{-1}(\lambda c_0)$. Thus

$$c(z, \theta + 0) = \begin{cases} c_0, & 0 \leq z \leq L(1 - \lambda) \\ \lambda^{-1}c_0, & L(1 - \lambda) \leq z \leq L \end{cases} \tag{3.4.12}$$

Also, the concentrations emerging as $\gamma_1(t)$ all come from the minus side of EC and

$$\gamma_1(t) = \lambda c_0 = \bar{\gamma}_1 \tag{3.4.13}$$

In the first half of the second cycle (segments FG and GH), we see that

$$\Gamma_2(t) = \begin{cases} \lambda^{-1}c_0, & \theta \leq t \leq \theta\left(1 + \dfrac{\lambda}{2}\right) \\ c_0, & \theta\left(1 + \dfrac{\lambda}{2}\right) \leq t \leq \dfrac{3}{2}\theta \end{cases}$$

Hence

$$\bar{\Gamma}_2 = \lambda(\lambda^{-1}c_0) + (1 - \lambda)c_0 = (2 - \lambda)c_0 \tag{3.4.14}$$

On the other hand, the concentrations on the minus side of HK all correspond to characteristics emanating from JK and so are $\bar{\gamma}_1 = \lambda c_0$. After the temperature jump,

$$c\left(z, \frac{3\theta}{2} + 0\right) = \lambda^2 c_0 \tag{3.4.15}$$

It follows by a repetition of the argument that

$$c(z, 2\theta + 0) = \lambda^{-1}c(z, 2\theta - 0) = \begin{cases} \lambda c_0, & 0 < z < L(1 - \lambda) \\ (2 - \lambda)\lambda^{-1}c_0, & L(1 - \lambda) < z < L \end{cases}$$

and

$$\gamma_2 = \bar{\gamma}_2 = \lambda^2 c_0 \tag{3.4.16}$$

By looking at a typical succession of cycles as shown in Fig. 3.12, we see that the concentration $\bar{\gamma}_{n-1}$ is carried to a $+/-$ transition, becoming $\lambda\bar{\gamma}_{n-1}$ on the minus side and thus giving

$$\gamma_n = \lambda\bar{\gamma}_{n-1} = \bar{\gamma}_n \tag{3.4.17}$$

where the subscript n denotes the nth cycle. This agrees with Eqs. (3.4.13)

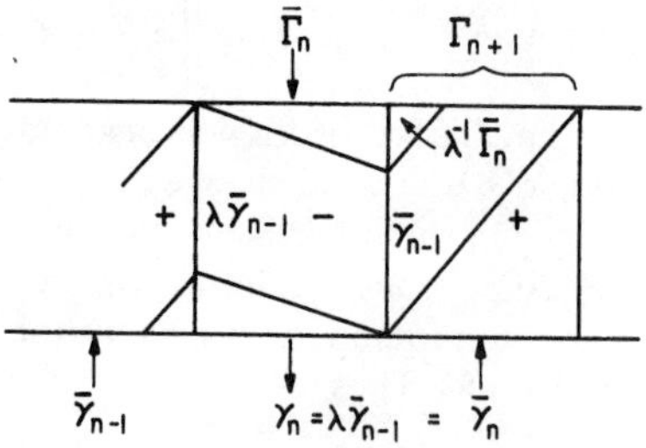

Figure 3.12

and (3.4.16) and gives

$$\bar{\gamma}_n = \lambda^n c_0 \tag{3.4.18}$$

On the other hand, $\bar{\Gamma}_{n+1}$ is the average of the fraction that comes from $\bar{\Gamma}_n$ suffering a $-/+$ transition to $\lambda^{-1}\bar{\Gamma}_n$ and a fraction coming from $\bar{\gamma}_{n-1}$ suffering two compensating transitions. The ratio of these fractions is λ to $(1 - \lambda)$, so

$$\begin{aligned} \bar{\Gamma}_{n+1} &= \lambda(\lambda^{-1}\bar{\Gamma}_n) + (1 - \lambda)\bar{\gamma}_{n-1} \\ &= \bar{\Gamma}_n + (\lambda^{-1} - 1)\bar{\gamma}_n \end{aligned} \tag{3.4.19}$$

It is not hard to see that

$$\bar{\Gamma}_n = (2 - \lambda^{n-1})c_0 \tag{3.4.20}$$

a formula agreeing with Eqs. (3.4.9) and (3.4.14). We observe that the *separation factor*

$$\alpha_n = \frac{\bar{\Gamma}_n}{\gamma_n} = 2\lambda^{-n} - \lambda^{-1} \tag{3.4.21}$$

increases with each cycle.

The arrangement derives its pumping action from the fact that the system is capable of storing more solute on the stationary phase at the lower temperature during that part of the cycle in which the flow is in the negative direction. The separation factor given by Eq. (3.4.21) shows a good agreement with experimental results over a reasonable range of concentration, but, as the concentration $\bar{\Gamma}_n$ increases, significant deviations are expected due to the nonlinear nature of the equilibrium relation (see Exercise 3.4.4).

Exercises

3.4.1. Consider the parametric pump when $L = V\theta(\mu_+ - \frac{1}{2}\mu_-)$ and show that

$$\bar{\gamma}_n = \lambda^n c_0, \qquad \bar{\Gamma}_n = c_0(3 - \lambda - \lambda^{n-2}), \qquad n > 2$$

3.4.2. If the parametric pump is so designated that there is an integer p such that

$$\mu_+ - (p - 1)\mu_- = \frac{2L}{V\theta}$$

show that

$$\frac{\bar{\gamma}_n}{c_0} = \lambda^n, \qquad \frac{\bar{\Gamma}_n}{c_0} = 2 + (p - 1)(1 - \lambda) - \lambda^{n-p}, \quad n > p$$

and find $\bar{\Gamma}_n$ for $n < p$.

3.4.3. Show that the parametric pump will ultimately get all the solute on the column into one of the reservoirs.

3.4.4. Note that at the end of every half-cycle the solute is redistributed in the following manner:

$$\varepsilon c_+ + (1 - \varepsilon)n_+ = \varepsilon c_- + (1 - \varepsilon)n_-$$

which represents a straight line of slope $-\varepsilon/(1 - \varepsilon)$ in the (c, n)-plane. Using the equilibrium lines for T_+ and T_-, draw a schematic diagram in the (c, n)-plane showing the variations in Γ_n as well as in γ_n. Can you construct a similar diagram for a nonlinear equilibrium relation (e.g., Langmuir isotherm) and thereby argue that the linear equilibrium assumption may lead to a significant error as the number of cycles increases? [cf. H.-K. Rhee and N. R. Amundson, *Ind. Eng. Chem. Fundam.* **9**, 303 (1970)].

3.5 Linear Equations with Variable Coefficients

The case of the linear equation with coefficients that are functions of the independent variables is scarcely more difficult. It will be worthwhile doing it in some detail for two independent variables since the geometry of the situation can be very clearly demonstrated. Let us first consider the strictly linear case so that

$$P(x, y)\frac{\partial z}{\partial x} + Q(x, y)\frac{\partial z}{\partial y} = R(x, y) \tag{3.5.1}$$

The characteristic differential equations are

$$\frac{dx}{ds} = P(x, y)$$

$$\frac{dy}{ds} = Q(x, y) \tag{3.5.2}$$

$$\frac{dz}{ds} = R(x, y)$$

In contrast to the case of constant coefficients, the characteristic direction σ in the (x, y)-plane will now vary from point to point since it is given by a function of x and y; that is,

$$\sigma = \frac{dy}{dx} = \frac{Q(x, y)}{P(x, y)} \tag{3.5.3}$$

The solution to this equation is a one-parameter family of curves, which replaces the family of parallel straight lines we would obtain in case of constant coefficients. This is shown in Fig. 3.13. (Compare this with, for example, Fig. 3.4.) These curves are the characteristics (or the characteristic ground curves), C, of Eq. (3.5.1).

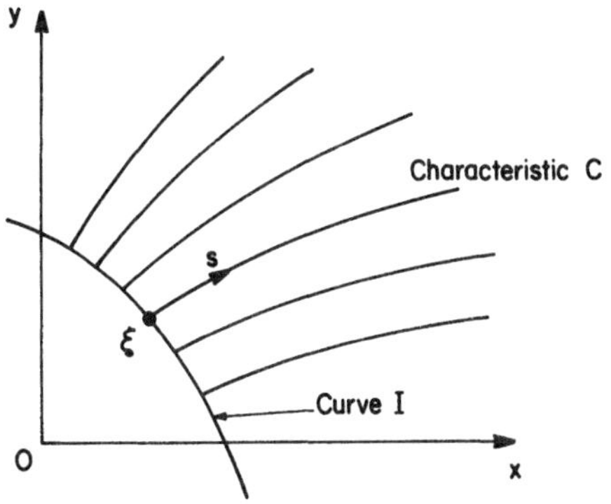

Figure 3.13

To be more specific with the solution, let us consider the initial data prescribed along an initial curve I. We will assume that I is not tangent to a characteristic and suppose that it is given parametrically, in terms of a parameter ξ, as

$$x = X(\xi), \qquad y = Y(\xi) \tag{3.5.4}$$

Then to give $z(x, y)$ at a point on I is equivalent to giving z as a function of ξ:

$$z = Z(\xi) = z(X(\xi), Y(\xi)). \tag{3.5.5}$$

If we take the intersection of I with a characteristic C as the point at which $s = 0$, we will then obtain the family of characteristics

$$x = x(s, \xi), \qquad y = y(s, \xi) \tag{3.5.6}$$

by integrating the two equations

$$\begin{aligned} \frac{dx}{ds} &= P(x, y), \qquad x(0, \xi) = X(\xi) \\ \frac{dy}{ds} &= Q(x, y), \qquad y(0, \xi) = Y(\xi) \end{aligned} \tag{3.5.7}$$

simultaneously. But now looking back at Eq. (3.5.2) we see that we have a third equation,

$$\frac{dz}{ds} = R(x, y), \qquad z(0, \xi) = Z(\xi) \tag{3.5.8}$$

for which the third initial condition comes from Eq. (3.5.5). The last equation can be integrated by quadratures once the previous two have been solved:

$$z(s, \xi) = Z(\xi) + \int_0^s R(x(s', \xi), y(s', \xi))ds' \tag{3.5.9}$$

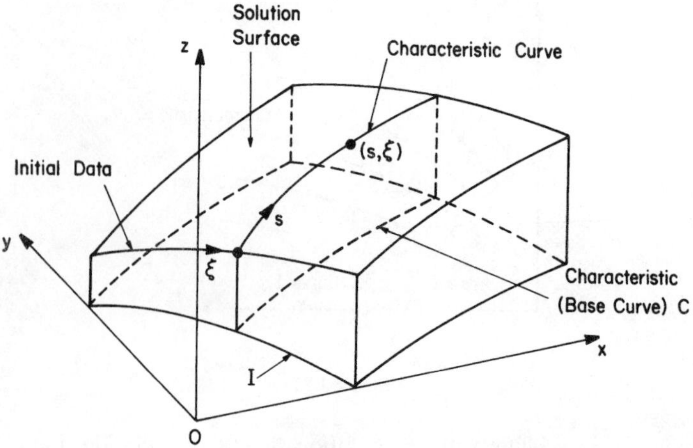

Figure 3.14

a formula that is the immediate analogue of Eq. (3.1.6) or (3.1.11). These generate a surface in three dimensions, $z(x, y)$, that manifestly satisfies both the equation and the initial condition. It may be possible to eliminate s and ξ and get the equations of the surface as $z = \phi(x, y)$ or as $\psi(x, y, z) = 0$, but it is not necessary to do this either from a conceptual or computational point of view. The construction of the solution surface is illustrated in Fig. 3.14.

We see that in the strictly linear case the characteristics C can be laid down once and for all, independently of the specification of the initial data that fixes the specific solution. This feature is not lost in the linear or semilinear case when R is a function of z, as well as of x and y. The only difference is that the differential equation is now of the form

$$\frac{dz}{ds} = R(x, y, z), \qquad z(0, \xi) = Z(\xi) \tag{3.5.10}$$

and cannot be integrated by quadratures, unless R is separable.

The solution surface is thus made up of a one-parameter family of curves in space all of which pass through the initial curve and satisfy the characteristic equations (3.5.7) and (3.5.8) or (3.5.10). It is instructive to see this in another way by writing the partial differential equation as

$$P(x, y)\frac{\partial z}{\partial x} + Q(x, y)\frac{\partial z}{\partial y} + R(x, y, z)\,(-1) = 0$$

(see Sec. 2.5). Now the normal to the solution surface $z(x, y)$ has direction cosines in ratio $\partial z/\partial x : \partial z/\partial y : -1$, and this equation states that such a normal is perpendicular to the direction given by the ratios $P:Q:R$. Hence if we solve the characteristic differential equations

$$\frac{dx}{ds} = P(x, y), \qquad \frac{dy}{ds} = Q(x, y), \qquad \frac{dz}{ds} = R(x, y, z) \tag{3.5.11}$$

we have a curve that is everywhere perpendicular to the normal to the solution surface, which is to say that it lies in the surface itself. The geometrical construction shows why the curve I cannot be a characteristic or tangent to a characteristic if the value of z is to be specified arbitrarily, for the same considerations as were outlined in the discussions of Fig. 3.2 apply.

An illustration of a linear equation is afforded by making the velocity of the cooling fluid in the heat-exchanger problem a function of time. Then

$$\frac{\partial T}{\partial t} + V(t)\frac{\partial T}{\partial x} = H(t)\,(T_a - T) \tag{3.5.12}$$

for H will generally be a function of V and so a function of t also. We have the characteristic differential equations

$$\begin{aligned} \frac{dt}{ds} &= 1 \\ \frac{dx}{ds} &= V(t) \\ \frac{dT}{ds} &= H(t)\,(T_a(t) - T) \end{aligned} \tag{3.5.13}$$

The first two equations give a family of characteristics C whose slope at time t is

$$\sigma = \frac{dt}{dx} = \frac{1}{V(t)} \tag{3.5.14}$$

These characteristics are horizontally parallel as shown in Fig. 3.15 because σ is a function of t alone. Also, the first equation shows that t and s differ only by a constant, so we could write

$$\frac{dT}{dt} = H(t)\,(T_a(t) - T) \tag{3.5.15}$$

for the third equation. If we let

$$U(t, t_0) = \int_{t_0}^{t} H(t')dt' \tag{3.5.16}$$

and

$$W(t, t_0) = \int_{t_0}^{t} V(t')\,dt' \tag{3.5.17}$$

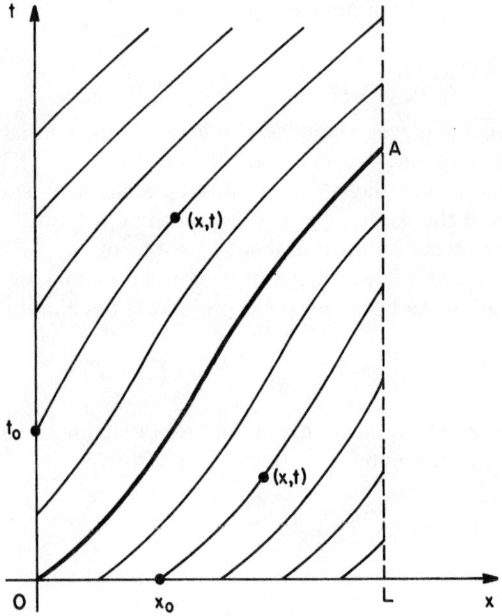

Figure 3.15

then the solution of Eq. (3.5.15) for which

$$T = T_i(t_0) \qquad \text{at } t = t_0 \tag{3.5.18}$$

is

$$T(t, t_0) = T_i(t_0) \exp[-U(t, t_0)] + \int_{t_0}^{t} T_a(t')H(t') \exp[-U(t, t')]\, dt' \tag{3.5.19}$$

As we will see shortly, the parameter t_0 can be related to x by using Eqs. (3.5.14) and (3.5.17).

Now suppose that, as before, T is specified as an initial distribution on the x-axis,

$$T(x, 0) = T_0(x) \tag{3.5.20}$$

and as the feed temperature on the t-axis,

$$T(0, t) = T_F(t) \tag{3.5.21}$$

We would again distinguish between a point (x, t) beneath the characteristic OA that passes through the origin and a point above. The former will lie on a characteristic through some point x_0 on the x-axis, $0 \leq x_0 \leq L$, so that integration of Eq. (3.5.14) from the point $(x_0, 0)$ yields

$$x = x_0 + W(t, 0) \tag{3.5.22}$$

The solution of Eq. (3.5.15) is given by Eq. (3.5.19), with $T_0(x_0)$ for $T_i(t_0)$ and $t_0 = 0$ in $U(t, t_0)$, as well as in the integral. But $x_0 = x - W(t, 0)$, and so we obtain the solution in the form

$$T(x, t) = T_0(x - W(t, 0)) \exp[-U(t, 0)] + \int_0^t T_a(t')H(t') \exp[-U(t, t')]\, dt' \tag{3.5.23}$$

For a point (x, t) above the characteristic OA, there must be a time t_0 for $x = 0$ such that

$$x = W(t, t_0) \tag{3.5.24}$$

and then Eq. (3.5.19) gives

$$T(x, t) = T_F(t_0) \exp[-U(t, t_0)] + \int_{t_0}^t T_a(t')H(t)' \exp[-U(t, t')]\, dt' \tag{3.5.25}$$

If Eq. (3.5.24) can be solved to give t_0 as a function of x and t, then this could be substituted in Eq. (3.5.25), but this is not necessary either conceptually or computationally.

As another example of a linear equation with variable coefficients, we will consider the chromatography of a single solute with cross-sectional area A varying in the direction of flow. The system is otherwise the same as that discussed in Sec. 3.3. The material balance now reads

$$\frac{\partial}{\partial z}\{\varepsilon V(z)A(z)c\} + \frac{\partial}{\partial t}\{A(z)[\varepsilon c + (1 - \varepsilon)n]\} = 0$$

or

$$Q\frac{\partial c}{\partial z} + A(z)\left\{\varepsilon\frac{\partial c}{\partial t} + (1 - \varepsilon)\frac{\partial n}{\partial t}\right\} = 0 \tag{3.5.26}$$

where

$$Q = \varepsilon V(z)A(z) = \text{constant volumetric flow rate} \tag{3.5.27}$$

If we define the dimensionless variables as

$$x = \frac{\int_{z_0}^{z} A(z')\, dz'}{\int_0^{z_0} A(z')\, dz'} \tag{3.5.28}$$

and

$$\tau = \frac{Qt}{\varepsilon \int_0^{z_0} A(z')\, dz'} \tag{3.5.29}$$

then Eq. (3.5.26) can be rewritten in the form

$$\frac{\partial c}{\partial x} + \frac{\partial c}{\partial \tau} + \frac{1 - \varepsilon}{\varepsilon} \frac{\partial n}{\partial \tau} = 0 \tag{3.5.30}$$

which is identical to Eq. (3.3.1) with unity for V. Consequently, all the arguments in Sec. 3.3 are equally applicable if the initial and feed data are compatible.

For an annular cylindrical bed, fed through a central channel and drained at the periphery, the initial and feed data would be specified as

$$\begin{aligned} &\text{at } t = 0, \qquad c = c_0(r) \\ &\text{at } r = r_0, \qquad c = c_F(t) \end{aligned} \tag{3.5.31}$$

Here the radial position r corresponds to z and r_0 denotes the radius of the central channel (see Exercise 1.2.2). If the flow is assumed purely radial, we have from Eqs. (3.5.28) and (3.5.29)

$$x = \left(\frac{r}{r_0}\right)^2 - 1 \tag{3.5.32}$$

and

$$\tau = \frac{Qt}{(\pi \varepsilon r_0^2 d)} \tag{3.5.33}$$

where d is the depth of the bed. The data can be rewritten as

$$\begin{aligned} &\text{at } \tau = 0, \qquad c = c_0\left(r_0\sqrt{1 + x}\right) = f(x) \\ &\text{at } x = 0, \qquad c = c_F\left(\frac{\pi r_0^2 d \varepsilon \tau}{Q}\right) = g(\tau) \end{aligned} \tag{3.5.34}$$

For a given set of data, therefore, we simply determine the solution in terms of x and τ and then recover the original variables by using Eqs. (3.5.32) and (3.5.33). This mode of operation is of potential interest in cases where high throughput rates and wide but shallow beds are desired. Radial flow geometry is also characteristic of one operating method used in paper chromatography.

For a spherical bed with spherical symmetry, we may put

$$x = \left(\frac{r}{r_0}\right)^3 - 1 \tag{3.5.35}$$

$$\tau = \frac{Qt}{\frac{4}{3}\pi\varepsilon r_0^3} \tag{3.5.36}$$

and rewrite the data in the form of Eq. (3.5.34).

Exercises

3.5.1. Write down the characteristic differential equations of Eq. (1.14.11) for the first optimization problem in Sec. 1.14. They should be equivalent to imposing the optimal policy $T = T_m(\xi)$ on the original process equations with one difference. Account for this difference.

3.5.2. Explore the theory of the linear equilibrium parapump when $V(t)$ and $T(t)$ are arbitrary periodic functions of time.

3.5.3. Consider the chromatography of a single solute with a linear equilibrium relation. If the cross-sectional area varies in the direction of flow, determine the characteristic direction σ in the (z, t)-plane. For an annular cylindrical bed, discuss the wave propagation in the r-direction. Can you formulate a bottleneck problem and show how the wave motion will be affected by the variation in cross-sectional area?

3.6 Linear Equations with *n* Independent Variables

The strictly linear equation with n independent variables may be written

$$\sum P_i(x_1, \ldots, x_n)p_i = R(x_1, \ldots, x_n) \tag{3.6.1}$$

where

$$p_i = \frac{\partial z}{\partial x_i}, \qquad i = 1, 2, \ldots, n \tag{3.6.2}$$

If $z = z(x_1, \ldots, x_n)$ is a hypersurface (i.e., an n-dimensional structure) in $(n + 1)$-dimensional space, $dz = \Sigma p_i\, dx_i$ or

$$\sum p_i\, dx_i - dz = 0$$

Thus the direction given by $\{p_1 : p_2 : \ldots : p_n : -1\}$ is orthogonal to any element $dx_1 : dx_2 : \ldots : dx_n : dz$ lying in the surface and therefore defines the direction normal to it. Now the differential equation

$$\sum p_i P_i - R = 0$$

may be interpreted as saying that the direction given by

$$\frac{dx_i}{ds} = P_i(x_1, \ldots, x_n), \qquad i = 1, 2, \ldots, n \tag{3.6.3}$$

$$\frac{dz}{ds} = R(x_1, \ldots, x_n) \tag{3.6.4}$$

is orthogonal to the normal to the solution surface. Hence the curves that satisfy these equations must lie in the surface. The solutions of Eqs. (3.6.3) give the characteristic ground curves, $x_i(s)$, and Eq. (3.6.4) can be solved by quadratures to give $z(s)$. The functions $x_i(s)$ and $z(s)$ provide the parametric form of a characteristic curve in $(n + 1)$-dimensional space.

As we saw in Sec. 2.1, we will need initial conditions specified on an $(n - 1)$-dimensional manifold. This may be conveniently described parametrically in terms of $(n - 1)$ parameters $\xi_1, \ldots, \xi_{n-1}$ by

$$x_i = X_i(\xi_1, \ldots, \xi_{n-1}), \qquad i = 1, 2, \ldots, n \tag{3.6.5}$$

If the value of z on this manifold is given by

$$z = Z(\xi_1, \ldots, \xi_{n-1}) \tag{3.6.6}$$

then Eqs. (3.6.3) and (3.6.4) can be solved with initial conditions $x_i = X_i$, $z = Z$ when $s = 0$. We may denote this $(n - 1)$-parameter family of solutions by

$$x_i = x_i(\xi_1, \ldots, \xi_{n-1}, s), \qquad i = 1, 2, \ldots, n \tag{3.6.7}$$

$$z = z(\xi_1, \ldots, \xi_{n-1}, s) \tag{3.6.8}$$

and we see that this now defines an n-dimensional structure in the $(n + 1)$-dimensional space of $\{x_1, \ldots, x_n, z\}$. In principle, the n parameters could be eliminated from the n equations (3.6.7) to give $z = z(x_1, \ldots, x_n)$. In practice, this may be neither necessary nor desirable.

For the linear or semilinear equation, we can again get the characteristic ground curves from Eqs. (3.6.3), but we must now solve

$$\frac{dz}{ds} = R(x_1, \ldots, x_n, z) = R(s, z) \tag{3.6.9}$$

as a differential equation. Here R can be regarded as a function of s and z since $x_1, \ldots, x_n$ are calculable functions of s from Eqs. (3.6.3). Again we have the linear feature that the characteristic ground curves can be laid down once and for all.

We would again run into trouble if the initial manifold were anywhere parallel to a characteristic direction. Now an arbitrary direction in the initial manifold is given by

$$dx_i = \sum_{p=1}^{n-1} \frac{\partial X_i}{\partial \xi_p} d\xi_p$$

The condition that no values of $d\xi_p$ can be found to make these proportional to P_i is that the determinant

$$\begin{vmatrix} \dfrac{\partial X_1}{\partial \xi_1} & \cdots & \dfrac{\partial X_1}{\partial \xi_{n-1}} & p_1 \\ \vdots & & \vdots & \vdots \\ \dfrac{\partial X_n}{\partial \xi_1} & \cdots & \dfrac{\partial X_n}{\partial \xi_{n-1}} & p_n \end{vmatrix} \tag{3.6.10}$$

should not vanish.

To illustrate the linear equation with three independent variables, we will look at a problem that starts by being linear but appears at first to be headed for nonlinearity. This is the second optimization problem of Sec. 1.14, where for $u(x, y, z)$ we have

$$u_z = \max_k \, [kx(1 + u_y - u_x) - \rho k^r y(1 + u_y)] \tag{3.6.11}$$

Apart from the maximization operation, this is a linear equation. For example, if the maximization decreed that k should be held constant, we could write the equation as

$$-kxu_x + (kx - \rho k^r y)u_y - u_z = -kx + \rho k^r y$$

Then, making the translations $u \leftrightarrow z$, $x \leftrightarrow x_1$, $y \leftrightarrow x_2$, $z \leftrightarrow x_3$, $u_x \leftrightarrow p_1$, $u_y \leftrightarrow p_2$, $u_z \leftrightarrow p_3$, we have the characteristic differential equations

$$\frac{dx}{ds} = -kx$$

$$\frac{dy}{ds} = kx - \rho k^r y$$

$$\frac{dz}{ds} = -1$$

$$\frac{du}{ds} = -kx + \rho k^r y$$

If we divide the first two by the third equation, we have

$$\frac{dx}{dz} = kx, \qquad \frac{dy}{dz} = -kx + \rho k^r y$$

which are analogous to the process equations

$$\frac{da}{dt} = -ka, \qquad \frac{db}{dt} = ka - \rho k^r b$$

The difference is that although x and y are the values of a and b, z is the duration of the process, and so $z = 0$ is the end of the operation, whereas $t = 0$ is the beginning. Since z and t run in opposite directions, we expect to see a change of sign. Of course, constant k is a very simple type of control policy, but we will see in Sec. 9.3 that the same characteristic equations can be obtained in the more general case where k is chosen to maximize the expression in the brackets of Eq. (3.6.11).

REFERENCES

3.2. The transient behavior of the heat exchanger is treated by the Laplace transformation in

W. H. Ray, "Forced Flow Heat Exchanger Dynamics," *Ind. Eng. Chem. Fundam.* **5**, 138–139 (1966).

Application of the pulsed reactor and graphs of the integrals of Exercise 3.2.5 are given in

W. H. Blanton, Jr., C. H. Byers, and R. P. Merrill, "Quantitative Rate Coefficients from Pulsed Microcatalytic Reactors," *Ind. Eng. Chem. Fundam.* **7**, 611–617 (1968).

3.3. Propagations of linear waves are clearly illustrated in

R. L. Streeter, *The Analysis and Solution of Partial Differential Equations*, Brooks/Cole Publishing Co., Monterey, Calif., 1973, Chapter 9.

3.4. The treatment of parametric pumping follows somewhat on the lines laid down in

R. L. Pigford, B. Baker, and D. E. Blum, "An Equilibrium Theory of the Parametric Pump," *Ind. Eng. Chem. Fundam.* **8**, 144–149 (1969).

R. Aris, "Equilibrium Theory of the Parametric Pump," *Ind. Eng. Chem. Fundam.* **8**, 603–604 (1969).

The effect of the nonlinear nature of the equilibrium relation is briefly discussed in

H.-K. Rhee and N. R. Amundson, "An Equilibrium Theory of the Parametric Pump," *Ind. Eng. Chem. Fundam.* **9**, 303–304 (1970).

3.5. The chromatography of a single solute with radial flow is treated in

L. Lapidus and N. R. Amundson, "Mathematics of Adsorption in Beds: III. Radial Flow," *J. Phys. Colloid Chem.* **54**, 821–829 (1950).

See also the last section of

H.-K. Rhee, R. Aris, and N. R. Amundson, "On the Theory of Multicomponent Chromatography," *Phil. Trans. Roy. Soc. Lond.* **A267**, 419–455 (1970).

Chromatographic Equations with Finite Rate Expressions

4

When a system consists of more than one phase and the interphase transfer rate is finite, the differential equation describing the system is found coupled with a rate expression, which may be linear or nonlinear (see Sec. 1.2). Some of these problems can be treated by using the Laplace transformation or other special techniques. The use of the Laplace transform is naturally restricted to linear systems or to those that can be linearized in some way. In this chapter we will first see that the solution determined by the Laplace transform matches up with that of the method of characteristics and discuss some problems in the chromatography of a single solute whose rate of partition is finite. The last section will be concerned with a model for the poisoning of the catalyst particles in fixed-bed reactors. Here we will see that the system of equations may be integrated by quadratures.

4.1 Solution by the Laplace Transformation

Normally, the Laplace transformation is associated with a variable whose range is from zero to infinity (or one whose range can be so extended), and very often this variable is the time. We will usually denote this variable by t and the transform variable by p defining the transform $\bar{f}(p)$ of a function $f(t)$ by

$$\bar{f}(p) = L[f(t)] = \int_0^\infty e^{-pt} f(t)\, dt \tag{4.1.1}$$

The inversion formula

$$f(t) = L^{-1}[\bar{f}(p)] = \frac{1}{2\pi i} \int_{\alpha - i\infty}^{\alpha + i\infty} e^{pt} \bar{f}(p)\, dp \tag{4.1.2}$$

where α is greater than the abscissa of convergence [i.e., the least value of Re p for which the integral (4.1.1) is defined], will not often be needed, as there exist very extensive tabulations of the Laplace transform and its inverse. (See the References.)

Let the equation to be solved be

$$a \frac{\partial u}{\partial x} + b \frac{\partial u}{\partial t} = c \tag{4.1.3}$$

where a and b may be functions of x and

$$c = h(x)u \tag{4.1.4}$$

Suppose further that the initial and boundary conditions are

$$\begin{aligned} u(x, 0) &= f(x) \\ u(0, t) &= g(t) \end{aligned} \tag{4.1.5}$$

Now

$$\int_0^\infty e^{-pt} \frac{\partial u}{\partial x}(x, t)\, dt = \frac{d\bar{u}}{dx}(x, p) \tag{4.1.6}$$

and

$$\begin{aligned} \int_0^\infty e^{-pt} \frac{\partial u}{\partial t}(x, t)\, dt &= [e^{-pt} u(x, t)]_0^\infty + p \int_0^\infty e^{-pt} u(x, t)\, dt \\ &= -u(x, 0) + p\bar{u}(x, p) \end{aligned} \tag{4.1.7}$$

for $u(x, t)$ must be less than $Ae^{\alpha t}$ for some finite value α if the transform

$$\bar{u}(x, p) = \int_0^\infty e^{-pt} u(x, t)\, dt \tag{4.1.8}$$

is to exist, and it follows that the integrated part vanishes at the upper limit if Re $p > \alpha$. Multiplying Eq. (4.1.3) by e^{-pt} and integrating from zero to infinity with respect to t gives

$$a(x)\frac{d\bar{u}}{dx}(x, p) + b(x)[-f(x) + p\bar{u}(x, p)] = h(x)\bar{u}(x, p)$$

or

$$a\frac{d\bar{u}}{dx} + (pb - h)\bar{u} = bf \tag{4.1.9}$$

In this equation, p is regarded as a parameter, and so we write a total derivative for $\bar{u}$ as a function of the one variable x. The resulting equation is a first-order linear ordinary differential equation

$$\frac{d\bar{u}}{dx} + q(x)\bar{u} = r(x) \tag{4.1.10}$$

where

$$q(x) = p\frac{b(x)}{a(x)} - \frac{h(x)}{a(x)} \tag{4.1.11}$$

$$r(x) = \frac{b(x)f(x)}{a(x)} \tag{4.1.12}$$

An initial condition for this ordinary equation is obtained by the transform of the condition at $x = 0$,

$$\bar{u}(0, p) = \bar{g}(p) = \int_0^\infty e^{-pt}g(t)\,dt \tag{4.1.13}$$

The solution of Eq. (4.1.10) can be obtained quite straightforwardly. Let

$$Q(x) = \int_0^x q(x')\,dx' \tag{4.1.14}$$

Then multiplying both sides by exp $Q(x)$ and integrating gives

$$[\bar{u}(x, p) \exp Q(x)]_0^x = \int_0^x r(x') \exp Q(x')\,dx'$$

or

$$\bar{u}(x, p) = \bar{g}(p) \exp[-Q(x)] + \int_0^x r(x') \exp\{-[Q(x) - Q(x')]\}\,dx' \tag{4.1.15}$$

From this, $u(x, t)$ can be obtained by inversion easily enough, but it is instructive to pick the formula apart a little to see how it connects with the method of characteristics. If

$$B(x) = \int_0^x \left[\frac{b(x')}{a(x')}\right] dx' \tag{4.1.16}$$

$$C(x) = \int_0^x \left[\frac{h(x')}{a(x')}\right] dx' \tag{4.1.17}$$

we see that

$$Q(x) = pB(x) - C(x) \tag{4.1.18}$$

and so

$$\bar{u}(x, p) = \bar{g}(p) \exp\{-pB(x) + C(x)\} + \int_0^x R(x, x') \exp\{-p[B(x) - B(x')]\} dx', \tag{4.1.19}$$

where

$$R(x, x') = r(x') \exp[C(x) - C(x')] \tag{4.1.20}$$

Now the inverse transform of $\exp(-pA)$ is

$$L^{-1}[\exp(-pA)] = \delta(t - A) \tag{4.1.21}$$

where $\delta(t - A)$ is the Dirac delta function centered at $t = A$ and is such that

$$\begin{aligned} \delta(t - A) &= 0, \qquad t \neq A \\ \int_0^\infty F(t)\delta(t - A)\, dt &= F(A) \end{aligned} \tag{4.1.22}$$

Since this is the only way in which p enters the integral in Eq. (4.1.19) and since the integration is a linear operation, the inverse transform of this part of $\bar{u}(x, p)$ is

$$\int_0^x R(x, x')\delta(t - B(x) + B(x'))\, dx' \tag{4.1.23}$$

If $t > B(x)$, the argument of the delta function is always strictly positive and so this integral vanishes. To find the inverse of the first term in Eq. (4.1.19), we need the convolution theorem for the Laplace transform; that is, if $\bar{f}(p)$ and $\bar{g}(p)$ are the transforms of $f(t)$ and $g(t)$, respectively, then

$$L^{-1}[\bar{f}(p)\bar{g}(p)] = \int_0^t f(t - \tau)g(\tau)\, d\tau = \int_0^t f(\tau)g(t - \tau)\, d\tau \tag{4.1.24}$$

Applying this to the first term and using Eq. (4.1.22), we see that it is the transform of

$$e^{C(x)} \int_0^t g(t - \tau)\delta(\tau - B(x))\, d\tau = g(t - B(x))\, e^{C(x)} \tag{4.1.25}$$

provided that $t > B(x)$; otherwise, it is zero. Therefore, the solution of Eq. (4.1.3) can be written as

$$u(x, t) = \begin{cases} \int_0^x R(x, x')\delta(t - B(x) + B(x'))\, dx', & t \leq B(x) \\ g(t - B(x))e^{C(x)} & t > B(x) \end{cases} \tag{4.1.26}$$

To see how this matches up with the solution by the method of characteristics, we will write Eq. (4.1.3) in the form

$$\frac{\partial u}{\partial x} + \frac{b}{a}\frac{\partial u}{\partial t} = \frac{h}{a} u$$

for which the characteristic differential equations are

$$\frac{dx}{ds} = 1, \qquad \frac{dt}{ds} = \frac{b(x)}{a(x)}, \qquad \frac{du}{ds} = \frac{h(x)}{a(x)} u \tag{4.1.27}$$

If the initial data are specified as $x = x_0$, $t = t_0$, and $u = u_0$ for $s = 0$, these may be integrated to give

$$\begin{aligned} x &= x_0 + s \\ t &= t_0 + B(s) \\ u &= u_0 e^{C(s)} \end{aligned} \tag{4.1.28}$$

For the characteristic emanating from the origin, we have $x_0 = t_0 = 0$, and thus

$$t = B(x) \tag{4.1.29}$$

If $t > B(x)$, the characteristics emanate from the t-axis, so $x_0 = 0$, $u_0 = g(t_0)$, and $t_0 = t - B(x) > 0$. The last equation of Eq. (4.1.28) gives

$$u(x, t) = g(t - B(x))e^{C(x)} \tag{4.1.30}$$

which is identical to Eq. (4.1.26) for $t > B(x)$. If $t \leq B(x)$, we have $t_0 = 0$, $u_0 = f(x_0)$, and $s = x - x_0$. From Eqs. (4.1.16) and (4.1.17),

$$\begin{aligned} B(s) &= B(x - x_0) = B(x) - B(x_0) \\ C(s) &= C(x - x_0) = C(x) - C(x_0) \end{aligned} \tag{4.1.31}$$

so

$$u(x, t) = f(x_0)e^{C(x) - C(x_0)} \tag{4.1.32}$$

where x_0 is determined in terms of x and t by the equation

$$B(x_0) = B(x) - t \tag{4.1.33}$$

We now have only to confirm that Eq. (4.1.32) is the same as the integral in

Eq. (4.1.26). For this purpose, we will write the integral as

$$I = \int_0^x \frac{R(x, x')}{B'(x')} \delta(B(x') - [B(x) - t])B'(x')\, dx$$

in which B' denotes the derivative of $B(x)$. Applying Eq. (4.1.33) and then Eqs. (4.1.16) and (4.1.20), one can immediately find that

$$I = \frac{R(x, x_0)}{B'(x_0)} = f(x_0)e^{C(x) - C(x_0)}$$

Therefore, Eq. (4.1.26) corresponds precisely to the solution that we can derive by the method of characteristics.

We may apply this without hesitation to the transient heat-exchanger problem of Sec. 3.2. If, for simplicity, we take T_a, T_0, and T_F to be constants and write

$$u(x, t) = \frac{T_a - T(x, t)}{T_a - T_F} \tag{4.1.34}$$

Eq. (3.2.14) can be written

$$V\frac{\partial u}{\partial x} + \frac{\partial u}{\partial t} = -Hu \tag{4.1.35}$$

and

$$u(x, 0) = f(x) = u_0 \tag{4.1.36}$$
$$u(0, t) = \frac{T_a - T_F}{T_a - T_F} = 1$$

Then, from Eq. (4.1.15),

$$\begin{aligned}\bar{u}(x, p) &= \frac{1}{p}\exp\left(-\frac{px}{V} - \frac{Hx}{V}\right) \\ &\quad + \frac{u_0}{V}\exp\left(-\frac{px}{V} - \frac{Hx}{V}\right)\int_0^x \exp\left(\frac{px'}{V} + \frac{Hx'}{V}\right) dx' \\ &= \frac{u_0}{p + H} + \left(\frac{1}{p} - \frac{u_0}{p + H}\right)\exp\left[-(p + H)\frac{x}{V}\right] \end{aligned} \tag{4.1.37}$$

This may be easily inverted to give an expression equivalent to Eq. (3.2.21).

4.2 Linear Chromatography

A more instructive example is afforded by the linear chromatography of a single solute whose rate of partition is proportional to the difference from equilibrium. For this we have c and n, denoting the concentrations in the mobile and stationary phases, respectively, and a mass balance gives

$$\varepsilon V \frac{\partial c}{\partial z} + \varepsilon \frac{\partial c}{\partial t} + (1 - \varepsilon) \frac{\partial n}{\partial t} = 0 \tag{4.2.1}$$

where ε = fractional void volume in the bed
V = interstitial velocity of mobile phase
z = distance from inlet
t = time

The partition between the two phases is represented by

$$\frac{\partial n}{\partial t} = k(\gamma c - n) \tag{4.2.2}$$

in which γ is the adsorption equilibrium constant [cf. Eq. (1.2.8)]. We will not attempt to render the equations more elegant by change of variable, although this obviously might be done, but will use the obviousness of this formulation to make the equations as physically immediate as possible. The initial and boundary conditions are

$$c(z, 0) = f(z), \qquad n(z, 0) = h(z) \tag{4.2.3}$$

$$c(0, t) = g(t) \tag{4.2.4}$$

conditions that we may wish to specialize later.

Denoting the Laplace transform with respect to the variable t by a bar over the function, we have

$$V \frac{d\bar{c}}{dz} + p\bar{c} + \nu p\bar{n} = f(z) + \nu h(z) \tag{4.2.5}$$

and

$$(p + k)\bar{n} - k\gamma\bar{c} = h(z) \tag{4.2.6}$$

with

$$\bar{c}(0, p) = \bar{g}(p) \tag{4.2.7}$$

in which

$$\nu = \frac{1 - \varepsilon}{\varepsilon}. \tag{4.2.8}$$

We observe that Eq. (4.2.6) is no longer a differential equation, but immediately gives

$$\bar{n} = \frac{k\gamma\bar{c}}{p + k} + \frac{h(z)}{p + k} \tag{4.2.9}$$

Substituting this in Eq. (4.2.5), we have

$$V\frac{d\bar{c}}{dz} + p\left(1 + \frac{\nu k\gamma}{p + k}\right)\bar{c} = f(z) + \frac{\nu k}{p + k}h(z) \tag{4.2.10}$$

Thus, letting

$$p\left(1 + \frac{\nu k\gamma}{p + k}\right) = P(p) \tag{4.2.11}$$

we have, from Eqs. (4.2.7) and (4.2.10),

$$\bar{c}(z, p) = \bar{g}(p)\exp\left[-\frac{zP(p)}{V}\right] + \frac{1}{V}\int_0^z \left[f(z') + \frac{\nu k}{p + k}h(z')\right]\exp\left[-\frac{(z - z')P(p)}{V}\right] dz' \tag{4.2.12}$$

From this equation it is evident that everything will turn on our ability to invert the two transforms $\exp[-zP(p)/V]$ and $k(p + k)^{-1}\exp[-zP(p)/V]$. Let us write

$$L^{-1}[e^{-zP(p)/V}] = F(z, t) \tag{4.2.13}$$

and

$$L^{-1}\left[\frac{k}{p + k}e^{-zP(p)/V}\right] = G(z, t) \tag{4.2.14}$$

Then applying the convolution theorem to Eq. (4.2.12) gives

$$c(z, t) = \int_0^t g(t - t')F(z, t')\,dt' + \frac{1}{V}\int_0^x [f(z - z')F(z', t) + \nu h(z - z')G(z', t)]\,dz' \tag{4.2.15}$$

The inversion of these transforms is a useful illustration of the syntax of the Laplace transform and of the use of a standard lexicon. Let us first make the manipulations easier by introducing some abbreviations. The expression $P(p)$ may be written

$$P(p) = p + \nu\gamma\frac{kp}{p + k} = p + \nu\gamma k - \nu\gamma k\frac{k}{p+k}$$

$$= k\left[\frac{P}{k} + \nu\gamma - \nu\gamma\left(1 + \frac{p}{k}\right)^{-1}\right]$$

The last move of taking out a factor of k is suggested by the observation that $P(p)$ is more simply written as a function of p/k and that the factor $k/(p + k)$ in the second transform is a function only of the same combination. Moreover, the factor (kz/V), which now appears outside the brackets when we write $zP(p)/V$, is dimensionless. This suggests the following abbreviations:

$$\xi = \frac{kz}{V}, \qquad \beta = \nu\gamma \tag{4.2.16}$$

Then

$$\bar{F} = \exp\left[-\xi\left(\frac{p}{k}\right) - \beta\xi + \beta\xi\left(1 + \left(\frac{p}{k}\right)\right)^{-1}\right] \tag{4.2.17}$$

$$\bar{G} = \left[1 + \left(\frac{p}{k}\right)\right]^{-1} \exp\left[-\xi\left(\frac{p}{k}\right) - \beta\xi + \beta\xi\left(1 + \left(\frac{p}{k}\right)\right)^{-1}\right] \tag{4.2.18}$$

As a first point of syntax, we should consider what it means to have p occurring only in the combination p/k. Table 4.1 shows some entries as they might appear in the introductory section of general formulas in a dictionary of transforms. (References are given to two or three of the standard works.) The first entry under "scale change" shows us that, if p appears with a factor of a, it corresponds to a change of the time scale by a factor of $1/a$. Putting $k = 1/a$, we see that if

$$F^* = L^{-1}\left[\exp\left(-\xi p - \beta\xi + \frac{\beta\xi}{1 + p}\right)\right] = F^*(\xi, t)$$

then

$$F(z, t) = kF^*(\xi, kt)$$

This suggests that we should transform the time variable as well and write

$$\eta - kt \tag{4.2.19}$$

TABLE 4.1

	$f(t) = L^{-1}[\bar{f}(p)]$	$L[f(t)] = \bar{f}(p)$	I	II	III
1. Scale change	$\frac{f(t/a)}{a}$	$\bar{f}(ap)$	4.1.4	I.21	1.3
2. Real shift	$f(t - a),\ t \geq a$; $0,\ t < a$	$e^{-ap}\bar{f}(p)$	4.1.4	I.11	1.5
3. Complex shift	$e^{-at}f(t)$	$\bar{f}(p + a)$	4.1.5	I.19	1.9
4. Real differentiation	$f'(t)$	$p\bar{f}(p) - f(0)$	4.1.8		1.29

I: A. Erdelyi, ed., *Tables of Integral Transforms*, McGraw-Hill, New York, 1954.

II: P. A. McCollum and B. F. Brown, *Laplace Transform Tables and Theorems*, Holt, Rinehart and Winston, New York, 1965.

III: G. E. Roberts and H. Kaufman, *Table of Laplace Transforms*, W. B. Saunders, Philadelphia, 1966.

Let us make these transformations in Eq. (4.2.15); then if c_0 is a characteristic concentration and we write

$$\begin{aligned} c(z, t) &= c_0\rho(\xi, \eta), \qquad & g(t) &= c_0\phi(\eta), \\ f(z) &= c_0\chi(\xi), \qquad & \nu h(z) &= c_0\psi(\xi) \end{aligned} \tag{4.2.20}$$

we have

$$\begin{aligned} \rho(\xi, \eta) = &\int_0^\eta \phi(\eta - \eta')F^*(\xi, \eta')d\eta' \\ &+ \int_0^\xi [\chi(\xi - \xi')F^*(\xi', \eta) + \psi(\xi - \xi')G^*(\xi', \eta)]d\xi' \end{aligned} \tag{4.2.21}$$

where

$$\begin{aligned} L(F^*) &= \exp\left(-\xi p - \beta\xi + \frac{\beta\xi}{1 + p}\right) \\ L(G^*) &= \frac{1}{1 + p}\, exp\left(-\xi p - \beta\xi + \frac{\beta\xi}{1 + p}\right) \end{aligned} \tag{4.2.22}$$

The exponential term $\beta\xi$ is independent of p and so just gives a multiplicative factor of $\exp(-\beta\xi)$. The term $-\xi p$ in the exponent is more interesting, as the second entry in Table 4.1 shows. In fact, if

$$\begin{aligned} L(F^+) &= \exp\left(\frac{\beta\xi}{1 + p}\right) \\ L(G^+) &= \frac{1}{1 + p}\exp\left(\frac{\beta\xi}{1 + p}\right) \end{aligned} \tag{4.2.23}$$

then

$$\begin{aligned} F^*(\xi, \eta) &= \begin{cases} e^{-\beta\xi}F^+(\xi, \eta - \xi), & \eta \geq \xi \\ 0, & \eta < \xi \end{cases} \\ G^*(\xi, \eta) &= \begin{cases} e^{-\beta\xi}G^+(\xi, \eta - \xi), & \eta \geq \xi \\ 0, & \xi < \xi \end{cases} \end{aligned} \tag{4.2.24}$$

Finally, there is the third rule given in Table 4.1, which shows that, if we can find the inverse transform of $\bar{f}(p)$, then we can find also that of $\bar{f}(p + 1)$. In fact, if

$$L(F^{**}) = \exp\left(\frac{\beta\xi}{p}\right)$$

and (4.2.25)

$$L(G^{**}) = p^{-1}\exp\left(\frac{\beta\xi}{p}\right)$$

then

$$F^*(\xi, \eta) = \begin{cases} e^{-\beta\xi-\eta+\xi}F^{**}(\xi, \eta - \xi), & \eta \geq \xi \\ 0, & \eta < \xi \end{cases}$$

and (4.2.26)

$$G^*(\xi, \eta) = \begin{cases} e^{-\beta\xi-\eta+\xi}G^{**}(\xi, \eta - \xi), & \eta \geq \xi \\ 0, & \eta < \xi \end{cases}$$

Substituting these back into Eq. (4.2.21), we have two formulas depending on whether η is greater or less than ξ. If $\eta \geq \xi$,

$$\rho(\xi, \eta) = \int_\xi^\eta \phi(\eta - \eta')e^{-\beta\xi-\eta'+\xi}F^{**}(\xi, \eta' - \xi)\, d\eta'$$

$$+ \int_0^\xi [\chi(\xi - \xi')F^{**}(\xi', \eta - \xi) + \psi(\xi - \xi')G^{**}(\xi', \eta - \xi)]e^{-\beta\xi'-\eta+\xi}d\xi' \qquad (4.2.27)$$

On the other hand, if $\eta \leq \xi$,

$$\rho(\xi, \eta) = \int_0^\eta [\chi(\xi - \xi')F^{**}(\xi', \eta - \xi)$$

$$+ \psi(\xi - \xi')G^{**}(\xi', \eta - \xi)]e^{-\beta\xi'-\eta+\xi}\, d\xi' \qquad (4.2.28)$$

The fact that the first integral has vanished altogether in the expression for ρ when $\xi > \eta$ is not surprising, since we would not expect the inlet conditions to have any effect as long as $z > Vt$.

Now G^{**} can be found quite readily from standard dictionaries. For example, Erdelyi (5.5.35) gives†

$$L^{-1}[p^{-\nu-1}e^{a/p}] = \left(\frac{t}{a}\right)^{\nu/2} I_\nu(2\sqrt{at}) \qquad \text{Re}\ \nu > -1 \qquad (4.2.29)$$

Thus putting $a = \beta\xi$ and $\nu = 0$ gives immediately that

$$G^{**}(\xi, \eta) = I_0(2\sqrt{\beta\xi\eta}) \qquad (4.2.30)$$

Unfortunately, the restriction on ν prevents us from putting $\nu = -1$ in Eq. (4.2.29) and so obtaining F^{**}. However, the rule for the Laplace transform of the derivative $f'(t)$ may be modified if we recall that the inverse transform of a constant is a delta function so that

$$L[f'(t) + f(0)\delta(t)] = p\bar{f}(p)$$

The delta function is no embarrassment since it will appear under an inte-

†Roberts and Kaufman, formula 3.31 of Section 2.

gration sign.† Thus

$$F^{**}(\xi, \eta) = \frac{\partial}{\partial\eta} I_0(2\sqrt{\beta\xi\eta}) + \delta(\eta) = \sqrt{\frac{\beta\xi}{\eta}} I_1(2\sqrt{\beta\xi\eta}) + \delta(\eta) \qquad (4.2.31)$$

Substituting from Eqs. (4.2.30) and (4.2.31) into (4.2.27) and (4.2.28), we have, for $\eta \geq \xi$,

$$\begin{aligned}
\rho(\xi,\eta) &= \phi(\eta-\xi)e^{-\beta\xi} \\
&\quad + \int_\xi^\eta \phi(\eta-\eta')e^{-\beta\xi-\eta'+\xi}\sqrt{\frac{\beta\xi}{\eta'-\xi}}I_1(2\sqrt{\beta\xi(\eta'-\xi)})\,d\eta' \\
&\quad + \int_0^\xi \chi(\xi-\xi')e^{-\beta\xi'-\eta+\xi}\sqrt{\frac{\beta\xi'}{\eta-\xi}}I_1(2\sqrt{\beta\xi'(\eta-\xi)})\,d\xi' \\
&\quad + \int_0^\xi \psi(\xi-\xi')e^{-\beta\xi'-\eta+\xi}I_0(2\sqrt{\beta\xi'(\eta-\xi)})\,d\xi' \\
&= \phi(\eta-\xi)e^{-\beta\xi} \\
&\quad + \int_0^{\eta-\xi} \phi(\eta-\xi-\eta'')e^{-\beta\xi-\eta''}\sqrt{\frac{\beta\xi}{\eta''}}I_1(2\sqrt{\beta\xi\eta''})\,d\eta'' \\
&\quad + \int_0^\xi \chi(\xi-\xi')e^{-\beta\xi'-(\eta-\xi)}\sqrt{\frac{\beta\xi'}{\eta-\xi}}I_1(2\sqrt{\beta\xi'(\eta-\xi)})\,d\xi' \\
&\quad + \int_0^\xi \psi(\xi-\xi')e^{-\beta\xi'-(\eta-\xi)}I_0(2\sqrt{\beta\xi'(\eta-\xi)})\,d\xi'
\end{aligned} \qquad (4.2.32)$$

For $\xi > \eta$, we have

$$\begin{aligned}
\rho(\xi,\eta) &= \int_0^\eta \chi(\xi-\xi')e^{-\beta\xi'-(\eta-\xi)}\sqrt{\frac{\beta\xi'}{\eta-\xi}}I_1(2\sqrt{\beta\xi'(\eta-\xi)})\,d\xi' \\
&\quad + \int_0^\eta \psi(\xi-\xi')e^{-\beta\xi'-(\eta-\xi)}I_0(2\sqrt{\beta\xi'(\eta-\xi)})\,d\xi'
\end{aligned} \qquad (4.2.33)$$

Note that in the first integral of this latter formula the argument of the delta function is always negative and it therefore vanishes.

These are rather fiercely complicated formulas, and we will get a better feel for them by considering some special cases. Consider first the saturation of a column initially innocent of solute by a stream of constant composition.

†Erdelyi, 5.5.31, or Roberts and Kaufman, 2.3.33.

For this case,

$$\phi = 1, \qquad \chi = \psi = 0 \tag{4.2.34}$$

so that ρ vanishes for $\xi > \eta$. For $\xi \leq \eta$, we have

$$\begin{aligned}
\rho(\xi, \eta) &= e^{-\beta\xi} + e^{-\beta\xi} \int_0^{\eta-\xi} e^{-\eta''} \sqrt{\frac{\beta\xi}{\eta''}} I_1(2\sqrt{\beta\xi\eta''})\, d\eta'' \\
&= e^{-\beta\xi} + e^{-\beta\xi}\left[e^{-\eta''} I_0(2\sqrt{\beta\xi\eta''})\right]_0^{\eta-\xi} \\
&\quad + e^{-\beta\xi} \int_0^{\eta-\xi} e^{-\eta''} I_0(2\sqrt{\beta\xi\eta''})\, d\eta'' \\
&= e^{-\beta\xi-(\eta-\xi)} I_0(2\sqrt{\beta\xi(\eta-\xi)}) \\
&\quad + e^{-\beta\xi-(\eta-\xi)} e^{\eta-\xi} \int_0^{\eta-\xi} e^{-\eta''} I_0(2\sqrt{\beta\xi\eta''})\, d\eta''
\end{aligned} \tag{4.2.35}$$

Let us now introduce an integral that has been defined and tabulated for this type of problem.†

$$\Phi(u, v) = e^u \int_0^u e^{-s} I_0(2\sqrt{vs})\, ds \tag{4.2.36}$$

This enjoys the following properties:

$$\Phi(u, v) = \sum_{0 \leq n < m} \frac{u^m v^n}{m!n!} \tag{4.2.37}$$

$$\Phi(u, v) + \Phi(v, u) + I_0(2\sqrt{uv}) = e^{u+v} \tag{4.2.38}$$

$$\Phi(0, v) = 0, \qquad \Phi(u, 0) = e^u - 1 \tag{4.2.39}$$

$$\lim_{u \to \infty} e^{-(u+v)}\, \Phi(u, v) = 1 \tag{4.2.40}$$

It is also useful to record here that

$$\Phi_u(u, v) = \Phi(u, v) + I_0(2\sqrt{uv}) \tag{4.2.41}$$

$$\Phi_v(u, v) = \Phi(u, v) - \sqrt{\frac{u}{v}}\, I_1(2\sqrt{uv}) \tag{4.2.42}$$

†The integral $\Phi(u, v)$ was tabulated by Brinkley and Brinkley (1947). It is essentially equivalent to the function

$$J(u, v) = 1 - e^{-v} \int_0^u e^{-s} I_0(2\sqrt{vs})\, ds$$

properties of which were fully explored by Goldstein (1953).

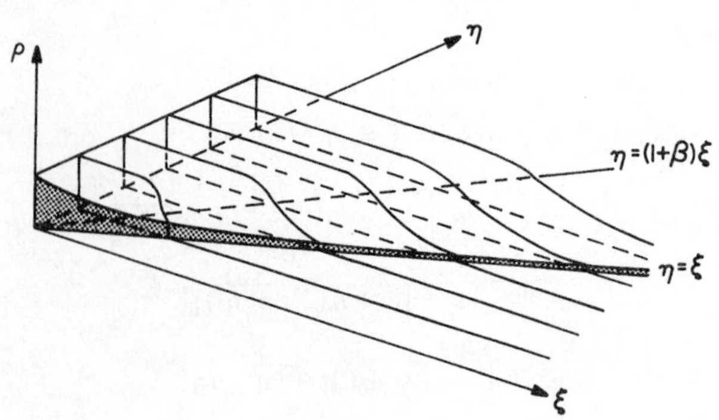

Figure 4.1

Then we see that $\beta\xi$ and $(\eta - \xi)$ play the role of v and u, respectively, and

$$\rho(\xi, \eta) = e^{-\beta\xi-(\eta-\xi)}[I_0(2\sqrt{\beta\xi(\eta - \xi)}) + \Phi(\eta - \xi, \beta\xi)]$$
$$= 1 - e^{-\beta\xi-(\eta-\xi)}\Phi(\beta\xi, \eta - \xi) \tag{4.2.43}$$

by use of Eq. (4.2.38).

Figure 4.1 shows the surface $\rho(\xi, \eta)$ with some of the sections by planes of constant η.

Such a figure can be constructed from tabulated values of Φ, but its general form is still far from evident. We might suspect that a "front" to the wave calculated in some average sense would move along the line $\eta = (1 + \beta)\xi$, since the total amount put on the column is $\varepsilon V c_0 t = (\varepsilon V c_0/k)\eta$, and the amount saturating the column in equilibrium with the concentration c_0 over a length z is $[\varepsilon c_0 + (1 - \varepsilon)\gamma c_0]z = (\varepsilon V c_0/k)(1 + \beta)\xi$. We note that the line $\eta = (1 + \beta)\xi$ is the characteristic C emanating from the origin [cf. with Eq. (3.3.6)]. We would also like it to be obvious that the front would get steeper as $k \to \infty$. All this, however, requires a rather sophisticated understanding of the integral Φ. The asymptotic behavior of $\Phi(u, v)$ is not simple, but the following formula, which is part of a much longer one due to Onsager, is useful:

$$\Phi(u, v) \sim \frac{1}{2}[1 - \operatorname{erf}(\sqrt{v} - \sqrt{u})]e^{u+v} - \frac{(u/v)^{1/4}}{1 + (u/v)^{1/4}} I_0(2\sqrt{uv}) + \cdots \tag{4.2.44}$$

We see that, when $\beta\xi$ and $\eta - \xi$ are both large, we can use the leading term and write

$$\rho(\xi, \eta) = \frac{1}{2}[1 + \operatorname{erf}(\sqrt{\eta - \xi} - \sqrt{\beta\xi})] \tag{4.2.45}$$

Now when $\beta\xi >> \eta - \xi$ (i.e., $\eta << (1 + \beta)\xi$), the argument of the error function is large and negative, so the error function itself is close to -1 and

ρ is very small. When $\eta >> (1 + \beta)\xi$, the argument of the error function is very large and ρ is close to 1. The "front" is thus located where the main change in ρ is taking place, in the neighborhood of $\eta = (1 + \beta)\xi$. Moreover, we see that as k gets larger, the region about the line $Vt = (1 + \beta)z$ in which $(\sqrt{\eta - \xi} - \sqrt{\beta\xi})$ is small continually contracts in size, so the front gets steeper, as we would expect.

For the elution of a column whose stationary phase is initially saturated at a concentration $h(z) = \gamma c_0$, we may take

$$\Phi(\eta) = \chi(\xi) = 0, \qquad \psi(\xi) = \beta$$

and then, from Eq. (4.2.32) or (4.2.33),

$$\rho(\xi, \eta) = \beta \int_0^{\zeta} e^{-\beta\xi' - (\eta - \xi)} I_0(2\sqrt{\beta\xi'(\eta - \xi)})\, d\xi'$$

where ζ is the lesser of ξ or η. Taking the case of $\eta > \xi$, we can make the substitutions $s = \beta\xi'$, $u = \beta\xi$, $v = \eta - \xi$, and obtain

$$\rho(\xi, \eta) = e^{-\beta\xi - (\eta - \xi)} \Phi(\beta\xi, \eta - \xi) \tag{4.2.46}$$

Finally, we may consider the injection of a short pulse of solute into a clean column. For this we suppose that

$$\chi = \psi = 0; \qquad \phi(\eta) = 0, \qquad \eta > \delta, \qquad \int_0^{\infty} \phi(\eta)\, d\eta = 1$$

where the last condition merely normalizes the amount injected in the interval $0 \le \eta \le \delta$. We need only consider the situation for $\eta > \xi > \delta$ for which Eq. (4.2.32) gives

$$\rho(\xi, \eta) = \phi(\eta - \xi)e^{-\beta\xi}$$

$$+ \int_{\eta - \xi - \delta}^{\eta - \xi} \phi(\eta - \xi - \eta'')e^{-\beta\xi - \eta''} \sqrt{\frac{\beta\xi}{\eta''}}\, I_1(2\sqrt{\beta\xi\eta''})\, d\eta'' = \phi(\eta - \xi)e^{-\beta\xi}$$

$$+ \int_0^{\delta} \phi(\eta')e^{-\beta\xi - (\eta - \xi) + \eta'} \sqrt{\frac{\beta\xi}{\eta - \xi - \eta'}}\, I_1(2\sqrt{\beta\xi(\eta - \xi - \eta')})\, d\eta'$$

The first term merely gives the pulse form moving down the column with the same speed as the carrier stream, but decaying exponentially as it does so. This part of the solution is never observed in practice. If we let $\delta \to 0$ and use the second mean-value theorem and the normalizing condition on the integral of ϕ, we have

$$\rho(\xi, \eta) = e^{-\beta\xi - (\eta - \xi)} \sqrt{\frac{\beta\xi}{\eta - \xi}}\, I_1(2\sqrt{\beta\xi(\eta - \xi)}) \tag{4.2.47}$$

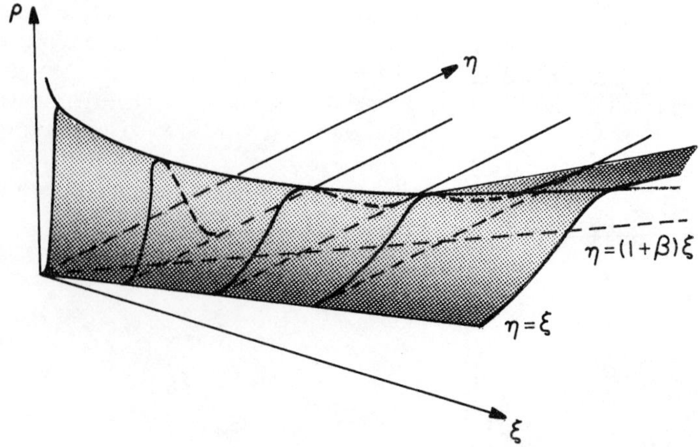

Figure 4.2

Now, when the argument of the Bessel function is large, the asymptotic expansion

$$I_1(x) \sim e^x(2\pi x)^{-1/2} \tag{4.2.48}$$

may be used to give

$$\rho(\xi, \eta) \sim \frac{1}{2\sqrt{\pi}}\left[\frac{\beta\xi}{(\eta - \xi)^3}\right]^{1/4} \exp[-(\sqrt{\eta - \xi} - \sqrt{\beta\xi})^2] \tag{4.2.49}$$

Here we see that the sharp pulse becomes peaked about the point where $\eta = (1 + \beta)\xi$, and we have the peak moving down the column with speed $V/(1 + \beta)$, as we should expect from Eqs. (3.3.6) and (3.3.9). We also see that as $k \to \infty$ the peak becomes sharper. Curves of ρ given by Eq. (4.2.47) are shown as the frame of a three-dimensional surface in Fig. 4.2.

Exercises

4.2.1. Show by termwise transformation of the series that

$$L[I_0(2\sqrt{at})] = p^{-1}e^{a/p}$$

$$L\left[\left(\frac{a}{t}\right)^{1/2} I_1(2\sqrt{at})\right] = e^{a/p} - 1$$

4.2.2. Obtain the concentration in the stationary phase from that in the mobile phase by inversion of Eq. (4.2.9). Use the solution for ρ given in Eq. (4.2.35).

4.2.3. Prove the properties of $\Phi(u, v)$ stated in Eqs. (4.2.37) to (4.2.42).

4.2.4. Goldstein (1953) uses the function

$$J(x, y) = 1 - e^{-y} \int_0^x e^{-\tau} I_0(2\sqrt{\tau y})\, d\tau$$

Find the corresponding properties for J and express the solution given by Eq. (4.2.43) in terms of this function.

4.2.5. Obtain the solution for a finite chromatogram; that is

$$f(z) = h(z) = 0$$

$$g(t) = \begin{cases} c_0, & 0 \leq t \leq T \\ 0, & t > T \end{cases}$$

4.3 Laplace Transformation as a Moment-generating Function

Although we have obtained the complete solution to the problem of linear chromatography, it was only possible to get real insight into the form of the solution by means of a far-from-obvious asymptotic expansion. When further details, such as diffusion in the mobile phase, are added to the linear model, the problem of inverting the Laplace transform becomes extremely difficult. There ought to be a way of getting some insight into the form of the solution without inverting the transform. We take our cue from the common practice in chromatography of injecting a small sample into the mobile phase at the entrance of the column during a short interval of time. Such a peak of solute moves down the column and spreads out, and the concentration is measured at some later point as a function of time. The peaklike form of the concentration wave suggests that we might describe it by means of its moments. The zeroth moment would measure the area under the curve, the first would give the mean residence time of the solute sample, and the second would give a measure of the variance of the residence times or the spread of the peak. The higher moments give a measure of the skewness, flatness, and so on, but in general these are not as important at the first two. If $c(z, t)$ denotes the concentration in the mobile phase at a point z as a function of t, we will denote the moments about the origin of t by

$$m_0 = m_0(z) = \int_0^\infty c(z, t)\, dt \tag{4.3.1}$$

$$m_1 = m_1(z) = \int_0^\infty t c(z, t)\, dt \tag{4.3.2}$$

$$m_2 = m_2(z) = \int_0^\infty t^2 c(z, t)\, dt \tag{4.3.3}$$

Then the mean residence time is

$$\mu = \frac{m_1}{m_0} \tag{4.3.4}$$

and the variance of residence times is

$$\sigma^2 = \left(\frac{m_2}{m_0}\right) - \left(\frac{m_1}{m_0}\right)^2 \tag{4.3.5}$$

Now the Laplace transform may be regarded as a moment-generating function, for if all the moments

$$M_n = \int_0^\infty t^n f(t)\, dt$$

of the function $f(t)$ exist, then

$$\begin{aligned} \bar{f}(p) &= \int_0^\infty e^{-pt} f(t)\, dt \\ &= \int_0^\infty \left[1 - pt + \frac{1}{2}p^2t^2 + \cdots + \frac{(-)^n}{n!}p^nt^n + \cdots\right] f(t)\, dt \qquad (4.3.6) \\ &= M_0 - M_1 p + \cdots + \frac{(-)^n}{n!} M_n p^n + \cdots \end{aligned}$$

The coefficients in this expansion can be evaluated by

$$M_n = (-)^n \bar{f}^{(n)}(0) \tag{4.3.7}$$

where $\bar{f}^{(n)}(p)$ is the nth derivative of $\bar{f}(p)$. This tells us that if the moments exist we can calculate them by means of the Laplace transform, but to proceed in the other direction and having found the Laplace transform to assert that the moments exist requires some condition on the Laplace transform.

If the Laplace transform exists for some value of the complex number p, then it is analytic in the half-plane to the right of the line through p parallel to the imaginary axis. In fact, if the Laplace transform converges in any half-plane Re $p > \sigma$, we refer to σ as the abscissa of convergence. It can be shown that in this half-plane

$$\int_0^\infty t^n e^{-pt} f(t)\, dt = (-)^n \bar{f}^{(n)}(p) \tag{4.3.8}$$

and it follows that if $\sigma < 0$ then $\bar{f}(p)$ is analytic at the origin, and we can calculate the moments by Eq. (4.3.7). In many important cases, however, $\sigma = 0$ and only a limited number of moments may exist, and more subtle conditions are required.

Let us illustrate this by the final-value theorem for the Laplace transform.

This asserts that if

$$\int_0^t f(t')dt' \to A, \qquad \text{as} \quad t \to \infty$$

then

$$\bar{f}(p) \to A, \qquad \text{as} \quad p \to 0+ \tag{4.3.9}$$

It is known as an *Abelian theorem*, being one of a class that states that a certain behavior of the transform follows from some feature of the behavior of the function. Converse theorems, in which the behavior of the function is deduced from that of the transform, are known as *Tauberian theorems* and usually involve supplementary conditions. Thus it is not true in general that if $\bar{f}(p) \to A$ as $p \to 0+$, then $\int_0^t f(t')\,dt' \to A$ as $t \to \infty$. It is true if the condition that $\bar{f}(p)$ is analytic at $p = 0$ is added, but this is a strong condition, and the interest and difficulty of Tauberian theorems lie in making such conditions as weak as possible. We will merely quote a theorem that is sufficient for most purposes.†

> THEOREM: If $\bar{f}(p)$ is analytic for Re $p > 0$ and if $(-)^n\bar{f}^{(n)}(p) \to A$ as $p \to 0+$, and if there exists a K such that $Kt + \int_0^t u^{n+1}f(u)\,du$ is a nondecreasing function in $0 < t < \infty$, then $M_n = A$.

Let us apply this to the chromatographic problem in which a sample specified by

$$c(0, t) = g(t) \tag{4.3.10}$$

is introduced to an empty column,

$$f(z) = h(z) = 0$$

Then, by Eqs. (4.2.11) and (4.2.12), we have

$$\bar{c}(z, p) = \bar{g}(p) \exp\left[\frac{-zP(p)}{V}\right] \tag{4.3.11}$$

$$P(p) = p\left(1 + \nu\gamma\frac{k}{p + k}\right) \tag{4.3.12}$$

We will suppose that $g(t)$ is a peaklike injection of a sample all of whose moments exist and write

$$m_n = \int_0^\infty t^n g(t)\,dt = (-)^n\bar{g}^{(n)}(0) \tag{4.3.13}$$

†It is an adaptation of the theorem proved by Widder (1946, p. 195).

Then clearly $\bar{g}(p)$ is analytic at the origin and so also is $\bar{c}(z, p)$; hence

$$m_n(z) = \int_0^\infty t^n c(z, t)\, dt = (-)^n \left[\frac{\partial^n}{\partial p^n}\bar{c}(z, p)\right]_{p=0} \tag{4.3.14}$$

In particular,

$$m_0(z) = \bar{c}(z, 0) = \bar{g}(0) = m_{00} \tag{4.3.15}$$

This just expresses the fact that the solute does not disappear, so the area under the peak wherever it is measured down the column is the same. It is convenient to normalize this to $m_0 = m_{00} = \bar{g}(0) = 1$ by a suitable change of concentration units.

For the first moment, we have

$$m_1(z) = -\left[\frac{\partial}{\partial p}\bar{c}(z, p)\right]_{p=0} = -\bar{g}'(0) + \frac{z}{V}P'(0)$$

or

$$m_1(z) - m_{10} = \frac{z}{V}(1 + \nu\gamma) = (1 + \beta)\frac{z}{V}$$

By taking the time $t = 0$ to coincide with m_{10}, the mean time of injection of the sample, we can make $m_{10} = 0$, and since $m_0 = 1$, we have

$$\mu = \frac{m_1(z)}{m_0} = (1 + \beta)\frac{z}{V} \tag{4.3.16}$$

Thus the mean residence of the solute in a column of length z is $(1 + \beta)z/V$, and we can speak of the peak as traveling with a speed $V/(1 + \beta)$. It is to be noticed that this only depends on the equilibrium constant γ and not on the rate of partition between the two phases, which is governed by the rate constant k. This agrees with what we were able to learn from the asymptotic expansion of the complete solution.

For the second moment, we have

$$m_2(z) = \left[\frac{\partial^2}{\partial p^2}\bar{c}(z,p)\right]_{p=0} = \bar{g}''(0) - \frac{z}{V}P''(0) - 2\frac{z}{V}P'(0)\bar{g}'(0) + \frac{z^2}{V^2}P'^2(0)$$

Since we have chosen the origin of time to make $m_{10} = -\bar{g}'(0) = 0$, and normalized the concentration to make $m_0 = m_{00} = 1$, the variance of the injected sample is

$$\sigma_0^2 = m_{20} = \bar{g}''(0)$$

This we cannot normalize in any way, but we may imagine, if the peak is sufficiently sharp, that σ_0^2 is very small. Then

$$\sigma^2 = m_2 - m_1^2 = \bar{g}''(0) - \frac{z}{V}P''(0) + \frac{z^2}{V^2}P'^2(0) - \frac{z^2}{V^2}P'^2(0)$$

or

$$\sigma^2 - \sigma_0^2 = -\frac{z}{V}P''(0) = 2\frac{\beta z}{kV} \tag{4.3.17}$$

Here, as we should expect, the variance is inversely proportional to the rate of partitioning between the phases, for the more rapidly equilibrium is established, the less chance the peak has to spread. It is a common feature of dispersing systems that the variance grows linearly with distance (see Exercise 4.3.1), and it allows us to define an effective dispersion coefficient for the system as

$$D_e = \frac{z^2(\sigma^2 - \sigma_0^2)}{2\mu^3} = \frac{\beta}{(1 + \beta)^3}\frac{V^2}{k} \tag{4.3.18}$$

This is the diffusion coefficient, which would give the same rate of growth of variance of a peak of solute moving with the same speed down an empty tube with no adsorption or partition taking place but only longitudinal diffusion imposed on plug flow.

It is clear that the Laplace transformation can be used to get a good physical feel for the form of the solution, and it may be used even when a complete inversion is out of the question. If we idealize the injection of the sample to make it a unit impulse (i.e., $\sigma_0 \to 0$), then $g(t) = \delta(t)$, $\bar{g}(p) = 1$, and

$$\bar{c}(z, p) = \exp\left[\frac{-zP(p)}{V}\right] = \exp\left[\left(-\mu p + \frac{1}{2}\sigma^2 p^2\right) + \tilde{\omega}(p)\right] \tag{4.3.19}$$

where $\tilde{\omega}(p)$ contains all the higher terms. Now from the standard tables

$$L\left[\frac{1}{\sigma\sqrt{2\pi}}e^{-(t-\mu)^2/2\sigma^2}\right] = \frac{1}{2}\operatorname{erf} c\left[\frac{\sigma}{\sqrt{2}}\left(p - \frac{\mu}{\sigma^2}\right)\right]e^{-\mu p + \sigma^2 p^2/2} \tag{4.3.20}$$

and if we consider this for long columns (i.e., σ and μ becoming large), then for $p < \mu/\sigma^2$, the complementary error function with negative argument rapidly approaches the value 2. Thus

$$L\left[c(z, t) - \frac{1}{\sigma\sqrt{2\pi}}e^{-(t-\mu)^2/2\sigma^2}\right]$$

$$= e^{-\mu p + (\sigma^2 p^2/2)}\left[e^{\tilde{\omega}(p)} - \frac{1}{2}\operatorname{erf} c\frac{\sigma}{\sqrt{2}}\left(p - \frac{\mu}{\sigma^2}\right)\right] \tag{4.3.21}$$

and the right side approaches zero as $p \to 0$ and $\mu/\sigma \to \infty$. We thus see that the peak tends to a Gaussian distribution of mean μ and variance σ.

Exercises

4.3.1. The diffusion of a solute in a stream of uniform speed V is governed by the equation

$$\frac{\partial c}{\partial t} = D\frac{\partial^2 c}{\partial x^2} - V\frac{\partial c}{\partial x}$$

If

$$c(0, t) = f(t), \qquad c(x, 0) = 0$$

and c remains bounded as $x \to \infty$, show that

$$\bar{c}(x, p) = \bar{f}(p) \exp[-xR(p)]$$

$$R(p) = -\frac{V}{2D} + \sqrt{\frac{V^2}{4D^2} + \frac{p}{D}}$$

Hence show that by appropriate normalization

$$\mu = \frac{x}{V}, \qquad \sigma^2 - \sigma_0^2 = \frac{2Dx}{V^3}$$

4.3.2. Lighthill and Whitham [*Proc. Roy. Soc. Lond.* **A229**, 281 (1955)] obtained a linearized equation for $\eta(x, t)$, the height of a flood wave at a point x at time t:

$$(gh_0 - v_0^2)\eta_{xx} - 2v_0\eta_{xt} - 2\lambda\eta_t - 3\lambda v_0 \eta_x = 0$$

subject to

$$\eta(0, t) = f(t), \qquad \eta(x, 0) = 0$$

In this equation, h_0 is the height of the undisturbed stream, v_0 the undisturbed stream speed, and $\lambda = gS_0/v_0$, where S_0 is the slope of the river bed. The quantity $F = v_0/(gh_0)^{1/2}$, the Froude number, is the remaining parameter of the system. Show that a humplike disturbance moves with a speed of $(3v_0/2)$ and that the variance of the hump that passes the point x is greater than that of $f(t)$ by

$$\frac{2x}{3gS_0} \frac{4 - F^2}{9F^2}$$

4.3.3. From the lexicon entry

$$L(e^{-t^2/4\alpha}) = \sqrt{\pi\alpha}\, e^{\alpha p^2} \operatorname{erf} c(p\sqrt{\alpha})$$

and the syntactical rules in Table 4.1, establish Eq. (4.3.20).

4.3.4. Find the skewness coefficient for the case of Exercise 4.3.1 when $\sigma_0 \to 0$. (The skewness coefficient is defined as the third moment about the mean divided by σ^3.)

4.4 Chromatography with a Langmuir Isotherm

There is a remarkable transformation discovered by Thomas that serves to linearize the equations for the chromatography of independent solutes even in the case where the isotherm is not linear. We again have the conservation equation,

$$\varepsilon V\frac{\partial c}{\partial z} + \varepsilon\frac{\partial c}{\partial t} + (1 - \varepsilon)\frac{\partial n}{\partial t} = 0 \tag{4.4.1}$$

but this time the partitioning is governed by the equation

$$\frac{\partial n}{\partial t} = k_1(N - n)c - k_2 n \tag{4.4.2}$$

We see that the equilibrium relation between c and n is

$$n = \frac{(k_1N/k_2)c}{1 + (k_1/k_2)c} = \frac{\gamma c}{1 + Kc} \tag{4.4.3}$$

where

$$\gamma = \frac{k_1N}{k_2}, \qquad K = \frac{k_1}{k_2} \tag{4.4.4}$$

[cf. Eqs. (1.2.6) and (1.2.7)]. When K is small or the concentration c is low, n is approximately γc, as in the previous case. The equations can be written

$$V\frac{\partial c}{\partial z} + \frac{\partial c}{\partial t} + \nu\frac{\partial n}{\partial t} = 0 \tag{4.4.5}$$

and

$$\frac{\partial n}{\partial t} = k_2[(\gamma - Kn)c - n] \tag{4.4.6}$$

giving as much in common as possible with the notation of the linear case, which could be obtained by setting $K = 0$ and dropping the suffix on k_2.

We will take the opportunity to introduce a transformation commonly encountered in the theory of chromatography,

$$x = \frac{z}{V}, \qquad y = \frac{1}{\nu}\left(t - \frac{z}{V}\right) \tag{4.4.7}$$

Since

$$V\frac{\partial}{\partial z} = \frac{\partial}{\partial x} - \frac{1}{\nu}\frac{\partial}{\partial y}, \qquad \frac{\partial}{\partial t} = \frac{1}{\nu}\frac{\partial}{\partial y}$$

we have

$$\frac{\partial c}{\partial x} + \frac{\partial n}{\partial y} = 0 \tag{4.4.8}$$

and

$$\frac{\partial n}{\partial y} = k[(\gamma - Kn)c - n] \tag{4.4.9}$$

where

$$k = \nu k_2$$

For a first boundary condition, we can specify the inlet concentration

$$c(0, t) = g(t)$$

so that in the new coordinates, where $t = \nu y$ at $z = 0$,

$$c(0, y) = g^*(y) = g(\nu y) \tag{4.4.10}$$

The second boundary condition is obtained by recognizing that the solute on the column at any point is unaffected until the front of the solvent just reaches that point (i.e., until y becomes positive). The idea here is that a certain distribution $n = h(z)$ has been laid down and the column drained of solvent. Then the front of the flow started at time $t = 0$ just reaches the point z at time z/V (i.e., when $y = 0$), and

$$c(x, 0) = h^*(x) = h(Vx) \tag{4.4.11}$$

Now the conservation equation, Eq. (4.4.8), may be satisfied identically if we set

$$n = \frac{\partial f}{\partial x}, \qquad c = -\frac{\partial f}{\partial y} \tag{4.4.12}$$

and look instead for the single function $f(x, y)$. Substituting in the equation for the rate of partitioning between the phases gives

$$\frac{\partial^2 f}{\partial y\, \partial x} + k\gamma\frac{\partial f}{\partial y} + k\frac{\partial f}{\partial x} - kK\frac{\partial f}{\partial y}\frac{\partial f}{\partial x} = 0 \tag{4.4.13}$$

This is a nonlinear equation, but it can be linearized by the substitution

$$f(x, y) = \frac{\gamma}{K}x + \frac{1}{K}y - \frac{1}{kK}\ln \phi(x, y) \tag{4.4.14}$$

for then

$$f_y = \frac{1}{K} - \frac{1}{kK}\frac{\phi_y}{\phi} \tag{4.4.15}$$

$$f_x = \frac{\gamma}{K} - \frac{1}{kK}\frac{\phi_x}{\phi} \tag{4.4.16}$$

$$f_{yx} = \frac{1}{kK}\frac{\phi_y\phi_x}{\phi^2} - \frac{1}{kK}\frac{\phi_{yx}}{\phi} \tag{4.4.17}$$

and so Eq. (4.4.13) yields

$$\frac{\partial^2\phi}{\partial y\, \partial x} = k^2\gamma\phi \tag{4.4.18}$$

From Eqs. (4.4.15), (4.4.12), and (4.4.10), we have the boundary condition

$$\frac{\phi_y(0, y)}{\phi(0, y)} = kKg^*(y) + k \tag{4.4.19}$$

and from Eqs. (4.4.16), (4.4.12), and (4.4.11),

$$\frac{\phi_x(x, 0)}{\phi(x, 0)} = \gamma k - kKh^*(x) \tag{4.4.20}$$

These are both linear equations that can be solved to give

$$\phi(0, y) = \exp\left[ky + kK\int_0^y g^*(y')\,dy'\right] = G(y) \tag{4.4.21}$$

$$\phi(x, 0) = \exp\left[\gamma kx - kK\int_0^x h^*(x')\,dx'\right] = H(x) \tag{4.4.22}$$

Since ϕ appears in a logarithm to give f, which we then differentiate to give c and n, it does not matter what value is given to $\phi(0, 0)$, and so we make it 1.

Applying the Laplace transform to Eqs. (4.4.18) and (4.4.21) with

$$\overline{\phi} = \overline{\phi}(x, p) = \int_0^\infty \exp(-py)\phi(x, y)\,dy \tag{4.4.23}$$

we have

$$p\frac{d\overline{\phi}}{dx} = \gamma k^2\overline{\phi} + H'(x) \tag{4.4.24}$$

and

$$\overline{\phi}(0, p) = \overline{G}(p) \tag{4.4.25}$$

Hence

$$\phi(x, p) = \overline{G}(p)\exp\left(\frac{\gamma k^2 x}{p}\right) + \frac{1}{p}\int_0^x H'(x')\exp\left[\frac{\gamma k^2(x - x')}{p}\right]dx' \tag{4.4.26}$$

and

$$\phi(x, y) = \int_0^y G(y - y')F_1(x, y')\,dy' + \int_0^x H'(x - x')\,F_2(x', y)\,dx' \tag{4.4.27}$$

where

$$L[F_1(x, y)] = \exp\left(\frac{\gamma k^2 x}{p}\right) \tag{4.4.28}$$

and

$$L[F_2(x, y)] = \frac{1}{p}\exp\left(\frac{\gamma k^2 x}{p}\right) \tag{4.4.29}$$

But by a rather simpler analysis than the one we did in the linear case, we can find F_1 and F_2 from the transform given in Eq. (4.4.26):

$$F_2(x, y) = I_0(2k\sqrt{\gamma y x}) \tag{4.4.30}$$

and

$$F_1(x, y) = k\sqrt{\frac{\gamma x}{y}}I_1\,(2k\sqrt{\gamma y x}) + \delta(y) \tag{4.4.31}$$

Hence

$$\phi(x, y) = G(y) + \int_0^y G(y - y')k\sqrt{\frac{\gamma x}{y'}}I_1\,(2k\sqrt{\gamma y' x})\,dy' + \int_0^x H'(x - x')I_0(2k\sqrt{\gamma y x'})\,dx'$$

or integrating the first integral by parts and using the fact that $G(0) = 1$,

$$\phi(x, y) = I_0(2k\sqrt{\gamma y x}) + \int_0^y G'(y - y')I_0(2k\sqrt{\gamma y' x})\,dy' + \int_0^x H'\,(x - x')I_0(2k\sqrt{\gamma y x'})\,dx' \tag{4.4.32}$$

Let us consider the two important cases of the saturation of an empty column and the elution of a saturated one. In the former case,

$$g^*(y) = c_0, \qquad h^*(x) = 0$$

and so, by Eqs. (4.4.21) and (4.4.22),

$$G(y) = \exp[k(1 + Kc_0)y], \qquad H(x) = \exp(\gamma k x) \tag{4.4.33}$$

Then Eq. (4.4.32) gives

$$\phi(x, y) = I_0(2k\sqrt{\gamma y x}) + k(1 + Kc_0)\exp[\mathrm{k}(1 + kc_0)y] \cdot \int_0^y \exp[-k(1 + Kc_0)y']I_0(2k\sqrt{\gamma x y'})\,dy' + \gamma k\exp(\gamma k x)\int_0^x \exp(-\gamma k x')I_0(2k\sqrt{\gamma y x'})\,dx' \tag{4.4.34}$$

$$= I_0(2k\sqrt{\gamma y x}) + \Phi\left(k(1 + kc_0)y, \frac{\gamma k x}{1 + Kc_0}\right)$$

$$+ \phi(\gamma k x, ky)$$

and substituting back in Eqs. (4.4.12), (4.4.15), and (4.4.16) gives

$$c(x, y) = c_0 \frac{I_0(2k\sqrt{\gamma y x}) + \Phi(k(1 + Kc_0)y, \gamma k x/(1 + Kc_0))}{I_0(2k\sqrt{\gamma y x}) + \Phi(k(1 + Kc_0)y, \gamma k x/(1 + Kc_0)) + \phi(\gamma k x, ky)} \tag{4.4.35}$$

and

$$n(x,y) = \frac{c_0}{1 + Kc_0} \cdot \frac{\Phi(k(1 + Kc_0)y, \gamma k x/(1 + Kc_0))}{I_0(2k\sqrt{\gamma y x}) + \Phi(k(1 + kc_0)y, \gamma k x/(1 + Kc_0)) + \Phi(\gamma k x, ky)} \tag{4.4.36}$$

Similarly, if the column is initially saturated with

$$g^*(y) = 0, \qquad h^*(x) = n_0 = \frac{\gamma c_0}{1 + Kc_0}$$

then

$$c(x, y) = c_0 \frac{\Phi(\gamma k x/(1 + Kc_0), k(1 + Kc_0)y)}{I_0(2k\sqrt{\gamma y x}) + \Phi(\gamma k x/(1 + Kc_0), k(1 + Kc_0)y) + \Phi(ky, \gamma k x)} \tag{4.4.37}$$

and

$$n(x,y) = \frac{\gamma c_0}{1 + Kc_0} \cdot \frac{I_0(2k\sqrt{\gamma y x}) + \Phi(\gamma k x/(1 + Kc_0), k(1 + Kc_0)y)}{I_0(2k\sqrt{\gamma y x}) + \Phi(\gamma k x/(1 + Kc_0), k(1 + Kc_0)y) + \Phi(ky, \gamma k x)} \tag{4.4.38}$$

Goldstein and Murray, among others, presented the most comprehensive studies of essentially the same problem by applying Heaviside's operational methods. (See H. Jeffreys, *Operational Methods in Mathematical Physics*, 2nd ed., Cambridge University Press, New York, 1931.) The problem was solved not only subject to constant initial and entry conditions but for a finite chromatogram (see Exercise 4.4.4) as well as for general entry conditions. The limits of these solutions were examined and proven to be identical to the corresponding solutions of the equilibrium model. The latter will be treated in the next two chapters.

Exercises

4.4.1. Establish Eqs. (4.4.37) and (4.4.38).

4.4.2. Show that the solutions for the linear case can be obtained from Eqs. (4.4.35) to (4.4.38).

4.4.3. Express the solution given by Eqs. (4.4.35) and (4.4.36) in terms of the function $J(x, y)$ (cf. Exercise 4.2.4).

4.4.4. Determine the solution subject to the conditions

$$\text{at } t = 0, \qquad c = n = 0$$

$$\text{at } z = 0, \qquad c = g(t) = \begin{cases} c_0, & 0 \leqslant t \leqslant T \\ 0, & t > T \end{cases}$$

4.5 Fixed-bed Adsorption with Recycle

In certain processing operations, a stream is taken from a process unit, fed to an adsorption column, and then recycled back to the original unit. As an example we would like to consider the simple model formulated by Cooney and Shieh (1972). Here a continuous stirred tank reactor is used for a first-order, reversible reaction $A \rightleftharpoons B$, and the exit stream is treated through an adsorbent bed to remove the desired product B. Unreacted component A is then recycled back to the reactor (see Fig. 4.3). Since the recycle stream remains low in B, the reverse reaction is suppressed and very high conversion of A can be obtained. We will assume that the reactor has a much larger capacity than the adsorber and component A is present in large excess so that the total concentration of A and B in the reactor remains nearly constant.†

For the adsorbent bed, we have

$$V\frac{\partial c}{\partial z} + \frac{\partial c}{\partial t} + \nu\frac{\partial n}{\partial t} = 0 \tag{4.5.1}$$

in which c and n denote the concentrations of B in the mobile and stationary phases, respectively, and ν is defined by Eq. (4.2.8). The partition between the two phases is given by

$$\frac{\partial n}{\partial t} = k(\gamma c - n) \tag{4.5.2}$$

A mass balance for B in the reactor gives

$$\theta\frac{dc_p}{dt} = c_R - c_P + \theta[k_r(c_T - c_P) - k'_r c_P] \tag{4.5.3}$$

†Cooney and Shieh (1972) suggested that the reaction rate might be assumed to be constant because the concentration of A would be nearly constant, but this is just a special case of the above.

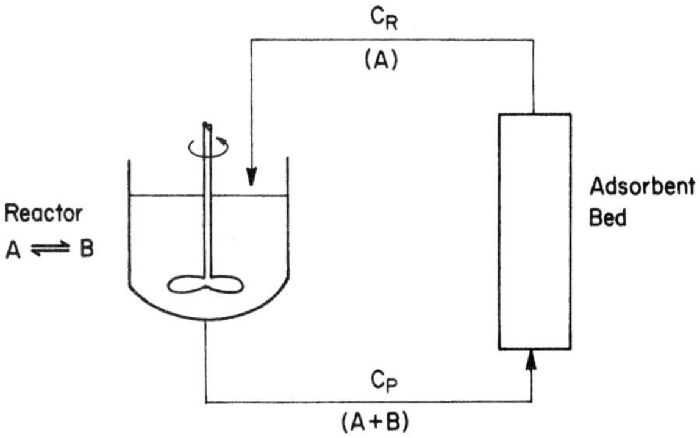

Figure 4.3

where c_P = concentration of B in the product stream from the reactor
c_R = concentration of B in the recycle stream
c_T = constant total concentration in the reactor
θ = mean residence time of the reactor
k_r, k_r' = rate constants of the forward and reverse reactions, respectively

Appropriate initial and boundary conditions are

$$
\begin{aligned}
c(z, 0) &= 0, \qquad n(z, 0) = n_0 = \gamma c_0 \\
c(0, t) &= c_P(t) \\
c(0, 0) &= c_P(0) = c_{P0}
\end{aligned}
\tag{4.5.4}
$$

The first pair of conditions implies that the bed is presaturated uniformly and drained of solvent.

We will introduce a coordinate transformation similar to those in Eq. (4.4.7) but with somewhat different scale factors:

$$
x = \frac{\nu\gamma k}{V}z, \qquad y = k\left(t - \frac{z}{V}\right) \tag{4.5.5}
$$

These new variables are now dimensionless. Equations (4.5.1) to (4.5.3) are then reduced to

$$
\gamma\frac{\partial c}{\partial x} + \frac{\partial n}{\partial y} = 0 \tag{4.5.6}
$$

$$
\frac{\partial n}{\partial y} = \gamma c - n \tag{4.5.7}
$$

$$
a\frac{dc_P}{dy} = c_R - bc_P + r \tag{4.5.8}
$$

where

$$a = \theta k, \qquad b = 1 + \theta(k_r + k_r'), \qquad r = \theta k_r c_T \tag{4.5.9}$$

The initial and boundary conditions become

$$\begin{aligned} n(x, 0) &= n_0 = \gamma c_0 \\ c(0, y) &= c_P(y) \\ c(0, 0) &= c_P(0) = c_{P0} \end{aligned} \tag{4.5.10}$$

Denoting the Laplace transform with respect to the variable y by a bar over the function, we have

$$\gamma \frac{d\bar{c}}{dx} + p\bar{n} = n_0 \tag{4.5.11}$$

$$(p + 1)\bar{n} - \gamma\bar{c} = n_0 \tag{4.5.12}$$

and

$$(ap + b)\,\bar{c}_P - \bar{c}_R = ac_{P0} + \frac{r}{p} \tag{4.5.13}$$

with

$$\bar{c}(0, p) = \bar{c}_P(p) \tag{4.5.14}$$

Substituting Eq. (4.5.12) in Eq. (4.5.11), we have

$$\frac{d\bar{c}}{dx} + \frac{p}{p + 1}\bar{c} = \frac{c_0}{p + 1} \tag{4.5.15}$$

which is solved subject to Eq. (4.5.14) to give

$$\bar{c}(x, p) = \bar{c}_P(p) \exp\left(-\frac{p}{p + 1}x\right) + \frac{c_0}{p}\left[1 - \exp\left(-\frac{p}{p + 1}x\right)\right] \tag{4.5.16}$$

For a bed of length L, we put

$$X = \frac{\nu\gamma kL}{V} \tag{4.5.17}$$

and observe that $\bar{c}(X, p) = \bar{c}_R$. Thus, using Eqs. (4.5.13) and (4.5.16), we have

$$\bar{c}_P(p) = \frac{c_0}{p} + \frac{Q + R/p}{ap + b - \exp\left[-\dfrac{pX}{p + 1}\right]} \tag{4.5.18}$$

where $\qquad Q = a(c_{p0} - c_0), \qquad R = r + (1 - b)c_0 \qquad$ (4.5.19)

Consequently, we have, from Eqs. (4.5.16) and (4.5.18),

$$\bar{c}(x, p) = \frac{c_0}{p} + \left(Q + \frac{R}{p}\right) \frac{\exp\left[-\dfrac{px}{p+1}\right]}{ap + b - \exp\left[-\dfrac{pX}{p+1}\right]}$$

or, by expanding the last factor in a series form,

$$\bar{c}(x,p) = \frac{c_0}{p} + \left(Q + \frac{R}{p}\right) \sum_{j=0}^{\infty} \left(\frac{1}{ap+b}\right)^{j+1} \exp\left[\frac{-p}{p+1}(jX + x)\right] \tag{4.5.20}$$

The inversion of this equation gives the solution not only for $c(x, y)$, $0 < x < X$, but for $c_P(y)$ at $x = 0$ and $c_R(y)$ at $x = X$.

Since the exponential factors in Eq. (4.5.20) can be written as $\exp[-(jX + x)] \cdot \exp[(jX + x)/(p + 1)]$, it is convenient to put

$$\bar{g}_j(x, p) = \exp[-(jX + x)] \left(\frac{1}{ap + b}\right)^{j+1} \exp\left(\frac{jX + x}{p + 1}\right) \tag{4.5.21}$$

and

$$\bar{G}(x,p) = \sum_{j=0}^{\infty} \bar{g}_j(x, p) \tag{4.5.22}$$

The inversion of $\bar{g}_j(x, p)$ can be obtained by applying the convolution theorem, Eq. (4.1.24), together with Eq. (4.2.29) and the third rule given in Table 4.1 (see Sec. 4.2). Thus we have

$$g_j(x, y) = \frac{\exp[-(jX + x)]}{j!a^{j+1}} \int_0^y (y - \eta)^j \cdot \exp\left[-\frac{b}{a}(y - \eta) - \eta\right] I_0(2\sqrt{(jX + x)\eta})\, d\eta \tag{4.5.23}$$

and

$$G(x, y) = \sum_{j=0}^{\infty} g_j(x, y) \tag{4.5.24}$$

Returning to Eq. (4.5.20), we observe

$$\bar{c}(x, p) = \frac{c_0}{p} + \left(Q + \frac{R}{p}\right)(p + 1)\bar{G}(x, p)$$

$$= \frac{c_0}{p} + Q(\bar{G} + p\bar{G}) + R\left(\bar{G} + \frac{\bar{G}}{p}\right) \tag{4.5.25}$$

and hence

$$c(x, y) = c_0 + Q\left[G(x, y) + \frac{\partial}{\partial y} G(x, y)\right] + R\left[G(x, y) + \int_0^y G(x, y')\,dy'\right] \tag{4.5.26}$$

This is the solution in series form. Under certain circumstances (e.g., if a is large and y/X is small) the series should converge very rapidly so that only the first term of the series may be required. Then

$$G(x, y) = g_0(x, y) = \frac{1}{a} e^{-x} \int_0^y \exp\left[-\frac{b}{a}y - \left(1 - \frac{b}{a}\right)\eta\right] I_0(2\sqrt{x\eta})\,d\eta$$

$$= \frac{1}{a\beta} \exp[-(x + y)] \exp(\beta y) \tag{4.5.27}$$

$$\int_0^{\beta y} \exp(-\beta\eta) I_0\left(2\sqrt{\frac{x}{\beta} \cdot \beta\eta}\right) d(\beta\eta)$$

$$= \frac{1}{a - b} \exp[-(x + y)]\, \Phi\left(\beta y, \frac{x}{\beta}\right)$$

where

$$\beta = 1 - \frac{b}{a} = 1 - \frac{1}{\theta k} - \frac{k_r + k_r'}{k} \tag{4.5.28}$$

and the function Φ is defined by Eq. (4.2.36). The derivative $\partial G/\partial y$ is determined by using Eqs. (4.2.38) and (4.2.41):

$$\frac{\partial G}{\partial y} = \frac{\partial g_0}{\partial y} = \frac{1}{a - b} \exp[-(x + y)] \left\{-\Phi\left(\beta y, \frac{x}{\beta}\right) + \frac{\partial \phi}{\partial y}\right\}$$

$$= -G(x, y) + \frac{\beta}{a - b} \tag{4.5.29}$$

$$\cdot \exp[-(x + y)] \left\{\exp\left(\beta y + \frac{x}{\beta}\right) - \Phi\left(\frac{x}{\beta}, \beta y\right)\right\}$$

For the integral term, we will switch the order of integration to obtain

$$\int_0^y G(x, y')\,dy' = \frac{\exp(-x)}{a - b} \int_0^y \exp(-y') \exp(\beta y')$$

$$\int_0^{\beta y'} \exp(-\beta\eta) I_0(2\sqrt{x\eta})\,d(\beta\eta)\,dy'$$

$$= \frac{\beta}{a - b} \exp(-x) \int_0^y \exp(-\beta\eta) \tag{4.5.30}$$

$$\cdot I_0 (2\sqrt{x\eta}) \int_\eta^y \exp[(\beta - 1)y'] \, dy' \, d\eta$$

$$= \frac{1}{a - b} \exp[-(x + y)]$$

$$\cdot \left\{ \frac{1}{\beta - 1} \Phi\left(\beta y, \frac{x}{\beta}\right) - \frac{\beta}{\beta - 1} \Phi(y, x) \right\}$$

Substituting these back into Eq. (4.5.26), we have

$$c(x, y) = c_0 + \frac{Q}{a} \exp[-(x + y)] \left\{ \exp\left(\frac{x}{\beta} + \beta y\right) \right.$$

$$\left. - \Phi\left(\frac{x}{\beta}, \beta y\right) \right\} + \frac{R}{b} \exp[-(x + y)] \left\{ \Phi(y, x) - \Phi\left(\beta y, \frac{x}{\beta}\right) \right\} \tag{4.5.31}$$

For $x = 0$ and $c_0 = 0$, we apply Eq. (4.2.39) in Eq. (4.5.31) to obtain

$$c(0, y) = c_{P0} \exp\left(-\frac{b}{a} y\right) + \frac{r}{b} \left\{ 1 - \exp\left(-\frac{b}{a} y\right) \right\} \tag{4.5.32}$$

This equation is identical to the solution of Eq. (4.5.8) with $c_R = 0$, so Eq. (4.5.31) may be valid up to the time of initial adsorbent bed breakthrough.

If we had assumed the reaction rate to be constant in Eq. (4.5.3), we would have had $b = 1$ and $R = r =$ the reaction rate, while others remain the same. Thus Eqs. (4.5.26) and (4.5.31) are still valid with $b = 1$ if we interpret R as the constant reaction rate.

Exercises

4.5.1. The concentration of B in the stationary phase can be obtained by using Eqs. (4.5.12) and (4.5.25). Show that

$$n(x, y) = n_0 + \gamma Q G(x, y) + \gamma R \int_0^y G(x, y') \, dy'$$

4.5.2. Apply the final-value theorem to $\bar{c}(x, p)$ and show that, in the limit as $y \to \infty$, the concentration $c(x, y)$ approaches to the equilibrium concentration of B in the reactor.

4.5.3. Express Eq. (4.5.31) in terms of the function $J(x, y)$ and use its properties, $J(x, 0) = \exp(-x)$ and $J(0, y) = 1$, to obtain Eq. (4.5.32).

4.5.4. Chang and others (1973) use the function

$$\Psi_j(u, v) = \exp(-v) \sum_{m=0}^{\infty} A_{jm}(u) v^m / (m!)^2$$

where the coefficient A's are calculated from the relations

$$A_{0m} = \exp(-u)u^m$$

$$A_{j0} = \frac{u^{j-1}}{(j-1)!} - A_{j-1,0}$$

$$A_{jm} = mA_{j,m-1} - A_{j-1,m}$$

Integrate Eq. (4.5.23) by parts and express the solution given by Eq. (4.5.26) in terms of the function Ψ_j.

4.6 Poisoning in Fixed-bed Reactors

One model for the poisoning of the catalyst particles in fixed beds considers that the poison is a strongly adsorbed foreign substance that completely blocks off the reaction sites on the catalyst surface. The outer layers of each catalyst particle, which we will take to be spherical, therefore form a poisoned mantle through which the reactants must diffuse before they can react within the unpoisoned core. Figure 4.4 shows the situation for a typical particle, the extent of the poisoning depending of course on the position of the particle in the bed and on the time. We make the pseudo-steady-state assumption that the core shrinks so slowly that at any time the concentration distribution through the mantle is governed by the steady-state diffusion equation. Thus, if $c(r)$ denotes the concentration of poison at radius r,

$$\frac{D}{r^2}\frac{\partial}{\partial r}\left(r^2\frac{\partial c}{\partial r}\right) = 0, \qquad a \geq r \geq b \tag{4.6.1}$$

The general solution of this equation is

$$c(r) = A + \frac{B}{r}$$

so in the notation of the figure

$$c_a = A + \frac{B}{a}, \qquad D\left(\frac{dc}{dr}\right)_s = -\frac{DB}{a^2}$$

and

$$c_b = A + \frac{B}{b}, \qquad D\left(\frac{dc}{dr}\right)_0 = -\frac{DB}{b^2}$$

Now at the surface of the particle the flux $D(dc/dr)_s$ into the particle is supplied by the transport across the boundary layer. If k_g is the mass-transfer coefficient for this,

$$D\left(\frac{dc}{dr}\right)_a = k_g(c_f - c_a)$$

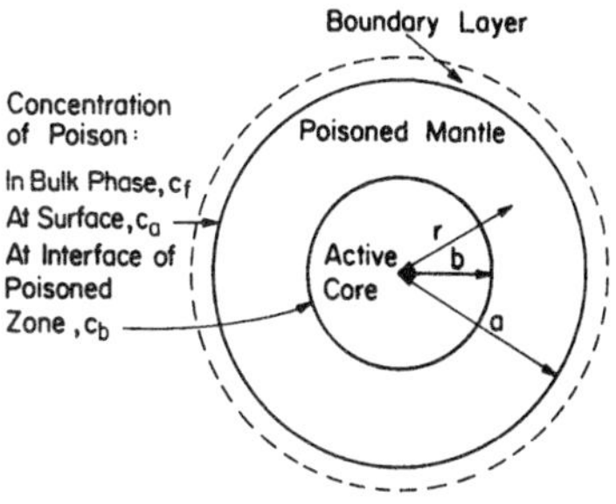

Figure 4.4

or

$$A + \frac{B}{a}\left(1 - \frac{D}{k_g a}\right) = c_f \tag{4.6.2}$$

Similarly, at the interface of the poisoned mantle and unpoisoned core, the flux of poison is equal to its rate of adsorption, or

$$D\left(\frac{dc}{dr}\right)_b = k_a c_b$$

Thus

$$A + \frac{B}{b}\left(1 + \frac{D}{k_a b}\right) = 0 \tag{4.6.3}$$

Solving for A and B and substituting back, we have that the diffusion rate of poison is

$$f = 4\pi b^2 D\left(\frac{dc}{dr}\right)_b = 4\pi b^2 k_a c_b = 4\pi b^2 k_a c_f \frac{\delta}{\delta + \rho(1 - \rho) + \beta\rho^2} \tag{4.6.4}$$

where $$\rho = \frac{b}{a}, \qquad \delta = \frac{D}{k_a a}, \qquad \beta = \frac{D}{k_g a}$$

The parameter β is a reciprocal Biot number for external mass transfer, and δ is a Damköhler number for adsorption.

This diffusion rate of poison is proportional to the rate at which the volume of the mantle grows. Thus

$$-4\pi b^2 \frac{db}{dt} = \lambda f$$

where λ is the reciprocal of the saturation concentration of the poison. Using Eq. (4.6.4) and the dimensionless variable $\rho = b/a$,

$$\frac{d\rho}{dt} = -\frac{\lambda k_a c_f}{a} \frac{\delta}{\delta + \rho(1 - \rho) + \beta\rho^2} \tag{4.6.5}$$

During the period from $t = 0$ to $t = t$, the total amount of poison that has passed into the mantle is

$$\int_0^t f\,dt = \int_b^a \frac{4\pi}{\lambda\rho} b^2\,db = \frac{4\pi a^3}{3\lambda}(1 - \rho^3)$$

The average concentration of poison per unit volume of pellet is thus $n = (1 - \rho^3)/\lambda$ and increases at a rate

$$\frac{dn}{dt} = -\frac{3\rho^2}{\lambda}\frac{d\rho}{dt} = \frac{3k_a c_f}{a}\frac{\delta\rho^2}{\delta + \rho(1 - \rho) + \beta\rho^2} \tag{4.6.6}$$

We now recognize that a given particle is at position z in the bed and that the mean concentration in it is therefore $n(z, t)$. Similarly, for that particle, c_f is $c(z, t)$, the concentration in the flowing stream at that position and time. Thus, for the poison in the bed, we have the chromatographic equation

$$V\frac{\partial c}{\partial z} + \frac{\partial c}{\partial t} + \nu\frac{\partial n}{\partial t} = 0 \tag{4.6.7}$$

If we take the independent variables

$$y = \alpha\left(t - \frac{z}{V}\right), \qquad x = \alpha' z \tag{4.6.8}$$

where α and α' have yet to be chosen, Eqs. (4.6.6) and (4.6.7) can be written

$$\alpha' V\frac{\partial c}{\partial x} + \nu\alpha\frac{\partial n}{\partial y} = 0 \tag{4.6.9}$$

and

$$\alpha\frac{\partial n}{\partial y} = \frac{3k_a}{a}\frac{\delta\rho^2}{\delta + \rho(1 - \rho) + \beta\rho^2}c \tag{4.6.10}$$

Now $\lambda n = 1 - \rho^3$ is a dimensionless form of the concentration in the pellet, so we should take

$$v(x, y) \equiv v = \lambda n = 1 - \rho^3 \tag{4.6.11}$$

and retain ρ as a convenient abbreviation for $(1 - v)^{1/3}$. If $c_0 u_0(y)$ is the inlet concentration in the mobile phase, we can also take

$$u(x, y) \equiv u = \frac{c}{c_0} \tag{4.6.12}$$

so Eq. (4.6.10) becomes

$$\alpha\frac{\partial v}{\partial y} = \frac{3k_a\lambda c_0}{a}\frac{\delta\rho^2 u}{\delta + \rho(1 - \rho) + \beta\rho^2}$$

and we have only to choose

$$\alpha = \frac{3k_a\lambda c_0}{a} \tag{4.6.13}$$

to obtain the form

$$\frac{\partial v}{\partial y} = \frac{\delta\rho^2 u}{\delta + \rho(1 - \rho) + \beta\rho^2} \tag{4.6.14}$$

But then by taking

$$\alpha' = \alpha\nu\frac{1}{\lambda c_0 V} = \frac{3\nu k_a}{aV} \tag{4.6.15}$$

we have

$$\frac{\partial u}{\partial x} + \frac{\partial v}{\partial y} = 0 \tag{4.6.16}$$

Equations (4.6.14) and (4.6.16) therefore give a pair of equations for the progressive poisoning of the bed. They are subject to the boundary conditions

$$u(x, 0) = v(x, 0) = 0 \tag{4.6.17}$$

$$u(0, y) = u_0(y) \tag{4.6.18}$$

and when the concentration in the feed is constant,

$$u_0(y) = 1 \tag{4.6.19}$$

When a first-order reaction is going on in the unpoisoned core of the pellet independently of the poisoning, its rate may be found by means of the effectiveness factor. Thus the rate of reaction in the pellet is

$$\frac{4}{3}\eta\pi a^3 k_r c_r(z, t)$$

where $c_r(z, t)$ is the concentration of the reactant disappearing by this first-order process, whose rate constant is k_r, and η is the effectiveness factor. For a spherical pellet poisoned to a fractional radius ρ, the effectiveness factor is

$$\eta = 3\phi^{-2}\left[\beta_r + \frac{1 - \rho}{\rho} + \frac{1}{\rho(\rho\phi \coth \rho\phi - 1)}\right]^{-1} \tag{4.6.20}$$

where ϕ = Thiele modulus = $A(k_r/D_r)^{1/2}$
β_r = $D_r/k_{gr}a$
D_r = diffusion coefficient of reactant in the pellet

For the reactant, we have the equation

$$\frac{\partial c_r}{\partial t} + V\frac{\partial c_r}{\partial z} = -\eta \nu k_r c_r \tag{4.6.21}$$

or

$$\frac{\partial c_r}{\partial x} = -\eta(x, y)\kappa c_r \tag{4.6.22}$$

where

$$\kappa = \frac{ak_r}{3k_a} \tag{4.6.23}$$

When $\eta \equiv 1$, we have

$$c_r = c_{r0}e^{-\kappa x}$$

but in general we have

$$c_r = c_{r0}\, exp\left[-\kappa \int_0^x \eta(x', y)\, dx'\right] \tag{4.6.24}$$

Thus

$$\frac{1}{x}\int_0^x \eta(x', y)\, dx' = A(x, y) \tag{4.6.25}$$

gives a measure of the total activity of the bed.

Now the Eqs. (4.6.14) and (4.6.16) are of the form

$$\frac{\partial u}{\partial x} = -g(v)u, \qquad \frac{\partial v}{\partial y} = g(v)u \tag{4.6.26}$$

and Bischoff (1969) has shown how these may be integrated by quadratures. For the boundary conditions

$$u(0, y) = u_0(y), \qquad v(x, 0) = v_0 \tag{4.6.27}$$

he sets

$$u(x, y) = u_0(y)\frac{v(x, y) - v_0}{v(0, y) - v_0} \tag{4.6.28}$$

So, from the second equation of (4.6.26),

$$\frac{\partial v}{\partial y} = g(v)u_0(y)\frac{v - v_0}{v(0, y) - v_0} \tag{4.6.29}$$

But at $x = 0$

$$\frac{d}{dy}v(0, y) = g[v(0, y)]u_0(y)$$

whence

$$\int_0^y u_0(y')\,dy' = \int_{v_0}^{v(0,y)} \frac{dw}{g(w)} \tag{4.6.30}$$

This gives an implicit relation between y and $v(0, y)$. From Eq. (4.6.28),

$$\frac{\partial u}{\partial x} = \frac{u_0(y)}{v(0, y) - v_0} \frac{\partial v}{\partial x}$$

and, by Eqs. (4.6.26) and (4.6.29), this gives

$$\frac{u_0(y)}{v(0, y) - v_0} \frac{\partial v}{\partial x} = -g(v)u_0(y)\frac{v - v_0}{v(0, y) - v_0}$$

or

$$\frac{1}{g(v)(v - v_0)} \frac{\partial v}{\partial x} = -1$$

Thus

$$x = -\int_{v(0,y)}^{v(x,y)} \frac{dw}{g(w)(w - v_0)} \tag{4.6.31}$$

and $v(0, y)$ is given by Eq. (4.6.30). This defines $v(x, y)$, and so also $u(x, y)$ by Eq. (4.6.28), implicitly in terms of x and y.

A fairly simple example is obtained when

$$g(v) = (1 + \alpha v)^{-1} \tag{4.6.32}$$

and $u_0(y) = 1$, $v_0 = 0$. Then Eq. (4.6.30) gives

$$y = \int_0^{v(0,y)} (1 + \alpha w)dw = v(0, y) + \tfrac{1}{2}\alpha[v(0, y)]^2$$

or

$$v(0, y) = \frac{1}{\alpha}(-1 + \sqrt{1 + 2\,\alpha y}) \tag{4.6.33}$$

Equation (4.6.31) gives

$$-x = \int_{v(0,y)}^{v(x,y)} \frac{1 + \alpha w}{w}\,dw = \ln\frac{v(x, y)}{v(0, y)} + \alpha[v(x, y) - v(0, y)] \tag{4.6.34}$$

Eliminating $v(0, y)$ from Eqs. (4.6.33) and (4.6.34) and from the relation (4.6.28) for this case, we have

$$\alpha v(x, y) = u(x, y)(-1 + \sqrt{1 + 2\alpha y}) \tag{4.6.35}$$

and

$$-x = \ln u(x, y) + (-1 + \sqrt{1 + 2\alpha y})[u(x, y) - 1] \tag{4.6.36}$$

Remarkably enough, the expression for the poisoned catalyst particle given by Eq. (4.6.14) can also be integrated in finite terms. Remembering that ρ is an abbreviation for $(1 - v)^{1/3}$, the equations are of the given form with

$$g(v) = \delta[\beta - 1 + (1 - v)^{-1/3} + \delta(1 - v)^{-2/3}]^{-1} \qquad (4.6.37)$$

$v_0 = 0$ and $u_0 = 1$. Thus, from Eq. (4.6.30), we have

$$\delta y = (\beta - 1)v(0, y) - \frac{3}{2}[1 - v(0, y)]^{2/3} + \frac{3}{2} - 3\delta[1 - v(0, y)]^{1/3} + 3\delta$$

or $\sigma_0(y) = 1 - \rho(0, y) = 1 - [1 - v(0, y)]^{1/3}$, the depth of penetration of the poison on the pellets at the inlet, is given by the cubic equation

$$(\beta - 1)\sigma_0^3 - 3(\beta - \tfrac{1}{2})\sigma_0^2 + 3(\beta + \delta)\sigma_0 - \delta y = 0 \qquad (4.6.38)$$

The relations between $\sigma(x, y) = 1 - \rho(x, y)$, $u(x, y)$, and $v(x, y)$ are

$$v(x, y) = 1 - [\rho(x, y)]^3 = \sigma(3 - 3\sigma + \sigma^2) \qquad (4.6.39)$$

and

$$u(x, y) = \frac{v(x, y)}{v(0, y)} = \frac{\sigma(3 - 3\sigma + \sigma^2)}{\sigma_0(3 - 3\sigma_0 + \sigma_0^2)} \qquad (4.6.40)$$

[cf. Eq. (4.6.28)], while Eq. (4.6.31) gives

$$\begin{aligned} -\delta x &= (\beta - 1) \ln u(x, y) \\ &\quad + (\delta + 1)\left[\ln \frac{\sigma}{\sigma_0} - \frac{1}{2} \ln \frac{3 - 3\sigma + \sigma^2}{3 - 3\sigma_0 + \sigma_0^2}\right. \\ &\quad \left. + \sqrt{3} \tan^{-1} \frac{2}{\sqrt{3}} \frac{\sigma_0 - \sigma}{1 + \frac{4}{3}(3 - 2\sigma)(3 - 2\sigma_0)}\right] \end{aligned} \qquad (4.6.41)$$

These are not pretty formulas, but they are quite tractable from a computational point of view, and they allow both the local effectiveness factor, η, and the average activity, A, to be calculated. Figure 4.5 shows the qualitative nature of the penetration $\sigma(x, y)$ and the resulting average activity for fixed β_r and several values of ϕ. Detailed figures may be found in the paper by Olson (1968).

Exercises

4.6.1. Establish the formulas (4.6.38), (4.6.39), (4.6.40), and (4.6.41). (They are modifications of those given by Bischoff.)

4.6.2. Bring the problem given in Exercise 1.9.2 within the ambit of this method.

4.6.3. Show that if $x >> \alpha y >> 1$, then the solution (4.6.34) is given approximately by $u(x, y) = \exp\{-[x - (2\alpha y)^{1/2}]\}$.

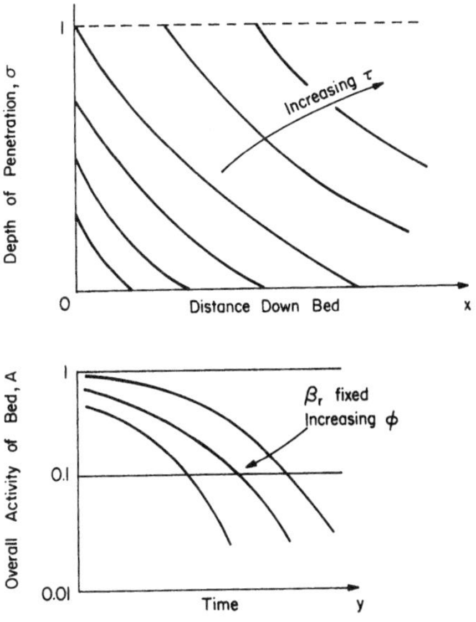

Figure 4.5

REFERENCES

4.1. There are a vast number of books of varying appeal and quality on the Laplace transform. For the mathematical theory,

D. W. Widder, *The Laplace Transform*, Princeton University Press, Princeton, N. J., 1946.

cannot be bettered. Another well-known mathematical treatise is

F. Doetsch, *Laplace Transformation*, Dover Publications, New York, 1943.

A popular introduction for the engineer is to be found in

R. V. Churchill, *Modern Operational Mathematics in Engineering*, McGraw-Hill, New York, 1958.

See also

M. Abramowitz and I. A. Stegun (eds.), *Handbook of Mathematical Functions*, National Bureau of Standards, Washington, D.C., 1964.

F. E. Nixon, *Handbook of Laplace Transformation*, 2nd ed., Prentice-Hall, Englewood Cliffs, N.J., 1965.

Of the easily accessible lexicons, three may be mentioned:

A. Erdelyi (ed.), *Tables of Integral Transforms*, Vol. I, McGraw-Hill, New York, 1954.

G. E. Roberts and H. Kaufman, *Table of Laplace Transforms*, W. B. Saunders, Philadelphia, 1966.

P. A. McCollum and B. F. Brown, *Laplace Transform Tables and Theorems*, Holt, Rinehart and Winston, New York, 1965.

4.2 and 4.4. The first application of Laplace transforms to the chromatographic equations was by

W. R. Marshall and R. L. Pigford, *Application of Differential Equations to Chemical Engineering Problems*, University of Delaware, Newark, Del., 1947.

See also

N. R. Amundson, "A Note on the Mathematics of Adsorption in Beds, I," *J. Phys. Colloid Chem.* **52**, 1153–1157 (1948).

This was followed by a second part in the same journal, **54**, 812–820 (1950), and a third, **54**, 821–829 (1950), concerning radial flow. The first two parts rederive and extend to more general boundary conditions the results of

H. C. Thomas, "Heterogeneous Ion-exchange in a Flowing System," *J. Amer. Chem. Soc.* **66**, 1664–1666 (1944).

H. C. Thomas, "Chromatography: A Problem in Kinetics," *Ann. N.Y. Acad. Sci.* **49**, 161–182 (1948).

where Riemann's method is used.

The most elegant presentation of the basic solutions is to be found in

S. Goldstein, "On the Mathematics of Exchange Processes in Fixed Columns: I. Mathematical Solutions and Asymptotic Expansions, *Proc. Roy. Soc. Lond.* **A219**, 151–171 (1953).

"II. The Equilibrium Theory as the Limit of the Kinetic Theory," *Proc. Roy. Soc. Lond.* **A219**, 171–185 (1953).

"III. The Solution for General Entry Conditions, and a Method of Obtaining Asymptotic Expansions" (with J. D. Murray), *Proc. Roy. Soc. Lond.* **A252**, 334–347 (1959).

"IV. Limiting Values, and Correction Terms, for the Kinetic-theory Solution with General Entry Conditions" (with J. D. Murray), *Proc. Roy. Soc. Lond.* **A252**, 348–359 (1959).

"V. The Equilibrium-theory and Perturbation Solutions, and Their Connexion with Kinetic-theory Solutions, for General Entry Conditions" (with J. D. Murray), *Proc. Roy. Soc. Lond.* **A252**, 360–375 (1959).

Part I contains an extensive study of the function $J(x, y)$ and also a usefully articulate bibliography.

The singular perturbation technique was applied to a more general problem by

J. D. Murray, "Singular Perturbations of a Class of Nonlinear Hyperbolic and Parabolic Equations," *J. Math. and Phys.* **47**, 111–133 (1968).

The work of Vermeulen and Hiester should also be mentioned:

N. K. Hiester and T. Vermeulen, "Elution Equations for Adsorption and Ion-exchange in Flow Systems," *J. Chem. Phys.* **16**, 1087–1089 (1948).

N. K. Hiester and T. Vermeulen, "Saturation Performance of Ion-exchange and Adsorption Columns," *Chem. Eng. Progr.* **48**, 505–516 (1952).

T. Vermeulen and N. K. Hiester, "Ion-exchange Chromatography of Trace Components," *Ind. Eng. Chem.* **44**, 636–651 (1952).

and there are many other references in this field.

Tables of the integral $\Phi(u, v)$ were prepared by

S. R. Brinkley, Jr., and R. F. Brinkley, "Table of the Probability of Hitting a Circular Target," *Math. Tables and Aids to Comp.* **2**, 221 (1947).

4.3. The use of the Laplace transform as a generating function has been exploited in several connections. For an application to linear kinematic waves, see

A. Klinkenberg and F. Sjenitzer, "Holding Time Distributions of the Gaussian Type," *Chem. Eng. Sci.* **5**, 258–270 (1956).

J. J. van Deemter, F. J. Zuiderweg, and A. Klinkenberg, "Longitudinal Diffusion and Resistance to Mass Transfer as Causes of Nonideality in Chromatography," *Chem. Eng. Sci.* **5**, 271–289 (1956).

R. Aris, "On the Dispersion of Linear Kinematic Waves," *Proc. Roy. Soc. Lond.* **A245**, 268–277 (1958).

M. Kocirik, "A Contribution to the Theory of Chromatography: Linear Nonequilibrium Chromatography with a Chemical Reaction of the First-order," *J. Chromatog.* **30**, 459–468 (1967).

For an elegant application to the analysis of kinetic experiments, see

M. Suzuki and J. M. Smith, "Kinetic Studies by Chromatography," *Chem. Eng. Sci.* **26**, 221–235 (1971).

J. Andrieu and J. M. Smith, "Gas–Liquid Reactions in Chromatographic Columns," *Chem. Eng. J.* **20**, 211–218 (1980).

4.5. The treatment follows on the lines laid down in

F. H. I. Chang, K. S. Tan, and I. H. Spinner, "Fixed Bed Sorption with Recycle: Analytic Solutions for Linear Models," *AIChE J.* **19**, 188–191 (1973).

Some results obtained by direct numerical integration are found in

D. O. Cooney and D.-F. Shieh, "Fixed Bed Sorption with Recycle," *AIChE J.* **18**, 245–247 (1972).

4.6. The method of solution follows

K. B. Bischoff, "General Solution of Equations Representing Effects of Catalyst Deactivation in Fixed-bed Reactors," *Ind. Eng. Chem. Fundam.* **8**, 665–668 (1969).

Some actual calculations done by direct numerical integration are given in

J. H. Olson, "Rates of Poisoning in Fixed-bed Reactors," *Ind. Eng. Chem. Fundam.* **7**, 185–188 (1968).

For further references, see Bischoff's paper.

5 Homogeneous Quasi-linear Equations

We already observed in Chapter 1 that many physical systems may be described by one or more homogenous, first-order partial differential equations. Although these are no more than special cases of quasi-linear systems, a homogeneous quasi-linear equation possesses all the characteristic features of a quasi-linear equation and can be treated in a much simpler manner. Indeed, solutions are usually constructed by using the characteristics and may contain discontinuities that are typical of quasi-linear systems. Thus solutions are better represented in terms of wave phenomena. In this chapter and the following two we will discuss in detail homogeneous equations with applications and the corresponding mathematical theory, whereas nonhomogeneous equations will be treated in Chapter 8.

We first discuss the simple structure of the solution and illustrate it somewhat in detail by treating a chromatographic problem with the Freundlich isotherm (Sec. 5.1). In Sec. 5.2 the notion of simple wave is introduced, and solutions are interpreted in terms of nonlinear waves. The result is then applied to the equilibrium chromatography of a single solute in Sec. 5.3. Here we consider not only the Langmuir isotherm (convex function) but also the BET isotherm (nonconvex function) and discuss various cases when discontinuities must be introduced. This naturally leads to Sec. 5.4, in which we treat discontinuities in solutions to formulate the conservation law and introduce the "entropy condition" required to ensure the uniqueness of the solution. On the basis of their different properties, discontinuities are classified into shocks, semishocks, and contact discontinuities. Section 5.5 again deals with the equilibrium chromatography to show how discontinuities come into the solutions of actual problems. By considering both the Langmuir isotherm and the BET isotherm, examples of shocks and semi-shocks are discussed. Application is extended to the problem of water flooding (Sec. 5.6), which is involved with a nonconvex flux function and so gives rise to a solution with a semishock. The final section is concerned with quasi-linear equations with more than two independent variables. The method of solution is illustrated by considering the development of a paper chromatogram.

5.1 Reducible Equations

The equation to be considered is

$$P(x, y, z)p + Q(x, y, z)q = P(x, y, z)\frac{\partial z}{\partial x} + Q(x, y, z)\frac{\partial z}{\partial y} = 0 \tag{5.1.1}$$

where P and Q are continuous functions of x, y, and z, and $P^2 + Q^2 \neq 0$. From Sec. 2.5, we find that the characteristic differential equations are given by

$$\frac{dx}{ds} = P(x, y, z) \tag{5.1.2}$$

$$\frac{dy}{ds} = Q(x, y, z) \tag{5.1.3}$$

$$\frac{dz}{ds} = 0 \tag{5.1.4}$$

The last equation implies that z remains constant along the characteristic curves.

If both P and Q are dependent on z only, Eq. (5.1.1) is called *reducible*,† and we note that conservation equations of many physical systems belong to this class of equations. From Eqs. (5.1.2) and (5.1.3), then, we have

$$\frac{dy}{dx} = \frac{Q(z)}{P(z)} = \sigma(z) \tag{5.1.5}$$

†The term *reducible* is adopted here in accordance with the mathematical theory for a pair of quasi-linear equations (see page 39 of Courant and Friedrichs, 1948). This will be the subject of Sec. 11.3 of the second volume.

which represents the characteristic direction in the (x, y)-plane. Now σ is a function of z, but z is invariant along each of the characteristic curves, so all the characteristic curves are necessarily straight and parallel to the (x, y)-plane. Therefore, the entire solution is given by a family of straight characteristic curves.

We will be mainly concerned with reducible equations in this and the next two chapters because complete solutions can be determined in a straightforward manner by virtue of the straightness of characteristic curves. In fact, if we require the solution surface to pass through a given initial curve I,

$$x = X(\xi), \qquad y = Y(\xi), \qquad z = Z(\xi) \tag{5.1.6}$$

we may solve Eqs. (5.1.4) and (5.1.5) with these values for x, y, and z when $s = 0$. Thus we obtain the characteristic curve in parametric form:

$$y - Y(\xi) = \sigma(z)\,\{x - \chi(\xi)\} \tag{5.1.7}$$

$$z = Z(\xi) \tag{5.1.8}$$

from which the parameter ξ may be eliminated to give the solution surface in the form

$$F(x, y, z) = 0 \tag{5.1.9}$$

The geometrical picture is given in Fig. 5.1. Here we just choose an arbitrary number of points along the curve I and use the value z_i specified there to evaluate $\sigma(z_i)$. We then draw a straight characteristic curve from each point in the direction $[1, \sigma(z_i), 0]$ so that all the characteristic curves are parallel to the (x, y)-plane. [Compare this with Eqs. (5.1.7) and (5.1.8).] The totality of these characteristic curves forms the solution surface in the (x, y, z)-space. The projection of characteristic curves onto the (x, y)-plane will give a family of straight characteristics C, as shown in Fig. 5.1. Their slopes are determined by $\sigma(z)$. Notice that, in contrast to the case of linear equations, it is not generally possible to lay down the characteristics C in the (x, y)-plane once and for all since their slopes vary from solution to solution according to the specified initial data.

As an example that can be easily pictured in a geometrical fashion, let us consider the elution of a chromatographic column when the adsorption equilibrium follows a Freundlich isotherm

$$n = \alpha' c^{\beta+1} \tag{5.1.10}$$

Equation (1.2.9) can then be adapted to give†

$$V\frac{\partial c}{\partial x} + (1 + \alpha c^{\beta})\frac{\partial c}{\partial t} = 0 \tag{5.1.11}$$

†Note that here x is used as the position variable.

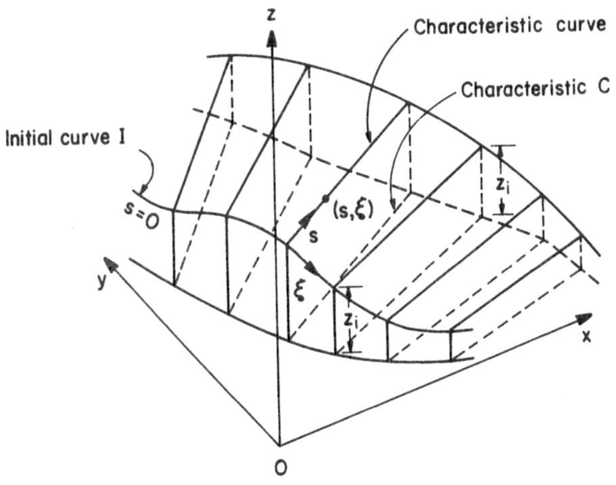

Figure 5.1

where α is a fixed parameter defined by

$$\alpha = \frac{1 - \varepsilon}{\varepsilon} \alpha'(\beta + 1) \tag{5.1.12}$$

Thus we have a reducible equation for the concentration of adsorbate at point x and time t. Let the initial distribution of solute on the column be†

$$c(x, 0) = \begin{cases} \left(\dfrac{x}{L}\right)^{\mu}, & 0 \le x \le L \\ 1, & x \ge L \end{cases} \tag{5.1.13}$$

and the inflowing stream be pure solvent:

$$c(0, t) = 0, \qquad t \ge 0 \tag{5.1.14}$$

The characteristic differential equations may be written as

$$\begin{aligned} \frac{dt}{dx} &= \frac{1 + \alpha c^{\beta}}{V} = \sigma(c) \\ \frac{dc}{ds} &= 0 \end{aligned} \tag{5.1.15}$$

Now, for a Freundlich isotherm, β will be negative, and hence the characteristic curve for zero concentration will be parallel to the t-axis. Since

†An initial distribution of the form $c(x, 0) = \lambda x^{\mu}$ can be reduced to this by absorbing suitable constant factors into α and c.

$c(0, t) = 0$, the t-axis itself is a characteristic curve (see Fig. 5.3). On the x-axis, let $\xi = x/L$ so that $x = L\xi$, $t = 0$, and $c = \xi^\mu$ on $s = 0$. Then, solving Eq. (5.1.15), we have the equations

$$\begin{aligned} t &= \frac{1}{V}(1 + \alpha c^\beta)(x - L\xi) \\ c &= \xi^\mu \end{aligned} \tag{5.1.16}$$

and these define the characteristic curves.

To find $c(x, t)$, we must eliminate ξ and solve the equation

$$Vt = (1 + \alpha c^\beta)(x - Lc^{1/\mu}) \tag{5.1.17}$$

for c or, alternatively, solve

$$x = L\xi + \frac{Vt}{1 + \alpha\xi^{\beta\mu}} \tag{5.1.18}$$

for ξ and substitute into the equation $c = \xi^\mu$. Although this cannot be done in closed form, we can still get a good idea of what the solution looks like from the disposition of the characteristic curves, as shown in Fig. 5.2. Since β is negative, $\sigma(c)$ decreases as c increases, and so the characteristic curves will fan out as we proceed in the x-direction. From this we see that the concentration profile is always monotonic increasing with x, but flattens as time goes on and more and more of the solute is washed out of the bed. The situation is shown in Fig. 5.3, in which we attempt to indicate the value of concentration c by the density of the shading. The time at which the exit concentration has decreased to a value c is given by putting $x = L$ in Eq. (5.1.17); that is,

$$\frac{Vt}{L} = (1 + \alpha c^\beta)(1 - c^{1/\mu}) \tag{5.1.19}$$

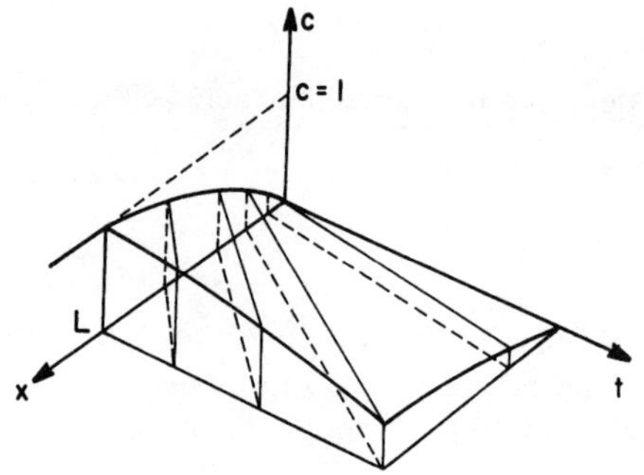

Figure 5.2

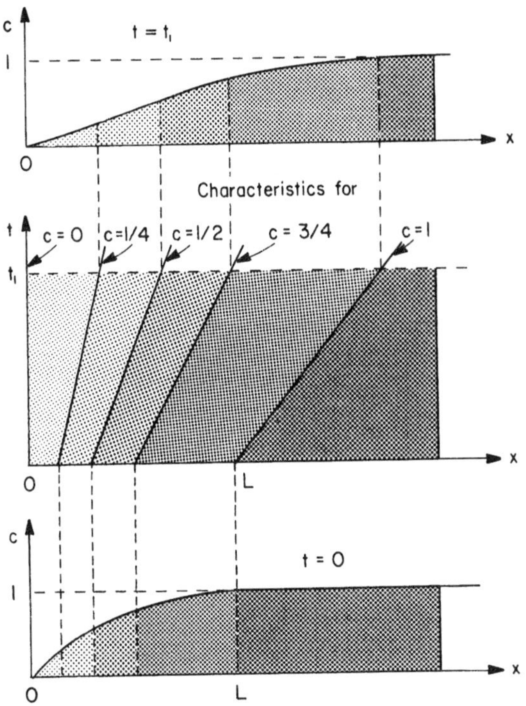

Figure 5.3

Initially, the amount of solute on the bed between $x = 0$ and $x = L$ is proportional to

$$\int_0^L c(x, 0)\, dx = L \int_0^1 \xi^\mu \, d\xi = \frac{L}{\mu + 1}$$

To find what fraction remains at time $\bar{t}$, we calculate

$$\phi = \frac{\mu + 1}{L} \int_0^L c(x, \bar{t})dx \tag{5.1.20}$$

For a fixed value of t (i.e., $t = \bar{t}$), we have, from Eq. (5.1.18),

$$\frac{dx}{L} = \left\{1 - \frac{V\bar{t}}{L} \frac{\alpha\beta\mu\xi^{\beta\mu - 1}}{(1 + \alpha\xi^{\beta\mu})^2}\right\}d\xi \tag{5.1.21}$$

Also, in changing the variable of integration from x to ξ, we need only integrate from $\xi = 0$ to $\xi = \rho$, the value corresponding to the exit concentration, $V\bar{t}/L = (1 + \alpha\rho^{\beta\mu})(1 - \rho)$. Thus

$$\phi = (\mu + 1)\int_0^\rho \left\{1 - \alpha\beta\mu(1 - \rho)(1 + \alpha\rho^{\beta\mu})\frac{\xi^{\beta\mu - 1}}{(1 + \alpha\xi^{\beta\mu})^2}\right\}\xi^\mu d\xi$$

$$= \rho^{\mu+1} + (\mu + 1)(1 - \rho)\left\{\rho^\mu - \mu\int_0^\rho \xi^{\mu-1}\frac{1 + \alpha\rho^{\beta\mu}}{1 + \alpha\xi^{\beta\mu}}d\xi\right\}$$

Let $\lambda = \rho^\mu$; then with $\lambda\xi' = \xi^\mu$,

$$\phi = \rho\lambda + (\mu + 1)(1 - \rho)\,\lambda\left\{1 - \int_0^1 \frac{(1 + \alpha\lambda^\beta)}{1 + \alpha\lambda^\beta(\xi')^\beta}d\xi'\right\} \tag{5.1.22}$$

Exercise

5.1.1. Reexamine Sec. 1.11 in the light of the characteristic differential equations that you can now write for the kinematic wave.

5.2 Simple Waves

Characteristic curves for homogeneous equations, Eq. (5.1.1), are parallel to the corresponding characteristics C in the (x, y)-plane since z remains unchanged along the curve. Thus it would be possible to determine the solution simply by tracing characteristics C from various points along the initial curve I in the (x, y)-plane. We expect this procedure to be particularly simple for reducible equations, because then all the characteristics become straight. An example of this may be found in Fig. 5.3. Thus we will consider the reducible equation

$$P(z)\frac{\partial z}{\partial x} + Q(z)\frac{\partial z}{\partial y} = 0 \tag{5.2.1}$$

with more detailed attention to the role of the characteristics C in relation to the solution.

We recall that every characteristic C carries a specific value of z, and so it represents a direction along which the state of z is propagated. This characteristic direction is given by

$$\frac{dy}{dx} = \frac{Q(z)}{P(z)} = \sigma(z) \tag{5.2.2}$$

which is clearly dependent on the state of z. If the initial data are continuous but have a discontinuous derivative (as at the point D on the initial curve I in Fig. 5.4), then the solution is given by a continuous surface at least in the neighborhood of the curve I, but the surface retains a corner along the characteristic curve through D. Thus an edge of this kind on the solution surface must have its projection along a characteristic C in the (x, y)-plane. This

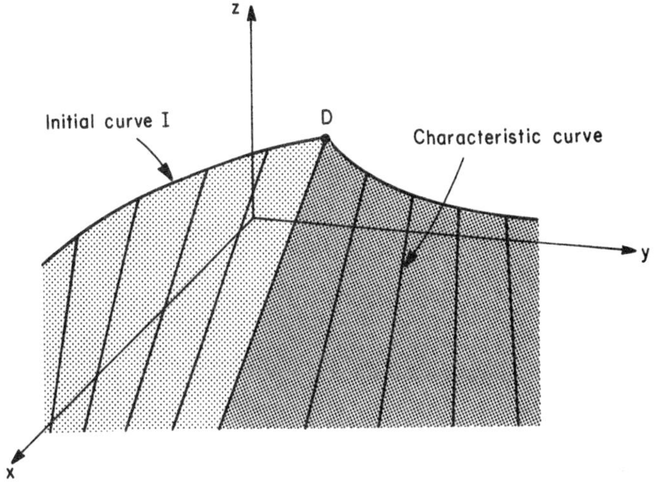

Figure 5.4

implies that discontinuities in derivatives are also propagated along characteristics in the (x, y)-plane.

The preceding arguments enable us to define the domain of dependence as well as the range of influence in a similar fashion as for linear systems (cf. Fig. 3.3 and the corresponding paragraph in Sec. 3.1). These are again determined by a single, straight characteristic C issuing from the point of interest, as shown in Fig. 5.5, but, unlike that for linear equations, the slope of such a characteristic is not fixed. It is rather dependent on the state of z at the point. Due to the existence of such domains of dependence and ranges of influence, the solution of a quasi-linear equation is not necessarily analytic: it may be composed of analytically different portions in different domains and thus the construction of solution by pieces is often possible.

Although we will eventually work with the most general type of initial data, including discontinuities in the variable z itself, we will consider at the moment piecewise continuously differentiable data and, for convenience, assume that the data are specified along the x-axis. Thus we may put the data in parametric form as

$$\text{at } y = 0, \qquad x = \xi \quad \text{and} \quad z = Z(\xi) \tag{5.2.3}$$

where $Z(\xi)$ is piecewise continuously differentiable. (Note that the initial curve I now coincides with the x-axis.) We then proceed to determine the solution piece by piece, treating separately each portion of the x-axis to which there correspond continuously differentiable data. Such a separate treatment will certainly lay a foundation for the solution as a whole.

Let us first suppose that z remains constant over a finite portion OA of the initial curve I; that is, $Z(\xi) = z_0$ for $0 \le \xi \le a$. If $\sigma(z_0) \ne 0$, characteristics

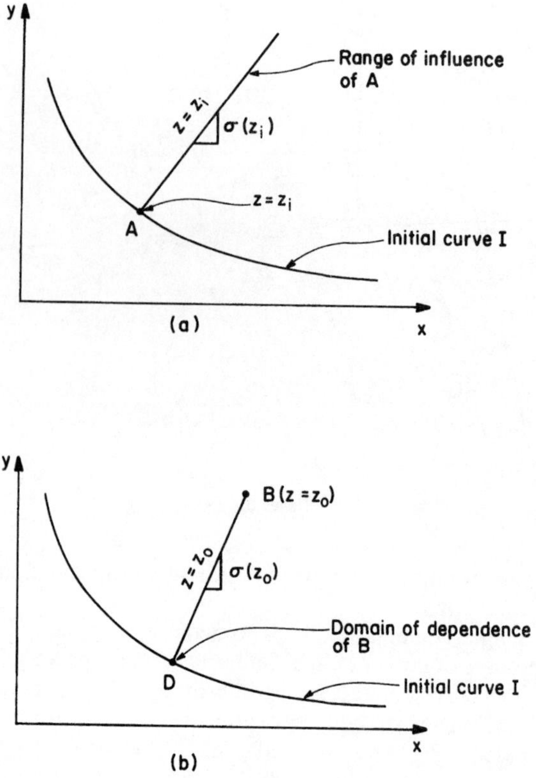

Figure 5.5

emanating from various points on OA are directed upward as shown in Fig. 5.6, and so the state z_0 is propagated through the domain. Therefore, unless the portion OA itself is characteristic, the solution in the region adjacent to that portion is given by a constant state that is the same as the initial state. It is obvious that all the characteristics are parallel and the region is necessarily bounded by characteristics.

Next we will consider a finite portion AB of the initial curve I over which z varies continuously from z_a at $\xi = a$ to z_b at $\xi = b$. Characteristics issuing from various points on AB have different slopes. Thus the region adjacent to the portion AB is covered by a family of straight, nonparallel characteristics C, as shown in Fig. 5.7. The solution in such a region will be called a *simple wave*† (see Courant and Friedrichs, 1948; Lax, 1957). At every point in a

†The term *simple wave* was coined by Courant and Friedrichs (1948) in relation to the solution of a quasi-linear system of two reducible equations. We will see in Chapter 11 of the second volume that these are really correspondent.

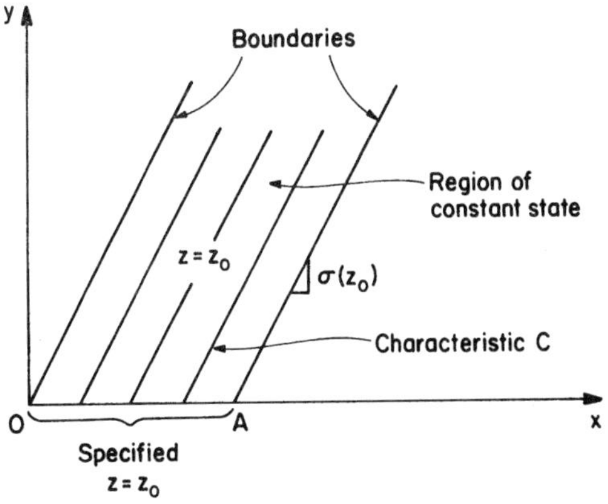

Figure 5.6

simple wave region, we can write

$$y = \sigma(z)\{x - \xi(z)\} \tag{5.2.4}$$

where $\xi(z)$ represents the inverse of the data $z = Z(\xi)$. If the portion AB is nowhere characteristic, $\sigma(z) \neq 0$ and Eq. (5.2.4) has a unique inversion; that is, z can be expressed as a function of x and y to give the solution. In fact, this is essentially what we have done for the example in the previous section; that is, Eq. (5.1.13) is equivalent to Eq. (5.2.4).

In Fig. 5.7 it is implied that σ decreases as ξ increases so that the characteristics fan out in the x-direction. The simple wave region expands continuously as the wave propagates. A simple wave of this kind is called an *expansion wave* (or a *rarefaction wave*; see Courant and Friedrichs, 1948). Clearly, for a simple wave to be expansive, we must have

$$\frac{d\sigma}{d\xi} = \frac{d\sigma}{dz}\frac{dZ}{d\xi} < 0 \qquad \text{for } a \leq \xi \leq b \tag{5.2.5}$$

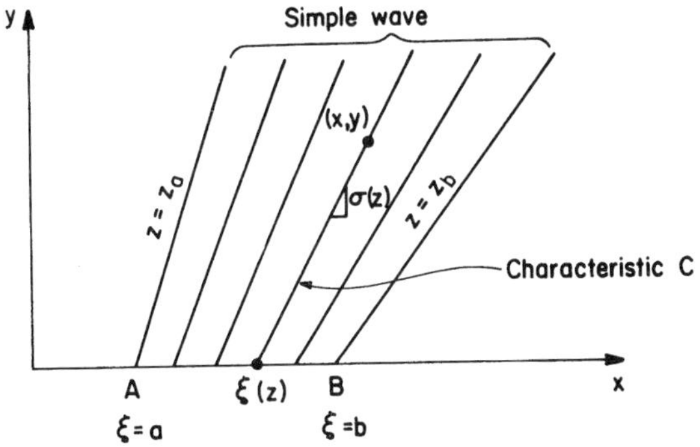

Figure 5.7

That is, the derivatives $d\sigma/dz$ and $dZ/d\xi$ must have opposite signs over the interval AB. Thus the nature of a simple wave depends not only on the nonlinear characteristic, $\sigma(z) = Q(z)/P(z)$, of the equation but on the manner the initial data, $z = Z(\xi)$, is prescribed. In fact, if the derivatives $d\sigma/dz$ and $dZ/d\xi$ have a common sign over the interval AB, that is,

$$\frac{d\sigma}{d\xi} = \frac{d\sigma}{dz}\frac{dZ}{d\xi} > 0 \qquad \text{for } a \leq \xi \leq b, \tag{5.2.6}$$

we expect that the characteristics become steeper as ξ increases, and so the simple wave region tends to be narrowed, as shown in Fig. 5.8(a). Such a wave is called a *compression wave* (or a *condensation wave*; see Courant and Friedrichs, 1948).

When the derivative $d\sigma/dz$ has a uniform sign for all z-values of interest, we say Eq. (5.2.1) is *strictly nonlinear* or *genuinely nonlinear* (Lax, 1957). For such a system, if the initial data $Z(\xi)$ show a monotonic increase or decrease with ξ, then the simple waves become strictly expansive or strictly compressive. But we cannot exclude the possibility that the initial data may not be a monotonic function of ξ or that the quasi-linear equation may not be strictly nonlinear. Under these circumstances, a simple wave can be partly expansive and partly compressive. Two cases are illustrated in Fig. 5.8(b) and (c).

In a compression wave region, the characteristics emanating from the initial curve tend to fan counterclockwise and thus eventually intersect one another in the (x, y)-plane. When this happens, z cannot be determined uniquely because z has different values along different characteristics. If z represents a physical quantity such as concentrations or temperature, this will pose a physically impossible situation. Such a difficulty can be resolved by allowing discontinuities in the solution itself, as we will see in Sec. 5.5.

With regard to expansion waves (see Fig. 5.7), we also notice that as the length of the portion AB decreases (i.e., as $b \to a$), the characteristics emanate from a neighborhood of the point A that gets smaller and smaller. In the limit $a = b$, we can think of an abrupt change in z from z_a to z_b at point A and the characteristics of all slopes between $\sigma(z_a)$ and $\sigma(z_b)$ fan out from a single point, A. The wave is then called a *centered* simple wave. In such a case the coordinate system can be translated to have the point A at the origin so that $\xi(z) = 0$ in Eq. (5.2.4). Then we have

$$\sigma(z) = \frac{y}{x} \tag{5.2.7}$$

and thus the state of z can be expressed as a function of y/x only. When we attempt to take the same limit for a compression wave, we find the characteristics cross over among themselves all at once at point A, and this would present a physically impossible situation from the beginning. Here again we will have to consider a discontinuous solution.

Suppose the initial data are prescribed by two different constant values

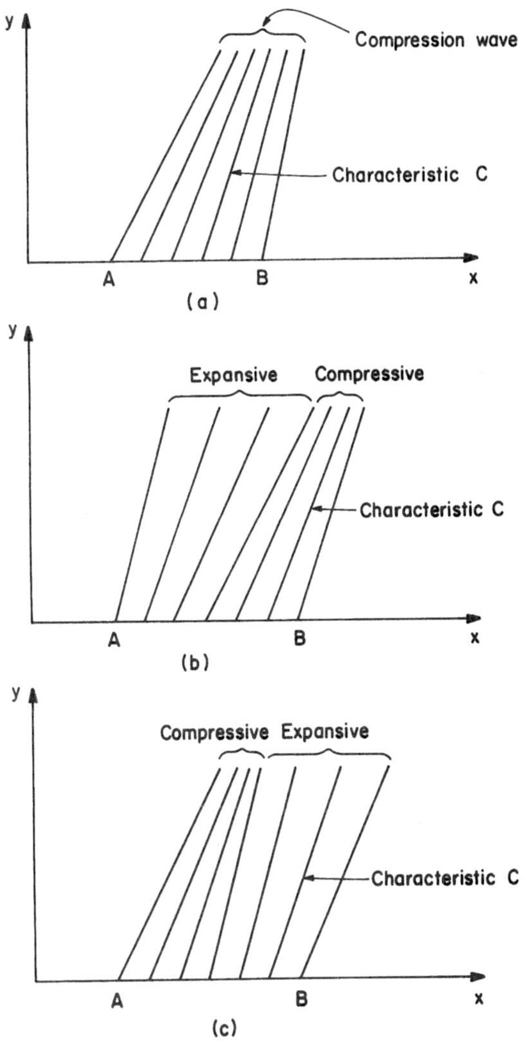

Figure 5.8

of z connected by a discontinuity at $x = 0$; that is,

$$\text{at } y = 0, \qquad z = \begin{cases} z_0, & x < 0 \\ z_1, & x > 0 \end{cases} \tag{5.2.8}$$

We will call this class of problems the *Riemann problem*, since such problems were first investigated by Riemann (see Courant and Friedrichs, 1948; Lax, 1957). If the differential equation is strictly nonlinear and if the characteristic

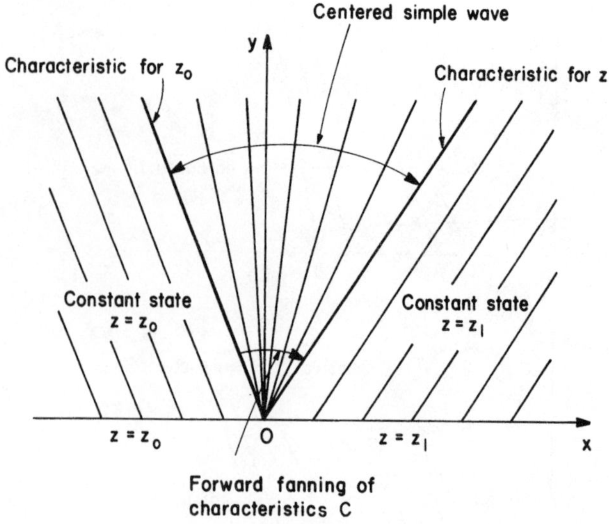

Figure 5.9

for the state z_0 lies on the left side of that for z_1, both emanating from the origin,† we see that the solution is composed of two constant states and a simple wave centered at the origin, as shown in Fig. 5.9. Hence, Eq. (5.2.7) holds in the simple wave region, and the solution is given by a function of y/x only. Indeed, this observation will be proved valid for systems of reducible equations and will bear a particular meaning in relation to their solutions, as we shall see in Chapter 13 of the second volume.

As an example, let us take the elution problem treated in the previous section. If $\mu \to 0$, the characteristics corresponding to values of c that are strictly less than 1 emanate from a small portion of the x-axis near the origin. In the limit $\mu = 0$, we may regard all concentrations in the range $0 \leq c \leq 1$ as being present at $x = t = 0$ and all the characteristics of slopes between $(1 + \alpha)/V$ and ∞ as fanning out from the origin. All the characteristics emanating from the x-axis are of slope $(1 + \alpha)/V$ (see Fig. 5.10). Thus the solution consists of a centered simple wave (above OD) and a constant state (beneath OD). In this case, Eq. (5.1.15) can be solved for c, giving

$$c(L, t) = \left\{\frac{(Vt/L) - 1}{\alpha}\right\}^{1/\beta}, \qquad \frac{Vt}{L} > 1 + \alpha \tag{5.2.9}$$

Recalling that β is negative, we see that this is a decreasing function of (Vt/L) having the value 1 at the point D. In fact, we now find that $c(x, t)$

†Note that if the characteristic for z_0 falls on the right side of that for z_1 we would face a physically impossble situation.

for the point (x, t) will be on the characteristic for which

$$\frac{t}{x} = \sigma(c) = \frac{1}{V}\{1 + \alpha[c(x, t)]^{\beta}\}$$

so that

$$c(x, t) = \left\{\frac{1}{\alpha}\left(\frac{Vt}{x} - 1\right)\right\}^{1/\beta} \tag{5.2.10}$$

provided that $Vt > (1 + \alpha)z$. If $Vt < (1 + \alpha)z$, then $c = 1$.

Exercises

5.2.1. Flooding near the mouth of a river has produced the following distribution of stage (cf. Exercise 1.11.2):

$$h(z, 0) = \begin{cases} h_0, & 0 \le z \le L_0 \\ h_0 + \alpha(z - L_0), & L_0 \le z \le L \end{cases}$$

Describe the decay of the flood waters if h is governed by

$$\frac{\partial h}{\partial t} + \beta h^{2/3}\frac{\partial h}{\partial z} = 0$$

(Note that z is an independent variable.)

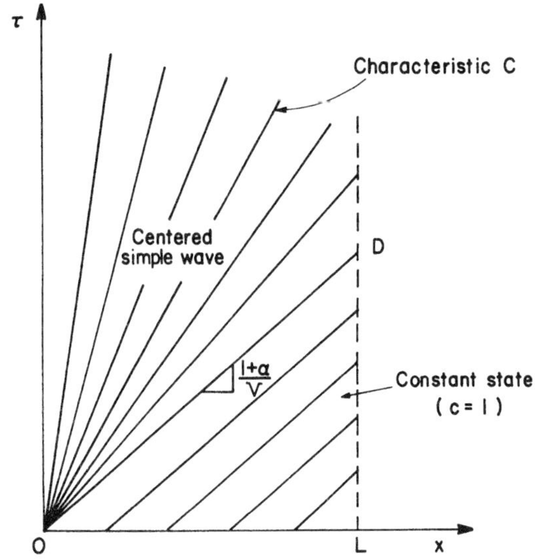

Figure 5.10

5.2.2. In the range $c_1 \leqslant c \leqslant c_2$, the isotherm for a certain ion exchanger can be represented by $f(c) = \alpha c^2$. Describe the saturation of a column when $c(x, 0) = c_1$, $c(0, t) = c_2$, and

$$V\frac{\partial c}{\partial x} + \{1 + \nu f'(c)\}\frac{\partial c}{\partial t} = 0$$

5.2.3. Due to some incident on the highway, the vehicle concentration k (vehicles per mile) at a particular moment ($t = 0$) shows a discontinuous distribution as follows:

$$k(z, 0) = \begin{cases} 150 & \text{for } z < 0, \\ 50 & \text{for } z > 0 \end{cases}$$

If the average speed v (miles per hour) is a linear function of k, the vehicle concentration is governed by Eq. (1.11.10). With $k_j = 200$ vehicles per mile and $v_m = 40$ mph, find the solution of this equation subject to the preceding initial condition. Discuss the traffic-flow situation in terms of the vehicle concentration k and the average speed v.

5.3 Equilibrium Chromatography of a Single Solute

To illustrate the application of simple wave solutions, we will consider somewhat in detail the single solute chromatography with the assumption of local equilibrium. Here the terms mobile phase and stationary phase will be used for the two phases in the column without prejudice as to whether they are solid, liquid, or gaseous. Likewise, the process transferring the solute from one phase to the other will be referred to as exchange, although it may be adsorption, partition, or absorption. If $c(z, t)$ and $n(z, t)$ are the concentrations at position z and time t in the mobile and stationary phases, respectively, we have, from Eq. (1.2.1),

$$\varepsilon V\frac{\partial c}{\partial z} + \varepsilon\frac{\partial c}{\partial t} + (1 - \varepsilon)\frac{\partial n}{\partial t} = 0 \qquad (5.3.1)$$

If we let

$$\nu = \frac{1 - \varepsilon}{\varepsilon} \qquad (5.3.2)$$

and introduce the dimensionless variables defined by

$$x = \frac{z}{Z}, \qquad \tau = \frac{Vt}{Z} \qquad (5.3.3)$$

where Z is a characteristic length for the column, Eq. (5.3.1) is rewritten as

$$\frac{\partial c}{\partial x} + \frac{\partial c}{\partial \tau} + \nu\frac{\partial n}{\partial \tau} = 0 \qquad (5.3.4)$$

The equilibrium isotherm

$$n = f(c) \tag{5.3.5}$$

is the relation that obtains between n and c at equilibrium. On physical grounds, we will require the condition

$$f'(c) > 0 \tag{5.3.6}$$

If we assume that the flow is sufficiently slow that equilibrium is maintained everywhere within the column at all times, we have

$$\frac{\partial c}{\partial x} + \{1 + \nu f'(c)\}\frac{\partial c}{\partial \tau} = 0 \tag{5.3.7}$$

which is a reducible equation. The characteristic direction is thus given by a function of c; that is,

$$\frac{d\tau}{dx} = 1 + \nu f'(c) \equiv \sigma(c) \tag{5.3.8}$$

In specifying the initial conditions, it would be unnatural to assume that the initial distributions of concentrations were not in equilibrium. Thus we have no need to specify $c(x, 0)$ and $n(x, 0)$ independently but may think of specifying either of the two or their combination, $c(x, 0) + \nu n(x, 0)$, whichever may be convenient. If we specify

$$c(x, 0) = F(x) \tag{5.3.9}$$

then it follows that $n(x, 0) = f(F(x))$. At the inlet of the column, $x = 0$, we can only specify the concentration of the solute in the mobile phase, say

$$c(0, \tau) = G(\tau) \tag{5.3.10}$$

The curve along which it is natural to give initial data is thus the pair of positive axes shown in Fig. 5.11, and since $f'(c)$ is positive and bounded for all values of c of physical interest, this initial curve I is never tangent to a

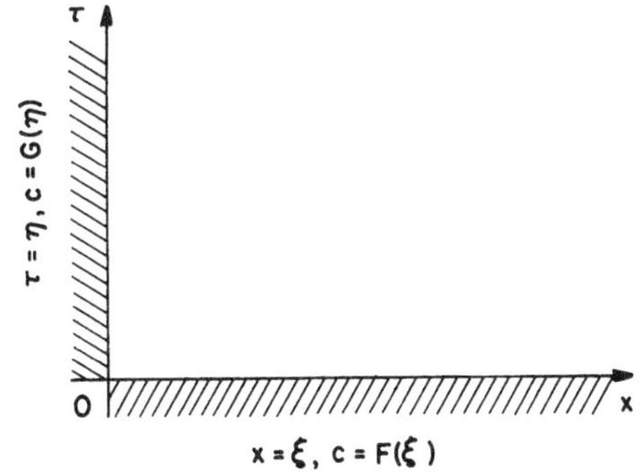

Figure 5.11

characteristic direction. An exception to this would have to be made at $c = 0$ if the Freundlich isotherm, $f(c) = \alpha c^{\beta}$, were used and β were less than 1, for then the characteristic direction would touch the τ-axis, as we have seen in Sec. 5.1. Frequently, we will find it convenient to put the data in parametric form as follows:

$$\begin{aligned} &\text{at } \tau = 0, \quad x = \xi \quad \text{and} \quad c = F(\xi) \\ &\text{at } x = 0, \quad \tau = \eta \quad \text{and} \quad c = G(\eta) \end{aligned} \tag{5.3.11}$$

First we will consider only convex isotherms for which we have

$$f''(c) < 0 \tag{5.3.12}$$

Since, then, $d\sigma/dc = \nu f''(c) < 0$, Eq. (5.3.7) is genuinely nonlinear. A particularly important example is afforded by the Langmuir isotherm [cf. Eq. (1.2.7)],

$$f(c) = \frac{\gamma c}{1 + Kc} \tag{5.3.13}$$

in which we have used the symbol γ for the product NK. In this case the characteristic direction is

$$\sigma(c) = 1 + \frac{\nu\gamma}{(1 + Kc)^2} \tag{5.3.14}$$

Suppose now that we have a nonuniformly adsorbed column, and this is to be eluted by feeding in a stream of pure solvent (or carrier). More specifically, let us consider a linear distribution of solute over a finite portion of the x-axis so that we have

$$\begin{aligned} &\text{at } \tau = 0, \quad x = \xi \quad \text{and} \quad c = \begin{cases} \dfrac{c_0}{\alpha}\,\xi, & 0 \le \xi \le \alpha \\ c_0, & \xi \ge \alpha \end{cases} \\ &\text{at } x = 0, \quad \tau = \eta \quad \text{and} \quad c = 0 \end{aligned} \tag{5.3.15}$$

We then proceed to construct the characteristics C in the (x, τ)-plane as shown in Fig. 5.12(b). All the characteristics emanating from the τ-axis are of slope $\sigma(0) = 1 + \nu\gamma$, whereas those from the x-axis for $\xi \ge \alpha$ are less steep with a slope $\sigma(c_0) = 1 + \nu\gamma/(1 + Kc_0)^2$. These two families of characteristics will give rise to two constant states, $c = 0$ and $c = c_0$, respectively. The characteristics emanating from the portion $0 \le \xi \le \alpha$ of the x-axis are of slopes between $\sigma(0)$ and $\sigma(c_0)$ and fan forward, giving an expansion wave. The layout of these characteristics may be better understood if we make use of the graph of the Langmuir isotherm, because the slope of the tangent to the isotherm, $f'(c)$, is a direct measure of the characteristic direction $\sigma(c)$ [cf. Eq. (5.3.8) and Fig. 5.12(a)]. The solution thus consists of two constant states separated by an expansion wave. For any point in this simple wave

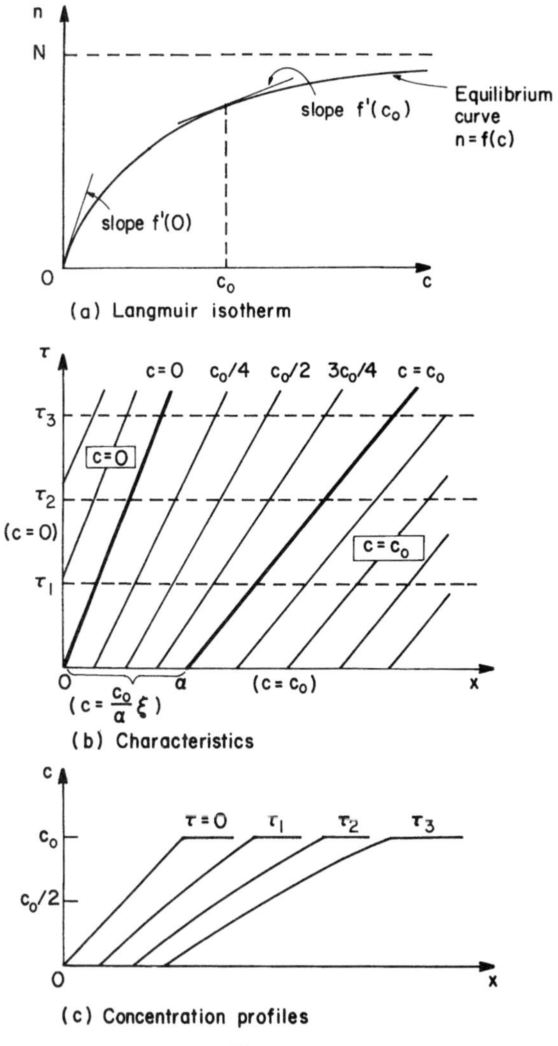

Figure 5.12

region, we can write

$$\tau = \left\{1 + \frac{\nu\gamma}{(1 + Kc)^2}\right\}\{x - \xi(c)\}$$

$$\xi(c) = \alpha\frac{c}{c_0} \tag{5.3.16}$$

or, a cubic equation for Kc,

$$(Kc)^3 + \left\{2 + \left(\frac{Kc_0}{\alpha}\right)(\tau - x)\right\}(Kc)^2 +$$

$$\left\{1 + \nu\gamma + \left(\frac{2Kc_0}{\alpha}\right)(\tau - x)\right\}(Kc) + \left(\frac{Kc_0}{\alpha}\right)\{\tau - x + \nu\gamma x\} = 0 \qquad (5.3.17)$$

and so the solution can be expressed as a function of x and τ. Although it is good to have the solution in a closed form, we can get a better idea of what the solution looks like by drawing the concentration profiles at successive times from the disposition of characteristics in the (x, τ)-plane. This procedure is illustrated in Fig. 5.12(b) and (c). The breakthrough curve from a column of finite length, say ξ_L, can be obtained easily by reading the concentration values at intersections of characteristics with a vertical line located at $\xi = \xi_L$.

We also notice that, the higher the state of concentration, the faster it is propagated through the column since the characteristic direction $\sigma(c)$ is just the reciprocal speed of propagation. Indeed, it is this behavior of the state of concentration that makes the simple wave expansive. Often, therefore, we will consider the solution in terms of wave propagation. This is an example of what we call a *nonlinear wave*, and we have now started to observe the difference in nature between nonlinear waves and linear waves, the latter having been treated in Sec. 3.3. Incidentally, the solvent flows through the column at the speed V or we have $d\tau/dx = 1$ for the flow of pure solvent, and clearly the solvent moves faster than any state of concentration does.

If we let α approach zero, we see that the characteristics emanating from the interval $0 \le \xi \le \alpha$ tend to translate to the left, maintaining their slopes. In the limit $\alpha = 0$, these characteristics all emanate from the origin; that is, the simple wave becomes centered there. The overall feature being similar to what we observe from Fig. 5.12, we show a three-dimensional view of the solution surface in Fig. 5.13. In practice, this represents the elution of a uniformly adsorbed column, which is a Riemann problem. In this case, we can write

$$\frac{\tau}{x} = \sigma(c) = 1 + \frac{\nu\gamma}{(1 + Kc)^2} \qquad (5.3.18)$$

for any point in the simple wave region so that

$$Kc = -1 + \sqrt{\frac{\nu\gamma}{\tau/x - 1}} \qquad (5.3.19)$$

which is a function of τ/x only. This fact was recognized in relation to the theory of chromatography as early as 1945 (see Walter, 1945; Offord and Weiss, 1949; Sillen, 1950) and was the basis of defining the throughput parameter T used by Vermeulen and co-workers (Hiester and Vermeulen, 1952; Klein, Tondeur, and Vermeulen, 1967; Tondeur and Klein, 1967).

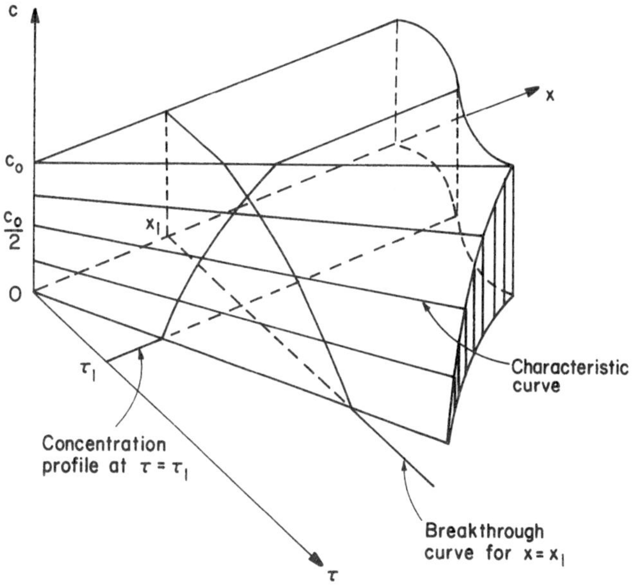

Figure 5.13

Next we will consider the saturation of a clean column. The initial data are

$$\begin{aligned} &\text{at } \tau = 0, \qquad c = 0 \quad \text{for} \quad x > 0 \\ &\text{at } x = 0, \qquad c = c_0 \quad \text{for} \quad \tau > 0 \end{aligned} \tag{5.3.20}$$

and thus we have another Riemann problem. The characteristics emanating from the x-axis ($\tau = 0$) have slope $\sigma(0) = 1 + \nu\gamma$, and these are steeper than those emanating from the τ-axis ($x = 0$), since the latter have slope $\sigma(c_0) = 1 + \nu\gamma/(1 + Kc_0)^2$. Figure 5.14 shows the confusing situation, which is compounded when we realize that there is a characteristic corresponding to every concentration between 0 and c_0 emanating from the origin. Note that these characteristics fan counterclockwise as we proceed in the x-direction. The point P lies on three characteristics, PQ, PR, and PO, corresponding to concentrations of c_0, 0, and $[-1 + \{\nu\gamma/(\tau/x - 1)\}^{1/2}]/K$ respectively. The solution at the time τ corresponding to P would look like the figure shown in the lower part of Fig. 5.14, where there are three values of c at the point P. Clearly, this is physically meaningless and some way must be found to circumvent the crossing of the characteristics. The trouble arises from the fact that the higher concentrations move faster than the lower ones. An advancing concentration wave with a sloping front, that is, a compression wave, gets progressively steeper as the higher concentration overtakes the lower, until the wave overtops itself. Figure 5.15 shows three stages: the first and second are physically acceptable, although the vertical tangent in

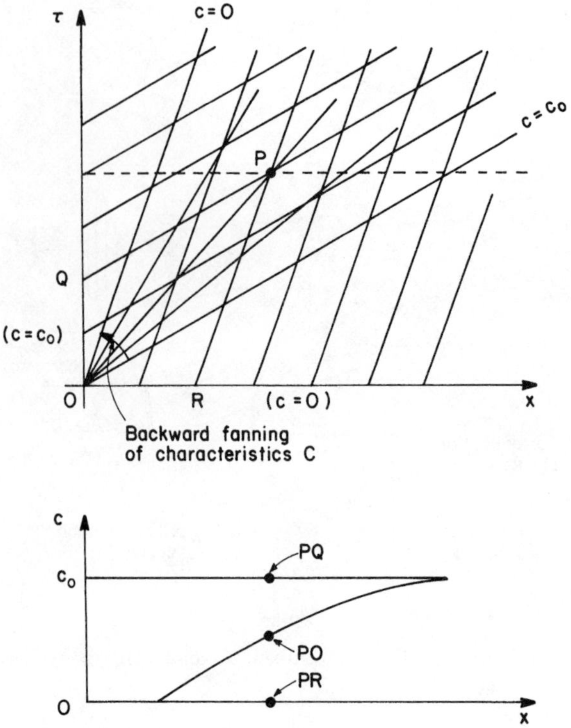

Figure 5.14

stage b shows that breaking is incipient. The concentration distribution in stage c has an unacceptable region where three values are present, and the only way to retain a physically meaningful solution is to introduce a discontinuity, as indicated by the broken line. In the case we are considering, there must be a discontinuity from the very first, for the initial data have a discontinuity at the origin and breaking takes place from the beginning. We will discuss discontinuous solutions in detail in Sec. 5.4 and complete the solution of the saturation problem in Sec. 5.5.

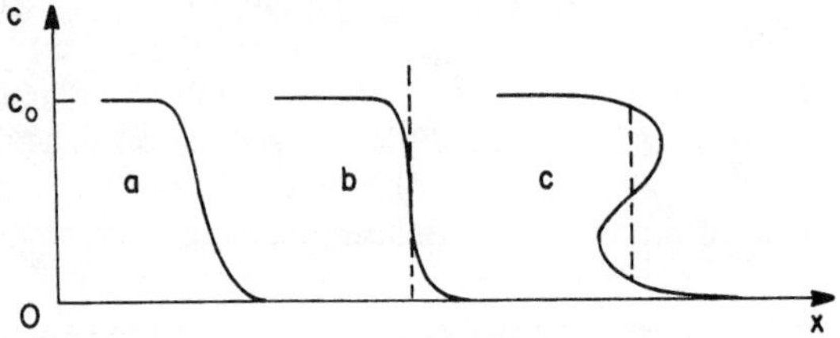

Figure 5.15

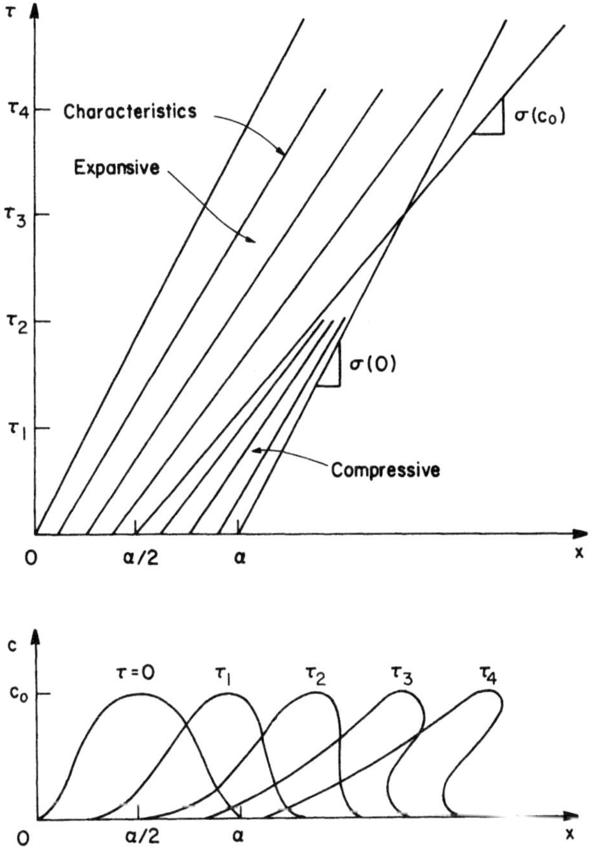

Figure 5.16

It is now apparent that, if we have a pulse-type distribution of solute with maximum of c_0 over a finite portion $(0 < x < \alpha)$ of the column at $\tau = 0$, as in Fig. 5.16, and introduce a stream of pure solvent, that is,

$$\begin{aligned} &\text{at } \tau = 0, \quad c = \begin{cases} f(x), & 0 < x < \alpha, \quad f(0) = 0, \quad f(\alpha) = 0 \\ 0, & x > \alpha \end{cases} \\ &\text{at } x = 0, \quad c = 0 \end{aligned} \tag{5.3.21}$$

we will have a simple wave that is expansive in the rear side and compressive in the front side. While it propagates, the wave form is continuously distorted and the compressive part of the wave ultimately breaks to give a triple-valued solution for $c(x, \tau)$. This is shown in Fig. 5.16. The breaking starts at $\tau = \tau_2$, when the profile of concentration first develops an infinite slope, and afterward we need to consider a discontinuity in the solution. This phenomenon is typical of nonlinear waves and in contrast to what we have observed for linear waves in Sec. 3.3 (compare Fig. 5.16 with Fig. 3.7).

At this point we will turn our attention to the equilibrium isotherm of sigmoid type, say the BET (Brunauer–Emmett–Teller) isotherm

$$\theta = f(\phi) = \frac{k\phi}{(1 - \phi)[1 + (k - 1)\phi]} \tag{5.3.22}$$

where θ is the surface coverage, ϕ the ratio of the partial pressure p to the vapor pressure P_0, and k a positive parameter. If N denotes the solid phase concentration of solute when the surface is fully covered by a monolayer, and if we let

$$\nu_1 = \frac{1 - \varepsilon}{\varepsilon} \frac{NRT}{P_0} \tag{5.3.23}$$

from Eq. (5.3.4) we have

$$\frac{\partial \phi}{\partial x} + \{1 + \nu_1 f'(\phi)\} \frac{\partial \phi}{\partial \tau} = 0 \tag{5.3.24}$$

so that the characteristic direction is

$$\frac{d\tau}{dx} = 1 + \nu_1 f'(\phi) = \sigma(\phi) \tag{5.3.25}$$

which is directly related to the slope of the tangent to the isotherm. Since the isotherm has an inflection point unless k is very small, $d\sigma/d\phi$ has a nonuniform sign and so Eq. (5.3.24) is not strictly nonlinear. Suppose the initial distribution of solute is linear as in Eq. (5.3.15), where ϕ_0 is larger than the concentration at the inflection point ϕ^*. When we elute this column with a stream of pure solvent, we immediately see that the solution is given by a simple wave that is expansive over the range $0 \le \phi \le \phi^*$ and compressive in the range from $\phi = \phi^*$ to $\phi = \phi_0$. The situation here will be very similar to that presented in the upper part of Fig. 5.16, although the initial distribution is strictly linear rather than of pulse type. [Examine the nature of the solution in the light of Eqs. (5.2.5) and (5.2.6)].

Let us consider the elution of a uniformly saturated column subject to the BET isotherm. This is a Riemann problem and, if we observe this as the limiting situation of the previous case, it is straightforward to produce the layout of characteristics in the (x, τ)-plane as in Fig. 5.17(b). The characteristics emanating from the τ-axis, where $\phi = 0$, have slope $\sigma(0) = (1 + \nu_1 k)$, whereas those emanating from the x-axis, where $\phi = \phi_0$, are of slope

$$\sigma(\phi_0) = 1 + \frac{\nu_1 k[1 + (k - 1)\phi_0^2]}{(1 - \phi_0)^2[1 + (k - 1)\phi_0]^2} \tag{5.3.26}$$

The latter can become steeper than the former if ϕ_0 approaches 1. Even more confusing is the behavior of characteristics issuing from the origin. As we traverse the change in ϕ from 0 to ϕ^* and then to ϕ_0, those characteristics

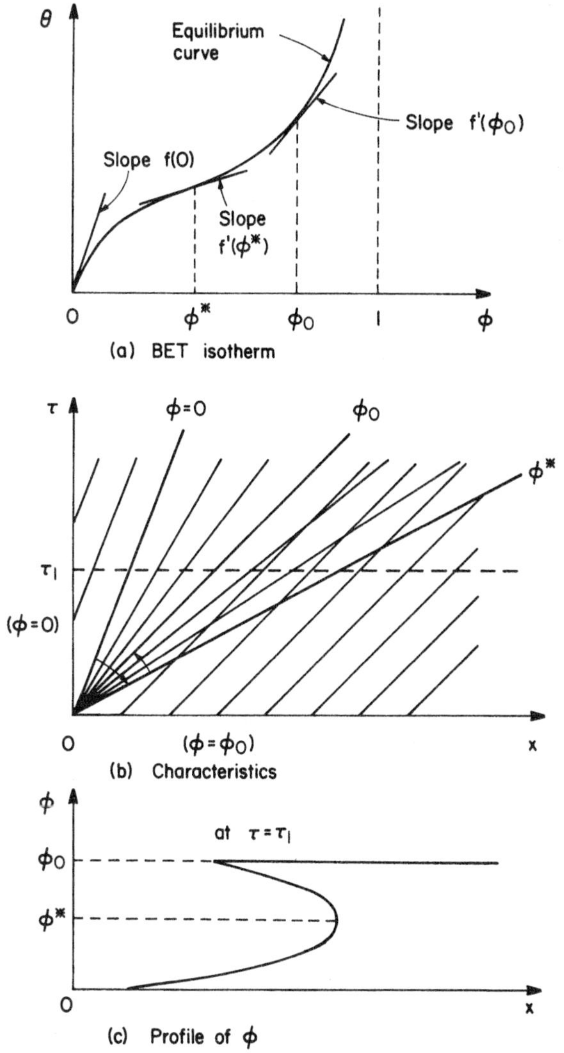

Figure 5.17

with slopes between $\sigma(0)$ and $\sigma(\phi^*)$ fan forward, but the remainder fan backward. Thus the profile of ϕ at any time $\tau = \tau_1$, will include a triple-valued zone as in Fig. 5.17(c), and here again a discontinuity must be introduced into the solution. The saturation of a clean column, for which the initial data are just the opposite of the previous case, would pose a similar difficulty. We now trace the change in ϕ from ϕ_0 to 0 and observe that, while the characteristics of slopes between $\sigma(\phi_0)$ and $\sigma(\phi^*)$ fan forward, the remainder fan backward. This backward fanning of characteristics makes the profile of ϕ triple valued so that we need to consider a discontinuous solution. These problems will be treated further in Sec. 5.5 after we discuss discontinous solutions in Sec. 5.4.

Exercises

5.3.1. For a single-solute chromatography with the Freundlich isotherm

$$f(c) = kc^{1/3}$$

determine the solution if the initial data are specified as

$$\text{at } \tau = 0, \qquad c = 1$$

$$\text{at } x = 0, \qquad c = \begin{cases} 1 - 0.875\tau & \text{for } 0 < \tau \leqslant 1 \\ 0.125 & \text{for } \tau > 1 \end{cases}$$

and the parameter values are given by

$$\nu = 1.5, \qquad k = 5(\text{mol/l})^{2/3}$$

Here c is in moles per liter. Obtain the breakthrough curve and discuss.

5.3.2. The radial chromatographic column is governed by the equation (see Exercise 1.2.2)

$$\frac{Q}{2\pi\varepsilon d}\frac{1}{r}\frac{\partial c}{\partial r} + \{1 + \nu f'(c)\}\frac{\partial c}{\partial t} = 0, \qquad a < r < b$$

where Q denotes the constant volumetric flow rate and d the depth of the bed. If

$$f(c) = \frac{\gamma c}{1 + Kc}$$

and

$$\begin{cases} c(r, 0) = c_0(r - a) & \text{for } a \leqslant r \leqslant b \\ c(a, t) = 0 & \text{for } t \geqslant 0 \end{cases}$$

discuss the progress of the elution and show that the disc will be cleaned off when a total volume of $(1 + \nu\gamma)\cdot$(free volume of the disc)$/Q$ has passed through. [Note that the coordinate transformation defined by Eqs. (3.5.28) and (3.5.29) may be applied to put the differential equation in the form of Eq. (5.3.7).]

5.3.3. Consider the saturation of a clean column for which the equilibrium is described by the BET isotherm. Thus the initial data are specified as in Eq. (5.3.20) and the equilibrium relation is given by Eq. (5.3.22). Can you show the layout of characteristics in the (x, τ)-plane, as well as an example of the profile of ϕ, and thereby argue that a discontinuity must be admitted in the solution from the beginning?

5.3.4. Answer the same question as in Exercise 5.3.3 when the equilibrium curve is sigmoid in a reversed fashion from the BET isotherm.

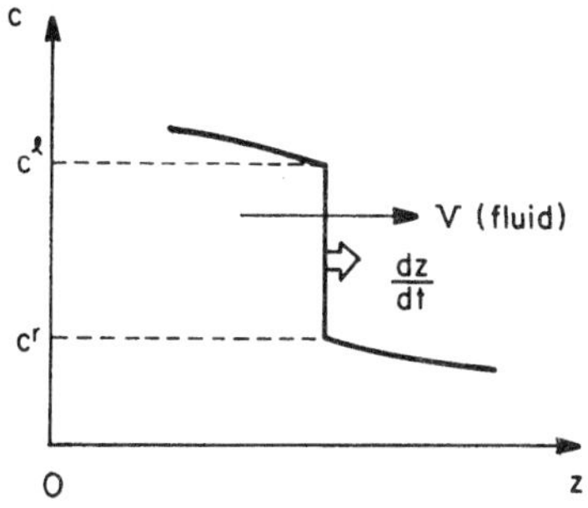

Figure 5.18

5.4 Discontinuities in Solutions

In this section we would like to study the nature of discontinuities in solutions of quasi-linear equations. For this purpose we will consider a physical system, the chromatography of a single solute with nonlinear equilibrium relations, mainly because the concept of discontinuous solutions is introduced on the basis of physical arguments. Suppose the concentration profile at an arbitrary time t contains a discontinuity as shown in Fig. 5.18. Note that if $c(z, t)$ is discontinuous, so is $n(z, t)$ because the two are uniquely related via the equilibrium relationship. At the position of the discontinuity, then, the partial differential equation (5.3.1) is no longer valid, and the conservation law must be formulated in the light of the fact that the discontinuity moves with such a speed that there is no accumulation of material or energy at the discontinuity.

We will adopt the superscripts l and r to denote, respectively, the left and right sides of a discontinuity in the z-direction. If we suppose the discontinuity moves with a speed dz/dt, then from the standpoint of an observer on the discontinuity the rate of solute transfer in the mobile phase through the discontinuity is

$$\varepsilon A c^l\left(V - \frac{dz}{dt}\right) - \varepsilon A c^r\left(V - \frac{dz}{dt}\right)$$

and this, when added to the rate of transfer of adsorbed solute in the reverse direction,

$$(1 - \varepsilon)A\frac{dz}{dt}(n^r - n^l)$$

must be zero, since there is no accumulation at the discontinuity. There results

$$V\frac{dt}{dz} = 1 + \nu\frac{n^l - n^r}{c^l - c^r} \tag{5.4.1}$$

We will use the square bracket [ ] to denote the *jump* in the quantity enclosed across a discontinuity. Since we are maintaining the hypothesis of equilibrium, the jump in n is related to that in c by

$$[n] = f(c^l) - f(c^r) = [f]$$

If we introduce the dimensionless variables, x and τ, defined by Eq. (5.3.3), we have†

$$\frac{d\tau}{dx} = 1 + \nu\frac{[f]}{[c]} \equiv \tilde{\sigma}(c^l, c^r) \tag{5.4.2}$$

which is often called a *jump condition* (see, for example, Lax, 1972). While this equation gives the propagation direction of the discontinuity in the (x, τ)-plane, we also see that $\tilde{\sigma}$ represents the reciprocal speed or slowness of the discontinuity. In this sense, Eq. (5.4.2) is equivalent to the well-known Rankine–Hugoniot relation in compressible fluid flow (Courant and Friedrichs, 1948). It is interesting to compare Eq. (5.4.2) with Eq. (5.3.8) in the light of Fig. 5.12(a). We see that the ratio $[f]/[c]$ is just the slope of the chord connecting two points on the isotherm for which $c = c^l$ and $c = c^r$, whereas $f'(c)$ is the slope of the tangent to the isotherm. We also notice that $\tilde{\sigma}$ is always greater than 1, so the mobile phase moves faster than a discontinuity and passes through it from the left to the right.

Given the state on one side of a discontinuity, say c^l, and its propagation speed $\tilde{\sigma}^{-1}$, we may determine the state c^r on the other side by using Eq. (5.4.2). However, we immediately find that an ambiguity still remains concerning the direction of jump in c, and thus Eq. (5.4.2) is not sufficient to give a unique state for c^r, which is required for a physically relevant solution. More generally, the solution of a quasi-linear equation is defined in the sense of weak solutions (see Chapter 7), and it is well known that the solution is not uniquely determined until we introduce an additional condition to regulate the direction of jumps across discontinuities. Such a condition is usually referred to as the *entropy condition* after the term that was used by Courant and Friedrichs (1948) in their pioneering work on supersonic flow and shock waves.

We now recall our observations in the previous section; that is, only when the characteristics fan backward as we proceed in the x-direction do we need to consider a discontinuity in the solution. Based on this, we are tempted to claim that the entropy condition is given by

$$\sigma(c^l) \leqslant \sigma(c^r) \tag{5.4.3}$$

where the equality sign is included to accommodate linear systems. This condition will be proved valid if the isotherm $f(c)$ is convex or sigmoid with

†This equation can be derived alternatively by considering the integral of $[\varepsilon c + (1 - \varepsilon)n]$ over a finite interval, which includes the point of a discontinuity (see Lax, 1972). This approach will be treated in Sec. 7.2.

only one inflection point, and these are the cases considered in the previous section. For more complex isotherms with more than one inflection point, however, there are cases when Eq. (5.4.3) fails to serve as a proper entropy condition, as we will see in the following.

In fact, Oleinik (1959) was the first to formulate the generalized entropy condition, which she called the "Condition E," and thus to establish the uniqueness theorem. In our notation the condition reads†

$$\tilde{\sigma}(c^l, c^r) \geqslant \tilde{\sigma}(c^l, c) \tag{5.4.4}$$

for every c between c^l and c^r. We then have the uniqueness theorem:

> THEOREM: A solution of Eq. (5.3.7) is unique if it satisfies at all points of discontinuity in c the jump condition, Eq. (5.4.2), as well as the entropy condition, Eq. (5.4.4).

Instead of presenting a lengthy proof, we will take a heuristic approach by considering various types of nonlinearity for the equilibrium isotherm $f(c)$. At the same time, we will interpret the implication of Eq. (5.4.4) and examine the characteristic features of discontinuities.

Shock. If the equilibrium relation $f(c)$ is given by a convex function, we have two different situations, as shown in Fig. 5.19. It is immediately apparent that the equality in Eq. (5.4.4) is never valid, whereas the inequality is equivalent to the conditions

$$\begin{aligned} c^l &> c^r, \qquad \text{if } f''(c) < 0 \\ c^l &< c^r, \qquad \text{if } f''(c) > 0 \end{aligned} \tag{5.4.5}$$

Note that Eq. (5.4.3) will give the same result. We also notice that, as we move along the straight chord from the point L to the point R, we see the equilibrium curve on the right side.

When we compare the slopes among the two tangents to the equilibrium curve and the chord connecting the two points L and R, we find the inequality

$$f'(c^l) < \frac{[f]}{[c]} < f'(c^r) \tag{5.4.6}$$

†For an alternative expression, see Lax (1971, 1973) and also Sec. 7.4 of this book. In general, the Condition E is expressed in terms of the propagation speed so that Eq. (5.4.4) should be written as

$$\frac{1}{\tilde{\sigma}(c^l, c^r)} \leqslant \frac{1}{\tilde{\sigma}(c^l, c)} \tag{5.4.4a}$$

As long as both $\tilde{\sigma}$ and σ have values of a common sign, the two inequalities are equivalent and the discussion of this section is correct. But if both $\tilde{\sigma}$ and σ can have positive as well as negative values, we must use Eq. (5.4.4a). For examples of the latter case, see Secs. 6.5 through 6.7.

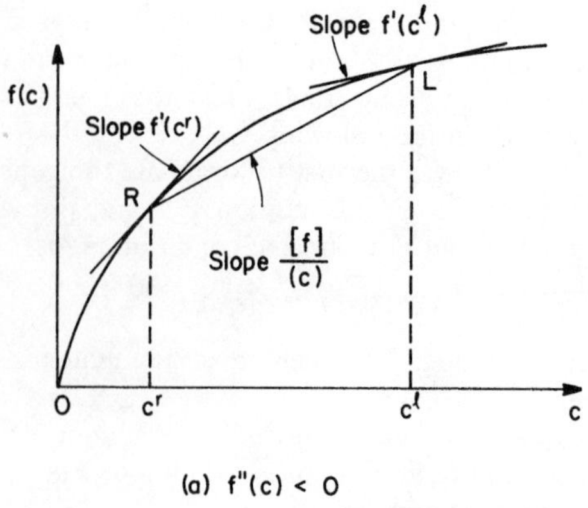

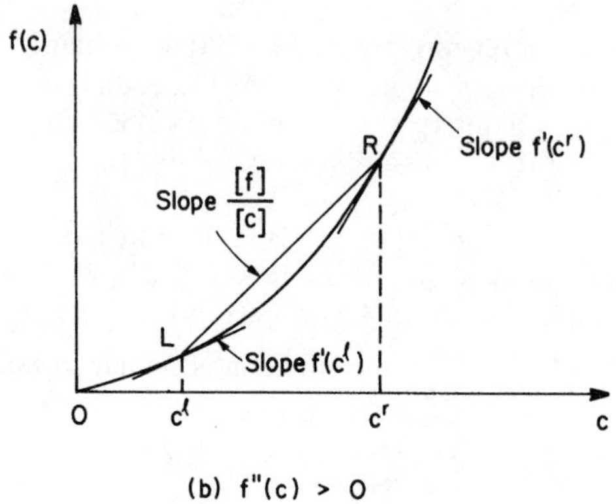

Figure 5.19

and so, from Eqs. (5.3.8) and (5.4.2),

$$\sigma(c^l) < \tilde{\sigma}(c^l, c^r) < \sigma(c^r) \tag{5.4.7}$$

According to Lax (1957)†, a discontinuity that satisfies this inequality condition is called a *shock* (or a *genuine shock*). For this reason, Eq. (5.4.7) is referred to as the *shock condition*. This implies that the discontinuities encountered in strictly nonlinear problems are all shocks.

Due to the condition in Eq. (5.4.7), the characteristics C for the state on either side of the shock impinge on the shock path S as shown in Fig. 5.20. Thus the state on the left side tends to propagate faster than the shock, while the state on the right side propagates with a lower speed. But the state on

†In fact, Lax used the propagation speed to define a shock, so Eq. (5.4.7) should be given with reciprocals of the quantities and with the inequality signs reversed. But as long as the three quantities have the common sign, the two are equivalent.

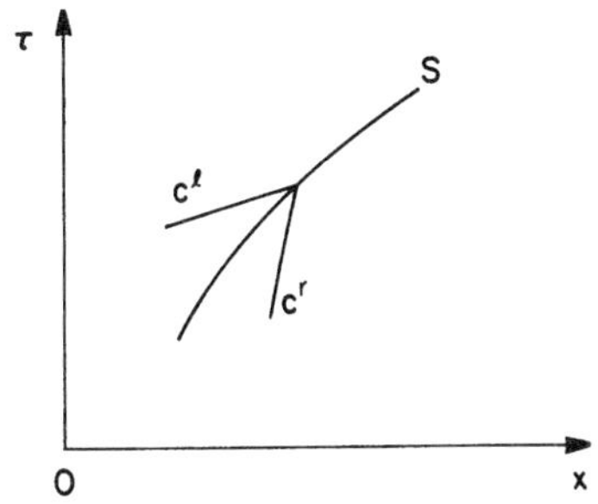

Figure 5.20

either side is not allowed to propagate beyond the shock path, so the shock remains sharp. In physical terms, this phenomenon is often referred to as the *self-sharpening tendency* of a shock.

Semishock. Suppose the equilibrium relation $f(c)$ changes its convexity just once at $c = c^*$; for example,

$$f''(c)\begin{cases} < 0 & \text{for } 0 \leq c < c^* \\ = 0 & \text{for } c = c^* \\ > 0 & \text{for } c > c^* \end{cases} \tag{5.4.8}$$

If $c^r < c^l < c^*$ or $c^* < c^l < c^r$, the situation is similar to that shown in Fig. 5.19(a) or (b), respectively, and the discontinuity connecting the states c^l and c^r is a shock. In fact, the state c^l can assume a slightly higher value (or a slightly lower value in the latter case) than c^* as long as the two points L and R can be connected by a straight chord without crossing the equilibrium curve. Note that such a crossing would violate the second law of thermodynamics.

Now we can think of two limiting situations in which the chord connecting the two points L and R is tangent to the equilibrium curve at the point L as shown in Fig. 5.21(a) and (b). Clearly, Eq. (5.4.4) implies that $c^l > c^r$ in the case of Fig. 5.21(a) and $c^l < c^r$ in the second case. Here again Eq. (5.4.3) leads to the same conclusion. In either case, however, we see the equilibrium curve on the right side as we move along the straight chord from L to R, which is the same as for a shock. Also, we notice that for both cases we have the condition

$$\sigma(c^l) = \tilde{\sigma}(c^l, c^r) < \sigma(c^r) \tag{5.4.9}$$

Thus the state on the left side propagates side by side with the same speed as the discontinuity, but the state on the right side tends to be overtaken by the discontinuity. A discontinuity of this type will be called a *semishock* because it is shocklike only on one side. There is a partially self-sharpening tendency with respect to the state on the right side. If c^l is higher than that

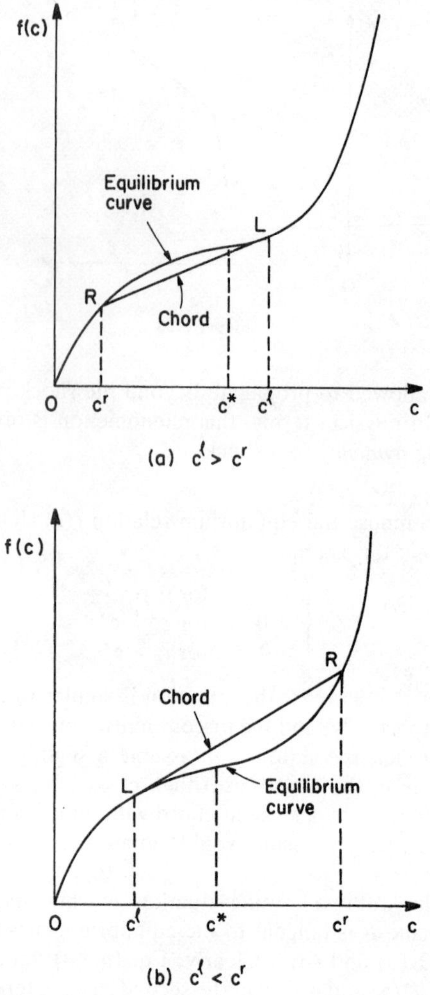

Figure 5.21

for the semishock in the case of Fig. 5.21(a), it may not be connected to c^r directly by a discontinuity. If the signs of $f''(c)$ in Eq. (5.4.8) are reversed, we expect to have semishocks for which we have

$$\sigma(c^l) < \tilde{\sigma}(c^l, c^r) = \sigma(c^r) \tag{5.4.10}$$

The existence of semishocks has been discussed by Liu (1974) in detail for a pair of reducible equations and by Jeffrey (1976) for systems of reducible

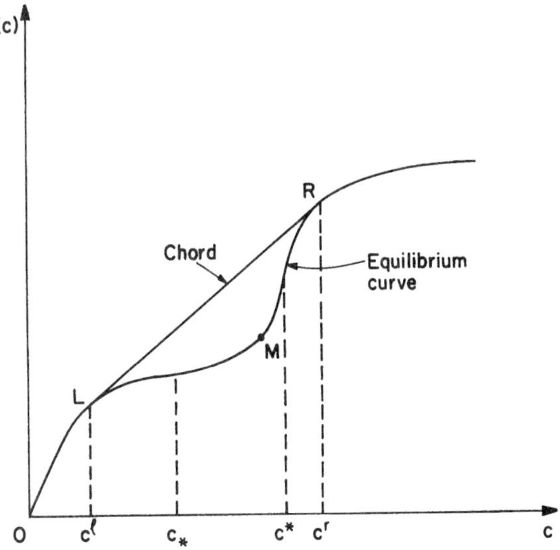

Figure 5.22

equations. These authors used the terms one-sided contact discontinuity and intermediate discontinuity, respectively.

Contact Discontinuity. Let us now consider a case when the equilibrium relation $f(c)$ has two inflection points, one at $c = c_*$ and the other at $c = c^*$, where $c_* < c^*$. An example is shown in Fig. 5.22, where we see three corners, L, M, and R. As long as both of the states c^l and c^r fall on the equilibrium curve below the upper corner R or above the lower corner L, the situation will look similar to the previous case. Thus we expect to have shocks as well as semishocks. But there is still another limiting situation for which the chord connecting the two points L and R is just tangent to the equilibrium curve at both ends as is depicted in Fig. 5.22. In this case, Eq. (5.4.4) implies $c^l < c^r$ so that here again we find the equilibrium curve on the right side as we proceed along the straight chord from L to R. In contrast to this, Eq. (5.4.3) would not give a unique direction of jump because the condition is satisfied with either $c^l > c^r$ or $c^l < c^r$. Therefore, we should not use Eq. (5.4.3) unless the isotherm is convex or sigmoid with one inflection point. It is obvious from Fig. 5.22 that we have

$$\sigma(c^l) = \tilde{\sigma}(c^l, c^r) = \sigma(c^r) \tag{5.4.11}$$

and so the states on both sides propagate with the same speed as the discontinuity. For this reason, we call such a discontinuity a *contact discontinuity* (see Lax, 1957; Liu, 1974; Jeffrey, 1976). In contrast to shocks, a contact discontinuity does not have a self-sharpening tendency.

We also notice that the preceding arguments remain valid even if the equilibrium curve contains more than two inflection points between L and R as long as the curve falls on the right side of the chord LR as we move along the chord from L to R. Since the same observation has been made in every case we have considered here, we are convinced that the entropy condition, Eq. (5.4.4), is equivalent to the following rule:

> RULE: If two states c^l and c^r are to be connected by a discontinuity, identify the chord connecting the two points L and R on the equilibrium curve for which $c = c^l$ and $c = c^r$, respectively. Then, for a system governed by Eq. (5.3.7), as we move along the chord from L to R, we must see the equilibrium curve on the right side.

If the variable coefficient appears with $\partial c/\partial x$ instead of $\partial c/\partial \tau$ in Eq. (5.3.7), we should traverse the chord from R to L to see the equilibrium curve on the right side (see Exercises 5.4.1 and 5.4.2). This rule will be found very useful when we treat actual problems in the next section.

Incidentally, if the equilibrium curve becomes linear between L and R and yet maintains a smooth connection at both ends (i.e., is continuously differentiable at both ends), the equality in Eq. (5.4.4) does hold and also Eq. (5.4.11). Thus both discontinuities with $c^l < c^r$ and with $c^l > c^r$ are admissible to the solution. This is not surprising at all because the system is essentially linear in this range of c-values. In this case, the system is said to be *linearly degenerate*. In fact, if we have a linear isotherm, and so a linear differential equation, discontinuities in the solution are all contact discontinuities and arise only when the initial data contain discontinuities. The direction of the jump is strictly determined by that in the initial data (see Sec. 3.1).

Exercises

5.4.1. By writing down a balance for the general kinematic wave (see Sec. 1.11), show that if the fluxes and concentrations on two sides of a discontinuity are q_1, q_2, k_1, and k_2, then its speed is $(q_1 - q_2)/(k_1 - k_2)$.

5.4.2. Derive the jump condition, Eq. (5.4.2), for a discontinuity in h in Exercise 5.2.1. Consider the entropy condition for this system and show that the discontinuities are shocks. How would you modify the Rule of this section in this case?

5.4.3. Consider the traffic-flow problem we treated in Exercise 5.2.3. If the vehicle concentration $k(z, t)$ contains a discontinuity, what are the conditions to be satisfied at the point of discontinuity? Show that, whenever the sum of the vehicle concentrations on both sides of a discontinuity equals the jamming concentration k_j, the discontinuity becomes stationary.

5.5 Discontinuous Solutions in Equilibrium Chromatography

We are now ready to complete the solutions of those problems left unfinished in Sec. 5.3. We would like to treat these in some detail in order to illustrate how to determine discontinuous solutions.

If the equilibrium relation is given by the Langmuir isotherm, Eq. (5.3.13), we have

$$[f] = \frac{\nu\gamma c^l}{1 + Kc^l} - \frac{\nu\gamma c^r}{1 + Kc^r} = \frac{\nu\gamma(c^l - c^r)}{(1 + Kc^l)(1 + Kc^r)}$$

and thus Eq. (5.4.2) gives

$$\tilde{\sigma}(c^l, c^r) = 1 + \frac{\nu\gamma}{(1 + Kc^l)(1 + Kc^r)} \tag{5.5.1}$$

for the propagation direction of a jump discontinuity (c^l, c^r). Since the isotherm is convex upward as shown in Fig. 5.19(a), the entropy condition requires that $c^l > c^r$ at a discontinuity, and we immediately see that Eq. (5.4.7) holds. This equation can also be obtained by comparing Eq. (5.3.14) with Eq. (5.5.1) under the condition $c^l > c^r$. Hence, all the discontinuities that arise in this problem are shocks.

For the saturation of a clean column, we have the initial data as in Eq. (5.3.20) and already know from the arguments in Sec. 5.3 that the solution would be involved with a discontinuity from the beginning. This discontinuity is definitely a shock and has states $c^l = c_0$ and $c^r = 0$, so the entropy condition is satisfied. The shock propagates in the direction

$$\tilde{\sigma}(c_0, 0) = 1 + \frac{\nu\gamma}{1 + Kc_0} \tag{5.5.2}$$

which is constant. Therefore, if we redraw the (x, τ)-plane of Fig. 5.14 in Fig. 5.23 and put in a straight line OS of slope $\tilde{\sigma}(c_0, 0)$ to represent the path of the shock, we can avoid any crossing of characteristics. For the region below OS will be covered by characteristics of slope $\sigma(0)$ like RS bearing the zero concentration from the x-axis, while above OS the characteristics QS will carry the uniform concentration c_0. The saturation wave will move into the column with a sharp front, as shown in the lower part of the figure. The front is a shock and remains sharp because of its self-sharpening tendency.

Suppose the equilibrium is governed by the BET isotherm, Eq. (5.3.22). Since the equilibrium curve is sigmoid with one inflection point, we expect to see shocks as well as semishocks in this problem. These discontinuities

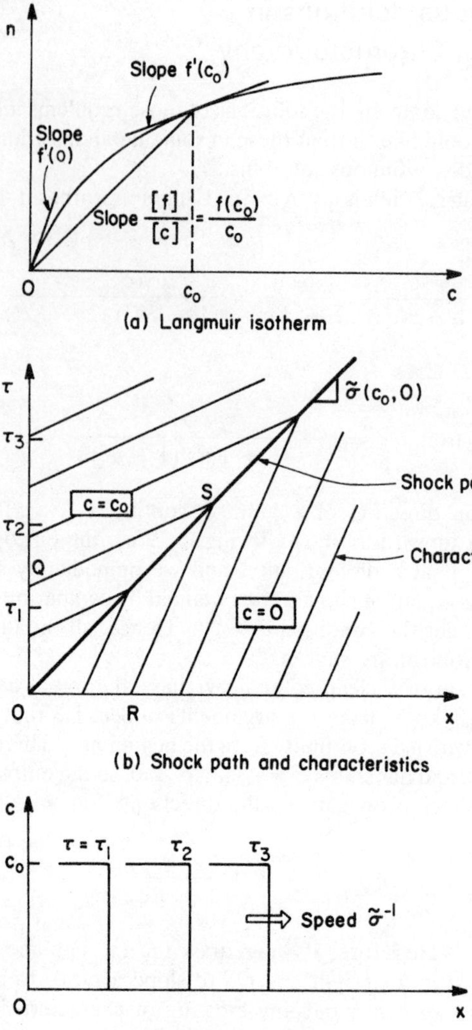

Figure 5.23

will propagate in the direction determined by Eqs. (5.3.22) and (5.4.2) with ν_1 for ν, as follows:

$$\tilde{\sigma}(\phi^l, \phi^r) = 1 + \frac{\nu_1[1 + (k-1)\phi^l\phi^r]}{(1-\phi^l)(1-\phi^r)[1+(k-1)\phi^l][1+(k-1)\phi^r]} \tag{5.5.3}$$

Note that Eq. (5.5.3) is symmetric with respect to the superscripts l and r. If the feed concentration ϕ_0 is low enough so that the corresponding point F on the equilibrium curve can be connected to the origin by a straight chord without crossing the curve, we see that the situation for the saturation of a clean bed remains the same as in the case with the Langmuir isotherm. Thus the saturation front is a shock and propagates in the direction

$$\tilde{\sigma}(\phi_0, 0) = 1 + \frac{\nu_1}{(1 - \phi_0)[1 + (k - 1)\phi_0]} \tag{5.5.4}$$

This is valid as long as the feed concentration ϕ_0 is less than the value

$$\phi_1 = \frac{1}{2}\frac{k - 2}{k - 1} \tag{5.5.5}$$

which is the concentration at the point where the chord through the origin becomes tangent to the equilibrium curve.

If the feed concentration ϕ_0 is larger than ϕ_1, the state ϕ_0 cannot be directly connected to the state $\phi = 0$ by a discontinuity. Recalling the behavior of those characteristics from the origin of the (x, τ)-plane (see Sec. 5.3), we notice that the characteristics for the states between ϕ_0 and ϕ_1 fan forward, and this would give a simple wave. The states $\phi^l = \phi_1$ and $\phi^r = 0$ can be connected by a discontinuity that propagates in the direction $\tilde{\sigma}(\phi_1, 0)$. This is a semishock because we have the relation $\sigma(\phi_1) = \tilde{\sigma}(\phi_1, 0) < \sigma(0)$ and so propagates with the same speed as the state ϕ_1. These arguments are presented in Fig. 5.24, from which we see that any crossing of characteristics is avoided. The saturation wave consists of two parts: one is the simple wave and the other the semishock. The two propagate side by side without interval, but the simple wave portion continues to expand. A solution of this kind is called a *combined wave*. Figure 5.25 shows the solution surface in three dimensions.

The elution with BET isotherm can be treated in a similar manner. Suppose the column is uniformly presaturated to a concentration ϕ_0. If this concentration is less than that at the inflection point (i.e., $\phi_0 < \phi^*$), the situation will be identical to the case with a Langmuir isotherm and so the solution is given by a centered simple wave. If the column concentration ϕ_0 is greater than ϕ^*, we observed in Sec. 5.3 that the characteristics emanating from the origin of the (x, τ)-plane and bearing the concentrations between ϕ^* and ϕ_0 would fan backward. Thus we expect to have a discontinuous solution. For this we pick the point I of concentration ϕ_0 on the equilibrium curve and draw the tangent to the curve (see the upper part of Fig. 5.26). The concentration ϕ_2 at this point of tangency is

$$\phi_2 = \frac{-1 + \sqrt{1 + (k - 2)\phi_0 - (k - 1)\phi_0^2}}{(k - 1)\phi_0} \tag{5.5.6}$$

which is real if $\phi_0 \leq 1$ and positive if $\phi_0 < (k - 2)/(k - 1)$. If we take ϕ^l

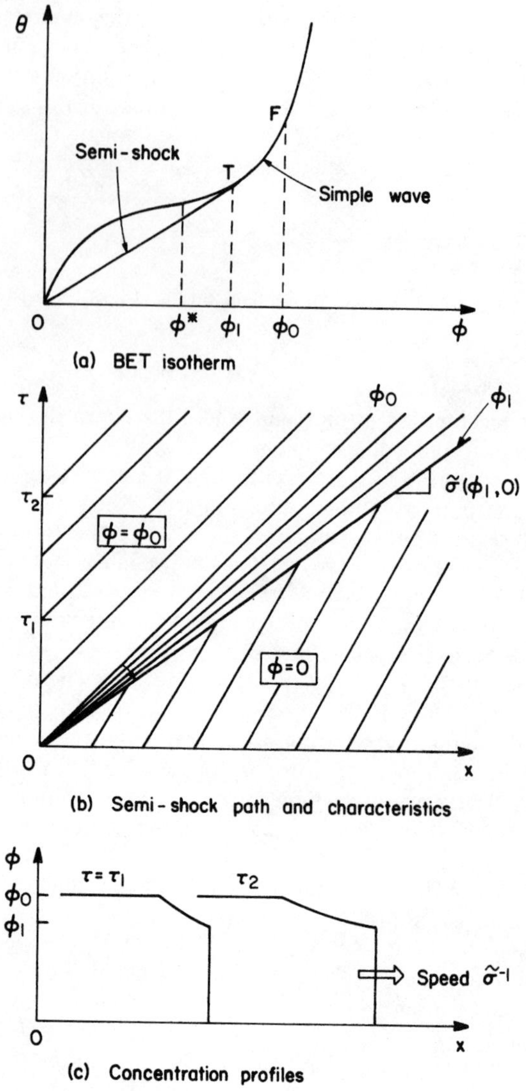

Figure 5.24

$= \phi_2$ and $\phi^r = \phi_0$, the entropy condition is satisfied so that the two states can be connected by a discontinuity. The variation of state from $\phi = 0$ to ϕ_2 will be continuous and give a simple wave. The discontinuity propagates in the direction $\tilde{\sigma}(\phi_2, \phi_0)$ and we have the relation $\sigma(\phi_2) = \tilde{\sigma}(\phi_2, \phi_0) < \sigma(\phi_0)$. Here we see another example of a semishock. The solution is given

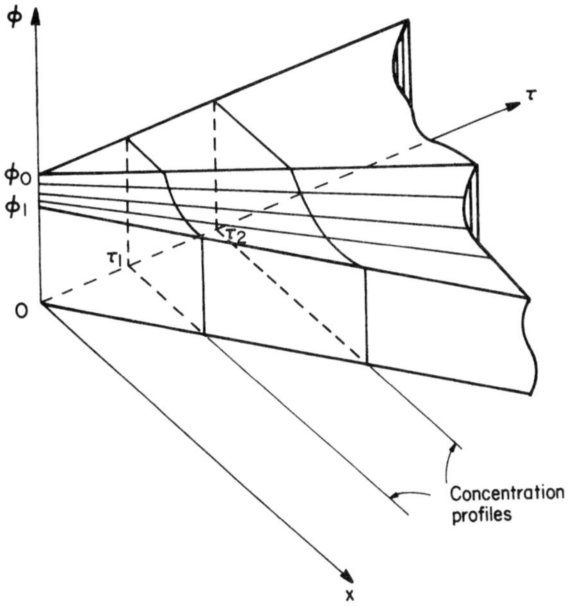

Figure 5.25

by a combined wave so that the elution wave remains partly diffuse and partly sharp. This is illustrated in Fig. 5.26. If $\phi_0 > (k - 2)/(k - 1)$, we can connect the point I and the origin of the equilibrium curve by a chord without touching or crossing the curve. This implies that, if we assign $\phi^l = 0$ and $\phi^r = \phi_0$, the entropy condition is satisfied (see Fig. 5.27). Hence the solution contains only the discontinuity, which is obviously a shock. This presents a case in which the elution wave is a shock.

Ion exchange with sigmoid isotherms was briefly discussed by Glueckauf (1947), who applied essentially the same approach to determine the partly diffuse and partly sharp boundaries. The existence of such combined waves was also confirmed experimentally with the system of H^+ and Cu^{2+} on Zeo-carb H. I.

Exercises

5.5.1. Assuming that the adsorption equilibrium follows a Freundlich isotherm, determine the solution for the saturation of a clean column.

5.5.2. Consider the equation of Exercise 5.2.2 but with the conditions $c(0, t) = c_1$ and $c(x, 0) = c_2$.

5.5.3. Discuss the saturation of a clean column, as well as the elution of a uniformly saturated column, when the adsorption equilibrium is described by the two-

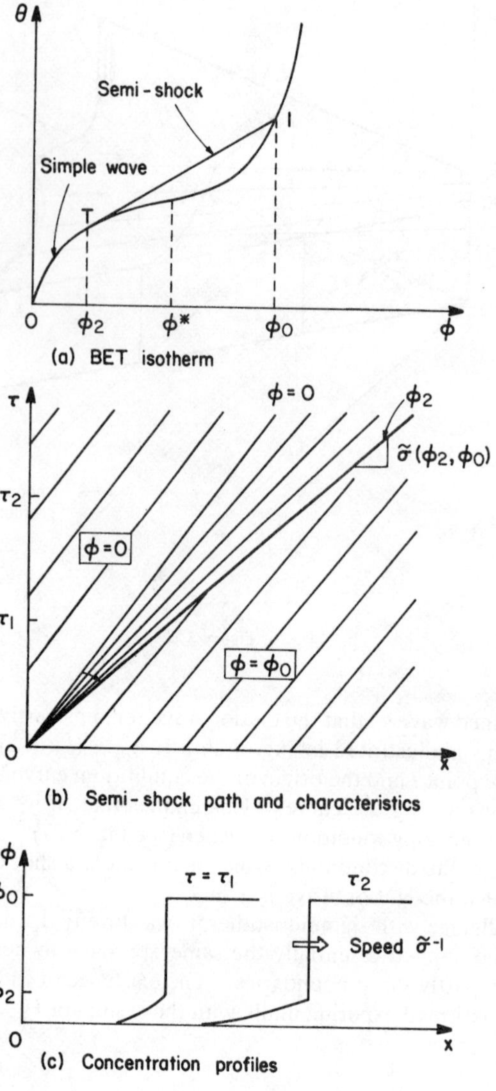

Figure 5.26

dimensional equation of state of van der Waals type; that is,

$$k_1 p = \frac{\theta}{1-\theta} e^{\theta/1-\theta} e^{-k_2\theta}$$

where p is the partial pressure of adsorbate, θ the surface coverage, and both k_1 and k_2 are positive constants.

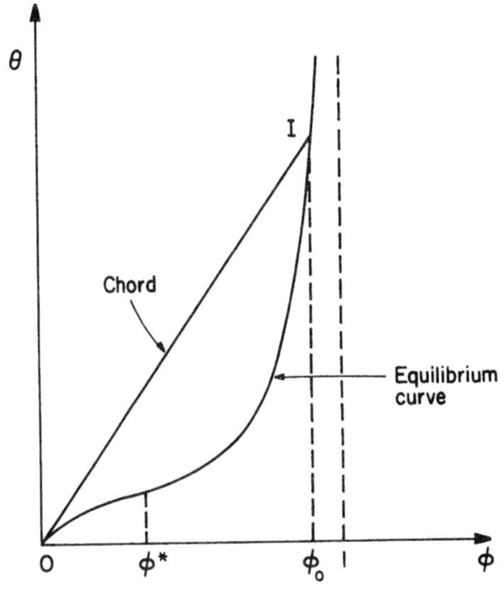

Figure 5.27

5.6 Water Flooding

Here we turn to another system for which the relationship between the flux and concentration is not purely convex or concave: the water flooding of oil reservoirs. If s and f are the volume fraction of water in the pore space and the fractional flow rate of water, respectively, at a distance z from the inlet at time t, we have, from Sec. 1.10, the equation

$$q_T \frac{\partial f}{\partial z} + \phi A \frac{\partial s}{\partial t} = 0 \tag{5.6.1}$$

Note that the subscript w used in Sec. 1.10 is dropped here for brevity. The fractional flow equation

$$f = F(s) \tag{5.6.2}$$

is the relation that determines f as a function of s. If we assume that the function $F(s)$ is continuously differentiable and introduce the dimensionless variables defined as

$$x = \frac{z}{L}, \qquad \tau = \frac{q_T t}{\phi A L} \tag{5.6.3}$$

where L denotes the length of the reservoir core, Eq. (5.6.1) can be rearranged to give the reducible equation

$$F'(s)\frac{\partial s}{\partial x} + \frac{\partial s}{\partial \tau} = 0 \tag{5.6.4}$$

We notice that τ represents the cumulative pore volume of water injected. Thus we have the characteristic direction

$$\frac{d\tau}{dx} = \frac{1}{F'(s)} = \sigma(s) \tag{5.6.5}$$

or the speed of propagation of the state s:

$$\frac{dx}{d\tau} = F'(s) = \frac{1}{\sigma(s)} \tag{5.6.6}$$

If the function $F(s)$ is nonlinear, we expect to see the characteristics crossing among themselves depending on the nature of the nonlinearity and the initial data. In such a case we must consider a discontinuity in the solution. The jump condition for the discontinuity can be formulated in the same manner as in Sec. 5.4 to give

$$\frac{d\tau}{dx} = \frac{[s]}{[F]} \equiv \tilde{\sigma}(s^l, s^r) \tag{5.6.7}$$

The entropy condition is the same as Eq. (5.4.4); that is,

$$\tilde{\sigma}(s^l, s^r) \geqslant \tilde{\sigma}(s^l, s) \tag{5.6.8}$$

for every s between s^l and s^r. With these conditions, the theorem in Sec. 5.4 is valid. We notice, however, that the variable coefficient $F'(s)$ appears with $\partial s/\partial x$ in the differential equation, and thus the rule of Sec. 5.4 must be modified as follows (see Exercises 5.4.2):

> RULE: If two states s^l and s^r are to be connected by a discontinuity, identify the chord connecting the two points L and R on the fractional flow curve for which $s = s^l$ and $s = s^r$, respectively. Then as we move along the chord from L to R we must see the fractional flow curve on the left side.

Let us now consider the actual situation of water flooding. The initial and feed conditions are specified as

$$\begin{aligned} &\text{at } \tau = 0, \qquad s = s_{wc} \\ &\text{at } x = 0, \qquad s = 1 \end{aligned} \tag{5.6.9}$$

the latter implying a pure water feed. We will take simple expressions for the relative permeability data [cf. Eq. (1.10.3)]

$$\begin{aligned} K_{ro} &= (K_{ro})_0(1 - \psi)^2 \\ K_{rw} &= (K_{rw})_0\psi^3 \end{aligned} \tag{5.6.10}$$

where

$$\psi = \frac{s - s_{wc}}{1 - s_{or} - s_{wc}} \tag{5.6.11}$$

and use Eq. (1.10.2) for the fractional flow equation. Thus we have

$$f = F(s) = \frac{\psi^3}{\psi^3 + \alpha(1 - \psi)^2} \tag{5.6.12}$$

in which

$$\alpha = \frac{\mu_w(K_{ro})_0}{\mu_o(K_{rw})_0} \tag{5.6.13}$$

The fractional flow curve contains an inflection point as sketched for some fixed value α in Fig. 5.28.

The characteristic direction in this case is given by

$$\sigma(s) = \frac{(1 - s_{or} - s_{wc})}{\alpha} \frac{[\psi^3 + \alpha (1 - \psi)^2]^2}{\psi^2(\psi - 1)(\psi - 3)} \tag{5.6.14}$$

and diverges at $\psi = 0$ and $\psi = 1$ or at $s = s_{wc}$ and $s = 1 - s_{or}$. This implies that the initial state $s = s_{wc}$ tends to remain stationary and also that the τ-axis is characteristic and so carries the state $s = 1 - s_{or}$ or $s = s_{wc}$. If we consider the variation of state from $s = 1$ to $s = s_{wc}$ as the limit of continuous monotonic variations, we can think of characteristics for states between $s = 1$ and $s = s_{wc}$ all emanating from the origin of the (x, τ)-plane. In the range $1 - s_{or} \leq s \leq 1$, we have $f = 1$ and $\sigma(s) = \infty$ so that the characteristics for states from $s = 1 - s_{or}$ to $s = 1$ are all vertical and coincide with the τ-axis.

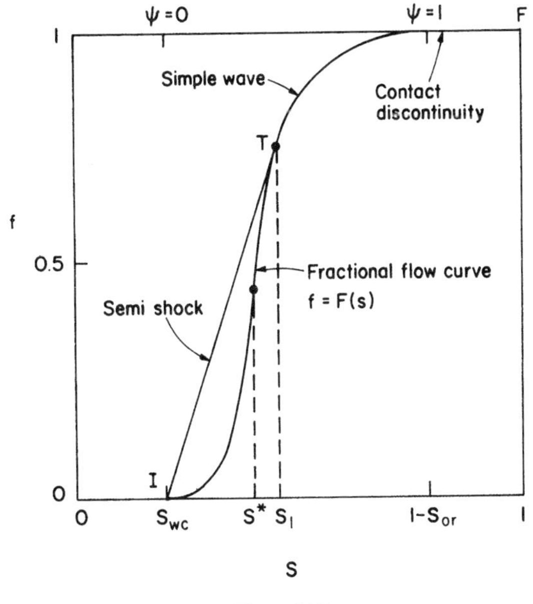

Figure 5.28

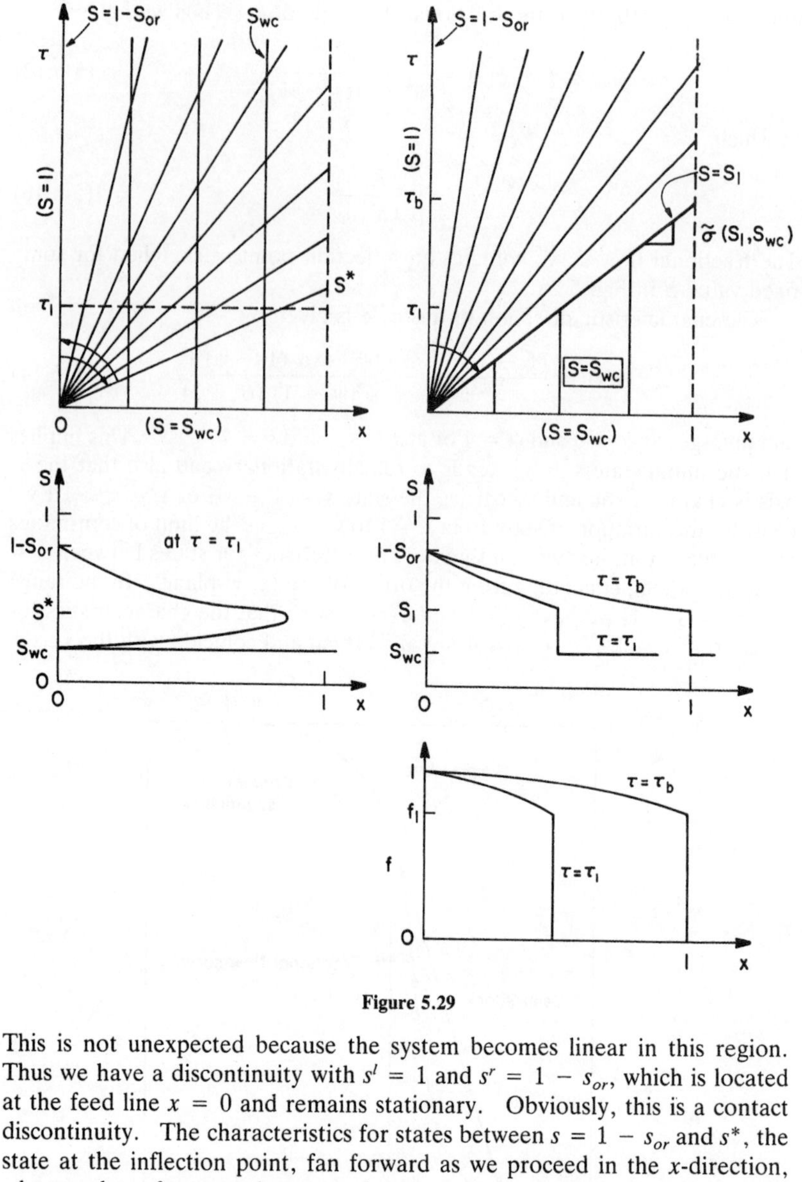

Figure 5.29

This is not unexpected because the system becomes linear in this region. Thus we have a discontinuity with $s^l = 1$ and $s^r = 1 - s_{or}$, which is located at the feed line $x = 0$ and remains stationary. Obviously, this is a contact discontinuity. The characteristics for states between $s = 1 - s_{or}$ and s^*, the state at the inflection point, fan forward as we proceed in the x-direction, whereas those for states between s^* and s_{wc} fan backward. The situation is shown on the left side of Fig. 5.29, where we also see the triple-valued profile of s. Here again we need to consider a discontinuity in the solution.

In view of the entropy condition or the Rule stated previously, we pick the point I of the initial state $s = s_{wc}$ in the fractional flow curve diagram and draw the tangent to the curve as shown in Fig. 5.28. The state s_1 at the point of tangency T is determined by the cubic equation

$$\psi_1^3 + 2\alpha\psi_1 - 2\alpha = 0 \tag{5.6.15}$$

If we assign $s^l = s_1$ and $s^r = s_{wc}$, the entropy condition is satisfied, and the jump condition gives

$$\tilde{\sigma}(s_1, s_{wc}) = \frac{s_1 - s_{wc}}{F(s_1)} = \frac{1}{2}(s_1 - s_{wc})(3 - \psi_1) \tag{5.6.16}$$

for which we have made use of Eq. (5.6.15). Hence, we see that the state varies continuously from $(1 - s_{or})$ to s_1 (simple wave part) and then jumps down from s_1 to s_{wc} discontinuously. Since we have the relation $\sigma(s_1) = \tilde{\sigma}(s_1, s_{wc}) < \sigma(s_{wc})$, the discontinuity is a semishock and it propagates at the same speed as the state s_1. The complete solution consists of a contact discontinuity ($s^l = 1$, $s_r = 1 - s_{or}$) located at the feed line $x = 0$, a simple wave ($1 - s_{or} \sim s_1$), a semishock ($s^l = s_1$, $s_r = s_{wc}$), and a constant state ($s = s_{wc}$). This solution structure is presented on the right side of Fig. 5.29 in comparison with the physically impossible situation on the left. We also notice that, due to the nature of the fractional flow curve, the profile of f will not contain a discontinuity at $x = 0$, and the state of flow on the right side of the semishock will be $f = 0$, which implies that the flow is of pure oil.

The breakthrough curve may be readily determined from Fig. 5.29 by reading the state along the vertical line at $x = 1$, but to make it meaningful as well as useful we have to do it for f or $(1 - f)$ as shown in Fig. 5.30. Note that τ_b denotes the time when the semishock reaches the exit at $x = 1$, and so $\tau_b = \tilde{\sigma}(s_1, s_{wc})$. Up to the moment $\tau = \tau_b$, we see pure oil coming out at the outlet of the reservoir core, but at $\tau = \tau_b$ the water phase suddenly makes its breakthrough, so the fraction of oil in the product stream drops down to a low value and afterward continues to decrease further. Since the dimensionless time τ is equivalent to the cumulative pore volume, we see that the amount of oil recovered, v_o, when expressed in terms of the pore volume,

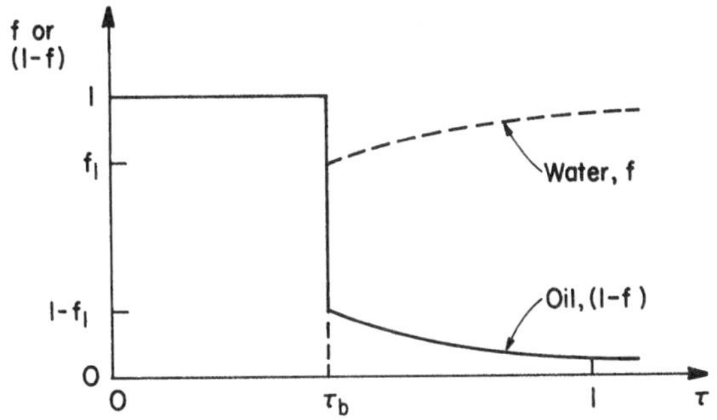

Figure 5.30

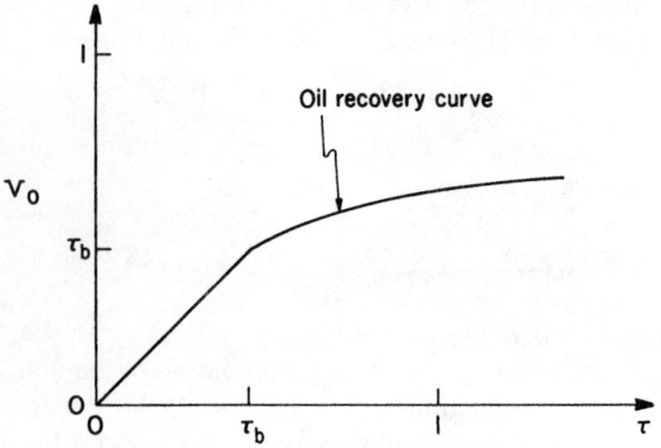

Figure 5.31

is given by the equations

$$v_o = \begin{cases} \tau & \text{for } \tau \leqslant \tau_b \\ \tau_b + \displaystyle\int_{\tau_b}^{\tau} (1 - f)\, d\tau & \text{for } \tau > \tau_b \end{cases} \tag{5.6.17}$$

or by the area under the breakthrough curve of oil, $(1 - f)$ versus τ. For $\tau > \tau_b$, we can find a characteristic reaching the exit from the origin for every τ, and thus the relation $\tau = \sigma(s)$ holds along the line $x = 1$. Now the second equation in Eq. (5.6.17) may be expressed in terms of s:

$$v_o(s) = \tilde{\sigma}(s_1, s_{wc}) + \int_{s_1}^{s} \{1 - F(\eta)\}\sigma'(\eta)\, d\eta, \qquad \text{for } \tau > \tau_b \tag{5.6.18}$$

The plot of v_o versus τ is called the oil recovery curve and is shown schematically in Fig. 5.31.

Exercises

5.6.1. Consider a core with the properties

$$K_{ro} = (1 - \psi)^3, \qquad K_{rw} = 0.25\psi^2$$

$$s_{wc} = 0.2, \qquad 1 - s_{or} = 0.7$$

$$\mu_w = 0.5\ cp, \qquad \mu_o = 40\ cp$$

and analyze the oil recovery by water flooding.

5.6.2. When the relative permeabilities are given by

$$K_{ro} = (K_{ro})_0(1 - \psi)^2(1 + 2\psi)$$

$$K_{rw} = (K_{rw})_0\psi^3$$

discuss the solution of water flooding (see Claridge and Bondor, 1974).

5.7 Quasi-linear Equation with n Independent Variables

We recall from the discussion in Sec. 2.1 that, if a particular solution $z(x_1, x_2, \ldots, x_n)$ of a partial differential equation is to be found, z must be specified on some $(n - 1)$-dimensional subspace of the $(n + 1)$-dimensional space of the x_i and z. The easiest way to do this is to define the $(n - 1)$-dimensional region parametrically by

$$x_i = X_i(\xi_1, \ldots, \xi_{n-1}) \tag{5.7.1}$$

and then z will be specified by

$$z = Z(\xi_1, \ldots, \xi_{n-1}) \tag{5.7.2}$$

Letting p_i denote $\partial z/\partial x_i$ and using the convention of summing on a repeated index from 1 to n, we can write the general quasilinear equation† as

$$\sum P_i(x_1, \ldots, x_n, z)p_i = R(x_1, \ldots, x_n, z) \tag{5.7.3}$$

By writing this as $\Sigma P_i p_i + R(-1) = 0$, we again see that the direction given by

$$\frac{dx_i}{ds} = P_i(x_1, \ldots, x_n, z) \tag{5.7.4}$$

$$\frac{dz}{ds} = R(x_1, \ldots, x_n, z) \tag{5.7.5}$$

is always perpendicular to the normal to the solution "surface" and hence lies in it. These equations again define the characteristics, and those that make up the particular solution that we are seeking form an $(n - 1)$-parameter family obtained by integrating the characteristic differential equations for $x_i(s, \xi_1, \ldots, \xi_{n-1})$ and $z(s, \xi_1, \ldots, \xi_{n-1})$ with the initial conditions

$$\begin{aligned} x_i(0, \xi_1, \ldots, \xi_{n-1}) &= X_i(\xi_1, \ldots, \xi_{n-1}) \\ z(0, \xi_1, \ldots, \xi_{n-1}) &= Z(\xi_1, \ldots, \xi_{n-1}) \end{aligned} \tag{5.7.6}$$

By eliminating the n quantities $\xi_1, \ldots, \xi_{n-1}$ and s between the $(n + 1)$ equations $x_i = x_i(s, \xi_1, \ldots, \xi_{n-1})$ and $z = z(s, \xi_1, \ldots, \xi_{n-1})$, we could obtain the solution in the form $z = z(x_1, \ldots, x_n)$, but this may not be possible in practice and is unnecessary either computationally or conceptually.

To illustrate this, let us consider the development of a paper chromatogram when there are no dispersive effects other than a weakly nonlinear isotherm. If the concentration of solute in the fluid phase is c and the equilibrium

†This equation can be reduced to an equivalent homogeneous linear differential equation with one additional independent variable x_{n+1}. This is done by making $z = x_{n+1}$ and regarding the solution in the implicit form, $\phi(x_1, \ldots, x_n, x_{n+1}) = \text{constant}$, as the new dependent variable. See Courant and Hilbert (1962). However, there is a difficulty in specifying the initial data.

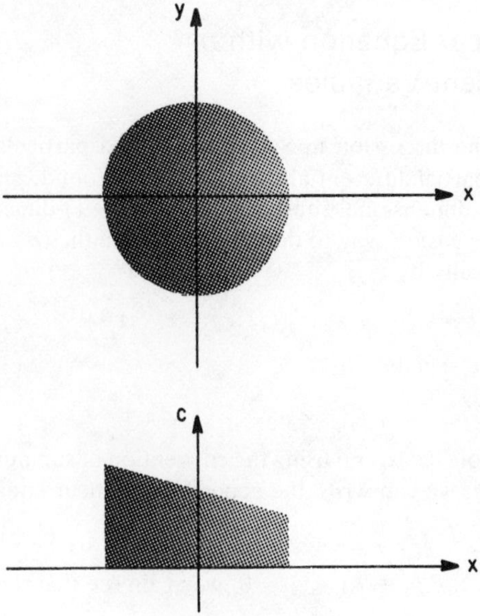

Figure 5.32

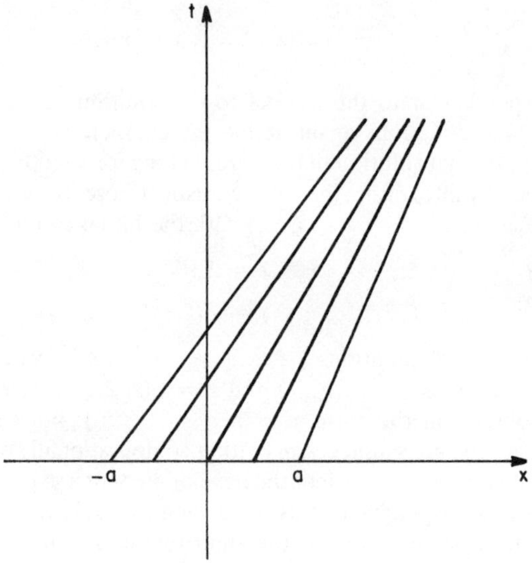

Figure 5.33

concentration on the paper $(\alpha c - \beta c^2)$, then the movement of the solute will be governed by

$$V_x \frac{\partial c}{\partial x} + V_y \frac{\partial c}{\partial y} + \frac{\partial c}{\partial t} + \frac{\partial}{\partial t}(\alpha c - \beta c^2) = 0$$

that is, by

$$V_x \frac{\partial c}{\partial x} + V_y \frac{\partial c}{\partial y} + (1 + \alpha - 2\beta c)\frac{\partial c}{\partial t} = 0 \tag{5.7.7}$$

Let us suppose that the solute has originally been put on within a circle of radius a about the origin. We may take $\xi_1 = r$ to be the radial distance from the origin and $\xi_2 = \theta$, the angular coordinate; then Eq. (5.7.1) will be

$$x = r \cos \theta, \qquad y = r \sin \theta, \qquad 0 \leqslant r \leqslant a, \quad 0 \leqslant \theta \leqslant 2\pi \tag{5.7.8}$$

and the specification of the initial distribution will be

$$c = C(r, \theta) \tag{5.7.9}$$

The characteristic differential equations are

$$\begin{aligned} \frac{dx}{ds} &= V_x, \qquad \frac{dy}{ds} = V_y \\ \frac{dt}{ds} &= 1 + \alpha - 2\beta c, \qquad \frac{dc}{ds} = 0 \end{aligned} \tag{5.7.10}$$

As in the previous cases, the characteristics are straight lines:†

$$x = r \cos \theta + V_x s \tag{5.7.11}$$

$$y = r \sin \theta + V_y s \tag{5.7.12}$$

$$t = [1 + \alpha - 2\beta C(r, \theta)]s \tag{5.7.13}$$

$$c = C(r, \theta) \tag{5.7.14}$$

The elimination of s, r, and θ is bedeviled by the fact that the characteristics can overlap. Let us simplify matters by taking $V_y = 0$, $V_x = V$, and $C(r, \theta) = f(r \cos \theta)$, where f is a decreasing function of its argument, say $b(a - r \cos \theta)$, where b is a constant. Figure 5.32 shows the initial distribution of solute in the spot. The characteristics will all lie in planes parallel to the y-axis, and we can picture them in a section by the plane $y = 0$ (see Fig. 5.33).

The characteristic emanating from the point at $x = x_0$ in the interval

†If $\beta = 0$, we have the uninteresting situation in which the spot simply moves as a circle with velocity components $V_x/(1 + \alpha)$ and $V_y/(1 + \alpha)$.

$[-a, a]$ of the x-axis is

$$x = x_0 + Vs$$
$$t = [1 + \alpha - 2\beta b(a - x_0)]s$$

or

$$Vt = [(1 + \alpha - 2\beta ba) + 2\beta bx_0](x - x_0) \tag{5.7.15}$$

To avoid any problem from the overlapping of characteristics and the formation of shocks—questions that will be more fully considered later—we will confine attention to times less than

$$\frac{(1 + \alpha - 4ab\beta)^2}{2b\beta V} \tag{5.7.16}$$

during which no overlapping can occur. If the isotherm is weakly nonlinear, β is small, so this may be quite a long interval of time. Because of the assumed initial distribution, we see that the diameter of the spot in the direction of motion contracts; in fact, at time t its diameter along the x-axis is

$$2a\left[1 - \frac{2b\beta Vt}{(1 + \alpha)(1 + \alpha - 4ab\beta)}\right] \tag{5.7.17}$$

The situation is shown in Fig. 5.34, where we attempt to indicate the value of c by the density of the shading.

The given initial distribution was designed to avoid shock fronts and was somewhat artificial. A more natural initial condition would be to make c constant within the initial circle, and in this case we could study the development of a diffuse rear boundary without running into any problem with shocks. If $C(r, \theta) = 1$† for $0 \leq r \leq a$, $0 \leq \theta \leq 2\pi$, and $C = 0$, $r > a$, we have to imagine all concentrations between 0 and 1 to be present on the boundary $r = a$. Confining attention to the rear boundary, $r = a$, $\pi/2 \leq \theta \leq 3\pi/2$, we have for the characteristic bearing the concentration c, $0 \leq c \leq 1$, the equations

$$\begin{aligned} x &= a \cos\theta + Vs \\ y &= a \sin\theta \\ t &= (1 + \alpha - 2\beta c)s \end{aligned} \tag{5.7.18}$$

Thus at time t the concentration contour of c is given by

$$\begin{aligned} x - \frac{Vt}{1 + \alpha - 2\beta c} &= a \cos\theta \\ y &= a \sin\theta \end{aligned} \tag{5.7.19}$$

† Again we are not losing any generality by taking the constant concentration to be unity.

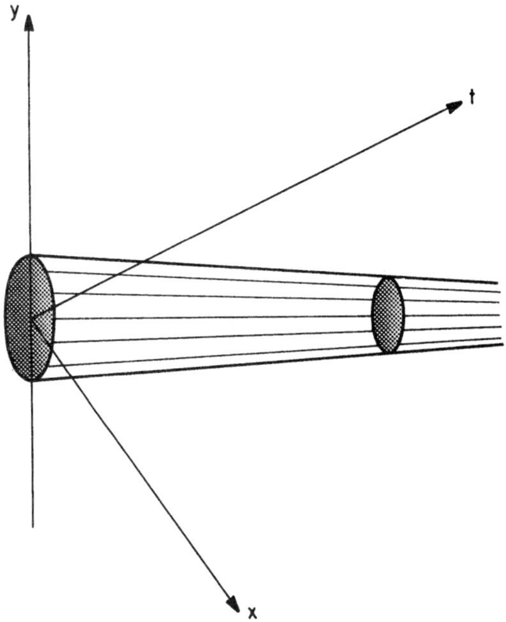

Figure 5.34

or the rear half of the circle

$$\left[x - \frac{Vt}{1 + \alpha - 2\beta c}\right]^2 + y^2 = a^2 \tag{5.7.20}$$

The blurring of the rear boundary is such that the contours of constant c are still circles, as shown in Fig. 5.35. However, we have not yet accounted for what should happen at the front of the spot.

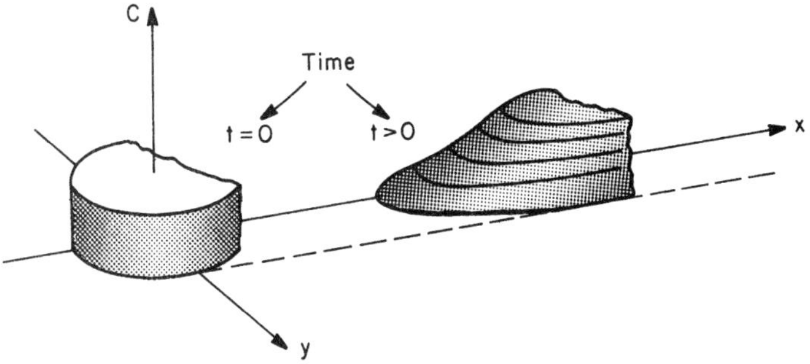

Figure 5.35

Exercises

5.7.1. Show that the envelope of the characteristics given by Eq. (5.7.15) is the parabola

$$Vt = \frac{1}{2}\beta b\left[(x - a) + \frac{1 + \alpha}{2\beta b}\right]^2 = \frac{[2\beta b(x - a) + (1 + \alpha)]^2}{8\beta b}$$

and that the first point of contact is by the characteristic emanating from $x_0 = -a$. Hence confirm the statement around Eq. (5.7.16).

5.7.2. Two solutes are adsorbed independently of one another following the isotherms $\alpha_1 c_1 - \beta_1 c_1^2$ and $\alpha_2 c_2 - \beta_2 c_2^2$, respectively, where β_1 and β_2 are sufficiently small. If they are put on the circle $x^2 + y^2 = a^2$ in concentration distributions $b_1(a - r\cos\theta)$ and $b_2(a - r\cos\theta)$, respectively, show that the time required for development of the chromatogram is at least

$$\frac{Vt}{2a} = \frac{(1 + \alpha_1)(1 + \alpha_2 - 4ab_2\beta_2)}{\alpha_1 - \alpha_2 + 4ab_2\beta_2}$$

What is meant by β_1 and β_2 being sufficiently small?

REFERENCES

5.1 and 5.2. The case of single homogeneous equations is briefly discussed, for example, in

P. D. Lax, "The Formation and Decay of Shock Waves," *Amer. Math. Monthly* **79**, 227–241 (1972).

P. D. Lax, "Hyperbolic Systems of Conservation Laws and the Mathematical Theory of Shock Waves," *Regional Conference Series in Applied Mathematics*, No. 11, SIAM, Philadelphia, Pa., 1973.

G. B. Whitham, *Linear and Nonlinear Waves*, Wiley-Interscience, New York, 1974.

Although these are concerned with pairs of or systems of equations, the following references are extremely helpful:

R. Courant and K. O. Friedrichs, *Supersonic Flow and Shock Waves*, Wiley-Interscience, New York, 1948.

P. D. Lax, "Hyperbolic Systems of Conservations Laws II," *Comm. Pure Appl. Math.* **10**, 537–566 (1957).

5.3. Equilibrium chromatography is the subject of many classical papers. See for example,

J. N. Wilson, "A Theory of Chromatography," *J. Amer. Chem. Soc.* **62**, 1583–1591 (1940).

D. DeVault, "The Theory of Chromatography," *J. Amer. Chem. Soc.* **65**, 532–540 (1943).

J. Weiss, "On the Theory of Chromatography," *J. Chem. Soc.* **1943**, 297–303 (1943).

J. W. Walter, "Multiple Adsorption from Solutions," *J. Chem. Phys.* **13**, 229–234 (1945).

E. Glueckauf, "Theory of Chromatography. Part II. Chromatograms of a Single Solute," *J. Chem. Soc.* **1947**, 1302–1308 (1947).

L. G. Sillen, "On Filtration through a Sorbent Layer: IV. The ψ Condition, a Simple Approach to the Theory of Sorption Columns," *Arkiv Kemi* **2**, 477–498 (1950).

S. Goldstein, "On the Mathematics of Exchange Processes in Fixed Columns: II. The Equilibrium Theory as the Limit of the Kinetic Theory," *Proc. Roy. Soc.* (London) **A219**, 171–185 (1953).

An extensive study in this direction is found in

H.-K. Rhee, Studies on the Theory of Chromatography, Ph.D. Thesis, University of Minnesota, 1968.

For the application of the throughput parameter, see Walter's 1945 paper and Sillen's 1950 paper, cited above, and

A. C. Offord and J. Weiss, "Chromatography with Several Solutes," *Disc. Faraday Soc.* **7**, 26–34 (1949).

N. K. Hiester and T. Vermeulen, "Saturation Performance of Ion-exchange and Adsorption Columns," *Chem. Eng. Progr.* **48**, 505–516 (1952).

G. Klein, D. Tondeur, and T. Vermeulen, "Multicomponent Ion Exchange in Fixed Beds: General Properties of Equilibrium Systems," *Ind. Eng. Chem. Fundam.* **6**, 339–351 (1967).

D. Tondeur and G. Klein, "Multicomponent Ion Exchange in Fixed Beds: Constant-separation-factor Equilibrium," *Ind. Eng. Chem. Fundam.* **6**, 351–361 (1967).

5.4 and 5.5. The concept of discontinuities and the formulation of the jump condition are commonly found in all the references cited above. For the generalized entropy condition and the uniqueness theorem, see

O. A. Oleinik, "Uniqueness and Stability of the Generalized Solution of the Cauchy Problem for a Quasi-linear Equation," *Uspekhi Mat. Nauk* (*N.S.*) **14**, 165–170 (1959); English translation in *Amer. Math. Soc. Transl.*, Ser. 2, **33**, 285–290 (1963).

Also see the second reference by Lax (1973) for Secs. 5.1 and 5.2, and

P. D. Lax, "Shock Waves and Entropy," in *Contributions to Nonlinear Functional Analysis*, edited by E. H. Zarantonello, Academic Press, New York, 1971, pp. 603–634.

Concerning the definition of a shock, see the last two references for Secs. 5.1 and 5.2: one by Courant and Friedrichs (1948) and the other by Lax (1957).

The problem of equilibrium relations or flux relations that are not purely convex or concave is discussed in

I. M. Gelfand, "Some Problems in the Theory of Quasilinear Equations," *Uspekhi Mat. Nauk* **14**, 87–158 (1959); English translation in *Amer. Math. Soc. Transl.*, Ser. 2, **29**, 295–381 (1963).

Z.-Q. Wu, "On the Existence and Uniqueness of the Generalized Solutions of the Cauchy Problem for Quasilinear Equations of First Order without Convexity Conditions," *Acta Math. Sinica* **13**, 515–530 (1963); English translation in *Chinese Math.* **4**, 561–577 (1964).

D. P. Ballou, "Solutions to Nonlinear Hyperbolic Cauchy Problems without Convexity Conditions," *Trans. Amer. Math. Soc.* **152**, 441–460 (1970).

For semishocks and contact discontinuities, see the discussions in

H.-K. Rhee and N. R. Amundson, "An Analysis of an Adiabatic Adsorption Column: Part I. Theoretical Development," *Chem. Eng. J.* **1**, 241–254 (1970).

T. P. Liu, "The Riemann Problem for General 2×2 Conservation Laws," *Trans. Amer. Math. Soc.* **199**, 89–112 (1974).

A. Jeffrey, *Quasilinear Hyperbolic Systems and Waves*, Pitman, London, 1976.

Combined waves in physical systems are to be found in the paper by Glueckauf (1947) cited above for Sec. 5.3.

5.6. The treatment of water flooding follows somewhat on the lines laid down in

S. E. Buckley and M. C. Leverett, "Mechanism of Fluid Displacement in Sands," *Trans. AIME* **146**, 107–116 (1942).

H. J. Welge, "A Simplified Method for Computing Oil Recovery by Gas or Water Drive," *Trans. AIME* **195**, 91–98 (1952).

W. L. Claridge and P. L. Bondor, "A Graphical Method for Calculating Linear Displacement with Mass Transfer and Continuously Changing Mobilities," *Soc. Pet. Eng. J.* **14**, 609–618 (Dec. 1974).

K. Y. Choi, "Studies on the Secondary and Tertiary Recoveries of Oil," M.S. Thesis, Seoul National University, Seoul, Korea, 1977.

5.7. The theory of quasi-linear equations with n independent variables is briefly discussed in

R. Courant and D. Hilbert, *Methods of Mathematical Physics: Vol. II. Partial Differential Equations*, Wiley-Interscience, New York, 1962.

See also the book by Whitman (1974) cited above for Secs. 5.1 and 5.2 and the references for Sec. 7.6.

Formation and Propagation of Shocks

6

The main theme of this chapter is the interaction of a shock with a simple wave. This we are going to describe in rather painful detail with a sequence of examples since some important ideas are introduced here concerning wave propagation. The conservation equation for equilibrium chromatography will be used for illustration throughout the chapter except for the last two sections, which are concerned with the subjects of traffic flow and sedimentation, respectively. The first section deals with the compression wave to show how it gives birth to a shock. In the following three sections, we will discuss the propagation of shocks while they are interacting with one or two simple waves in various manners. The moving-bed adsorber gives an example of backward propagating waves, and its analysis in a finite domain is of practical importance (Sec. 6.5). In Sec. 6.6, traffic-flow problems are treated with the simplest form of the flux curve. Some important features are elucidated in detail with illustrative examples. The final section on sedimentation takes up the case where the flux–concentration relationship is not purely convex. A complete analysis is given for quadratic, cubic, and quartic relations.

6.1 Formation of a Shock

In Secs. 5.2 and 5.3, we learned that in the region of a compression wave the characteristics would eventually cross one another, and over much of the (x, τ)-plane three characteristics could pass through the same point (see Figs. 5.8 and 5.16). Since it is impossible for more than one concentration to be present at one point and instant, such a solution is not physically acceptable and a discontinuity must be introduced that allows the solution to be one-valued. An arbitrary line could be drawn on one side of which the greatest value of concentration would be taken and on the other the least value. This would give a one-valued and piecewise smooth solution, but it would not necessarily be physically acceptable. To be physically acceptable, the discontinuity that is introduced to make the solution one-valued must move with a speed that satisfies the jump condition, Eq. (5.4.2), and the entropy condition, Eq. (5.4.4).

Looking again at the layout of characteristics in the (x, τ)-plane, we see that the family of characteristics has an envelope and that the crossing of characteristics takes place only inside the envelope, as shown in Fig. 6.1. The point at which the envelope takes the lowest ordinate is the point at which characteristics begin to intersect one another, and so the discontinuity or shock must initiate there. Hence, it is expedient to determine the equation of the envelope and use it to determine where a discontinuity is to be first introduced.

Here we will confine our discussion to a closed interval of the x- or τ-axis along which the initial or boundary data $[F(x)$ or $G(\tau)]$ are specified by a continuously differentiable function. We also assume that outside the interval the data remain constant without discontinuities at both end points and further that one-sided derivatives are defined at both ends. For the case of

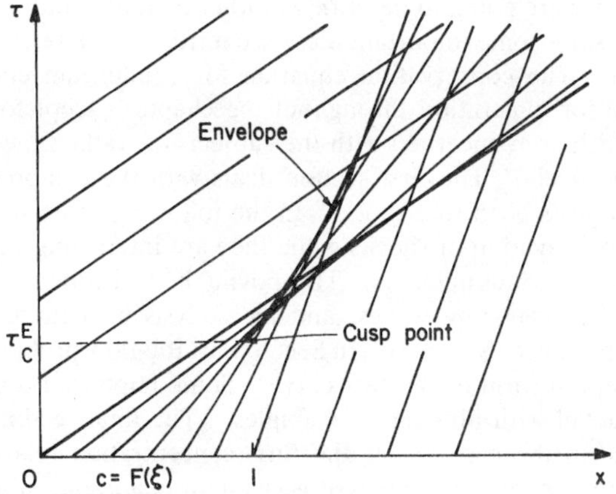

Figure 6.1

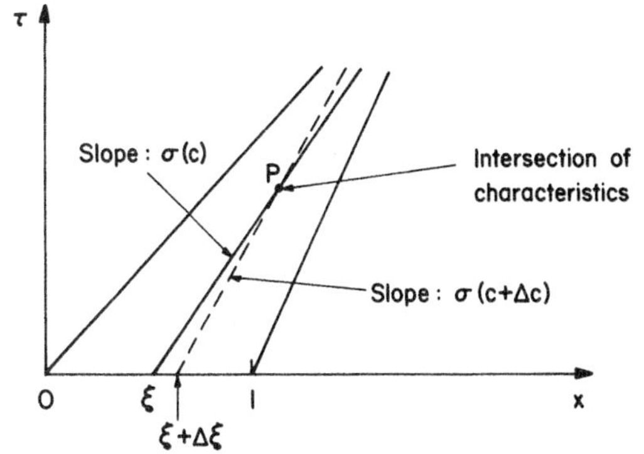

Figure 6.2

chromatography, the data may then be described for convenience of illustration as

$$\begin{aligned}
&\text{at}\quad \tau = 0, \quad x = \xi \quad \text{and} \quad c = F(\xi), \qquad \text{for} \quad 0 \le \xi \le 1 \\
&\qquad\qquad\qquad\qquad\qquad\quad c = c_1 = F(1), \qquad \text{for} \quad \xi \ge 1 \qquad (6.1.1) \\
&\text{at}\quad x = 0, \quad \tau = \eta \quad \text{and} \quad c = c_0 = F(0), \qquad \text{for} \quad \eta \ge 0
\end{aligned}$$

Note that both c_0 and c_1 are constant and that, since we are interested in a compression wave, the derivatives $d\sigma/dc$ and $dF/d\xi$ must have a common sign. Let us now consider the situation depicted in Fig. 6.2. Here the characteristics emanating from two adjacent points ξ and $\xi + \Delta\xi$ on the x-axis, respectively, intersect at point P. If we let $\Delta c = F(\xi + \Delta\xi) - F(\xi)$, the two characteristics will have the slopes $\sigma(c)$ and $\sigma(c + \Delta c)$, respectively, and thus we can immediately write two equations at P:

$$\begin{aligned}
\tau &= \sigma(c)\,\{x - \xi\} \\
\tau &= \sigma(c + \Delta c)\,\{x - (\xi + \Delta\xi)\}
\end{aligned} \qquad (6.1.2)$$

from which the point of intersection is determined as

$$x = \xi + \frac{\sigma(c + \Delta c)}{[\sigma(c + \Delta c) - \sigma(c)]/\Delta\xi}$$

$$\tau = \frac{\sigma(c)\sigma(c + \Delta c)}{[\sigma(c + \Delta c) - \sigma(c)]/\Delta\xi}$$

In the limit as $\Delta\xi$ approaches zero, we have

$$\begin{aligned}
x^E &= \xi + \sigma(c)\left\{\frac{d\sigma}{dc}\frac{dF}{d\xi}\right\}^{-1} \\
\tau^E &= \{\sigma(c)\}^2\left\{\frac{d\sigma}{dc}\frac{dF}{d\xi}\right\}^{-1}
\end{aligned} \qquad (6.1.3)$$

Since these equations represent the locus of the intersection between two adjacent characteristics, they give a parametric representation of the *envelope of characteristics* in terms of the parameter ξ with the correspondence $c = F(\xi)$. This point is denoted by the superscript E. We also notice that Eq. (6.1.3) can be determined by applying the method described in Sec. 0.10 (see Exercise 6.1.1).

If the compression wave is based on the feed data, we have

$$\begin{aligned} &\text{at} \quad \tau = 0, \quad x = \xi \quad \text{and} \quad c = c_0 = G(0), \qquad \text{for} \quad \xi \geq 0 \\ &\text{at} \quad x = 0, \quad \tau = \eta \quad \text{and} \quad c = G(\eta), \qquad \text{for} \quad 0 \leq \eta \leq 1 \\ &\qquad\qquad\qquad\qquad\qquad\qquad c = c_1 = G(1), \qquad \text{for} \quad \eta \geq 1 \end{aligned} \tag{6.1.4}$$

and the derivatives $d\sigma/dc$ and $dG/d\eta$ must have opposite signs. A similar procedure will lead us to the parametric representation of the envelope in terms of the parameter η:

$$\begin{aligned} x^E &= \left\{ -\frac{d\sigma}{dc}\frac{dG}{d\eta} \right\}^{-1} \\ \tau^E &= \eta + \sigma(c) \left\{ -\frac{d\sigma}{dc}\frac{dG}{d\eta} \right\}^{-1} \end{aligned} \tag{6.1.5}$$

Clearly, the envelopes can take a variety of shapes depending on the nature of the nonlinearity $d\sigma/dc$ as well as the function $F(\xi)$ or $G(\eta)$. We note, however, that an envelope necessarily encloses an angular region inside which all the crossing of characteristics take place or, we may say, the characteristics fan backward as we proceed in the x-direction. Thus a unique continuation of the solution through the region is mathematically impossible, so a discontinuity will first be formed at the point where τ^E becomes minimum. In fact, this happens at the cusp point (x_c^E, τ_c^E) of the envelope, and so we have the time

$$\tau = \tau_c^E = \min_{0 \leq \xi \leq 1} \tau^E \tag{6.1.6}$$

for the birth of a discontinuity. Henceforth, the jump across the discontinuity will grow while it propagates along a path that is necessarily located inside the region enclosed by the envelope, for otherwise we see that characteristics on one side or the other of the path still cross one another there. It then follows that at every point along the path we always find two characteristics impinging on the path, one from the left side and the other from the right. Therefore, a discontinuity originating from a compression wave is necessarily a shock (see Fig. 5.20).

To illustrate this, we will consider once again the Langmuir isotherm

$$n = f(c) = \frac{\gamma c}{1 + Kc} \tag{6.1.7}$$

for which

$$\sigma(c) = 1 + \frac{\nu\gamma}{(1 + Kc)^2} \tag{6.1.8}$$

and a linear distribution of solute in the initial bed; that is,

$$F(\xi) = c_0 - (c_0 - c_1)\xi \tag{6.1.9}$$

where $c_0 > c_1$. From Eq. (6.1.3), we have

$$\begin{aligned} x^E &= \xi + \frac{(1 + Kc)\{\nu\gamma + (1 + Kc)^2\}}{2\nu\gamma K(c_0 - c_1)} \\ \tau^E &= \frac{\{\nu\gamma + (1 + Kc)^2\}^2}{2\nu\gamma K(c_0 - c_1)(1 + Kc)} \end{aligned} \tag{6.1.10}$$

and, after differentiating the second equation with respect to ξ and making some rearrangements,

$$\frac{d\tau^E}{d\xi} = \frac{d\tau^E}{dc}\frac{dF}{d\xi} = \frac{3}{2\nu\gamma}\left\{1 + \frac{\nu\gamma}{(1 + Kc)^2}\right\}\left\{\frac{\nu\gamma}{3} - (1 + Kc)^2\right\} \tag{6.1.11}$$

It is obvious that τ^E becomes minimum for $Kc = -1 + \sqrt{\nu\gamma/3}$, but because we are interested in the closed interval, $0 \leqslant \xi \leqslant 1$, we must pay attention to the end points. In fact, if $1 + Kc_0 < \sqrt{\nu\gamma/3}$, $d\tau^E/d\xi$ remains positive throughout the interval, and so τ^E attains its minimum at $\xi = 0$ or for $c = c_0$. This implies that a shock will be formed along the characteristic emanating from the origin and carrying the concentration $c = c_0$. The cusp point is determined by putting $\xi = 0$ and $c = c_0$ in Eq. (6.1.10). After its formation the shock will propagate with c^l fixed at c_0 and c^r decreasing from c_0 to c_1 continuously, as shown in Fig. 6.3(a), where the numbers on the profiles of c represent the time sequence, that is, 0 at $\tau = 0$, 1 at $\tau = \tau_c^E$, 2 at $\tau > \tau_c^E$, and 3 at the time when the shock attains its full strength. Since the reciprocal speed is given by

$$\tilde{\sigma}(c_0, c) = 1 + \frac{\nu\gamma}{(1 + Kc_0)(1 + Kc)} \tag{6.1.12}$$

[see Eq. (5.5.1)], the shock is decelerating. On the other hand, if $1 + Kc_1 > \sqrt{\nu\gamma/3}$, the sign of $d\tau^E/d\xi$ is uniformly negative, and thus a shock is incipient along the characteristic emanating from the point $\xi = 1$. The cusp point is given by Eq. (6.1.10) with $\xi = 1$ and $c = c_1$. While it propagates, c^r is fixed at c_1 and c^l increases from c_1 to c_0 so that the shock is now accelerating. This is illustrated in Fig. 6.3(b) with the same scheme as in part (a). Finally, if $1 + Kc_1 < \sqrt{\nu\gamma/3} < 1 + Kc_0$, τ^E reaches its minimum when $Kc = -1$

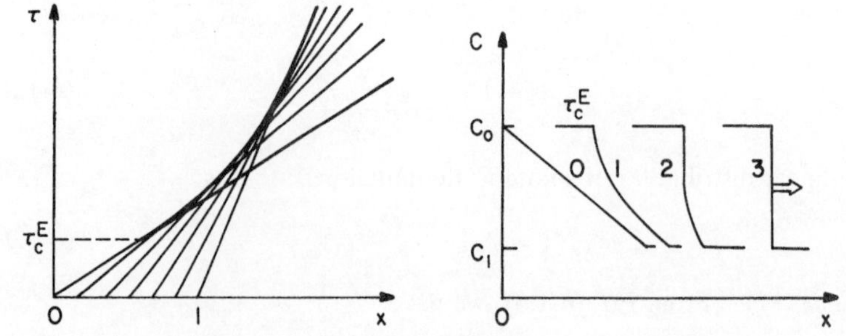

(a) Case when $1 + Kc_0 < \sqrt{\nu\gamma/3}$

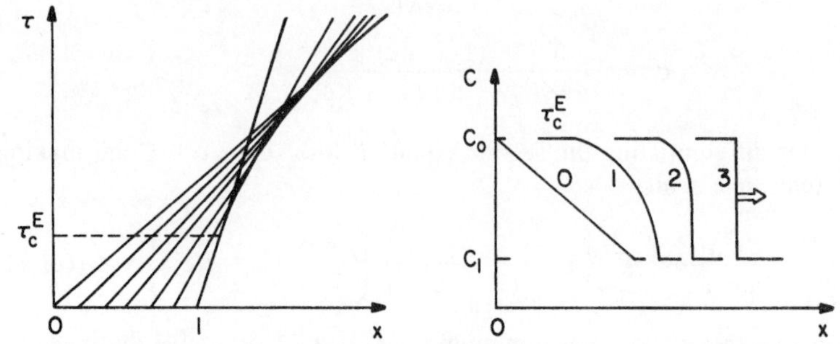

(b) Case when $1 + Kc_1 > \sqrt{\nu\gamma/3}$

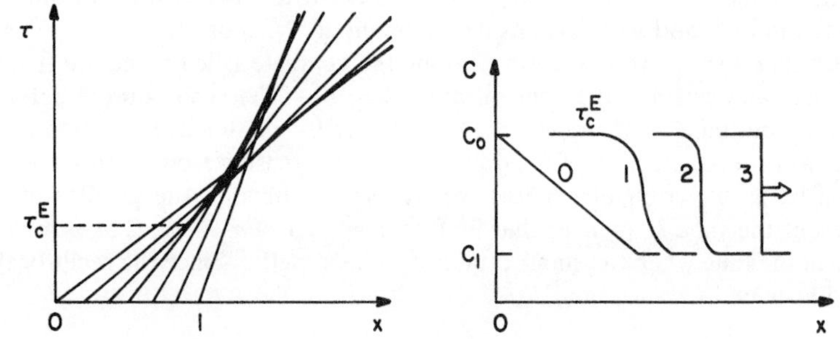

(c) Case when $1 + Kc_1 < \sqrt{\nu\gamma/3} < 1 + Kc_0$

Figure 6.3

$+ \sqrt{\nu\gamma/3}$. Thus a shock is formed at some point along the characteristic carrying this concentration. The characteristic emanates from the point

$$\xi = \frac{Kc_0 - 1 + \sqrt{\nu\gamma/3}}{K(c_0 - c_1)} \tag{6.1.13}$$

on the x-axis. We then find the cusp point from Eq. (6.1.10) as

$$x_c^E = \frac{Kc_0 - 1 + \frac{5}{3}\sqrt{\nu\gamma/3}}{K(c_0 - c_1)}$$
$$\tau_c^E = \frac{8}{3}\frac{\sqrt{\nu\gamma/3}}{K(c_0 - c_1)} \tag{6.1.14}$$

In this case both c^l and c^r vary in opposite directions, so the propagation of the shock is more complicated. This is schematically shown in Fig. 6.3(c) together with the evolution of profiles in time, $0 \to 1 \to 2 \to 3$.

Returning to Eq. (6.1.3), we notice that

$$\frac{d\sigma}{d\xi} = \frac{d\sigma}{dc}\frac{dF}{d\xi} > 0$$

for a compression wave. Therefore, if $d\sigma/d\xi$ is a nonincreasing function of ξ (i.e., $d^2\sigma/d\xi^2 \le 0$), then τ^E is monotonically increasing, so a shock will form on the characteristic emanating from the origin. If we make a plot of σ versus ξ by using the relation $\sigma = \sigma(F(\xi))$, we obtain a concave upward curve or a straight line and this gives us some geometrical feeling. To examine the case when such a curve lies below the diagonal, let us consider a special situation in which all the characteristics with bases on the interval $0 \le \xi \le 1$ intersect one another at a single point.† This implies that a shock of full strength will suddenly appear at the point. The profile of σ versus ξ is given by the equation

$$\sigma_B = \frac{\sigma_0}{1 - (1 - \sigma_0/\sigma_1)\xi}, \qquad 0 \le \xi \le 1 \tag{6.1.15}$$

where $\sigma_0 = \sigma(c_0)$ and $\sigma_1 = \sigma(c_1)$. The curve is a portion of a hyperbola and concave upward, as shown in Fig. 6.4. We will call this the *base curve*, as designated by the subscript in Eq. (6.1.15).

Suppose now that the initial data $c = F(\xi)$ are specified so that the curve $\sigma = \sigma(F(\xi))$ may be plotted. With σ_0 and σ_1 known, the base curve $\sigma_B(\xi)$ can be drawn on the same plane. In one of the typical situations, we see the $\sigma(\xi)$ curve may entirely lie above the $\sigma_B(\xi)$ curve and remain concave upward, as shown in Fig. 6.5. If we choose an arbitrary point ξ', $0 < \xi' < 1$, on the x-axis, the characteristic C' issuing from ξ' is steeper than the line $\overline{P\xi'}$, and so C' intersects $\overline{OP}$ at an intermediate point Q. Let us then consider a new base curve $\sigma'_B(\xi)$ over the interval $0 \le \xi \le \xi'$. Since $(\partial/\partial\sigma_1)/(\partial^2\sigma_B/\partial\xi^2) > 0$ from Eq. (6.1.15), we can show that the $\sigma'_B(\xi)$ curve falls between the two original curves over the interval $0 < \xi < \xi'$. For an arbitrary point ξ'' in this

†For specific examples, refer to Fig. 6.38 and also Lax (1972, 1973).

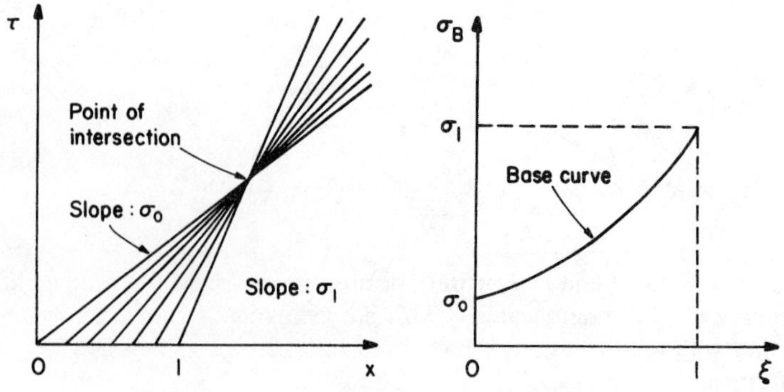

Figure 6.4

interval, the characteristic C'' is steeper than $\overline{Q\xi''}$ so that C'' intersects $\overline{OQ}$ at an intermediate point R. This leads us to conclude that all the crossings among the characteristics take place above the line OP. Therefore, a shock will first be formed on the characteristic emanating from the origin.

Figure 6.6 presents other typical situations that are of interest. In case (a), the $\sigma(\xi)$ curve falls below the $\sigma_B(\xi)$ curve, and by applying the same argument as before we can show that a shock will form on the characteristic for $c = c_1$. In cases (b) and (c), the $\sigma(\xi)$ curve crosses the $\sigma_B(\xi)$ curve, so we may regard these as combinations of the previous two cases. Clearly, a shock will form on the characteristic issuing from the point $\xi = \bar{\xi}$ under the circumstances of case (b), whereas the characteristics from end points will separately give birth to shocks in case (c).

When the compression wave is based on the feed data [see Eq. (6.1.4)], the geometrical argument becomes rather simple, because from Eq. (6.1.5)

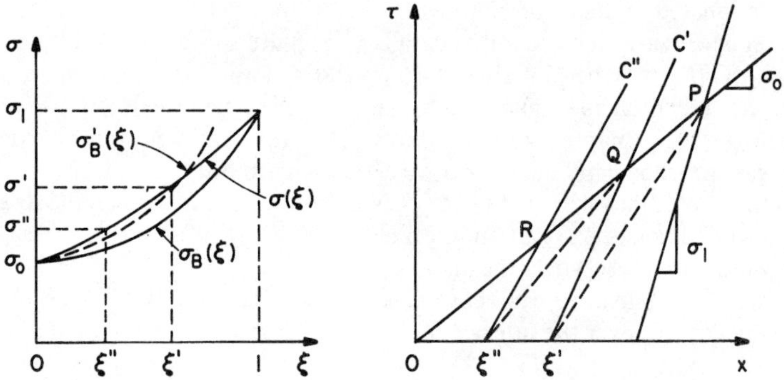

Figure 6.5

we have

$$\frac{d\tau^E}{d\eta} = \sigma\left(\frac{d\sigma}{d\eta}\right)^{-2}\frac{d^2\sigma}{d\eta^2} \tag{6.1.16}$$

Thus the derivatives $d\tau^E/d\eta$ and $d^2\sigma/d\eta^2$ have a common sign, and this suggests a convenient method of discerning the characteristic from which a shock will be formed. If the feed data are plotted in the form of a $\sigma = \sigma(G(\eta))$ curve, they might exhibit one of the four forms shown in Fig. 6.7. In case (a), $d^2\sigma/d\eta^2 < 0$ and the minimum of τ^E is found at $\eta = \eta_2$, and a shock will form on the characteristic for $c = G(\eta_2)$. Similarly, in case (b), a shock will form on the characteristic for $c = G(\eta_1)$. In case (c), τ^E attains its minimum at the point of inflection, and the characteristic from $\eta = \bar{\eta}$ will give birth to a shock. In case (d), the inflection point corresponds to a maximum of τ^E, so a shock will first be formed on the characteristic from the end point with the greatest slope $d\sigma/d\eta$, and then a second shock will form on the characteristic from the other end point.

For quasi-linear equations in N space dimensions, the initial-value problem may be written in the form

$$\frac{\partial u}{\partial t} + \nabla \cdot \mathbf{f}(u) = \frac{\partial u}{\partial t} + \sum_{j=1}^{N} f_j'(u)\frac{\partial u}{\partial x_j} = 0 \tag{6.1.17}$$

$$u(\mathbf{x}, 0) = u_0(\mathbf{x}) \tag{6.1.18}$$

where the flux $\mathbf{f}$ is a vector-valued function of u with N elements, $f_1, f_2, \ldots, f_N$, and x denotes the position vector. For this problem, Conway (1977) has treated the formation of shocks and proved that if†

$$\mathbf{f}''(u_0(\mathbf{x})) \cdot \nabla u_0(\mathbf{x}) < 0 \tag{6.1.19}$$

then the solution cannot be continued as a smooth function after a finite time. Thus, if $\mathbf{f}$ is genuinely nonlinear and $u_0(\mathbf{x})$ vanishes outside a finite interval of x [i.e., $u_0(\mathbf{x})$ has compact support], the solution must develop singularities as t increases. Conway has also given an expression for the time when a discontinuity must be introduced.

Exercises

6.1.1. The first equation of Eq. (6.1.2),

$$\tau = \sigma(c)[x - \xi]$$

represents a family of straight lines in the (x, τ)-plane if c is defined as a differentiable function of the parameter ξ [i.e., $c = F(\xi)$]. By applying the theorem of Sec. 0.10, determine the envelope of the family of the lines.

†Compare Eq. (6.1.19) to Eq. (5.2.6), noting that in Eq. (6.1.17) the nonlinearity appears in the derivatives with respect to the space variables.

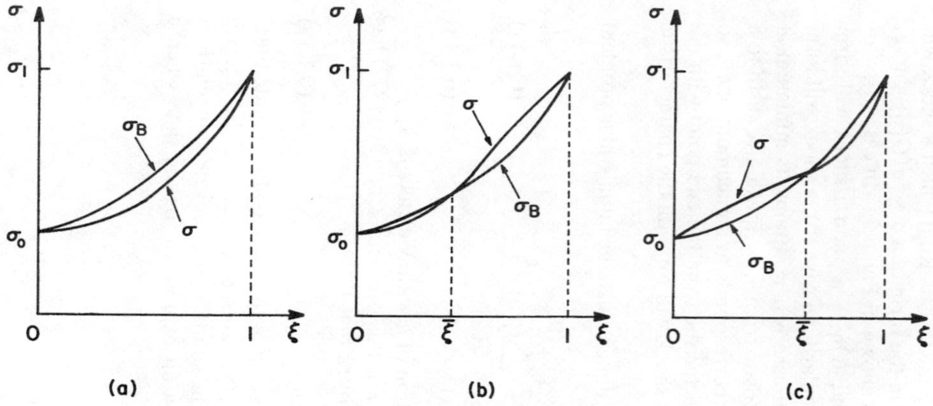

Figure 6.6

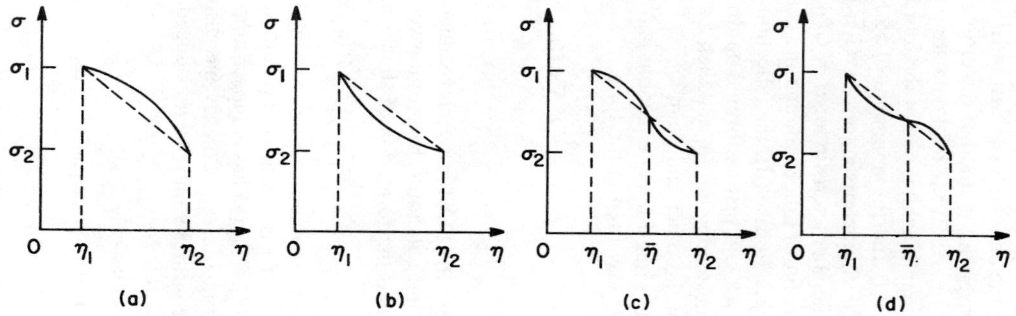

Figure 6.7

6.1.2. Derive Eq. (6.1.5).

6.1.3. Discuss the formation of a shock when the equilibrium relation is given by the Langmuir isotherm and the initial data are specified as in Eq. (6.1.1) with (a) $F(\xi) = c_0(1 - \xi^2)$ and (b) $F(\xi) = c_0(1 - \xi)^2$, both for $0 \le \xi \le 1$.

6.1.4. The following equilibrium relation holds in single-solute chromatography:

$$f(c) = kc^{3/2}$$

If the data are prescribed as

$$\begin{aligned} &\text{at } \tau = 0, && c = \alpha + \beta x, && \text{for } 0 \le x \le 1 \\ & && c = \alpha + \beta, && \text{for } x \ge 1 \\ &\text{at } x = 0, && c = \alpha, && \text{for } \tau > 0 \end{aligned}$$

with $\alpha, \beta > 0$, determine the point (x_0, τ_0) at which a shock starts to appear.

6.1.5. Consider chromatography with a Langmuir isotherm, where $K = 5$ l/mol and $\gamma = 2.5$. The initial data are prescribed as in Eq. (6.1.1) with $c_0 = 0.5$, $c_1 = 0.1$ mol/l, and linear function $F(\xi)$. Apply the base curve method to discern the characteristic on which a shock will form for (a) $\varepsilon = \frac{1}{6}$, (b) $\varepsilon = \frac{1}{3}$, and (c) $\varepsilon = \frac{1}{2}$. Compare the results with those from the envelope method (cf. Rhee, 1968).

6.1.6. In the radial chromatography considered in Exercise 5.3.2, let $c(r, 0) = c_0(r)$ be a monotonic decreasing function of r in $a \le r \le b$. If $c(b, t) = 0$, show that there will be no crossing of characteristics provided that

$$-c_0'(a) < \frac{1}{\nu\gamma K}\frac{a}{b^2 - a^2}\{\nu\gamma[1 + Kc_0(a)] + [1 + Kc_0(a)]^3\}$$

6.2 Saturation of a Column

In this section we wish to solve the problem of saturating an initially clean column with a feed in which the feed concentration eventually rises to a steady value. We will consider the Langmuir isotherm so that the characteristic direction is given by

$$\sigma(c) = 1 + \frac{\nu\gamma}{(1 + Kc)^2} \tag{6.2.1}$$

The bed is initially clean so that

$$\text{at } \tau = 0, \qquad c = 0 \tag{6.2.2}$$

and, for simplicity, we will take the inlet feed to be of a linearly increasing concentration over the period $0 \le \tau \le \alpha$ and thereafter to be held constant at the value c_0. Thus

$$\text{at } x = 0, \quad \tau = \eta \quad \text{and } c = \begin{cases} G(\eta) = \dfrac{c_0}{\alpha}\eta, & \text{for } 0 \le \eta \le \alpha \\ G(\alpha) = c_0, & \text{for } \eta \ge \alpha \end{cases} \tag{6.2.3}$$

All the characteristics emanating from the x-axis have slope $(1 + \nu\gamma)$ and those emanating from the τ-axis above $\tau = \alpha$ will have slope $1 + \nu\gamma/(1 + Kc_0)^2$. The characteristic emanating from the point $\tau = \eta$, $0 \le \eta \le \alpha$, has a slope

$$\sigma\left(\frac{c_0\eta}{\alpha}\right) = 1 + \frac{\nu\gamma}{(1 + Kc_0\eta/\alpha)^2} \tag{6.2.4}$$

and since this is decreasing with η, we have a compression wave based on the interval $0 \le \eta \le \alpha$ of the τ-axis (see Fig. 6.8). Thus we expect a shock to be formed. The envelope of characteristics is determined by introducing Eqs. (6.2.1) and (6.2.2) into Eq. (6.1.5):

$$\begin{aligned} x^E &= \frac{\alpha}{2\nu\gamma Kc_0}\left(1 + \frac{Kc_0}{\alpha}\eta\right)^3 \\ \tau^E &= \eta + \frac{\alpha}{2\nu\gamma Kc_0}\left(1 + \frac{Kc_0}{\alpha}\eta\right)\left\{\nu\gamma + \left(1 + \frac{Kc_0}{\alpha}\eta\right)^2\right\} \end{aligned} \tag{6.2.5}$$

These are both monotone increasing with η, so τ^E takes its minimum at $\eta = 0$. This is in contrast to the case with a linear initial distribution of solute in the

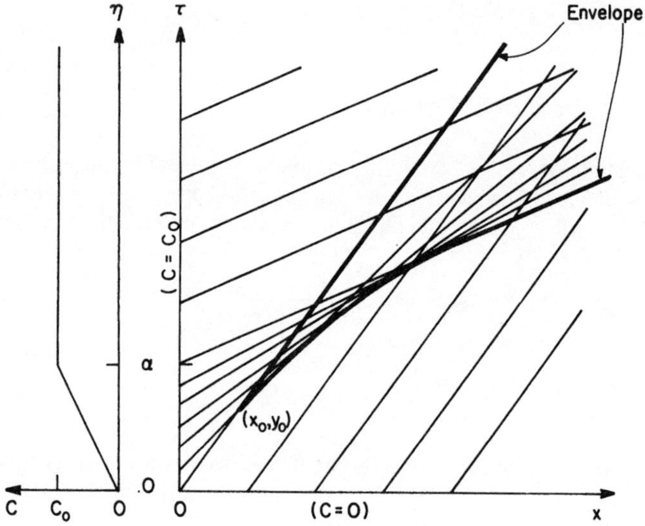

Figure 6.8

bed, which we discussed in the previous section. The cusp point of the envelope falls on the characteristic emanating from the origin ($\eta = 0$), as shown in Fig. 6.8, and has the coordinates

$$x_0 = x_c^E = \frac{\alpha}{2\nu\gamma K c_0}$$
$$\tau_0 = \tau_c^E = \frac{\alpha(1 + \nu\gamma)}{2\nu\gamma K c_0} \tag{6.2.6}$$

Clearly, we have $c = 0$ at this point.

This implies that a shock will form at the point (x_0, τ_0) and will have a zero concentration on the right side, while the concentration on the other side will rise as it propagates. From Eq. (5.5.1) we have the reciprocal speed of propagation,

$$\tilde{\sigma}(c, 0) = 1 + \frac{\nu\gamma}{1 + Kc} \tag{6.2.7}$$

where c denotes the concentration on the left side, and thus we find the shock to be accelerating. We expect a picture such as Fig. 6.9 to develop, where A is the point of inception of the shock. Suppose that P is a typical point (x, τ) on the shock path and that c is the concentration on the left side of the shock. This concentration must have been borne on a characteristic from the point on the τ-axis where $\eta = G^{-1}(c) = \alpha c/c_0$. Thus the point P must lie on

$$\tau = \eta(c) + \sigma(c)x \tag{6.2.8}$$

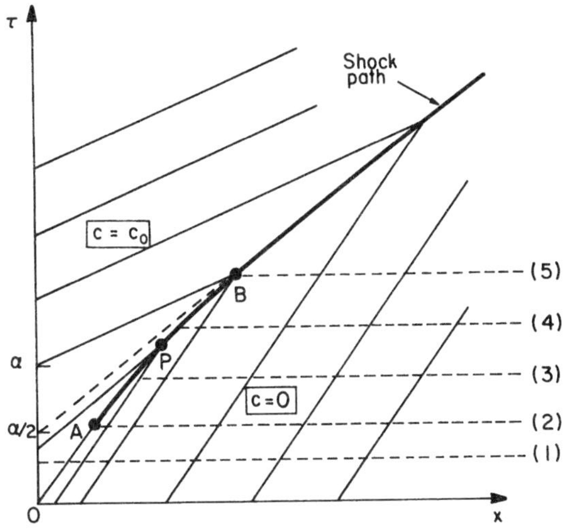

Figure 6.9

Note that here η is considered as a function of c. But if the shock has $c^l = c$ and $c^r = 0$, it moves with the reciprocal speed

$$\frac{d\tau}{dx} = \tilde{\sigma}(c, 0) \tag{6.2.9}$$

We could eliminate c between these two equations and so obtain a differential equation for the path of the shock. However, it is more convenient to treat c as a parameter along the shock path and differentiate Eq. (6.2.8) with respect to c:

$$\frac{d\tau}{dc} = \frac{d\eta}{dc} + \frac{d\sigma}{dc}x + \sigma\frac{dx}{dc}$$

But, by Eq. (6.2.9),

$$\frac{d\tau}{dc} = \tilde{\sigma}\frac{dx}{dc}$$

so that, on substituting for $d\tau/dc$,

$$(\tilde{\sigma} - \sigma)\frac{dx}{dc} - \frac{d\sigma}{dc}x = \frac{d\eta}{dc} \tag{6.2.10}$$

This is a first-order differential equation for x with the independent variable c. After this equation is solved subject to the initial condition

$$x = x_0 \quad \text{at} \quad c = 0 \tag{6.2.11}$$

we can determine τ from Eq. (6.2.8) to give a parametric representation of the shock path.

In the present case, we substitute Eqs. (6.2.1), (6.2.3), and (6.2.7) in Eq. (6.2.10) to obtain

$$\frac{c}{(1 + Kc)^2}\frac{dx}{dc} + \frac{2x}{(1 + Kc)^3} = \frac{\alpha}{\nu\gamma K c_0} \tag{6.2.12}$$

This equation can be made exact by multiplying throughout by K^2c, so that

$$\frac{d}{dc}\left\{x\left(\frac{Kc}{1 + Kc}\right)^2\right\} = \frac{\alpha K}{\nu\gamma c_0}c$$

and, after integration subject to Eq. (6.2.11),

$$x\left(\frac{Kc}{1 + Kc}\right)^2 = \frac{\alpha K c^2}{2\nu\gamma c_0}$$

or

$$x = \frac{\alpha}{2\nu\gamma K c_0}(1 + Kc)^2 \tag{6.2.13}$$

Substituting in Eq. (6.2.8) for τ, we have

$$\tau = \frac{\alpha c}{c_0} + \left\{1 + \frac{\nu\gamma}{(1 + Kc)^2}\right\}\frac{\alpha(1 + Kc)^2}{2\nu\gamma Kc_0} \tag{6.2.14}$$

$$= \frac{\alpha}{2\nu\gamma Kc_0}\{\nu\gamma(1 + 2Kc) + (1 + Kc)^2\}$$

From Eq. (6.2.13), we have

$$Kc = -1 + \left(\frac{2\nu\gamma Kc_0 x}{\alpha}\right)^{1/2} \tag{6.2.15}$$

and thus the jump in c (or the shock strength) grows like $\sqrt{x}$. If we eliminate c between Eqs. (6.2.13) and (6.2.14), we have

$$\tau - x + \frac{\alpha}{2Kc_0} = \frac{\alpha(1 + Kc)}{Kc_0} = \left(\frac{2\alpha\nu\gamma}{Kc_0}x\right)^{1/2} \tag{6.2.16}$$

so the path of the shock is a part of a parabola.

The shock is fully developed when c has reached its maximum value of c_0. This is at point B of Fig. 6.9, whose coordinates are

$$x = \frac{\alpha}{2\nu\gamma Kc_0}(1 + Kc_0)^2$$

$$\tau = \frac{\alpha(1 + \nu\gamma)}{2\nu\gamma Kc_0}(1 + Kc_0)^2 - \frac{\alpha Kc_0}{2} \tag{6.2.17}$$

Thereafter the shock travels on a straight path since it is always between concentrations c_0 and 0. This path is tangent to the curved path APB at B because it has a slope of $1 + \nu\gamma/(1 + Kc_0)^2$, which is the same as the slope of the curve at B [cf. Eqs. (6.2.13) and (6.2.14)]. Since it must go through B, its equation is

$$\tau = \frac{\alpha}{2} + \left(1 + \frac{\nu\gamma}{1 + Kc_0}\right)x \tag{6.2.18}$$

That is, its intercept on the τ-axis is $\alpha/2$. If we let α approach zero, so that the full concentration c_0 is initiated at time $\tau = 0$, then the shock starts immediately and proceeds along the straight line

$$\tau = \left(1 + \frac{\nu\gamma}{1 + Kc_0}\right)x \tag{6.2.19}$$

Figure 6.10 shows the distribution of solute along the column at various instants as marked in Fig. 6.9:

(1) Before the shock has formed
(2) At the inception of the shock
(3) During the development of the shock before $\tau = \alpha$

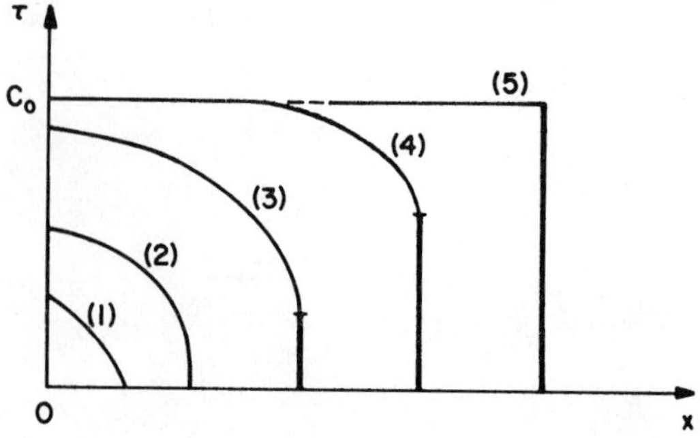

Figure 6.10

(4) During the development of the shock after $\tau = \alpha$
(5) When the shock is just fully developed or any instant thereafter

A three-dimensional view of the solution surface is given in Fig. 6.11.

When the compression wave is based on the initial data $c = F(\xi)$, rather than on the feed data $c = G(\eta)$, we should first discern the characteristic from which a shock will form (see Fig. 6.3). If the shock starts to form on one of the end characteristics, then the state on one side of the shock remains fixed, say $c^l = c_0 =$ fixed, and the state on the other side of c varies contin-

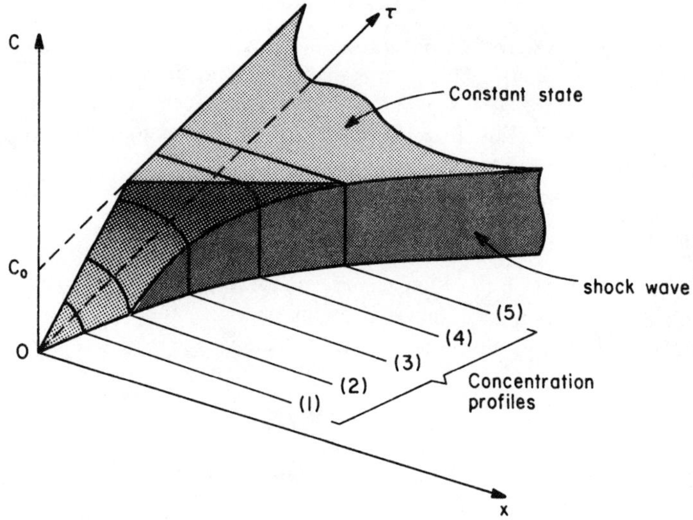

Figure 6.11

uously, maintaining the correspondence to the parameter ξ through $c = F(\xi)$. Thus, at a typical point (x, τ) on the shock path, we can write the equations

$$\tau = \sigma(c)\{x - \xi(c)\} \tag{6.2.20}$$

$$\frac{d\tau}{dx} = \tilde{\sigma}(c_0, c)$$

where $\xi(c) = F^{-1}(c)$. Now, treating c as the parameter running along the shock path and eliminating τ from Eq. (6.2.20), we have

$$\{\tilde{\sigma}(c_0, c) - \sigma(c)\}\frac{dx}{dc} - \frac{d\sigma}{dc}x = -\frac{d}{dc}\{\sigma(c)\xi(c)\} \tag{6.2.21}$$

which is a differential equation for x subject to the initial condition

$$x = x_0 \quad \text{at} \quad c = c_0$$

where x_0 is the abscissa at the point of inception of the shock. In the case of the Langmuir isotherm, Eq. (6.2.21) can be made exact so that the solution may be expressed in terms of c in a closed form (see Exercise 6.2.2). If, however, the shock is first formed on an intermediate characteristic, as we see in Fig. 6.3(c), the states on both sides of the shock vary simultaneously while it propagates, and thus the analysis is more involved. We will discuss this case in Sec. 6.4.

Exercises

6.2.1. Consider a column that is uniformly presaturated to an equilibrium concentration c_0. This column is to be further saturated with a feed in which the feed concentration rises linearly to a steady value c_1 over a period of time α. If the equilibrium relation is given by the Langmuir isotherm, analyze in detail the time evolution of the system.

6.2.2. For a chromatographic column subject to the Langmuir isotherm

$$f(c) = \frac{\gamma c}{1 + Kc}$$

we have the data

$$c(0, \tau) = c_0$$

$$c(x, 0) = \begin{cases} c_0 - \dfrac{(c_0 - c_1)x}{\alpha}, & 0 \le x \le \alpha \\ c_1, & x > \alpha \end{cases}$$

where $c_0 > c_1$ and $1 + Kc_0 < \sqrt{\nu\gamma/3}$. Determine the point at which a shock will first form and the path along which the shock will propagate after its birth.

6.3 Development of a Finite Chromatogram

Let us consider the case where a stream of constant concentration c_1 is fed into the clean bed during the interval $0 < \tau < \alpha$ and, thereafter, the feed stream consists of pure solvent. In this case the initial and boundary conditions are

$$\text{at} \quad \tau = 0, \qquad c = 0$$

$$\text{at} \quad x = 0, \qquad c = \begin{cases} c_1, & 0 \le \tau \le \alpha \\ 0, & \tau > \alpha \end{cases} \tag{6.3.1}$$

Here again we will consider the Langmuir isotherm. Then there is a shock formed from the very start as the state of concentration c_1 moves into the bed, and this shock moves along the straight line

$$\tau = \left(1 + \frac{\nu\gamma}{1 + Kc_1}\right)x \tag{6.3.2}$$

At the point $\tau = \alpha$, all concentrations between $c = 0$ and $c = c_1$ are present, and the characteristics fan out to give the centered simple wave

$$\tau = \alpha + \left\{1 + \frac{\nu\gamma}{(1 + Kc)^2}\right\}x, \qquad 0 \le c \le c_1 \tag{6.3.3}$$

See Fig. 6.12. Now the characteristic bearing the maximum concentration c_1 has a smaller slope in the (x, τ)-plane than that of the shock, so they intersect at a point A. This is the point

$$x_0 = \frac{\alpha}{\nu\gamma Kc_1}(1 + Kc_1)^2$$
$$\tau_0 = \frac{\alpha}{\nu\gamma Kc_1}(1 + Kc_1)(1 + \nu\gamma + Kc_1) \tag{6.3.4}$$

Beyond this point the simple wave continuously overtakes the shock from behind so that the state on the left side of the shock decreases, while we have a fixed state $c = 0$ on the right. If P is a typical point (x, τ) on the shock path and if c is the concentration on the left side of the shock, we have

$$\tau = \alpha + \left\{1 + \frac{\nu\gamma}{(1 + Kc)^2}\right\}x \tag{6.3.5}$$

since c must have been borne on a characteristic from the point $\tau = \alpha$ on the τ-axis. On the other hand, the slope of the shock path is given by

$$\frac{d\tau}{dx} = 1 + \frac{\nu\gamma}{1 + Kc} \tag{6.3.6}$$

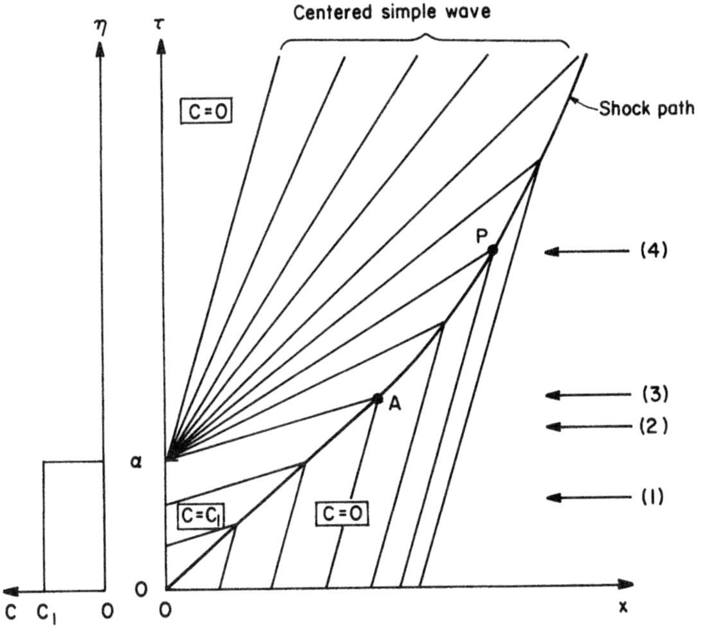

Figure 6.12

and since c decreases with time, the shock is continuously decelerating. Now, from Eq. (6.3.5),

$$\frac{\nu\gamma(\tau - x - \alpha)}{x} = \left(\frac{\nu\gamma}{1 + Kc}\right)^2 \tag{6.3.7}$$

and, from Eq. (6.3.6),

$$\frac{d}{dx}(\tau - x - \alpha) = \frac{\nu\gamma}{1 + Kc} \tag{6.3.8}$$

Hence,

$$\frac{d}{dx}(\tau - x - \alpha) = \left\{\frac{\nu\gamma(\tau - x - \alpha)}{x}\right\}^{1/2} \tag{6.3.9}$$

and, integrating from the point $A(x_0, \tau_0)$, where the interaction begins, we have

$$\begin{aligned}(\tau - x - \alpha)^{1/2} - (\nu\gamma x)^{1/2} \\ = (\tau_0 - x_0 - \alpha)^{1/2} - (\nu\gamma x_0)^{1/2} = -(\alpha K c_1)^{1/2}\end{aligned} \tag{6.3.10}$$

in which Eq. (6.3.4) has been introduced. Thus the shock path is a portion of a parabola.

We also notice that, from Eq. (6.3.10),

$$\left\{\frac{\nu\gamma(\tau - x - \alpha)}{x}\right\}^{1/2} = \nu\gamma - \left(\frac{\alpha\nu\gamma Kc_1}{x}\right)^{1/2}$$

and, by combining this with Eq. (6.3.7),

$$Kc = \frac{1}{-1 + \left(\dfrac{\nu\gamma x}{\alpha Kc_1}\right)^{1/2}} \tag{6.3.11}$$

This implies that the jump across the shock, c, decays like $1/\sqrt{x}$ for large time,† so the interaction goes on indefinitely. Since the rear end characteristic of the centered simple wave bears a zero concentration, its equation is

$$\tau - \alpha = (1 + \nu\gamma)x$$

and, from Eq. (6.3.10), we have

$$\tau - \alpha = (1 + \nu\gamma)x - 2(\alpha\nu\gamma Kc_1 x)^{1/2} + \alpha Kc_1$$

For a fixed time, therefore, the distance Δx between the shock path and the end characteristic is given by

$$(1 + \nu\gamma)\Delta x = 2(\alpha\nu\gamma Kc_1 x)^{1/2} - \alpha Kc_1 \tag{6.3.12}$$

This implies that the length of the chromatogram tends to grow like $\sqrt{x}$ for large time.‡ Note that x denotes the position of the shock.

Figure 6.13 shows the progress of the chromatographic band along the bed. It is drawn for a series of times as marked in Fig. 6.12:

(1) $\tau < \alpha$, the feed concentration is still c_1.

(2) $\alpha < \tau < \tau_0$, the feed is now pure solvent but the point of continuity of concentration c_1 has not yet caught up with the shock.

(3) $\tau = \tau_0$, the shock will now begin to diminish in intensity.

(4) $\tau > \tau_0$, this form is typical for all later times.

Figure 6.14 shows the solution surface in three dimensions.

In the preceding, Eq. (6.3.5) is in a particularly simple form so that the solution scheme applied turns out to be simple also. To study the general situation of this kind, let us take the data prescribed as follows:

$$\text{at} \quad \tau = 0, \qquad c = c_0$$

$$\text{at} \quad x = 0, \qquad \tau = \eta$$

$$\text{and} \quad c = \begin{cases} c_1 = G(\eta_1), & 0 \leq \eta < \eta_1 \\ G(\eta), & \eta_1 \leq \eta \leq \eta_2 \\ c_2 = G(\eta_2), & \eta > \eta_2 \end{cases} \tag{6.3.13}$$

†This has been proved in a general context by Lax (1957, 1972). See also Sec. 7.9 for the general treatment.

‡See the above footnote.

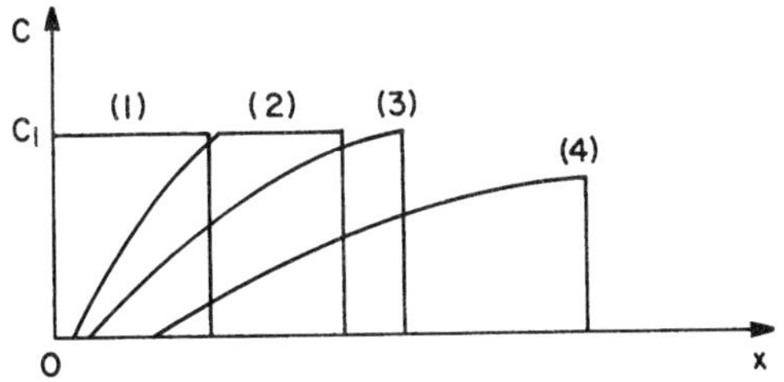

Figure 6.13

where $c_1 > c_0$, c_2 and $G(\eta)$ is a monotone decreasing function. If we consider a convex isotherm, that is, if $f''(c) < 0$, we have a shock formed from the origin with $c^l = c_1$ and $c^r = c_0$. This shock propagates along the line

$$\tau = \tilde{\sigma}(c_1, c_0)x \tag{6.3.14}$$

The data specified over the portion $\eta_1 \leq \tau \leq \eta_2$ of the τ-axis give rise to an expansion wave as depicted in Fig. 6.15. The characteristic issuing from the point $\tau = \eta_1$ on the τ-axis bears the concentration c_1, and its equation is

$$\tau = \eta_1 + \sigma(c_1)x \tag{6.3.15}$$

Since $\tilde{\sigma}(c_1, c_0) > \sigma(c_1)$, this characteristic intersects the shock path at the point $A(x_0, \tau_0)$, where

$$x_0 = \frac{\eta_1}{\tilde{\sigma}(c_1, c_0) - \sigma(c_1)}$$
$$\tau_0 = \frac{\tilde{\sigma}(c_1, c_0)\eta_1}{\tilde{\sigma}(c_1, c_0) - \sigma(c_1)} \tag{6.3.16}$$

Starting from this point, the concentration on the left side of the shock decreases continuously due to the interaction with the expansion wave, and so the shock is decelerating. The state on the right side is fixed at $c = c_0$. Suppose that P is a typical point (x, τ) on the shock path and that c is the concentration on the left side of the shock. Then we see that this concen-

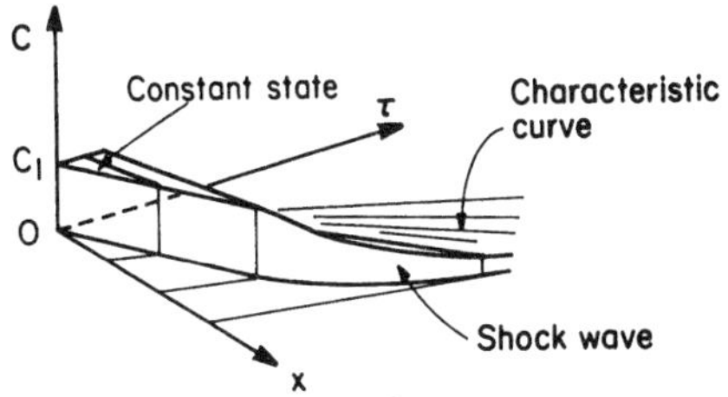

Figure 6.14

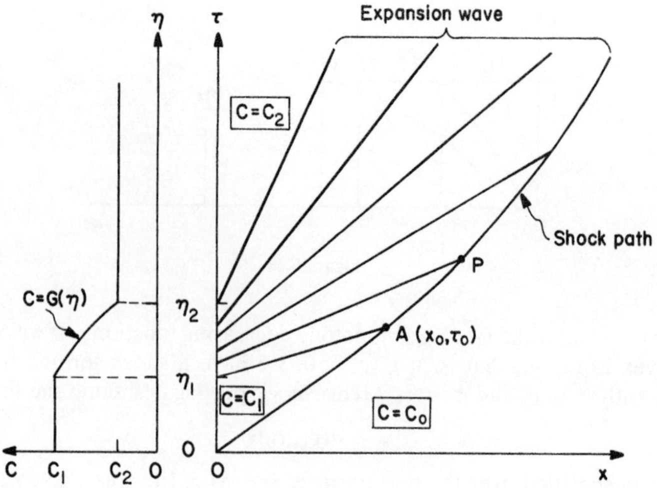

Figure 6.15

tration must have been borne on a characteristic from the point on the τ-axis where the feed concentration is c, that is, $\eta = G^{-1}(c)$. At the point P, therefore, we can write two equations:

$$\tau = \eta(c) + \sigma(c)x \tag{6.3.17}$$

and

$$\frac{d\tau}{dx} = \tilde{\sigma}(c, c_0)x \tag{6.3.18}$$

Now treating c as a parameter running along the shock path, we may differentiate Eq. (6.3.17) with respect to c and substitute for $d\tau/dc$ in Eq. (6.3.18). Thus we have

$$\{\tilde{\sigma}(c, c_0) - \sigma(c)\}\frac{dx}{dc} - \frac{d\sigma}{dc}x = \frac{d\eta}{dc} \tag{6.3.19}$$

which is essentially the same as Eq. (6.2.10). The initial condition is

$$x = x_0 \quad \text{at} \quad c = c_1 \tag{6.3.20}$$

Once x is determined in terms of c, τ is given by Eq. (6.3.17), so we have a parametric representation of the shock path.

To illustrate this, we will consider the Langmuir isotherm for which

$$\sigma(c) = 1 + \frac{\nu\gamma}{(1 + Kc)^2} \tag{6.3.21}$$

and

$$\tilde{\sigma}(c, c_0) = 1 + \frac{\nu\gamma}{(1 + Kc_0)(1 + Kc)} \tag{6.3.22}$$

From Eq. (6.3.19), we have

$$\frac{c - c_0}{(1 + Kc)^2}\frac{dx}{dc} + \frac{2(1 + Kc_0)}{(1 + Kc)^3}x = \frac{1 + Kc_0}{\nu\gamma K}\frac{d\eta}{dc} \tag{6.3.23}$$

and this can be made exact by multiplying throughout by $K^2(c - c_0)$ to give

$$\frac{d}{dc}\left\{\left[\frac{K(c - c_0)}{1 + Kc}\right]^2 x\right\} = \frac{1 + Kc_0}{\nu\gamma} K(c - c_0)\frac{d\eta}{dc}$$

or, after integration with Eq. (6.3.20),

$$x = x_0\left(\frac{1 + Kc}{1 + Kc_1}\frac{c_1 - c_0}{c - c_0}\right)^2 + \frac{1 + Kc_0}{\nu\gamma}\left\{\frac{1 + Kc}{K(c - c_0)}\right\}^2 \int_{c_1}^{c} K(c - c_0)\frac{d\eta}{dc}\,dc \tag{6.3.24}$$

Now, from Eq. (6.3.16),

$$\begin{aligned} x_0 &= \frac{\eta_1}{\nu\gamma}\frac{(1 + Kc_0)(1 + Kc_1)^2}{K(c_1 - c_0)} \\ \tau_0 &= \frac{\eta_1}{\nu\gamma}\frac{(1 + Kc_1)\{\nu\gamma + (1 + Kc_0)(1 + Kc_1)\}}{K(c_1 - c_0)} \end{aligned} \tag{6.3.25}$$

and hence

$$x = \frac{1 + Kc_0}{\nu\gamma}\left\{\frac{1 + Kc}{K(c - c_0)}\right\}^2 \left\{K(c_1 - c_0)\eta_1 + \int_{c_1}^{c} K(c - c_0)\frac{d\eta}{dc}\,dc\right\} \tag{6.3.26}$$

If the simple wave is centered (i.e., $\eta_1 = \eta_2 = \alpha$), then the integral term vanishes so that

$$x = \frac{\alpha}{\nu\gamma}K(c_1 - c_0)(1 + Kc_0)\left\{\frac{1 + Kc}{K(c - c_0)}\right\}^2 \tag{6.3.27}$$

which would reduce to Eq. (6.3.11) with $c_0 = 0$. We also notice from Eq. (6.3.26) that x diverges in the limit as $c \to c_0$. This implies that if $c_2 \leq c_0$ the interaction between the simple wave and the shock continues indefinitely. If $c_2 > c_0$, however, the interaction will be over when $c = c_2$, and thereafter the shock will propagate with a constant speed with fixed states on both sides, $c^l = c_2$ and $c^r = c_0$.

In the case when $G(\eta)$ is linear, we have

$$\frac{c - c_1}{c_2 - c_1} = \frac{\eta - \eta_1}{\eta_2 - \eta_1}$$

or

$$\eta(c) = G^{-1}(c) = \eta_1 + (\eta_2 - \eta_1)\frac{c - c_1}{c_2 - c_1} \tag{6.3.28}$$

and, by substituting in Eq. (6.3.26),

$$x = \frac{1 + Kc_0}{\nu\gamma}\left\{\frac{1 + Kc}{K(c - c_0)}\right\}^2 \left\{K(c_1 - c_0)\eta_1 - \frac{\eta_2 - \eta_1}{c_2 - c_1}\frac{Kc_1^2}{2}\left(1 - \frac{c}{c_1}\right)\left(1 + \frac{c}{c_1} - 2\frac{c_0}{c_1}\right)\right\} \tag{6.3.29}$$

If $c_0 = 0$, this equation reduces to

$$x = \frac{Kc_1}{\nu\gamma}\left(1 + \frac{1}{Kc}\right)^2 \left\{\eta_1 - \frac{c_1}{2}\frac{\eta_2 - \eta_1}{c_2 - c_1}\left[1 - \left(\frac{c}{c_1}\right)^2\right]\right\} \tag{6.3.30}$$

For large times, the term $(c/c_1)^2$ becomes negligible and thus the shock strength tends to zero like $1/\sqrt{x}$ [cf. Eq. (6.3.11)]. The concentration profiles at successive times will be similar to those in Fig. 6.13 except for the nonvanishing concentrations, that is, c_0 in the front and c_2 in the tail.

If the expansion wave is based on the initial distribution of solute $c = F(\xi)$, we have the equation

$$\tau = \sigma(c)\{x - \xi(c)\} \tag{6.3.31}$$

instead of Eq. (6.3.17), where $\xi(c) = F^{-1}(c)$, so the term $d\eta/dc$ in Eq. (6.3.19) should be replaced by $-(d/dc)\{\sigma(c)\xi(c)\}$ [cf. Eq. (6.2.21)].

Exercises

6.3.1. From Eqs. (6.3.5) and (6.3.6), derive an equation that is equivalent to Eq. (6.3.23) and solve it to express x as a function of c. Use this solution to confirm Eqs. (6.3.10) and (6.3.11).

6.3.2. A column is of length X and a sample is put on for duration α. How soon can another sample of the same size be put on if their chromatograms are not to overlap? Assume that the equilibrium relation is given by the Langmuir isotherm.

6.3.3. Put $\alpha = c_0/c_1$ in Eq. (6.3.1) and consider the limit as $\alpha \to 0$ (or $c_1 \to \infty$), maintaining $c_0 = \alpha c_1 =$ fixed. This is equivalent to an impulse introduced at $\tau = 0$. Could you apply this limit to Eqs. (6.3.4), (6.3.10), (6.3.11) and (6.3.12) so that the propagation of an impulse can be analyzed? Discuss if these expressions are relevant. How does the time evolution of the system in this case compare with that for a square-wave input?

6.3.4. An initially clean column of voidage 0.4 is being saturated with a feed stream having $c = 0.1$ mol/l. The adsorption equilibrium is described by the Langmuir isotherm with $\gamma = 5$ and $K = 5$ l/mol. If at $\tau = 5$ the feed concentration drops to $c = 0.05$ mol/l, analyze the time evolution of the system, showing clearly the (x, τ)-plane portrait.

6.3.5. If we have $c_0 = c_2 = 0$ and

$$G(\eta) = c_1 \sin \frac{\pi}{2}\left(\frac{\eta_2 - \eta}{\eta_2 - \eta_1}\right), \qquad \eta_1 \le \eta \le \eta_2$$

in Eq. (6.3.13), determine the shock path in terms of the parameter c. Can you argue that the shock strength c decays like $1/\sqrt{x}$ for large time?

6.3.6. Suppose that the equilibrium relation is given by

$$f(c) = kc^{3/2}$$

Given the data

$$\text{at } \tau = 0, \qquad c = \begin{cases} c_0\left(1 - \dfrac{x}{\alpha}\right), & \text{for } 0 < x \le \alpha \\ 0, & \text{for } x > \alpha \end{cases}$$

$$\text{at } x = 0, \qquad c = \begin{cases} c_0, & \text{for } 0 \le \tau \le \beta \\ 0, & \text{for } \tau > \beta \end{cases}$$

sketch the portrait of the solution in the (x, τ)-plane and analyze the interaction between the shock and the simple wave. Draw the concentration profiles at various instants.

6.3.7. Consider single-solute chromatography with a Langmuir isotherm when the initial data are prescribed as follows:

$$\text{at} \quad \tau = 0, \qquad c = \begin{cases} c_0\dfrac{x}{\alpha}, & 0 \le x \le \alpha \\ c_0, & \alpha \le x \le \beta \\ 0, & x \ge \beta \end{cases}$$

$$\text{at} \quad x = 0, \qquad c = 0, \qquad \tau \ge 0$$

Sketch the portrait of solution in the (x, τ)-plane and find the parametric representation of the shock path.

6.4 Propagation of a Pulse

With a purely convex isotherm, suppose that the initial condition is specified by a pulse-type distribution of solute. We then expect to have a simple wave that is partly expansive and partly compressive (see Sec. 5.3; in particular,

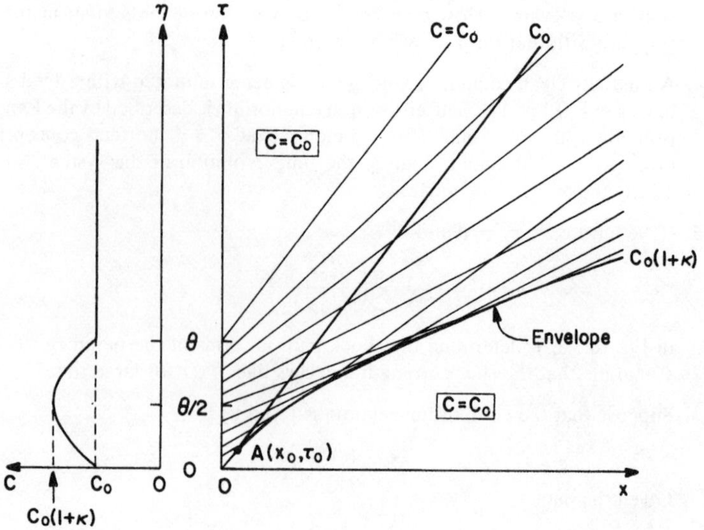

Figure 6.16

Fig. 5.16). The compressive part of the wave will eventually give birth to a shock. While it propagates, the strength of the shock will first grow and then gradually be weakened due to its interaction with the simple wave.

To examine the situation in detail, we will consider the Langmuir isotherm and the initial data prescribed as follows:

$$\text{at} \quad \tau = 0, c = c_0$$
$$\text{at} \quad x = 0, \tau = \eta$$
$$c = \begin{cases} G(\eta) = c_0\left(1 + K\sin\dfrac{\pi}{\theta}\eta\right), & 0 < \eta < \theta \\ c_0, & \eta > \theta \end{cases} \tag{6.4.1}$$

where both κ and θ are positive constants. Thus we have a half-cycle sinusoidal signal in the feed concentration. The feed data for $0 \le \eta \le \theta/2$ will give a compression wave, whereas those for $\theta/2 \le \eta \le \theta$ yield an expansion wave, as shown in Fig. 6.16. The characteristics in the region of the compression wave form an envelope, and its equation may be given in a parametric form by using Eqs. (6.1.5) and (6.1.8):

$$\begin{aligned} x^E &= \frac{\theta(1 + Kc)^3}{2\pi\nu\gamma Kc_0\kappa\cos(\pi\eta/\theta)} \\ \tau^E &= \eta + \frac{\theta(1 + Kc)\{\nu\gamma + (1 + Kc)^2\}}{2\pi\nu\gamma Kc_0\kappa\cos(\pi\eta/\theta)} \end{aligned} \tag{6.4.2}$$

for $0 \le \eta \le \theta/2$, where c is related to η by Eq. (6.4.1). Both x^E and τ^E are monotone increasing functions of η, and so we have $\eta = 0$ (or $c = c_0$) at the cusp point. This implies that a shock will be formed at the point $A(x_0, \tau_0)$, where

$$x_0 = \frac{\theta(1 + Kc_0)^3}{2\pi\nu\gamma K c_0 \kappa}$$
$$\tau_0 = \frac{\theta(1 + Kc_0)\{\nu\gamma + (1 + Kc_0)^2\}}{2\pi\nu\gamma K c_0 \kappa} \qquad (6.4.3)$$

and the concentration at the point is c_0.

While it is interacting with the compression wave, the shock accelerates because the concentration on the left side increases continuously from c_0 to $c_0(1 + \kappa)$, the maximum value of c at the peak. After the shock attains a maximum strength $c_0\kappa$, it starts to interact with the expansion wave, and the concentration on the left side now decreases so that the shock will have to decelerate. This is schematically shown in Fig. 6.17. Here again the concentration on the right side of the shock remains fixed at c_0. Hence, if we let c stand for the concentration on the left side at an arbitrary point (x, τ) on the shock path, it is immediately apparent that we can write Eq. (6.3.19) together with Eqs. (6.3.21) and (6.3.22). In the present case, we have

$$\eta = G^{-1}(c) = \frac{\theta}{\pi} \sin^{-1}\left(\frac{c - c_0}{c_0\kappa}\right) \qquad (6.4.4)$$

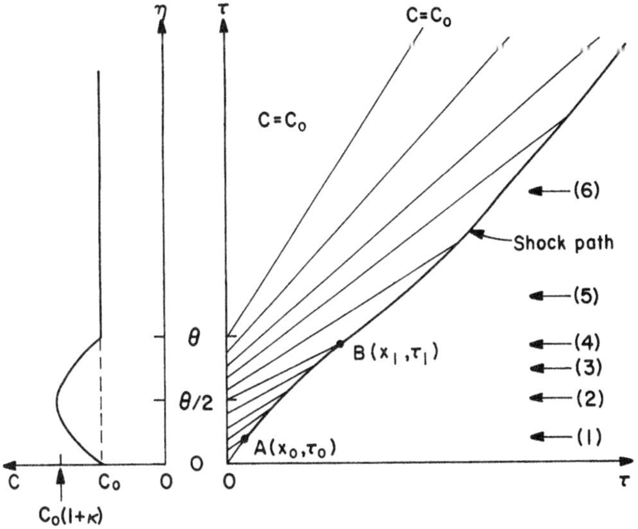

Figure 6.17

and thus, from Eq. (6.3.23),

$$\frac{c - c_0}{(1 + Kc)^2}\frac{dx}{dc} + \frac{2(1 + Kc_0)}{(1 + Kc)^3}x = \frac{\theta(1 + Kc_0)}{\pi\nu\gamma K}\frac{1}{\sqrt{(c_0\kappa)^2 - (c - c_0)^2}}$$

or

$$\frac{d}{dc}\left\{\left[\frac{K(c - c_0)}{1 + Kc}\right]^2 x\right\} = \frac{\theta(1 + Kc_0)}{\pi\nu\gamma}\frac{K(c - c_0)}{\sqrt{(c_0\kappa)^2 - (c - c_0)^2}} \tag{6.4.5}$$

The pertinent initial condition is

$$x = x_0 \quad \text{at} \quad c = c_0 \tag{6.4.6}$$

so by integrating Eq. (6.4.5) with this condition, we have

$$x = \frac{\theta K c_0 \kappa(1 + Kc_0)}{\pi\nu\gamma}\left\{\frac{1 + Kc}{K(c - c_0)}\right\}^2\left\{1 - \sqrt{1 - \left(\frac{c - c_0}{c_0\kappa}\right)^2}\right\} \tag{6.4.7}$$

We then use Eq. (6.3.17) to determine τ:

$$\tau = \frac{\theta}{\pi}\sin^{-1}\left(\frac{c - c_0}{c_0\kappa}\right) + \left\{1 + \frac{\nu\gamma}{(1 + Kc)^2}\right\}x \tag{6.4.8}$$

These equations give the parametric representation of the shock path. In the limit as c approaches c_0, Eq. (6.4.7) yields two limits, x_0 and ∞, and the latter implies that the interaction goes on indefinitely. In other words, the shock becomes weaker and weaker in strength but never vanishes completely. The acceleration of the shock is reversed when c reaches its maximum value of $c_0(1 + \kappa)$. This is at the point B in Fig. 6.17, whose coordinates are

$$\begin{aligned} x_1 &= \frac{\theta(1 + Kc_0)\{1 + Kc_0(1 + \kappa)\}^2}{\pi\nu\gamma K c_0 \kappa} \\ \tau_1 &= \frac{\theta}{2} + \frac{\theta(1 + Kc_0)\{\nu\gamma + [1 + Kc_0(1 + \kappa)]^2\}}{\pi\nu\gamma K c_0 \kappa} \end{aligned} \tag{6.4.9}$$

In Fig. 6.18 we present the distribution of solute along the column at various instants as marked in Fig. 6.17:

(1) $\tau = \tau_0$, at the inception of the shock
(2) $\tau = \theta/2$, when the peak is just introduced
(3) $\theta/2 < \tau < \tau_1$, during the acceleration of the shock
(4) $\tau = \tau_1$, when the shock attains its maximum strength
(5), (6) $\tau > \tau_1$, when the shock continues to decelerate; these forms are typical for all later times

Here we clearly see how the shock, after inception, first grows and then decays. As the shock decays with time, the rear side (expansion wave) becomes more and more diffuse. Notice that the details of the original pulse

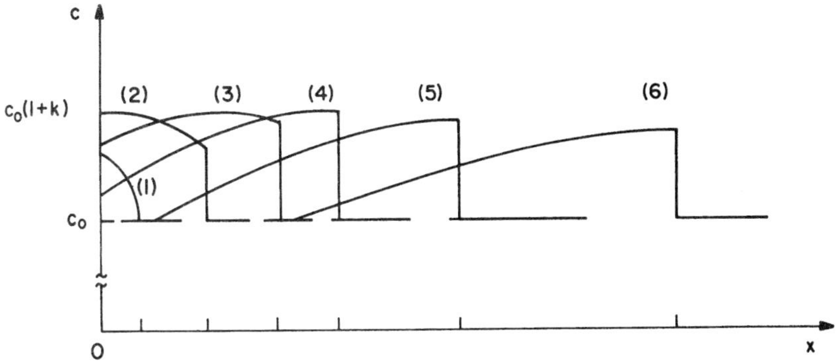

Figure 6.18

have been lost. While the intensity of the pulse gets weaker indefinitely, its band continues to broaden.

Next, we will turn to the problem in which the states on both sides of a shock vary at the same time. This is the situation we expect to obtain when a shock propagates between two simple waves. If x and τ are the coordinates of an arbitrary point on the shock path, we may regard both of them as functions of c^l and c^r. Thus we can write

$$\frac{d\tau}{dx} = \tilde{\sigma}(c^l, c^r) = \frac{(\partial\tau/\partial c^l)(dc^l/dc^r) + (\partial\tau/\partial c^r)}{(\partial x/\partial c^l)(dc^l/dc^r) + (\partial x/\partial c^r)} \tag{6.4.10}$$

or

$$\frac{dc^l}{dc^r} = -\frac{\tilde{\sigma}(\partial x/\partial c^r) - (\partial\tau/\partial c^r)}{\tilde{\sigma}(\partial x/\partial c^l) - (\partial\tau/\partial c^l)} = \psi(c^l, c^r) \tag{6.4.11}$$

The right side being given as a function of c^l and c^r, we have now a differential equation the solution of which would give the relationship between c^l and c^r along the shock path. The initial condition can always be specified in the form

$$c^l = c^* \quad \text{at} \quad c^r = c_* \tag{6.4.12}$$

where c^* and c_* are the concentrations on the left and right sides of the shock, respectively, when they just start to change simultaneously.

Suppose that both simple waves are based on the feed data prescribed along the τ-axis (see Fig. 6.19). At an arbitrary point (x, τ) on the shock path, we always have two characteristics, one from the point $(0, \eta')$ bearing $c = c^l$ and the other from the point $(0, \eta'')$ carrying $c = c^r$. The equations of these characteristics are

$$\begin{aligned} \tau - \eta' &= \sigma(c^l)x \\ \tau - \eta'' &= \sigma(c^r)x \end{aligned} \tag{6.4.13}$$

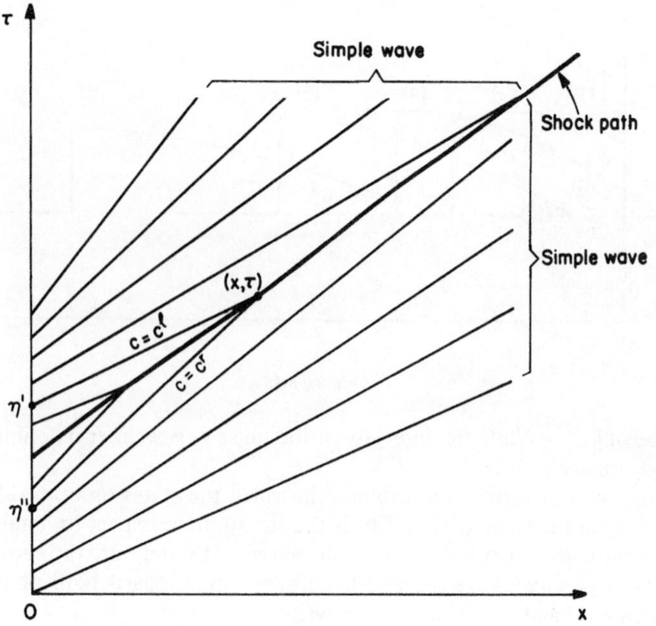

Figure 6.19

Since (x, τ) is an arbitrary point, we may regard η' or η'' as a function of c^l or c^r, respectively. Each of these functions is just the inverse of the corresponding portion of the feed data. Solving Eq. (6.4.13) for x and τ, we have

$$x = \frac{\eta'(c^l) - \eta''(c^r)}{\sigma(c^r) - \sigma(c^l)}$$

$$\tau = \frac{\sigma(c^r)\eta'(c^l) - \sigma(c^l)\eta''(c^r)}{\sigma(c^r) - \sigma(c^l)} \tag{6.4.14}$$

The partial derivatives in Eq. (6.4.11) are now determined in terms of c^l and c^r and substituted in Eq. (6.4.11) to give

$$\frac{dc^l}{dc^r} = \frac{\tilde{\sigma} - \sigma^l}{\tilde{\sigma} - \sigma^r} \frac{(\eta' - \eta'')(d\sigma^r/dc^r) + (\sigma^r - \sigma^l)(d\eta''/dc^r)}{(\eta' - \eta'')(d\sigma^l/dc^l) + (\sigma^r - \sigma^l)(d\eta'/dc^l)} \tag{6.4.15}$$

where $\tilde{\sigma} = \tilde{\sigma}(c^l, c^r)$, $\sigma^l = \sigma(c^l)$, $\sigma^r = \sigma(c^r)$, $\eta' = \eta'(c^l)$, and $\eta'' = \eta''(c^r)$. If this equation is solved subject to Eq. (6.4.12), we have the pair (c^l, c^r) at every point along the shock path. It is then possible to trace the shock path by substituting the pair (c^l, c^r) in Eq. (6.4.14).

If the simple waves are based on the initial data specified along the

x-axis, we can write two equations,

$$\begin{aligned} \tau &= \sigma(c^l)\{x - \xi'(c^l)\} \\ \tau &= \sigma(c^r)\{x - \xi''(c^r)\} \end{aligned} \tag{6.4.16}$$

at an arbitrary point (x, τ) on the shock path, where ξ' and ξ'' are the inverse functions of the initial data prescribed on the left and right sides of the shock, respectively. Thus

$$\begin{aligned} x &= \frac{\sigma^r\xi'' - \sigma^l\xi'}{\sigma^r - \sigma^l} \\ \tau &= \sigma^l\sigma^r\frac{\xi'' - \xi'}{\sigma^r - \sigma^l} \end{aligned} \tag{6.4.17}$$

in which $\sigma^l = \sigma(c^l)$ and $\sigma^r = \sigma(c^r)$, and by substituting in Eq. (6.4.11) we obtain

$$\frac{dc^l}{dc^r} = \frac{\tilde{\sigma} - \sigma^l}{\tilde{\sigma} - \sigma^r}\,\frac{\sigma^l(\xi'' - \xi')(d\sigma^r/dc^r) - \sigma^r(\sigma^r - \sigma^l)(d\xi''/dc^r)}{\sigma^r(\xi'' - \xi')(d\sigma^l/dc^l) - \sigma^l(\sigma^r - \sigma^l)(d\xi'/dc^l)} \tag{6.4.18}$$

This equation is subject to the initial condition, Eq. (6.4.12). Following the same procedure as before, we should be able to determine the concentrations on both sides of the shock and to locate the shock path. When we have a compression wave and if a shock is to be formed on one of the intermediate characteristics [see Fig. 6.3(c)], then we expect the states on both sides to vary simultaneously as the shock propagates. Thus Eqs. (6.4.17) and (6.4.18) are needed, but the initial condition is troublesome because we have $c^l = c^r$, and so $\tilde{\sigma} = \sigma^l = \sigma^r$ at the point of shock formation. One may certainly try to generate the solution in a series form. For numerical integration, it seems reasonable to use -1 as the limiting value of dc^l/dc^r at the point of shock formation.

We are now naturally led to a problem with periodic data. Suppose we have a train of half-cycle sinusoidal signals instead of a single signal in Eq. (6.4.1) and in Fig. 6.17. [See Fig. 6.20(a).] Up to the point of shock inception, the situation remains similar to that in Fig. 6.17 except for the expansion wave in front of the shock. Once the shock is formed, however, we see that the concentrations on both sides increase due to the simultaneous interactions with two simple waves. Hence, the shock accelerates faster than the case with a single signal. This is so until the concentration on the left side reaches a maximum value of $c_0(1 + \kappa)$. Thereafter, the shock will gradually decelerate and asymptotically attains a constant speed because the states on both sides change in opposite directions and asymptotically approach a common value. For large time, therefore, we can imagine the picture of the solution as shown in Fig. 6.21.

According to Lax (1957, 1972), the shock strength in a periodic wave

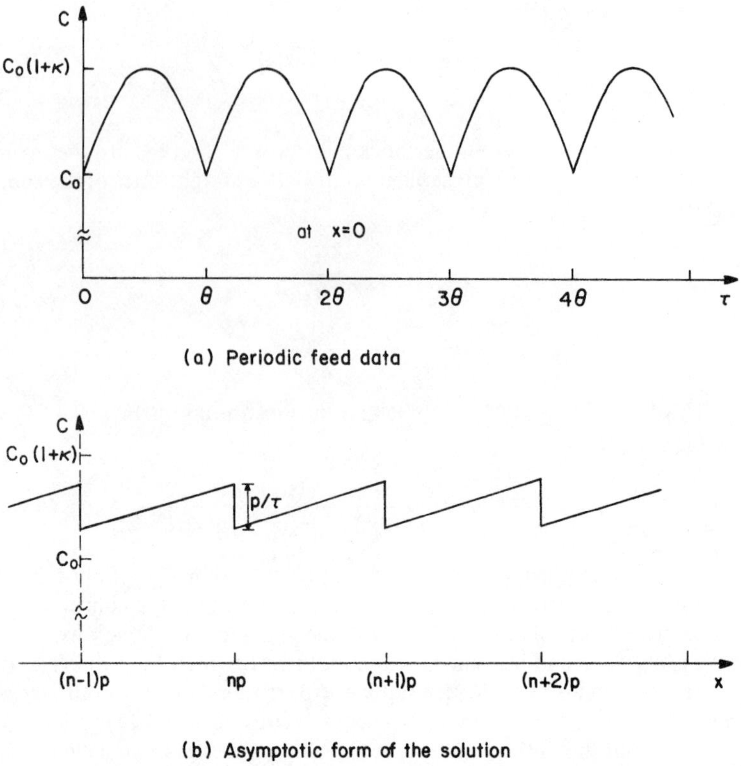

Figure 6.20

decays like $1/x$ for large time.† We may say equivalently that the decay is like $1/\tau$ for large τ because the shock path tends to be linear. Whitham (1974) showed that the asymptotic solution between successive shocks is linear in x with slope $1/\tau$. Thus the asymptotic form of the entire profile is the sawtooth wave as shown in Fig. 6.20(b). Here the period p is dependent on the ultimate slope, $\bar{\sigma}_\infty$, of the shock path (i.e., $p = \theta/\bar{\sigma}_\infty$). But the ultimate slope may be determined only through the detailed analysis of earlier parts. Lax (1957, 1972) as well as Whitham (1974) also proved that, if there is a single hump in the data so that a single shock is to appear in a later stage, the shock strength decays like $1/\sqrt{x}$ for large time [cf. Eqs. (6.3.11) and (6.3.12), and also Eq. (6.3.30)]. We find it rather paradoxical since the solution for a periodic wave, which represents a more severe initial disturbance, would decay faster than the solution in a single hump. This was noticed by Lax (1972), so we may call it the *Lax paradox*.

†The original proof was for the equation $(\partial u/\partial t) + (\partial f/\partial x) = 0$, where $f''(u) > 0$, and for the initial condition $u(x, 0)$.

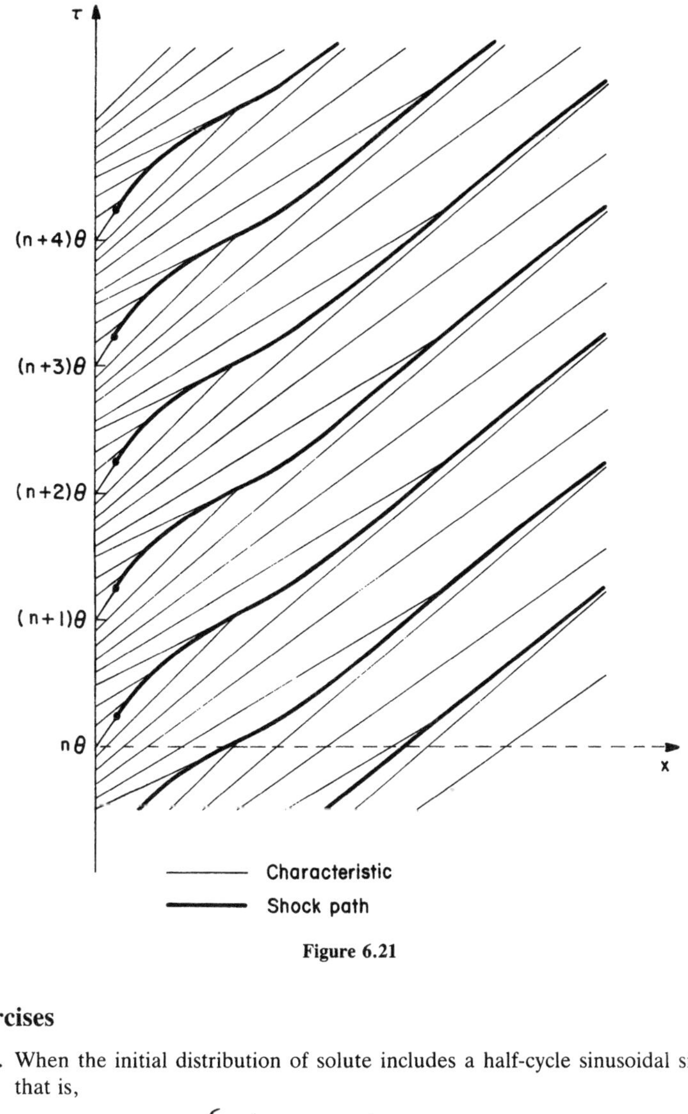

Figure 6.21

Exercises

6.4.1. When the initial distribution of solute includes a half-cycle sinusoidal signal, that is,

$$\text{at}\quad \tau = 0, \qquad c = \begin{cases} c_0\left(1 + \kappa \sin\dfrac{\pi}{\theta}x\right), & 0 \le x \le \theta \\ c_0, & x > \theta \end{cases}$$

$$\text{at}\quad x = 0, \qquad c = c_0$$

analyze the propagation of the wave through the column and discuss how the wave form changes. Use the Langmuir isotherm for the equilibrium relation.

6.4.2. Discuss the overlapping of two finite chromatograms (see Exercise 6.3.2).

6.4.3. A gravity water wave propagates in such a way that the water surface elevation h above the undisturbed level is governed by

$$\frac{\partial h}{\partial t} + \left(1 + \frac{3}{2}h\right)\frac{\partial h}{\partial x} = 0, \qquad -\infty < x < \infty, \qquad t > 0$$

if h and its derivatives remain small. An appropriate initial condition is

$$h(x, 0) = \begin{cases} h_0(1 + \cos \pi x), & |x| \le 1 \\ 0, & |x| > 1 \end{cases}$$

Conduct an analysis of the wave propagation. Also, discuss the case with $h_0 < 0$. What are the major differences between the two cases?

6.4.4. Derive Eqs. (6.4.15) and (6.4.18).

6.4.5. For a sawtooth wave as the initial distribution of solute, integrate Eq. (6.4.18) numerically with the Langmuir isotherm and locate the shock paths in the (x, τ)-plane. Also, examine the decay of shocks. Assume that the wave is initially between $c = 0.1$ and $c = 0.5$ mol/l with a period of unity and that the isotherm parameters are $\gamma = 2.5$ and $K = 5$ l/mol.

6.5 Analysis of a Countercurrent Adsorber

When the adsorbent phase moves countercurrently against the fluid phase, we need to consider the convection through the solid phase. For a single-solute case, we have Eq. (1.5.1) in the form

$$V_f\frac{\partial c}{\partial z} - \frac{1-\varepsilon}{\varepsilon}V_s\frac{\partial n}{\partial z} + \frac{\partial c}{\partial t} + \frac{1-\varepsilon}{\varepsilon}\frac{\partial n}{\partial t} = 0 \tag{6.5.1}$$

in which the subscripts f and s are used with the interstitial velocity V to denote the fluid and solid phases, respectively. We will introduce the dimensionless parameters and variables defined as

$$\nu = \frac{1-\varepsilon}{\varepsilon}, \qquad \mu = \frac{(1-\varepsilon)V_s}{\varepsilon V_f} \tag{6.5.2}$$

$$x = \frac{z}{L}, \qquad \tau = \frac{V_f t}{L} \tag{6.5.3}$$

where L is the total length of the column. Note that ν denotes the volume ratio between the two phases, whereas μ is the ratio of the volumetric flow rates. Now Eq. (6.5.1) becomes

$$\frac{\partial c}{\partial x} - \mu\frac{\partial n}{\partial x} + \frac{\partial c}{\partial \tau} + \nu\frac{\partial n}{\partial \tau} = 0 \tag{6.5.4}$$

Since we are concerned here with the equilibrium case, we have the isotherm

$$n = f(c) \tag{6.5.5}$$

so that, on substituting in Eq. (6.5.4), we obtain the quasi-linear equation

$$\{1 - \mu f'(c)\}\frac{\partial c}{\partial x} + \{1 + \nu f'(c)\}\frac{\partial c}{\partial \tau} = 0 \tag{6.5.6}$$

The characteristic direction is

$$\sigma(c) = \frac{d\tau}{dx} = \frac{1 + \nu f'(c)}{1 - \mu f'(c)} \tag{6.5.7}$$

which is fixed since c remains invariant in this direction. Thus the characteristics are straight lines. It is evident that for large values of μ the characteristics can be directed backward with negative slopes. Since

$$\frac{\partial \sigma}{\partial \mu} = \frac{1 + \nu f'(c)}{\{1 - \mu f'(c)\}^2} f'(c) > 0 \tag{6.5.8}$$

we see that the characteristic for a specific value of c would rotate counterclockwise as μ increases and for $\mu = 1/f'(c)$ the characteristic becomes vertical (see Fig. 6.22). We also observe that

$$\frac{d\sigma}{dc} = \frac{(\nu + \mu)f''(c)}{\{1 - \mu f'(c)\}^2} \tag{6.5.9}$$

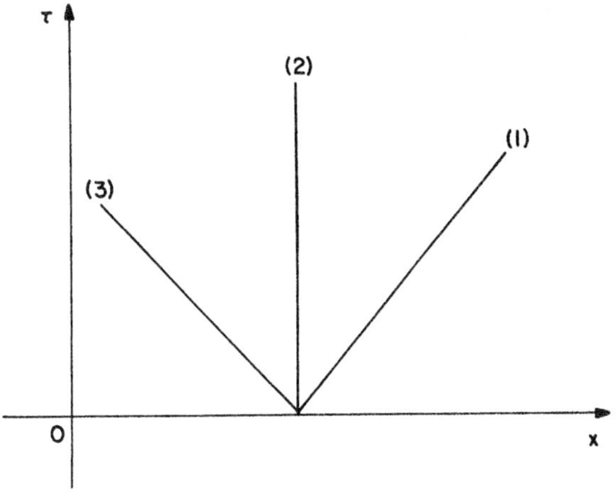

(1) Characteristic for small μ

(2) Characteristic for $\mu = 1/f'(c)$

(3) Characteristic for large μ

Figure 6.22

That is, the derivative $d\sigma/dc$ takes the same sign as $f''(c)$.† This implies that the structure of the simple wave solution is independent of the velocity of the solid phase μ. For example, if c increases in the x-direction for $f''(c) < 0$, then we have an expansion wave. But the region of the simple wave rotates counterclockwise as μ increases.

If the solution contains a discontinuity, it propagates in the direction

$$\tilde{\sigma}(c^l, c^r) = \frac{d\tau}{dx} = \frac{1 + \nu([f]/[c])}{1 - \mu([f]/[c])} \tag{6.5.10}$$

which has been formulated by applying the same procedure as in the fixed-bed case (see Sec. 5.4). Here again we see the possibility of having a discontinuity that remains stationary ($\tilde{\sigma} = \infty$) or that propagates backward ($\tilde{\sigma} < 0$). The derivative of $\tilde{\sigma}$ with respect to the ratio $[f]/[c]$ has positive sign, so the entropy condition essentially remains unaffected by the parameter μ, but, because $\tilde{\sigma}$ can be of negative sign, we have to state the entropy condition as follows: At all points of discontinuity in the solution the inequality

$$\tilde{\sigma}(c^l, c^r)^{-1} \leq \tilde{\sigma}(c^l, c)^{-1} \tag{6.5.11}$$

must be satisfied for all c between c^l and c^r. [Compare this with Eq. (5.4.4).] Then both the theorem and rule in Sec. 5.4 are valid. In the case of a convex isotherm (e.g., the Langmuir isotherm), Eq. (6.5.11) gives

$$f'(c^l) < \frac{[f]}{[c]} < f'(c^r) \tag{6.5.12}$$

so that, from Eqs. (6.5.7) and (6.5.10), we have

$$\sigma(c^l)^{-1} > \tilde{\sigma}(c^l, c^r)^{-1} > \sigma(c^r)^{-1} \tag{6.5.13}$$

The discontinuity moves faster than the state c^r but with a lesser speed than the state c^l. This is the very definition of a shock, and thus every discontinuity in a system with a convex isotherm is a shock.

Here we wish to consider the problem with uniform initial state of the bed and constant feed concentrations from both sides. Thus the initial condition is

$$\text{at} \quad \tau = 0, \qquad c = c_0 \tag{6.5.14}$$

and there is no need to specify the initial adsorbed concentration since $n_0 = f(c_0)$. At both boundaries there is no accumulation of solute so that, from Eqs. (1.5.5) and (1.5.6), we have

$$\text{at} \quad x = 0, \qquad c_a - c_{0+} = \mu(n_a - n_{0+}) \tag{6.5.15}$$

$$\text{at} \quad x = 1, \qquad c_{1-} - c_b = \mu(n_{1-} - n_b) \tag{6.5.16}$$

†If $-V_s > V_f$ in case of co-current contact, we have $\nu + \mu < 0$ and $d\sigma/dc$ takes the opposite sign of $f''(c)$.

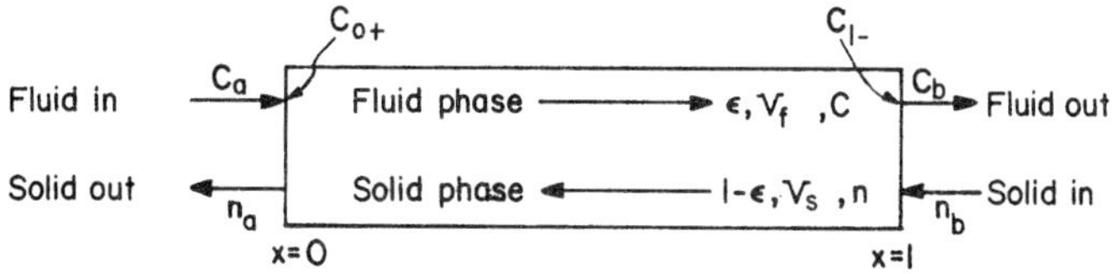

Figure 6.23

where the subscripts a and b refer to the inlet sides of the fluid and solid streams, respectively, and the subscripts $0+$ and $1-$ represent the states just inside the boundaries, as shown in Fig. 6.23. Both c_a and n_b are assumed constant. These boundary conditions can be satisfied identically by the continuity of c and n. If not satisfied identically, they serve to determine the outputs n_a and c_b by

$$n_a = n_{0+} + \frac{c_a - c_{0+}}{\mu} \tag{6.5.17}$$

$$c_b = c_{1-} + \mu(n_b - n_{1-}) \qquad 6.5.18)$$

This implies that there can be a discontinuity in c at either boundary. Such a discontinuity is different from the shock since it has nothing to do with the jump condition, Eq. (6.5.10), or the entropy condition, Eq. (6.5.11), and will be called a *boundary discontinuity*, or B.D. in abbreviation.

Now we will investigate in detail the time evolution of the adsorption process for μ values in various ranges. For simplicity we will confine ourselves to those cases in which we have $f''(c) < 0$. Thus all the discontinuities appearing inside the column are shocks and the entropy condition requires that $c^l > c^r$ at every point on the shock path.

Let us first suppose that the state of the incoming solid stream is the same as that of the initial bed, so

$$n_b = n_0, \qquad c_{be} = f^{-1}(n_b) = c_0 \tag{6.5.19}$$

where c_{be} is the equilibrium concentration corresponding to n_b. If $c_a < c_{be}$, we expect to see a simple wave centered at the origin, so for low values of μ the solution in the (x, τ)-diagram will look like that given in Fig. 6.24(a). In fact, the structure of the solution remains unchanged as long as the characteristic for c_a is directed forward, or if

$$0 \le \mu < \frac{1}{f'(c_a)} = \mu^* \tag{6.5.20}$$

which is called the *lower range of* μ. Since the whole range of influence is forward-facing, we have $c_{0+} = c_a$, and Eq. (6.5.15) becomes trivial to give $n_a = n_{ae} = f(c_a)$. In this sense, we say that the boundary at $x = 0$ remains active in its data-transmitting character. There is no discontinuity at $x = 0$ and equilibrium is maintained across the boundary. On the other hand, the nature of the other boundary at $x = 1$ is completely different. For $\tau < \tau_1$ [see Fig. 6.24(a)], the situation at $x = 1$ remains intact since the head of the simple wave has not reached there yet. Over the period from τ_1 to τ_2, however, we see that c_{1-} (and accordingly n_{1-}) decreases gradually from c_0 (or n_b) to c_a (or n_{ae}). The boundary bears a discontinuity across which equilibrium is not maintained, and the state of c_b is given by Eq. (6.5.18). Beyond the instant $\tau = \tau_2$, where

$$\tau_s = \tau_2 = \frac{1 + \nu/\mu_*}{1 - \mu/\mu_*} \tag{6.5.21}$$

no further change is observed, so the system reaches a steady state for which we have

$$\begin{aligned} n_{as} &= n_{ae} = f(c_a) \\ c_s &= c_a, \qquad 0 < x < 1 \\ c_{bs} &= c_a + \mu(n_b - n_{ae}) \end{aligned} \tag{6.5.22}$$

For the variation of c_b with time, one can get a better feeling by looking at Fig. 6.24(b), in which the boundary discontinuity at $x = 1$ is represented by the line segment of slope $1/\mu$. Clearly, c_b decreases with time. Figure 6.25 shows the profiles of c at various instants.

If μ is further increased so that it falls in the range

$$\mu_* \le \mu < \frac{1}{f'(c_{be})} = \mu^* \tag{6.5.23}$$

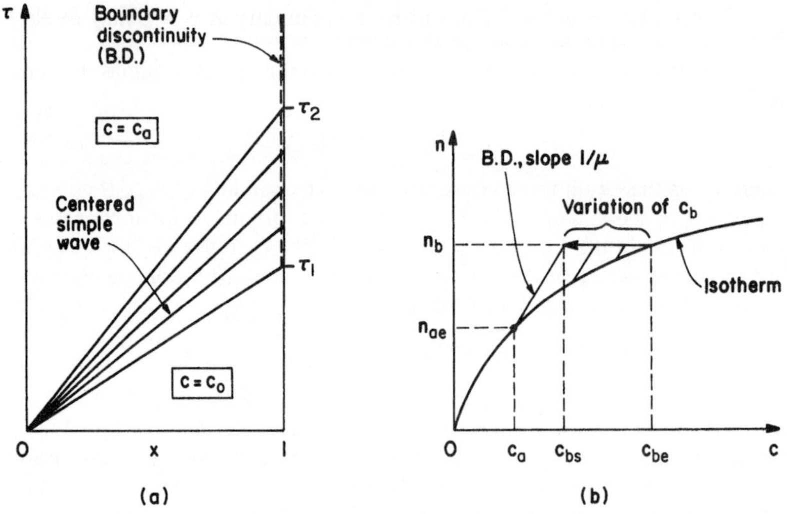

Figure 6.24

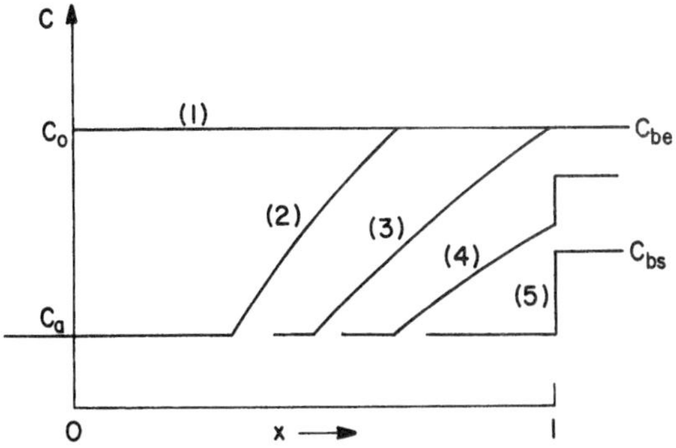

Figure 6.25

we always find a concentration c_∞, $c_a \le c_\infty < c_{be}$, for which σ becomes infinitely large. This implies that

$$f'(c_\infty) = \frac{1}{\mu} \tag{6.5.24}$$

and in case of the Langmuir isotherm we have $Kc_\infty = -1 + \sqrt{\mu\gamma}$. The characteristic for $c = c_\infty$ then becomes vertical so that the state c_∞ remains stationary at $x = 0$. Since $c_{0+} = c_\infty$ for $\tau > 0$, we have a boundary discontinuity† at $x = 0$ from the beginning, and n_a is given by Eq. (6.5.17) with $c_{0+} = c_\infty$ and $n_{0+} = n_\infty = f(c_\infty)$. Here the boundary at $x = 0$ may be said to be partially active in its data-transmitting capability. Inside the column we observe the characteristics for all concentrations from c_∞ to c_{be} fanning forward, so the profile of concentration tends to flatten out indefinitely. See Fig. 6.26(a). The ultimate state is obviously $c = c_\infty$, $0 < x < 1$, but such a state is not attained at a finite time. The situation at $x = 1$ is similar to that which we found with the lower range of μ, except for the fact that c_{1-} approaches c_∞ only asymptotically. For $\tau > \tau_1$, therefore, we have a boundary discontinuity at $x = 1$, the jump across which gradually grows to reach the maximum value $(c_{bs} - c_\infty)$, as shown in Fig. 6.26(b). At steady state, attained only at $\tau = \infty$, we have

$$\begin{aligned} n_{as} &= n_\infty + \frac{c_a - c_\infty}{\mu} \\ c_s &= c_\infty, \qquad 0 < x < 1 \\ c_{bs} &= c_a + \mu(n_b - n_\infty) \end{aligned} \tag{6.5.25}$$

†Note that the boundary at $x = 0$ is characteristic, and so we can specify there only $c = c_\infty$ to have the continuity in c. Otherwise, we expect to have a discontinuity at $x = 0$.

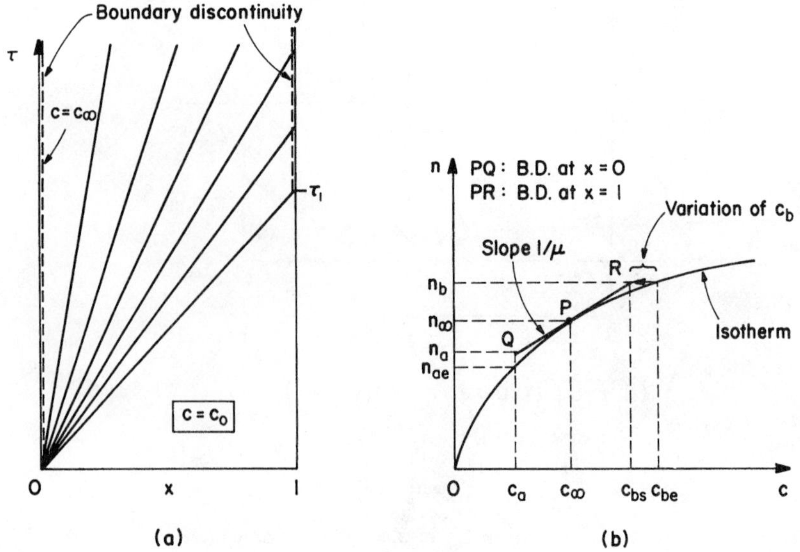

Figure 6.26

The range of μ given by Eq. (6.5.23) is called the *transition range*.

Finally, we come to the *upper range of* μ:

$$\mu \geq \mu^* \tag{6.5.26}$$

In this case the characteristics for all concentrations between c_a and c_{be} are directed backward. Thus the feed data we specify at $x = 0$ cannot penetrate the boundary, whereas the data at $x = 1$ are fully transmitted into the column. In other words, we have a passive boundary at $x = 0$ and an active one at $x = 1$. It is immediately clear that the state inside the column is $c = c_{be}$ and $n = n_b$, $0 < x < 1$, and Eq. (6.5.16) becomes trivial to yield $c_b = c_{be}$. There is no discontinuity at $x = 1$ and equilibrium is maintained across the boundary. At $x = 0$, however, the boundary bears a discontinuity from the beginning, since we have $c_{0+} = c_{be}$. The situation is shown in Fig. 6.27. The whole system attains a steady state immediately, which is

$$\begin{aligned} n_{as} &= n_b + \frac{c_a - c_{be}}{\mu} \\ c_s &= c_{be}, \qquad 0 < x < 1 \\ c_{bs} &= c_{be} = f^{-1}(n_b) \end{aligned} \tag{6.5.27}$$

The information on steady states is summarized in Table 6.1, as well as in Fig. 6.28. If we exclude the transition range, there are two distinct steady states, one with a discontinuity at $x = 0$ and the other with a discontinuity

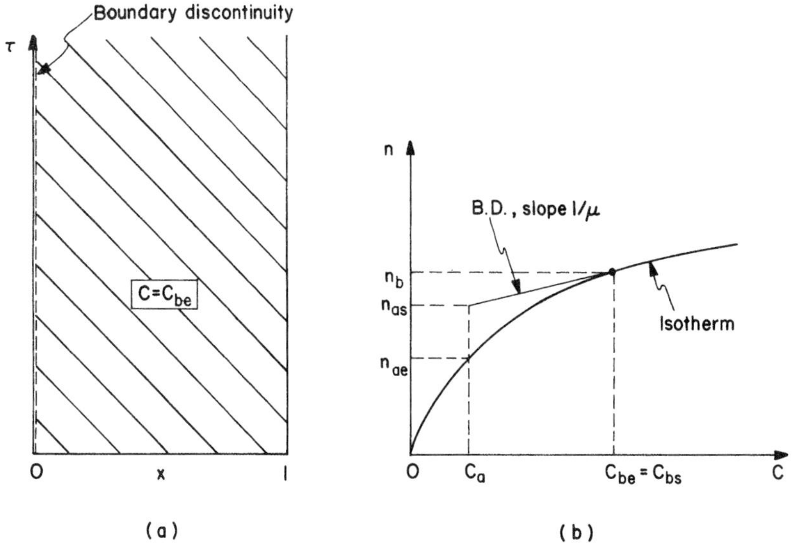

Figure 6.27

at $x = 1$, and these correspond to the lower range and the upper range of μ, respectively. Therefore, the particular steady state that would be attained is indeed dependent on the range to which the specific value of μ belongs. The process we have considered would be most likely the elution of a preadsorbed solid material for which we desire to have low n_{as}, high c_{bs}, and low τ_s. From Fig. 6.28 it would seem best to choose μ^*, but if it is more important to have the lowest possible n_{as}, one should select a μ value slightly less than μ_*. Though it may not be very definite, this type of approach can provide valuable information for design purposes.

TABLE 6.1

	Boundary discontinuity		Steady state			
Range of μ	$x = 0$	$x = 1$	τ_s	n_{as}	c_s, $0 < x < 1$	c_{bs}
Lower range $0 \le \mu < \mu^*$	No	Yes	τ_2	n_{ae}	c_a	Eq. (6.5.22)
Transition range $\mu_* \le \mu < \mu^*$	Yes	Yes	∞	Eq. (6.5.25)	c_∞	Eq. (6.5.25)
Upper range $\mu \ge \mu^*$	Yes	No	0	Eq. (6.5.27)	c_{be}	c_{be}

1. τ_s: The time at which the steady state is attained.
2. $\mu_* = 1/f'(c_a)$; $\mu^* = 1/f'(c_{be})$.

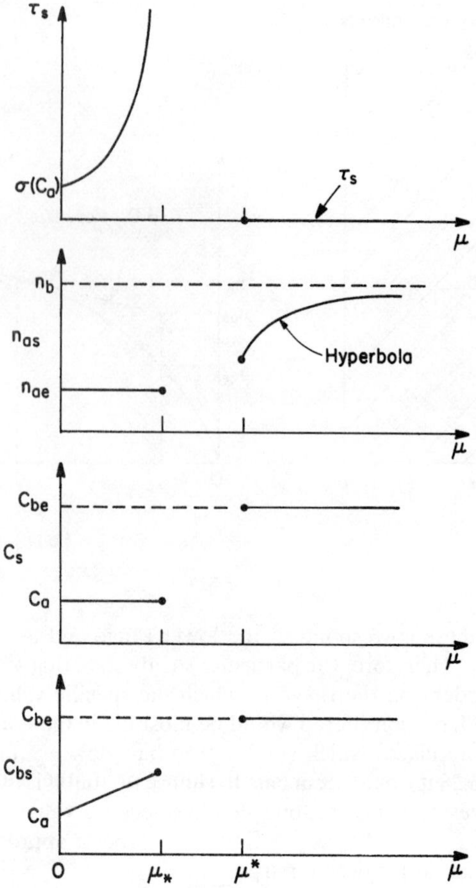

Figure 6.28

If $c_a > c_0 = c_{be}$, we expect to see a shock propagating from the origin instead of the simple wave in the previous case. The shock maintains a constant strength of $(c_a - c_{be})$ and propagates in the direction

$$\tilde{\sigma}(c_a, c_{be}) = \frac{1 + \nu[(n_{ae} - n_b)/(c_a - c_{be})]}{1 - \mu[(n_{ae} - n_b)/(c_a - c_{be})]} \tag{6.5.28}$$

which is constant. Thus we see that the transition range of μ shrinks to a single value μ_c such that

$$\mu_c = \mu_* = \mu^* = \frac{c_a - c_{be}}{n_{ae} - n_b} \tag{6.5.29}$$

In the lower range, $0 \leq \mu < \mu_c$, the shock propagates through the column to reach the other end at

$$\tau_s = \frac{1 + \nu/\mu_c}{1 - \mu/\mu_c} \tag{6.5.30}$$

and, thereafter, we have the steady-state situation with a discontinuity at $x = 1$. For $\mu = \mu_c$, the shock becomes stationary at $x = 0$ and $c = c_{be}$ for $0 < x < 1$. Although c is discontinuous at $x = 0$, equilibrium is maintained across the boundary. In the upper range, $\mu > \mu_c$, the boundary at $x = 0$ becomes passive in data-transmitting capability and bears a boundary discontinuity with $c_{0+} = c_{be}$. (For the steady-state consideration, see Exercise 6.5.2.)

In general, the initial state of the column (c_0 and n_0) may not be the same as the state of the incoming fluid stream (c_a and n_{ae}) nor the same as that of the incoming solid stream (c_{be} and n_b). Then we have to consider two waves, one emanating from the origin and the other from the point (0, 1). If they are both simple waves, the analysis is very much the same as in the previous case, with μ_* and μ^* defined by Eqs. (6.5.20) and (6.5.23), respectively. If the data are prescribed in such a way that one is a shock and the other a simple wave, then the two waves may meet within the column and interact with each other. The interaction can be analyzed by applying the procedure discussed in Sec. 6.3.

The case of two shock waves is of special interest and worth discussing here in some detail. Suppose we have $c_a > c_0 > c_{be}$ so that the entropy condition is satisfied at both ends of the initial line. We will use μ_c defined by Eq. (6.5.29) to divide the μ spectrum. If $0 \le \mu < \mu_c$ (the lower range of μ), the shock S_a from the origin propagates forward since $\tilde{\sigma}(c_a, c_0) > \tilde{\sigma}(c_a, c_{be}) > 0$. For small values of μ, the shock S_b from the other side may not come into the column at all, but, as μ increases, $\tilde{\sigma}(c_0, c_{be})$ becomes negative, and thus S_b propagates backward into the column. The two shocks then meet, combine, and continue as a single shock. This phenomenon is called the *confluence of shocks* (see, for example, Whitham, 1974). The combined shock of strength $(c_a - c_{be})$ propagates in the direction $\tilde{\sigma}(c_a, c_{be}) > 0$ and so eventually reaches the boundary at $x = 1$. This happens at $\tau = \tau_s$, where τ_s is given by Eq. (6.5.30). Thereafter, the system is in a steady state with a boundary discontinuity at $x = 1$. The steady state is given by Eq. (6.5.22). The situation is depicted in Fig. 6.29, while Fig. 6.30 shows the distribution of solute at successive times. Note that c_{bs} is higher than c_{be} and this is not favored.

Suppose now we have $\mu = \mu_c$. The two shocks, S_a and S_b, meet and combine as before, but here the combined shock S becomes stationary because $\tilde{\sigma}(c_a\, c_{be}) = \infty$. Both boundaries are active, and their influences are balanced within the column by the presence of the stationary shock. There are no boundary discontinuities at either boundary, so the equilibrium is maintained across both boundaries, as shown in Fig. 6.31. For the purpose of saturating the adsorbent phase, this is the best result that we would expect to attain.

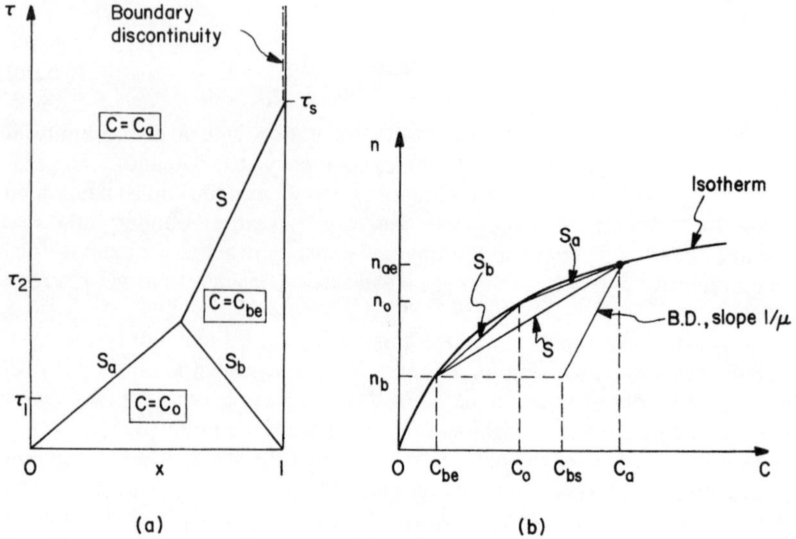

Figure 6.29

If $\mu > \mu_c$ (the upper range of μ), the combined shock S (or the shock S_b for higher values of μ) has a negative speed and thus propagates backward to reach the boundary at $x = 0$ at $\tau = \tau_s$, as shown in Fig. 6.32. For $\tau \geq \tau_s$, we have $c = c_{be}$, $0 < x \leq 1$, with a boundary discontinuity at $x = 0$. This is the steady state, which is the same as given by Eq. (6.5.27).

The steady-state values are plotted against the parameter μ in Fig. 6.33. For saturation, one would desire to have high n_{as}, low c_{bs}, and low τ_s. These requirements are best met if $\mu = \mu_c$.

Exercises

6.5.1. Derive Eq. (6.5.10) and discuss the geometrical meaning of the entropy condition, Eq. (6.5.11), in the diagram of the equilibrium curve.

6.5.2. When the state of the initial bed is identical to that of the incoming solid stream and when $c_a > c_0 = c_{be}$, analyze the time evolution of the exchange process for μ values in various ranges. Draw the (x, τ) diagram, the (n, c) diagram, and the (c, x) diagram in each case.

6.5.3. Consider a countercurrently moving bed adsorption problem. If the adsorption equilibrium follows a Langmuir-type isotherm and if the initial and boundary conditions are specified as

$$\text{at } \tau = 0, \quad c = c_0$$

$$\text{at } x = 0, \quad c = c_a$$

$$\text{at } x = 1, \quad n = n_b$$

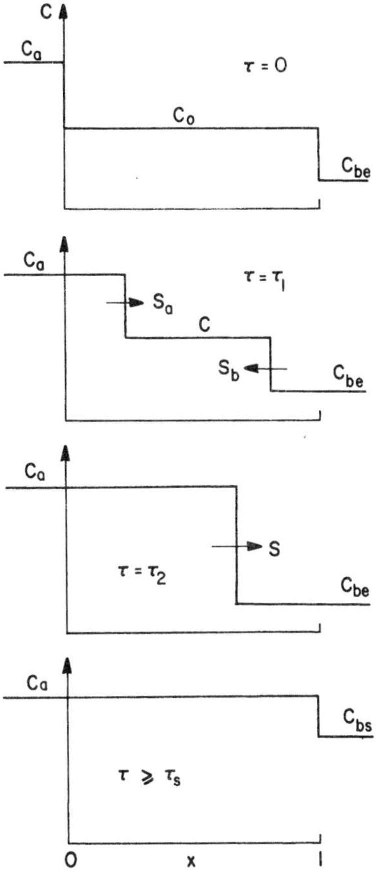

Figure 6.30

where $c_0 > c_a > c_{be} = f^{-1}(n_b)$, divide the μ-spectrum into suitable ranges and discuss the solution for a μ value in each range. Determine the steady-state values of c, n_a, and c_b as a function of μ, respectively.

6.6 Analysis of Traffic Flow

In this section we would like to apply the basic ideas to traffic flow problems and see how effective a simple approach can be in the treatment of complex situations. Let us consider the flow of vehicles along a roadway. If k is the concentration of vehicles in number per kilometer and q is the flux of vehicles along the roadway in number per hour, then the equation for the conservation

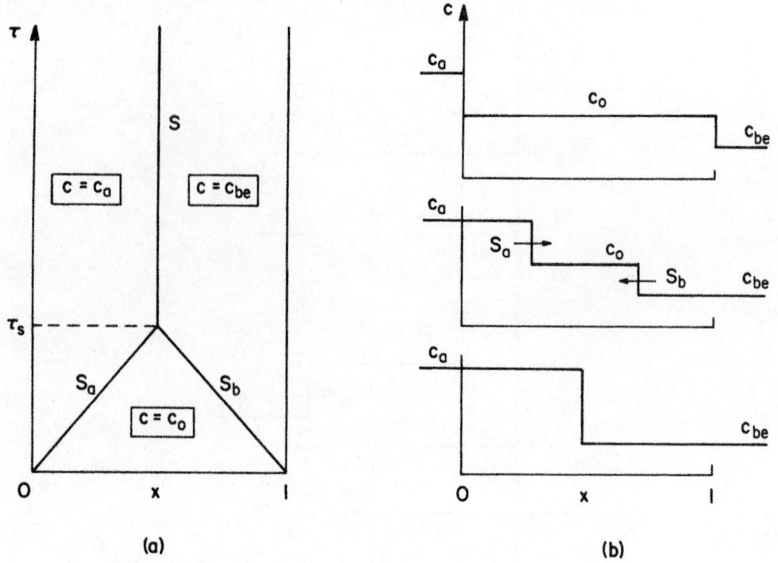

Figure 6.31

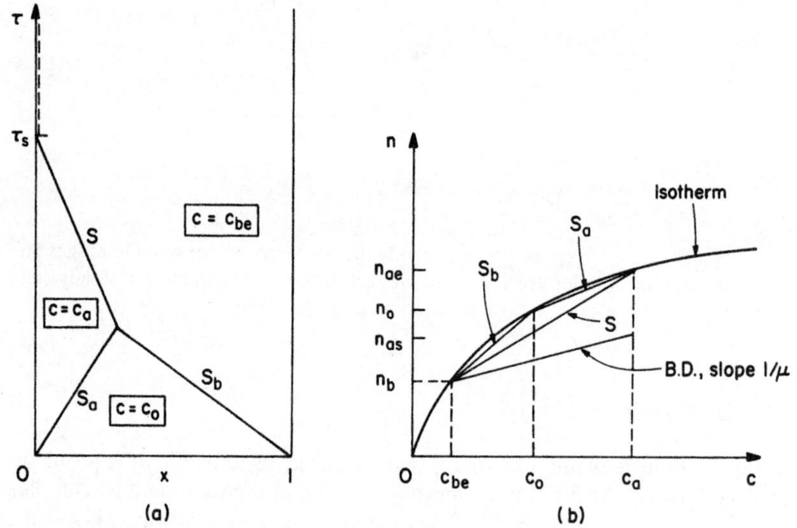

Figure 6.32

of vehicles at a distance z from the intersection at time t is

$$\frac{\partial k}{\partial t} + \frac{\partial q}{\partial z} = 0 \tag{6.6.1}$$

(see Sec. 1.11). The flux here is the product of k and the average vehicle velocity v in kilometers per hour; that is,

$$q = kv \tag{6.6.2}$$

and it is clear from our highway experience that the velocity v will be a decreasing function of k, which starts from a finite maximum value at $k = 0$ and decreases to zero as $k \to k_j$, the value for which all movement stops. Thus the flux can be expressed as a function of the vehicle concentration k.

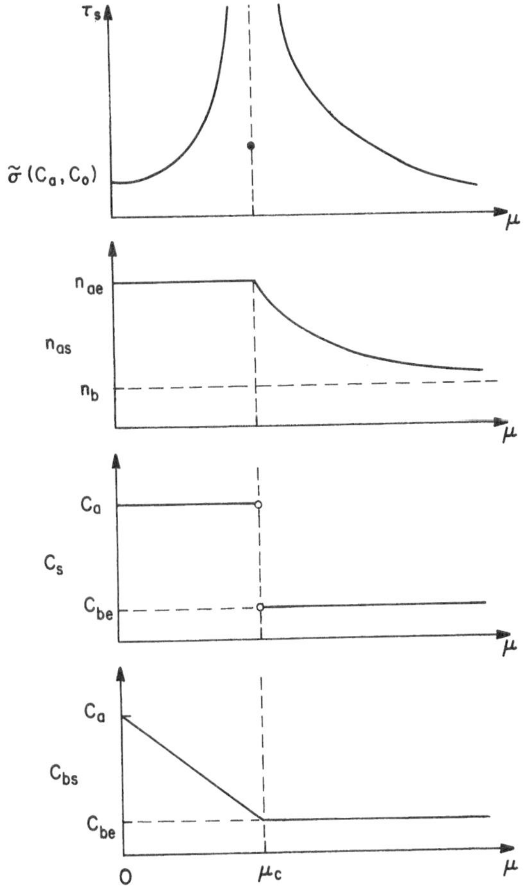

Figure 6.33

For example, for the Lincoln Tunnel in New York City, it has been determined that

$$q = ck \ln \frac{k_j}{k} \tag{6.6.3}$$†

For our purpose here, we will take the simplest relation between v and k,

$$v = v_m\left(1 - \frac{k}{k_j}\right) \tag{6.6.4}$$

where v_m is the maximum velocity on the uncluttered highway, and then the flux is expressed in the form

$$q(k) = (v_m k_j)\left(\frac{k}{k_j}\right)\left(1 - \frac{k}{k_j}\right) \tag{6.6.5}$$

which obtains the maximum value $v_m k_j/4$ when $k/k_j = \frac{1}{2}$. Let us define dimensionless variables as

$$\rho = \frac{k}{k_j}, \qquad w = \frac{v}{v_m}, \qquad \phi = \frac{q}{v_m k_j} \tag{6.6.6}$$

so that

$$w(\rho) = 1 - \rho \tag{6.6.7}$$

and

$$\phi(\rho) = \rho(1 - \rho) \tag{6.6.8}$$

By using the characteristic length L (e.g., the distance between intersections), we will make the independent variables dimensionless:

$$x = \frac{z}{L}, \qquad \tau = \frac{v_m t}{L} \tag{6.6.9}$$

Then Eq. (6.6.1) is rewritten in the form

$$\frac{\partial \rho}{\partial \tau} + \frac{\partial \phi}{\partial x} = 0$$

or

$$\frac{\partial \rho}{\partial \tau} + \phi'(\rho)\frac{\partial \rho}{\partial x} = 0 \tag{6.6.10}$$

The characteristic direction is given by

$$\frac{d\tau}{dx} = \frac{1}{\phi'(\rho)} = \frac{1}{1 - 2\rho} = \sigma(\rho) \tag{6.6.11}$$

†Note that the logarithmic expression does not give a finite value for the vehicle velocity v at $k = 0$.

along which the variable ρ remains constant. Thus the speed of propagation of the state ρ is

$$\frac{dx}{d\tau} = \phi'(\rho) = 1 - 2\rho = \lambda(\rho) \tag{6.6.12}$$

Since this speed is less than the vehicle velocity w, the vehicle moves faster than disturbances. The characteristics will be directed forward if $0 \le \rho < \frac{1}{2}$ and backward if $\frac{1}{2} < \rho \le 1$ (see Fig. 6.34). For $\rho = \frac{1}{2}$, the characteristic becomes vertical in the (x, τ)-plane and so the state $\rho = \frac{1}{2}$ remains stationary relative to the road.

If the solution includes discontinuities in ρ, we have to consider the jump condition, which can be readily formulated in the form

$$\frac{dx}{d\tau} = \frac{[\phi]}{[\rho]} = 1 - \rho^l - \rho^r = \tilde{\lambda}(\rho^l, \rho^r) \tag{6.6.13}$$

This is the propagation speed of the discontinuity (ρ^l, ρ^r), and its reciprocal gives the propagation direction $\tilde{\sigma}(\rho^l, \rho^r)$. Clearly, the speed $\tilde{\lambda}(\rho^l, \rho^r)$ is less than either of the local vehicle velocities, $w(\rho^l)$ and $w(\rho^r)$, and so vehicles cross the discontinuity from the left to the right side. Recalling the observation with Eq. (6.6.12), we notice that not only disturbances but discontinuities propagate against the flow of traffic so that a driver is facing them ahead. Since the flux curve $\phi(\rho)$ is convex, the entropy condition is given by the shock condition, which reads in this case

$$\lambda(\rho^l) > \tilde{\lambda}(\rho^l, \rho^r) > \lambda(\rho^r) \tag{6.6.14}$$

This is satisfied if

$$\rho^l < \rho^r \tag{6.6.15}$$

Hence, every discontinuity arising in this problem is a shock, across which the vehicle concentration ρ increases while the vehicle velocity $w(\rho)$ decreases.

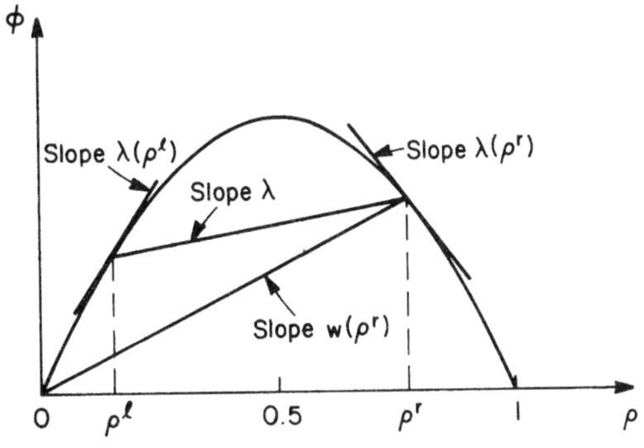

Figure 6.34

This implies that vehicles decelerate across a shock, which is consistent with our experience.

The character of the vehicle movement on the roadway is definitely going to depend on the initial distribution of vehicles at $\tau = 0$. We will assume that the initial data are specified on the whole x-axis. To illustrate what can happen, we will first begin with simple examples and then proceed to treat the traffic-light problem.

Suppose that the initial condition is given by

$$\rho(x, 0) = \begin{cases} 1, & x < 0 \\ 0, & x > 0 \end{cases} \tag{6.6.16}$$

Here we have a roadway free of vehicles in the front and congested with bumper-to-bumper vehicles behind. In practice, we will encounter this situation when the traffic light located at $x = 0$ has just turned green from red.

Since ρ decreases in the x-direction, the characteristics for ρ in the range from $\rho = 1$ to $\rho = 0$ fan clockwise, and so we have a simple wave centered at the origin as depicted in Fig. 6.35. The state ρ propagates forward if $0 \le \rho < \frac{1}{2}$ and backward if $\frac{1}{2} < \rho \le 1$, while the state $\rho = \frac{1}{2}$ is stationary at $x = 0$. For an arbitrary point $D(x, \tau)$, where the state is ρ, we can draw the characteristic OD and write the equation

$$\frac{x}{\tau} = \lambda(\rho) = 1 - 2\rho$$

or

$$\rho(x, \tau) = \frac{1}{2}\left(1 - \frac{x}{\tau}\right) \tag{6.6.17}$$

which is the solution in the region of the centered simple wave. Outside this region we have two constant states, $\rho = 0$ and $\rho = 1$. Here again the solution is given by a function of x/τ only. The profiles of the vehicle concentration at successive times are shown in Fig. 6.35.

Let us now examine the movement of individual vehicles by tracing their path in the (x, τ)-plane. The first car located at $x = 0$ immediately attains the maximum speed 1 and moves with the same speed along the line OA, which carries the state $\rho = 0$. The car initially positioned, for example, at $x = -\frac{1}{2}$ will remain there until the disturbance originated from the origin O reaches the car along the characteristic OB. This occurs at $L(-\frac{1}{2}, \frac{1}{2})$. From this point the car will accelerate first rapidly and then gradually, and its speed asymptotically approaches the maximum value 1. The vehicle path $LMDN$ can be determined by writing, for an arbitrary point D, the equation

$$\frac{dx}{d\tau} = w(\rho) = 1 - \rho \tag{6.6.18}$$

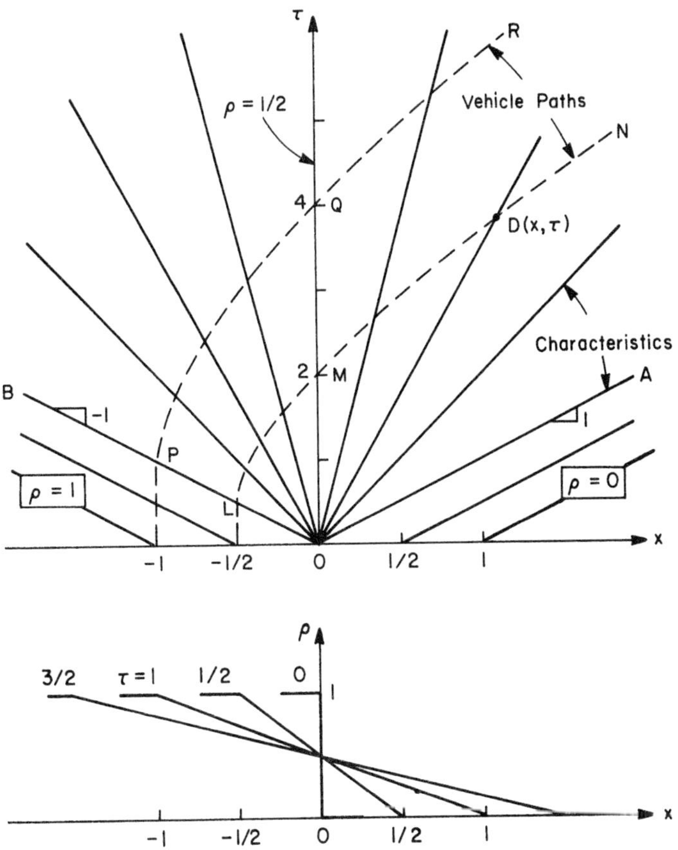

Figure 6.35

and combining this with Eq. (6.6.17) to get

$$\frac{dx}{d\tau} - \frac{x}{2\tau} = \frac{1}{2}$$

or

$$\frac{d}{d\tau}\left(\frac{x}{\sqrt{\tau}}\right) = \frac{1}{2\sqrt{\tau}}$$

Integrating from the starting point $(-a, a)$, we obtain

$$x = \tau - 2\sqrt{a\tau} \tag{6.6.19}$$

and so the path is given by a portion of a parabola. For the path LMN, $a = \frac{1}{2}$ so that the particular vehicle will cross the intersection ($x = 0$) at $\tau = 2$.

Since $a = 1$ at point P, we see that the path intersects the τ-axis at $\tau = 4$. This implies that, if the last vehicle in the block, $-1 \leq x < 0$, is to pass through the intersection, the green period of the traffic light must exceed this value of 4. In terms of actual parameters, this critical value is expressed as $t_G = 4L/v_m$, where L is the length of the block.

If the initial vehicle distribution is of the form

$$\rho(x, 0) = \begin{cases} \rho_0, & x < 0 \\ \rho_0 - (\rho_0 - \rho_1)x, & 0 \leq x \leq 1 \\ \rho_1, & x > 1 \end{cases} \tag{6.6.20}$$

where $\rho_0 > \rho_1$, then we again have a simple wave that is not centered but based on the portion (0, 1) of the x-axis. Outside the region of this simple wave we have two constant states, $\rho = \rho_0$ and $\rho = \rho_1$. For every point (x, τ) in the simple wave region, we have a characteristic with intercept ξ on the x-axis (see Fig. 6.36), and thus

$$x - \xi = (1 - 2\rho)\tau \tag{6.6.21}$$

But

$$\rho = \rho_0 - (\rho_0 - \rho_1)\xi$$

from Eq. (6.6.20), and eliminating ξ gives

$$\rho(x, \tau) = \frac{\rho_0 + (\rho_0 - \rho_1)(\tau - x)}{1 + 2(\rho_0 - \rho_1)\tau} \tag{6.6.22}$$

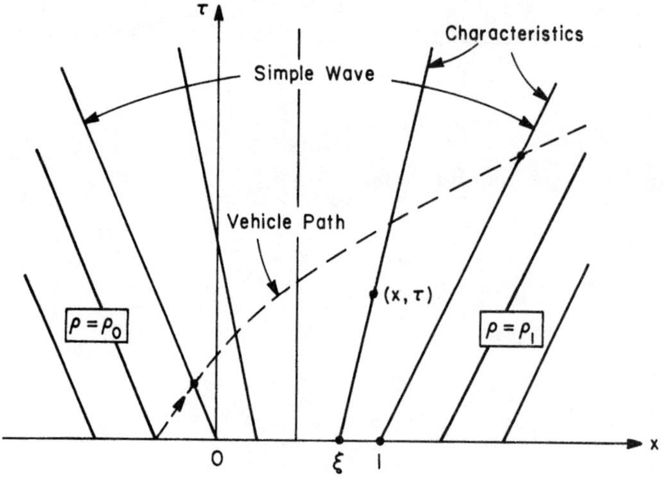

Figure 6.36

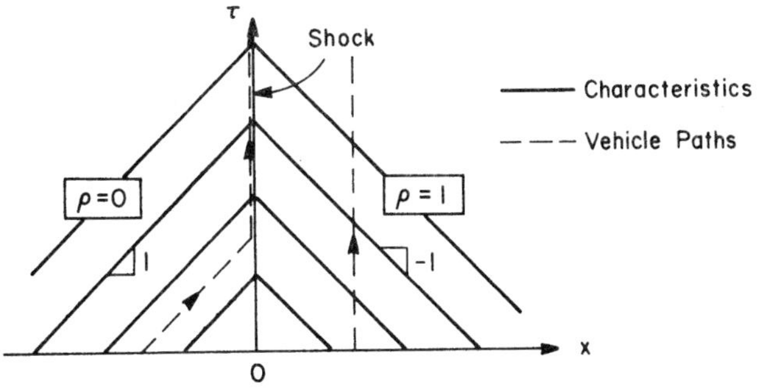

Figure 6.37

Note that if $\rho_0 < \rho_1$ we should have trouble at $\tau = 1/[2(\rho_1 - \rho_0)]$. The vehicle path can be determined by substituting Eq. (6.6.22) in Eq. (6.6.18) and integrating.

Next we will consider the initial condition

$$\rho(x, 0) = \begin{cases} 0, & x < 0 \\ 1, & x > 0 \end{cases} \tag{6.6.23}$$

This is the reversed situation of Eq. (6.6.16) and is expected to arise if the traffic light located at $x = 0$ turns red after all the vehicles have passed through the intersection.† The shock condition (6.6.15) is satisfied with $\rho^l = 0$ and $\rho^r = 1$, so we have a shock of strength 1 and this shock will be stationary since its speed is $\tilde{\lambda}(0, 1) = 0$. Therefore, the initial distribution of vehicles will remain unchanged. A car may approach the shock with the speed 1, but it will have to come to a complete stop at the intersection because the speed is zero on the other side of the shock, as shown in Fig. 6.37.

A more interesting feature is observed when the initial vehicle distribution has a positive slope. For illustration, let us consider the case with

$$\rho(x, 0) = \begin{cases} \rho_0, & x < 0 \\ \rho_0 + (\rho_1 - \rho_0)x, & 0 \le x \le 1 \\ \rho_1, & x > 1 \end{cases} \tag{6.6.24}$$

where $\rho_0 < \rho_1$. The characteristics emanating from points in the interval $0 < x < 1$ of the x-axis fan counterclockwise, and so we have here a compression wave. To determine the envelope of characteristics, we rearrange Eq.

†Note that the condition (6.6.23) can be expressed equivalently as

$$w(x, 0) = \begin{cases} 1, & x < 0 \\ 0, & x > 0 \end{cases}$$

and so there is no movement of vehicles beyond the traffic light.

(6.1.3) in the form

$$x^E = \xi - \lambda\left\{\frac{d\lambda}{d\rho}\frac{d}{d\xi}\rho(\xi, 0)\right\}^{-1}$$
$$\tau^E = -\left\{\frac{d\lambda}{d\rho}\frac{d}{d\xi}\rho(\xi, 0)\right\}^{-1} \tag{6.6.25}$$

by using the relation $\lambda(\rho) = 1/\sigma(\rho)$. Now ξ is the inverse of the initial data; that is,

$$\xi = \frac{\rho - \rho_0}{\rho_1 - \rho_0}, \qquad \rho_0 \leq \rho \leq \rho_1 \tag{6.6.26}$$

And since

$$\frac{d\lambda}{d\rho} = -2, \qquad \frac{d}{d\xi}\rho(\xi, 0) = \rho_1 - \rho_0$$

we have

$$x^E = \frac{1 - 2\rho_0}{2(\rho_1 - \rho_0)}$$
$$\tau^E = \frac{1}{2(\rho_1 - \rho_0)} \tag{6.6.27}$$

This represents a fixed point in the (x, τ)-plane, and this implies that all the crossing of characteristics necessarily takes place at a single point; thus a shock of full strength $(\rho_1 - \rho_0)$ will appear at this point suddenly. This is shown in Fig. 6.38. Here the profile of ρ at fixed τ remains linear till the shock formation because, from Eqs. (6.6.21) and (6.6.26), we have

$$\rho(x, \tau) = \frac{\rho_0 - (\rho_1 - \rho_0)(\tau - x)}{1 - 2(\rho_1 - \rho_0)\tau} \qquad \text{for} \quad \tau < \tau^E \tag{6.6.28}$$

in the region of the compression wave. We notice that with $\rho_0 > \frac{1}{2}$, the shock is formed on the left side of the intersection (i.e., $x^E < 0$).

Since the states on both sides, $\rho^l = \rho_0$ and $\rho^r = \rho_1$, remain fixed, the shock propagates from the beginning with constant speed given by

$$\tilde{\lambda}(\rho_0, \rho_1) = 1 - (\rho_0 + \rho_1) \tag{6.6.29}$$

As long as $(\rho_0 + \rho_1) < 1$, the shock propagates forward, and so the congested traffic will be cleared away. On the other hand, if $(\rho_0 + \rho_1) > 1$, the shock is directed backward and a traffic crawl will develop. In the special case with $\rho_0 = 0$ and $\rho_1 = 1$, we have $\tilde{\lambda} = 0$ and the shock is stationary at $x = \frac{1}{2}$. As illustrated by the vehicle paths in Fig. 6.38, the vehicles decelerate through the compression wave region as well as across the shock.

As a more realistic problem, let us consider a roadway with two traffic lights, one at $x = 0$ and another at $x = 1$, and suppose that the vehicles are

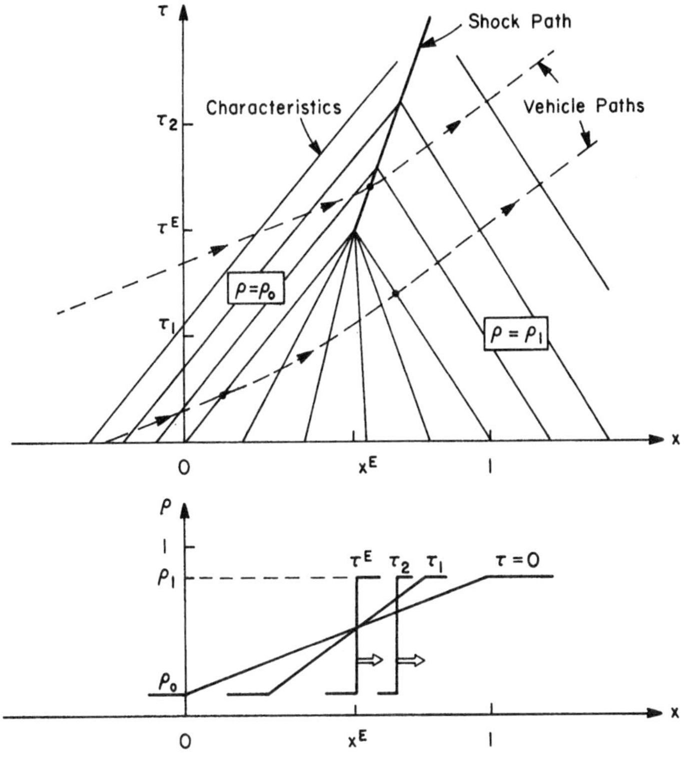

Figure 6.38

initially distributed as

$$\rho(x, 0) = \begin{cases} \rho_0, & x < 0 \\ \rho_1, & 0 < x < 1 \\ \rho_0, & x > 1 \end{cases} \tag{6.6.30}$$

where

$$\rho_1 > \frac{1}{2} > \rho_0, \qquad \rho_0 + \rho_1 > 1 \tag{6.6.31}$$

Now both the traffic lights turn green at the same time.

From the previous examples it is clear that we have a shock starting from the origin (0, 0) and a simple wave centered at the point (1, 0) as depicted in Fig. 6.39. The shock will propagate backward, since $\tilde{\lambda}(\rho_0, \rho_1) = 1 - (\rho_0 + \rho_1) < 0$, and its path is a straight line

$$x = \{1 - (\rho_0 + \rho_1)\}\tau$$

until it meets the leftmost characteristic AB of the simple wave for which we

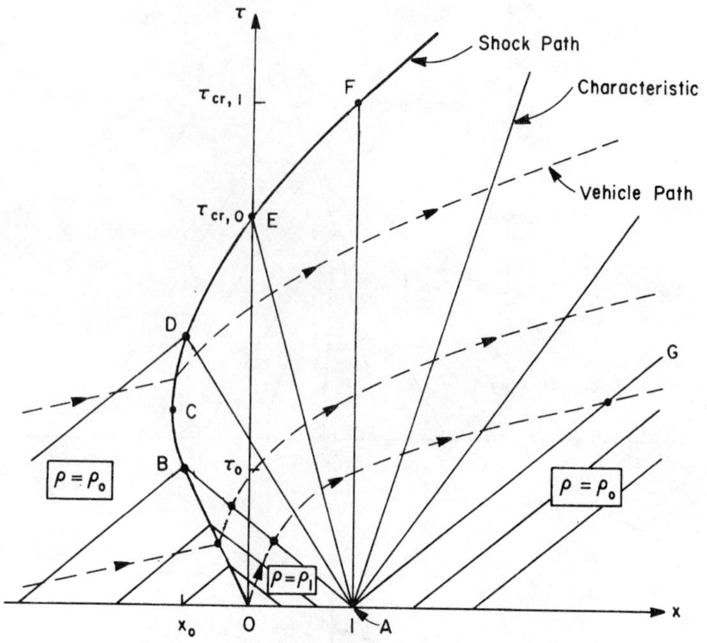

Figure 6.39

have the equation

$$x = 1 + (1 - 2\rho_1)\tau$$

The intersection B of the two lines has the coordinates

$$x_0 = \frac{1 - (\rho_0 + \rho_1)}{\rho_1 - \rho_0}$$
$$\tau_0 = \frac{1}{\rho_1 - \rho_0} \qquad (6.6.32)$$

From this point the shock interacts with the simple wave, and the state on the right side gradually decreases. Thus the shock accelerates and its path becomes curved as marked by the curve $BCDEF$ in Fig. 6.39.

We will take an arbitrary point $D(x, \tau)$ on the curved path and write the equation

$$\frac{dx}{d\tau} = 1 - (\rho_0 + \rho) \qquad (6.6.33)$$

where ρ is the state on the right side and so is carried along the characteristic AD:

$$x = 1 + (1 - 2\rho)\tau \qquad (6.6.34)$$

Treating ρ as the parameter running along the shock path, we have

$$\frac{dx}{d\rho} = (1 - \rho_0 - \rho)\frac{d\tau}{d\rho}$$

from Eq. (6.6.33) and

$$\frac{dx}{d\rho} = (1 - 2\rho)\frac{d\tau}{d\rho} - 2\tau$$

from Eq. (6.6.34). Hence

$$\frac{d\tau}{d\rho} = -\frac{2\tau}{\rho - \rho_0}$$

Integrating from the starting point B, where $\rho = \rho_1$, and using Eq. (6.6.32), we obtain

$$\tau = \tau_0\left(\frac{\rho_1 - \rho_0}{\rho - \rho_0}\right)^2 = \frac{\rho_1 - \rho_0}{(\rho - \rho_0)^2} \tag{6.6.35}$$

and, by substituting into Eq. (6.6.34),

$$x = 1 + (\rho_1 - \rho_0)\frac{1 - 2\rho}{(\rho - \rho_0)^2} \tag{6.6.36}$$

Clearly, both x and τ diverge in the limit as $\rho \to \rho_0$, and thus the interaction will go on indefinitely. The inversion of Eq. (6.6.35) gives

$$\rho - \rho_0 = \sqrt{\frac{\rho_1 - \rho_0}{\tau}} \tag{6.6.37}$$

so the shock strength $(\rho - \rho_0)$ decays like $1/\sqrt{\tau}$. [Compare with Eq. (6.3.11).]

Equations (6.6.35) and (6.6.36) form a parametric representation of the shock path, but the parameter ρ can be eliminated to give

$$x = 1 - 2\sqrt{(\rho_1 - \rho_0)\tau} + (1 - 2\rho_0)\tau \tag{6.6.38}$$

so the path is a part of a parabola. The width Δx of the simple wave region is determined by subtracting Eq. (6.6.38) from $x' = 1 + (1 - 2\rho_0)\tau$, the equation of the rightmost characteristic AG. Hence,

$$\Delta x = x' - x = 2\sqrt{(\rho_1 - \rho_0)\tau} \tag{6.6.39}$$

which asserts that the width grows like $\sqrt{\tau}$. [Compare with Eq. (6.3.12).] The simple wave solution in this region is determined by the family of characteristics for which we have

$$x - 1 = (1 - 2\rho)\tau$$

and so

$$\rho = \frac{1}{2}\left(1 - \frac{x}{\tau} + \frac{1}{\tau}\right)$$

In the limit as $\tau \to \infty$, we obtain the asymptotic expression

$$\rho - \frac{1}{2} \sim -\frac{x}{2\tau}, \qquad (1 - 2\rho_0)\tau - \Delta x < x < (1 - 2\rho_0)\tau \qquad (6.6.40)$$

where the range of x spans the simple wave region for a given time τ. Hence the simple wave solution tends to depend on the ratio x/τ only. The nature of the asymptotic behavior we have observed here was proved in a general context by Lax (1957, 1972) and Whitham (1974). We will come back to this subject in Sec. 7.9 to discuss the asymptotic behavior for the case of an arbitrary convex flux curve.

The shock will start to turn forward when $dx/d\tau = 0$ or $\rho = 1 - \rho_0$ so that the turning point C has, from Eqs. (6.6.35) and (6.6.36), the coordinates

$$x = \frac{1 - \rho_0 - \rho_1}{1 - 2\rho_1}$$
$$\tau = \frac{\rho_1 - \rho_0}{(1 - 2\rho_0)^2} \qquad (6.6.41)$$

When the shock crosses the intersection at $x = 0$, we have

$$\rho = \rho_1 - \sqrt{(\rho_1 - \rho_0)(\rho_1 + \rho_0 - 1)}$$

from Eq. (6.6.36), and thus

$$\tau = (\sqrt{\rho_1 - \rho_0} - \sqrt{\rho_1 + \rho_0 - 1})^{-2} = \tau_{cr,0} \qquad (6.6.42)$$

The shock then proceeds to pass through the block $0 < x < 1$ and crosses the intersection at $x = 1$ when $\rho = \frac{1}{2}$, and so

$$\tau = \frac{\rho_1 - \rho_0}{(\frac{1}{2} - \rho_0)^2} = \tau_{cr,1} \qquad (6.6.43)$$

While the shock further continues to move forward, its speed tends to approach asymptotically to a limiting value of $(1 - 2\rho_0)$. From this observation on the shock propagation, it is immediately apparent that, if the cluttered traffic is to pass through the intersections, the green periods at $x = 0$ and $x = 1$ must be set larger than $\tau_{cr,0}$ and $\tau_{cr,1}$, respectively. In the specific case with $\rho_0 = 0$ and $\rho_1 = 1$, the shock will first remain stationary at $x = 0$, start to move forward at $\tau = 1$, and pass the intersection at $x = 1$ when $\tau = 4$. Thus the green period t_G (in terms of the actual parameters) must exceed L/v_m at $x = 0$ and $4L/v_m$ at $x = 1$, respectively.

Finally, we will take the standard traffic problem in a congested area where there is a traffic-control situation at each intersection. Let us consider one cell that has a traffic queue of length α stopped and behind it a section clear of vehicles of length $\beta = (1 - \alpha)$. The latter includes not only the portion within the same block but also the cross street so that the distance between the two traffic lights is made equal to 1. For the sake of convenience, we will put the origin of the coordinates at the rear end of the traffic queue, and thus the traffic lights are located at $x = -(1 - \alpha)$ and $x = \alpha$. The initial condition is then expressed in the form

$$\rho(x, 0) = \begin{cases} 1, & x < -\beta \\ 0, & -\beta < x < 0 \\ 1, & 0 < x < \alpha \\ 0, & x > \alpha \end{cases} \tag{6.6.44}$$

where $\alpha + \beta = 1$.

Suppose both the traffic lights turn green. It is obvious that there will be a shock at $x = 0$ and two simple waves, one centered at $x = -\beta$ and the other centered at $x = \alpha$. Since $\tilde{\lambda}(0, 1) = 0$, the shock will remain stationary until either of the disturbances at $x = -\beta$ and at $x = \alpha$ reaches it. The question is then which disturbance would meet the shock first.

Let us suppose for the moment that $\alpha < \frac{1}{2}$ so that the shock must first interact with the simple wave centered at $x = \alpha$, as shown in Fig. 6.40. The vertical shock path intersects the leftmost characteristic at C, for which we have $\tau = \alpha$ since $x = \alpha - \tau$ is the equation for AC. From this point the state on the right side begins to decrease so that the shock is accelerated and its path is directed forward. For an arbitrary point $D(x, \tau)$, we can write the equation of the shock speed,

$$\frac{dx}{d\tau} = \tilde{\lambda}(0, \rho^r) = 1 - \rho^r$$

and also the equation of the characteristic AD:

$$x - \alpha = (1 - 2\rho^r)\tau$$

Eliminating ρ^r and putting the differential equation into the exact form, we get

$$\frac{d}{d\tau}\left(\frac{x - \alpha}{\sqrt{\tau}}\right) = \frac{1}{2\sqrt{\tau}}$$

which can be directly integrated from $C(0, \alpha)$ to give

$$x - \alpha = \tau - 2\sqrt{\alpha\tau} \tag{6.6.45}$$

This is the equation for the curved portion CDE of the shock path.

Sooner or later the simple wave centered at $x = -\beta$ will come to interact with the shock. Indeed, this happens from the point E, at which the rightmost characteristic BE expressed by the equation

$$x + \beta = \tau$$

intersects the shock path. Therefore, the point E has the coordinates

$$x_0 = \frac{(1 - 2\alpha)^2}{4\alpha} \qquad \tau_0 = \frac{1}{4\alpha} \tag{6.6.46}$$

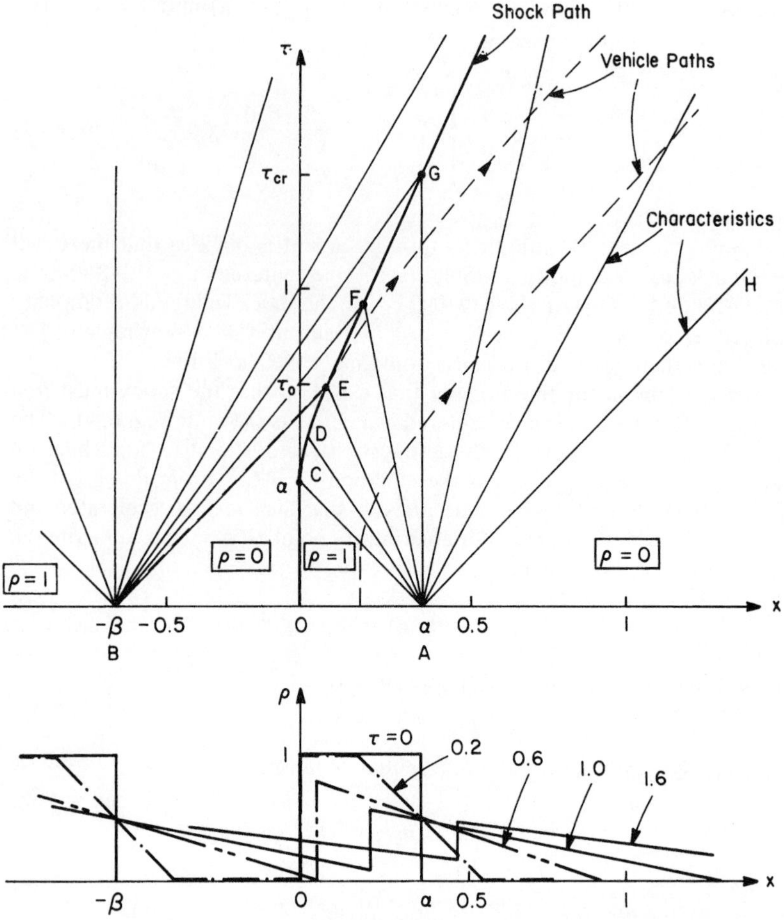

Figure 6.40

Beyond τ_0 we see that the states on both sides of the shock vary simultaneously, ρ^l increasing and ρ^r decreasing with time. But this can be resolved without much difficulty since we can write, at an arbitrary point F on the shock path, the equation of the shock speed,

$$\frac{dx}{d\tau} = \tilde{\lambda}(\rho^l, \rho^r) = 1 - \rho^l - \rho^r \tag{6.6.47}$$

and the equations of the characteristics AF and BF:

$$\begin{aligned} x - \alpha &= (1 - 2\rho^r)\tau \\ x + \beta &= (1 - 2\rho^l)\tau \end{aligned} \tag{6.6.48}$$

It is straightforward to eliminate both ρ^l and ρ^r and obtain

$$\frac{dx}{d\tau} = \frac{2x + 1 - 2\alpha}{2\tau}$$

Integrating from the point $E(x_0, \tau_0)$ and using Eq. (6.6.46), we get the linear equation with positive slope:

$$\tau = \frac{x}{1 - 2\alpha} + \frac{1}{2} \tag{6.6.49}$$

This implies that the shock continues to propagate forward with a constant speed of $(1 - 2\alpha)$, as illustrated by the straight shock path EFG in Fig. 6.40. In particular, the shock will pass through the intersection at $x = \alpha$ when $\tau = \tau_{cr} = 1/[2(1 - 2\alpha)]$, so the green period there must be greater than $t_G = L/[2(1 - 2\alpha)v_m]$.

Incidentally, the shock strength is readily found from Eq. (6.6.48); that is,

$$\rho^r - \rho^l = \frac{1}{2\tau} \tag{6.6.50}$$

so the strength decays like $1/\tau$. This nature of the shock propagation has been proved in general for large time by Lax (see Sec. 6.4). If the initial distribution of vehicles within the block were repetitive along the roadway, we would see shocks of the same nature at an interval of 1 in the x-direction. But the slope of the shock path at the later stage is equal to $1/(1 - 2\alpha)$, so the period of the shock appearance at a fixed position is $1/(1 - 2\alpha)$, just twice the critical value of the green period.

If $\alpha = \frac{1}{2}$, we must have $x = 0$ from Eq. (6.6.49), and this implies that the shock will always remain stationary at $x = 0$ while its strength decays according to Eq. (6.6.50). This is expected from the beginning, because the portrait in the (x, τ)-plane is symmetric with respect to the shock path.

Now we will briefly examine the case with $\alpha > \frac{1}{2}$ so that the traffic queue is longer than one-half block. Clearly, the shock at $x = 0$ will interact first with the simple wave centered at $x = -\beta$. During this interaction the shock will be directed backward with ρ^l increasing. We can apply the same procedure that leads to Eq. (6.6.45) and, noting that here ρ^r is the variable parameter, we would get

$$x + 1 - \alpha = 2\sqrt{(1 - \alpha)\tau} - \tau \tag{6.6.51}$$

for the shock path. This path is then reached by the boundary characteristic of the simple wave centered at $x = \alpha$, and the intersection is easily located at

$$x = x_1 = -\frac{(2\alpha - 1)^2}{4(1 - \alpha)}, \qquad \tau = \tau_1 = \frac{1}{4(1 - \alpha)} \tag{6.6.52}$$

Beyond this moment the shock will have the states on both sides changing simultaneously, but Eqs. (6.6.47) and (6.6.48) are equally valid here, so by using Eq. (6.6.52) for the initial condition we can find the linear equation of negative slope,

$$\tau = \frac{1}{2} - \frac{x}{2\alpha - 1} \tag{6.6.53}$$

for the shock path at the later stage. Hence the shock will continue to propagate backward with a constant speed of $(2\alpha - 1)$. Obviously, the traffic situation under these circumstances would be catastrophic.

Exercises

6.6.1. Consider Eq. (6.6.10) with the flux expression

$$\phi(\rho) = -\rho \log \rho$$

and the initial data

$$\rho(x, 0) = \begin{cases} 1, & x < 0 \\ 0.01, & x > 0 \end{cases}$$

Draw the portrait of the solution in the (x, τ)-plane and the profiles of ρ at various times. Find the equation for the path of the car initially located at $x = -1$ and suggest a proper value for the green period at $x = 0$.

6.6.2. Given the data $v_m = \frac{50}{3}$ km/h and $k_j = 50$ vehicles/km for the quadratic flux relation and the initial distribution of vehicles

$$k(z, 0) = \begin{cases} 40\ \text{km}^{-1}, & z < 0 \\ 30\ \text{km}^{-1}, & 0 < z < 0.5\ \text{km} \\ 15\ \text{km}^{-1}, & z > 0.5\ \text{km} \end{cases}$$

construct the solution first in the distance–time plane with typical vehicle paths emanating from each section of the initial line, and then the profiles of the vehicle concentration at various times. In particular, show that the zone with 30 vehicles/km moves with a constant speed without changing its width.

6.6.3. With the same flux relation as in Exercise 6.6.2, suppose the initial distribution of vehicles is given by

$$k(z, 0) = \begin{cases} 15\ \text{km}^{-1}, & z < 0 \\ 30\ \text{km}^{-1}, & 0 < z < 0.5\ \text{km} \\ 40\ \text{km}^{-1}, & z > 0.5\ \text{km} \end{cases}$$

and analyze the traffic flow for $t > 0$.

6.6.4. Eliminate ρ between Eqs. (6.6.33) and (6.6.34) to obtain a differential equation in x and τ, and show that the solution is identical to Eq. (6.6.38).

6.6.5. Solve the system of equations, Eqs. (6.6.47) and (6.6.48), for $\alpha > \frac{1}{2}$ and for $\alpha = \frac{1}{2}$.

6.6.6. Consider the traffic flow equation (6.6.10) when the flux relationship is given by

$$\phi(\rho) = -\rho \log \rho$$

where log denotes the common logarithms, and the initial distribution of vehicles is

$$\rho(x, 0) = \begin{cases} 0.01, & x < 0 \\ 1, & 0 < x < 1 \\ 0.01, & x > 1 \end{cases}$$

If both the traffic signs at $x = 0$ and $x = 1$ change from red to green, analyze the traffic flow in the (x, τ)-plane. Show that, when the shock interacts with the simple wave, the shock path may be expressed as

$$\tau = \frac{99}{97} \exp\left\{\int_1^\rho \frac{(100\mu - 1)\, d\mu}{\mu(\ln \mu - 100\mu + 5.6)}\right\}$$

$$x = 1 - \frac{1 + \ln \rho}{2.3}\tau$$

Note that $\rho = 0.1$ at $x = 1$ and thus suggest the critical value of the green period at $x = 1$.

6.6.7. With the flux relationship given in Exercise 6.6.6, consider the initial condition

$$\rho(x, 0) = \begin{cases} 0.1, & x < 0 \\ 0.01 + 0.99x, & 0 < x < 1 \\ 1.0, & x > 1 \end{cases}$$

Determine the point where a shock will be first formed, and formulate a differential equation governing the shock propagation.

6.6.8. For the shock path *EFG* in Fig. 6.40, show that Eq. (6.4.18) reduces to $d\rho^l/d\rho^r = -1$, which is subject to the condition, $\rho^l = 0$ when $\rho^r = 2\alpha$. Use this result in Eq. (6.6.48) to obtain Eq. (6.6.49).

6.7 Theory of Sedimentation

We come now to a physical theory for which the relationship between flux and concentration is not necessarily convex: the theory of the settling of sediments and the subsidence of slimes. When a single particle falls in a large volume of fluid, it acquires a terminal velocity, v_0, governed by Stoke's law or a suitable modification of it. However, when a large number of particles fall together in a suspension, their motions interact with one another, and the velocity of any one particle depends on the concentration of particles surrounding it. Let

n = number of particles per unit volume
f = downward flux of particles, that is, the number crossing a horizontal plane of unit area per unit time
$v(n)$ = mean velocity of particles when the concentration is n

Then

$$f = f(n) = nv(n) \tag{6.7.1}$$

If we make the z-axis vertical and downward with the origin at the free surface of the fluid† and make the usual number balance over a layer between z and $z + dz$, we arrive at the equation

$$\frac{\partial f}{\partial z} + \frac{\partial n}{\partial t} = 0 \tag{6.7.2}$$

where t is the time (see Exercise 1.11.5). It is assumed that the area of the sedimentation column is constant, although this restriction can be removed (see Exercise 6.7.1).

Many correlations have been proposed for the flux–concentration relationship, of which a useful compilation is given by Shannon and others (1963). We do not wish to go into the physics of this phenomenon but to draw attention to three types of relationship that give interesting effects in the solution of Eq. (6.7.2). At low concentrations the velocity of fall decreases from the Stokes velocity by an amount that is linear in concentration:

$$v(n) = v_0(1 - bn) \tag{6.7.3}$$

If this simple linear relationship persisted over the whole range of concentrations, the constant b would have to be the reciprocal of n_m, the greatest possible concentration or the concentration at which the suspension is close packed and thus stationary. We would thus have

$$v(n) = v_0\left(1 - \frac{n}{n_m}\right) \tag{6.7.4}$$

and

$$f(n) = nv(n) = (v_0 n_m)\left(\frac{n}{n_m}\right)\left(1 - \frac{n}{n_m}\right) \tag{6.7.5}$$

where n_m = maximum possible concentration and $v_0 n_m$ = maximum conceivable flux. The maximum concentration is attained in the final state of a settled suspension, but the flux $v_0 n_m$ is not physically attainable, being only a convenient upper bound. Indeed, it suggests that the variables might be made dimensionless by defining

$$u = \frac{n}{n_m} \tag{6.7.6}$$

$$\phi(u) = \frac{f(n)}{v_0 n_m} \tag{6.7.7}$$

†This choice of the coordinate is found consistent with the previous discussions.

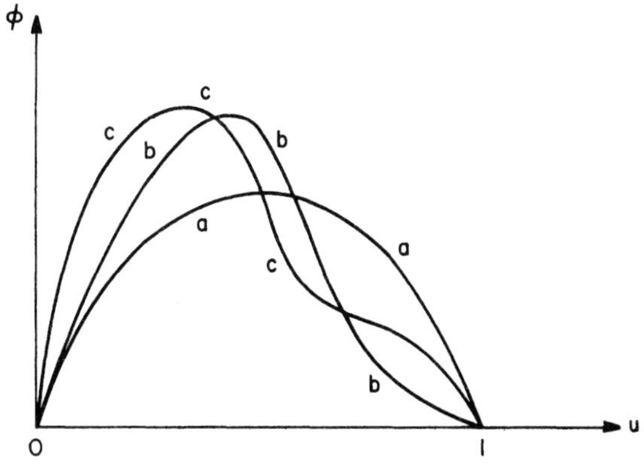

Figure 6.41

Then the first relation would be

$$\phi(u) = u(1 - u) \tag{6.7.8}$$

This is parabola a of Fig. 6.41, which is a convex curve with a maximum dimensionless flux of $\frac{1}{4}$ when $u = \frac{1}{2}$.

In practice, the flux–concentration relationship does not follow this simple relationship, for at higher concentrations the velocity falls off more rapidly than linearly. Two additional types of curve has been observed, which are sometimes called singly concave (curve b in Fig. 6.41) and doubly concave (curve c). These names are not entirely accurate but serve to distinguish the cases of (a) no points of inflection, (b) one point of inflection, and (c) two points of inflection.† A polynomial relationship between ϕ and u should be capable of expressing this, and Shannon has used a polynomial of degree 4. The simplest expressions that would represent the forms of curves b and c are cubic and quartic, respectively. Since $\phi(u) = 0$ at $u = 0$ and $u = 1$, and the limit of $\phi(u)/u$ as $u \to 0$ is 1, it is convenient to write these expressions as

$$\phi(u) = u(1 - u)\left(1 - \frac{u}{3\alpha - 1}\right) \tag{6.7.9}$$

and

$$\phi(u) = u(1 - u)\left\{1 + \frac{1 - 2(\alpha + \beta)}{1 - 2(\alpha + \beta) + 6\alpha\beta} u + \frac{u^2}{1 - 2(\alpha + \beta) + 6\alpha\beta}\right\} \tag{6.7.10}$$

†There are cases when the flux curve is doubly concave and doubly convex so that it has three points of inflection. See Shannon and Tory (1965).

where α and β are the locations of the inflection points. For the cubic expression in Eq. (6.7.9) to be positive throughout $0 \leq u \leq 1$, we must have

$$\frac{2}{3} \leq \alpha \leq 1 \tag{6.7.11}$$

and more complicated restrictions will obtain for α and β in Eq. (6.7.10). Clearly, these polynomial expressions are limited in their ability to fit experimental data, but they will show the characteristics of the several cases, especially when the settled sediment is incompressible.

We will consider the problem of the settling of a suspension in a column of height H initially filled with a suspension of uniform concentration u_0. Then it is very natural to make the independent variables dimensionless by taking

$$x = \frac{z}{H} \tag{6.7.12}$$

$$\tau = \frac{v_0 t}{H} \tag{6.7.13}$$

so that we have, from Eq. (6.7.2),

$$\frac{\partial \phi}{\partial x} + \frac{\partial u}{\partial \tau} = 0$$

or

$$\phi'(u)\frac{\partial u}{\partial x} + \frac{\partial u}{\partial \tau} = 0 \tag{6.7.14}$$

The initial and boundary conditions are

$$\begin{aligned} &\text{at} \quad \tau = 0, \qquad u = u_0 \\ &\text{at} \quad x = 0, \qquad u = 0 \\ &\text{at} \quad x = 1, \qquad \phi = 0 \end{aligned} \tag{6.7.15}$$

The vanishing flux condition at the bottom implies that $u = 1$ at $x = 1$ (see Fig. 6.41), and thus we can complete the data in Eq. (6.7.15) into one specifying u on the x-axis as follows:

$$u(x, 0) = \begin{cases} 0, & x \leq 0 \\ u_0, & 0 < x < 1 \\ 1, & x \geq 1 \end{cases} \tag{6.7.16}$$

The suspension will finally settle to a height of u_0 from the bottom and concentration $u = 1$ so that the solution $u(x, \tau)$ tends to

$$u_s(x) = \begin{cases} 0, & 0 \leq x < 1 - u_0 \\ 1, & 1 - u_0 < x \leq 1 \end{cases} \tag{6.7.17}$$

which is the steady state of the system.

From Eq. (6.7.14), we observe that the propagation speed of a specific concentration u is given by

$$\frac{dx}{d\tau} = \phi'(u) = \lambda(u) \tag{6.7.18}$$

which is the reciprocal of the characteristic direction $\sigma(u)$. Since u remains invariant while it propagates in the characteristic direction, the speed is constant and the propagation path is a straight line. We also notice that the speed $\lambda(u)$ is identical to the slope of the tangent to the flux curve so that it takes the positive sign (downward motion) for low concentrations, but becomes negative (upward motion) for higher concentrations. On the other hand, a discontinuity in u moves with speed

$$\frac{dx}{d\tau} = \frac{\phi(u^l) - \phi(u^r)}{u^l - u^r} = \tilde{\lambda}(u^l, u^r) \tag{6.7.19}$$

Note that this is the reciprocal of the propagation direction $\tilde{\sigma}(u^l, u^r)$. We see that $\tilde{\lambda}(u^l, u^r)$ is just the slope of the chord connecting the two points on the flux curve for which $u = u^l$ and $u = u^r$. Hence the speed can be positive, zero, or negative depending on the states on both sides of the discontinuity. Here again we need to put the entropy condition in the form of Eq. (6.5.11); that is, at all points of discontinuity in u the inequality

$$\tilde{\lambda}(u^l, u^r) \leq \tilde{\lambda}(u^l, u) \tag{6.7.20}$$

must be satisfied for all u between u^l and u^r. And yet the rule established in Sec. 5.6 remains valid. Thus, if two states u^l and u^r are to be connected by a discontinuity, we first identify the chord connecting the two points L and R on the flux curve for which $u = u^l$ and $u = u^r$, respectively, and move along the chord from L to R. Then we must see the flux curve on the left side. It is now clear from Fig. 6.41 that with a parabolic flux curve *a* we expect to have only shocks, because here the preceding entropy condition leads to the shock condition

$$\lambda(u^l) > \tilde{\lambda}(u^l, u^r) > \lambda(u^r) \tag{6.7.21}$$

or equivalently $u^l < u^r$, whereas, in the case of cubic and quartic relations (*b* and *c*), we can have both shocks and semishocks, for the entropy condition now admits a discontinuity satisfying the shock condition (6.7.21) as well as one satisfying either the condition $\lambda(u^l) > \tilde{\lambda}(u^l, u^r) = \lambda(u^r)$ or the condition $\lambda(u^l) = \tilde{\lambda}(u^l, u^r) > \lambda(u^r)$. Note that the last condition can be met only in the case of a quartic relation and that there is no possibility of having a contact discontinuity even with a quartic relation.

Let us first consider the simplest flux–concentration relationship, Eq. (6.7.8), for which

$$\lambda(u) = 1 - 2u \tag{6.7.22}$$

and

$$\tilde{\lambda}(u^l, u^r) = 1 - u^l - u^r \tag{6.7.23}$$

We observe that the slope of the characteristic varies from 1 when $u = 0$ to $(1 - 2u_0)^{-1}$ when $u = u_0$ and then to -1 when $u = 1$. Considering the initial condition, Eq. (6.7.16), we immediately find that the characteristics emanating from the origin will fan backward, and the same is true with those emanating from the point $x = 1$ on the x-axis. This implies that the shock condition (6.7.21) is satisfied both at $x = 0$ and $x = 1$. Thus we have two shocks from the beginning: one (S_a) starting from the top ($x = 0$) with $u^l = 0$ and $u^r = u_0$, and the other (S_b) from the bottom ($x = 1$) with $u^l = u_0$ and $u^r = 1$. The two remain sharp while they propagate as shocks. The situation is shown in Fig. 6.42. The shock S_a, representing the top of the suspension, moves with speed $(1 - u_0)$, and so OQ is the line

$$x = (1 - u_0)\tau \tag{6.7.24}$$

The speed of the shock S_b is $-u_0$, and thus PQ is the line

$$x = 1 - u_0\tau \tag{6.7.25}$$

The propagation of S_b along this line represents the growth of the layer of stationary sediment. The two meet together at Q and combine to give a new shock S with $u^l = 0$ and $u^r = 1$, which becomes stationary [cf. Eq. (6.7.23)]. The point of confluence Q is determined from Eqs. (6.7.24) and (6.7.25) to yield $\tau = 1$ and $x = 1 - u_0$. Thus the whole process of sedimentation is completed in a time H/v_0 whatever the initial concentration may be, provided that it is uniform.

The problem with the more general initial condition $u(x, 0) = u_0(x)$ can be solved without much difficulty since $\phi(u)$ is convex. We notice that in this case the top of the suspension will not fall with uniform velocity. For if its position is x and the concentration immediately below it is u, we have

$$\frac{dx}{d\tau} = \frac{\phi(u)}{u} = \tilde{\lambda}(0, u) \tag{6.7.26}$$

by Eq. (6.7.19), and u is given by $u_0(\xi)$, the concentration carried from the initial point ξ along the characteristic

$$x = \xi - \phi'(u_0(\xi))\tau \tag{6.7.27}$$

If ξ is regarded as a parameter along the shock path, we can express x and τ as functions of ξ; that is, we would rewrite $dx/d\tau$ as $(dx/d\xi)/(d\tau/d\xi)$ and differentiate Eq. (6.7.27) with respect to ξ, and then solve for $dx/d\xi$ and $d\tau/d\xi$ to find

$$\frac{dx}{d\xi} = \frac{\phi(u_0(\xi))\{1 - \phi''(u_0(\xi))u_0'(\xi)\tau\}}{u_0(\xi)\phi'(u_0(\xi)) - \phi(u_0(\xi))}$$

$$= -\frac{\{1 - u_0(\xi)\}\{1 + 2u_0'(\xi)\tau\}}{u_0(\xi)} \tag{6.7.28}$$

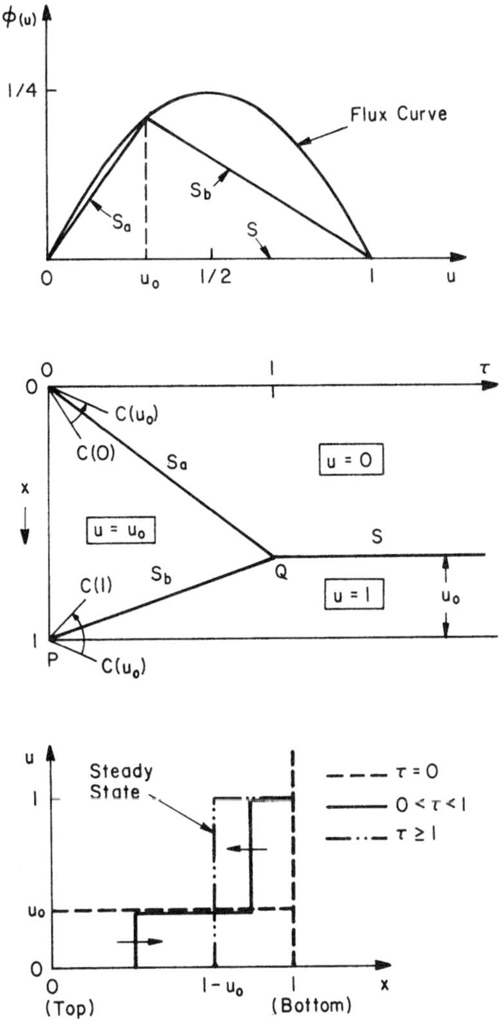

Figure 6.42

and

$$\frac{d\tau}{d\xi} = \frac{u_0(\xi)\{1 - \phi''(u_0(\xi))u_0(\xi)\tau\}}{u_0(\xi)\phi'(u_0(\xi)) - \phi(u_0(\xi))} = -\frac{1 + 2u_0(\xi)\tau}{u_0(\xi)} \tag{6.7.29}$$

These equations will be valid provided that the characteristics do not cross, that is, if

$$\phi''(u_0(\xi))u_0'(\xi)\tau \neq 1 \tag{6.7.30}$$

The parametric equations take care of this possibility rather well for such a point is a critical point of this pair of equations. The first form of the equations given in Eqs. (6.7.28) and (6.7.29) is generally valid; the second form is specialized for $\phi(u) = u(1 - u)$.

Turning now to the cubic relation given by Eq. (6.7.9), we will write this as

$$\phi(u) = u(1 - u)(1 - \gamma u), \qquad \gamma = \frac{1}{3\alpha - 1}, \qquad \frac{1}{2} < \gamma < 1 \tag{6.7.31}$$

so

$$\lambda(u) = \phi'(u) = 1 - 2(1 + \gamma)u + 3\gamma u^2 \tag{6.7.32}$$

$$\tilde{\lambda}(u^l, u^r) = 1 - (1 + \gamma)(u^l + u^r) + \gamma\{u^l u^l + u^l u^r + u^r u^r\} \tag{6.7.33}$$

A uniform initial distribution of sediments, $u(x, 0) = u_0$ for $0 < x < 1$, will again give rise to the shock S_a, which represents the top surface of the suspension, for with $u^l = 0$ we see the shock condition is satisfied for any value of $u_0 = u^r$. The shock moves downward with constant speed

$$\frac{dx}{d\tau} = \tilde{\lambda}(0, u_0) = (1 - u_0)(1 - \gamma u_0) \tag{6.7.34}$$

Due to the nature of the flux curve here, the structure of the solution will vary depending on the range of u to which the initial concentration u_0 belongs. These ranges are best illustrated by Fig. 6.43, in which a is the abscissa of the point at which the tangent to the curve at $u = 1$ meets the curve and α is the abscissa of the inflection point. Thus we will consider three different ranges of u_0: (1) the lower range, $0 < u_0 \leq a$; (2) the middle range, $a < u_0 < \alpha$; and (3) the upper range, $\alpha \leq u_0 < 1$. For the cubic relation (6.7.31), it can be shown that

$$a = \frac{1 - \gamma}{\gamma} \quad \text{and} \quad \alpha = \frac{1 + \gamma}{3\gamma} \tag{6.7.35}$$

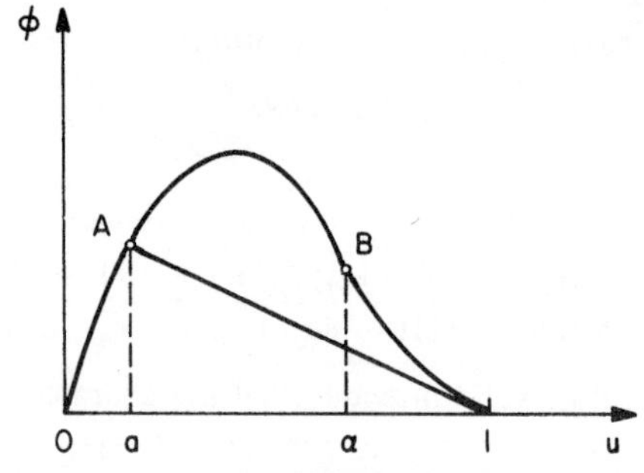

Figure 6.43

First, suppose that u_0 falls on the lower range (i.e., $0 < u_0 \leq a$). The shock S_a, representing the linear decrease of the upper surface, moves downward along the path OQ (see Fig. 6.44):

$$x = \tilde{\lambda}(0, u_0)\,\tau = \frac{\phi(u_0)}{u_0}\tau = (1 - u_0)(1 - \gamma u_0)\tau \qquad (6.7.36)$$

Note that here and in the sequel the expression involved with the function $\phi(u)$ is generally valid, while the explicit form is specialized for the cubic relation. Now the characteristics from the bottom end fan counterclockwise and then clockwise as we proceed in the x-direction, but the connection between the states $u = u_0$ and $u = 1$ is accomplished by a single discontinuity, which is a shock. Thus the picture is similar to the second part of Fig. 6.42. This shock S_b represents the linear buildup of the bottom layer and moves upward along the path PQ:

$$x - 1 = \tilde{\lambda}(u_0, 1)\tau = \frac{\phi(u_0)}{u_0 - 1}\tau = -u_0(1 - \gamma u_0)\tau \qquad (6.7.37)$$

The two shocks meet at Q, for which we have

$$\begin{aligned} \tau = \tau_0 &= \{\tilde{\lambda}(0, u_0) - \tilde{\lambda}(u_0, 1)\}^{-1} = \frac{u_0(1 - u_0)}{\phi(u_0)} = \frac{1}{1 - \gamma u_0} \\ x = x_0 &= 1 - u_0 \end{aligned} \qquad (6.7.38)$$

and the combined shock S becomes stationary, so the settling is complete. Figure 6.44 shows the situation in the (x, τ)-plane, and the concentration profile is similar to the last part of Fig. 6.42.

If $a < u_0 < \alpha$ (middle range), the situation at the top remains the same but, although the layout of the characteristics from the bottom end would be similar to that in the previous case, there is no way to connect the two points $(u_0, \phi(u_0))$ and $(1, 0)$ on the flux curve by a straight chord without crossing the flux curve. Consider a typical point D, $(u_0, \phi(u_0))$, on AB of the flux curve as marked in the first diagram of Fig. 6.45. By drawing the tangent from D to the flux curve, we locate the point of tangency E and find that the states $u^l = u_0$ and $u^r = u_1$ are connected by a semishock, where u_1 is the abscissa of the point E. On the right side, the semishock (S_b) is joined by a centered simple wave through which the connection varies from u_1 to 1.

The concentration u_1 is determined by the equation

$$\phi'(u_1) = \frac{\phi(u_0) - \phi(u_1)}{u_0 - u_1} \qquad (6.7.39)$$

and for the cubic relation (6.7.31) we find

$$u_1 = \frac{1}{2}\left(\frac{1 + \gamma}{\gamma} - u_0\right) \qquad (6.7.40)$$

The shock path OQ for S_a is given by Eq. (6.7.36), whereas the path of the

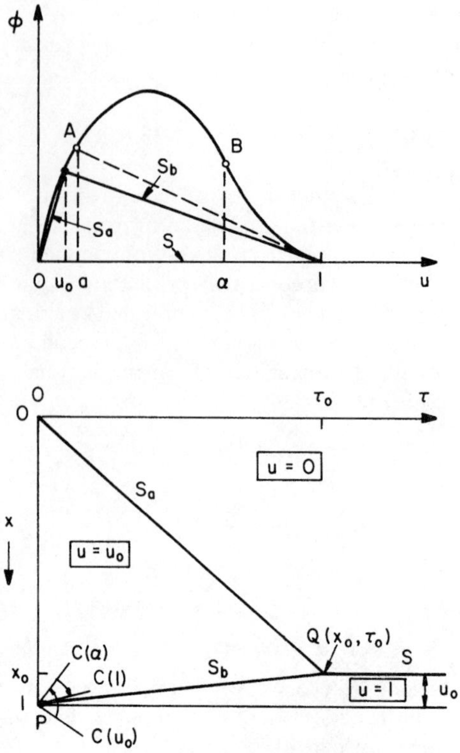

Figure 6.44

semishock S_b has the equation

$$x - 1 = \phi'(u_1)\tau = \frac{\phi(u_0) - \phi(u_1)}{u_0 - u_1}\tau$$

$$= \{1 - (1 + \gamma)(u_0 + u_1) + \gamma(u_0^2 + u_0u_1 + u_1^2\}\tau$$

so that their intersection Q has the coordinates

$$\tau_0 = \frac{u_0}{\phi(u_0) - u_0\phi'(u_1)} = \frac{u_0(u_1 - u_0)}{u_1\phi(u_0) - u_0\phi(u_1)} = \frac{4\gamma}{(1 + \gamma - \gamma u_0)^2}$$

$$x_0 = \frac{\phi(u_0)}{u_0}\tau_0 = (1 - u_0)(1 - \gamma u_0)\tau_0 \tag{6.7.41}$$

in which we have used Eq. (6.7.40) to obtain the last expression for τ_0.

From this point the resulting shock, representing the falling of the upper surface, interacts with the simple wave, and the shock path QR can, therefore,

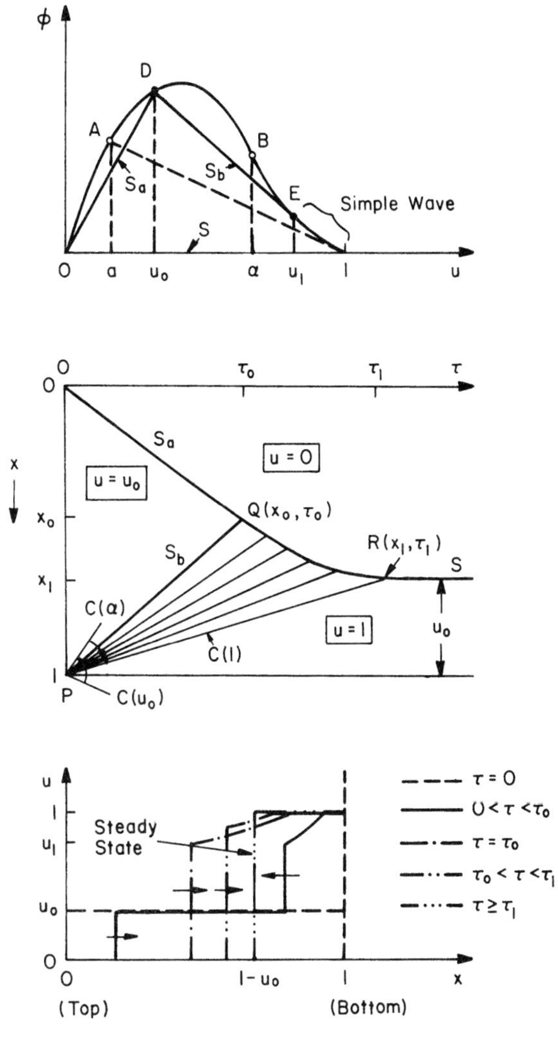

Figure 6.45

be described by the two equations

$$\frac{dx}{d\tau} = \tilde{\lambda}(0, u) = \frac{\phi(u)}{u} = 1 - (1 + \gamma)u + \gamma u^2$$
$$x - 1 = \lambda(u)\tau = \phi'(u)\tau = \{1 - 2(1 + \gamma)u + 3\gamma u^2\}\tau \tag{6.7.42}$$

where u is the concentration borne by one of the characteristics from the bottom. Treating u as a parameter along the shock path, we can differentiate the latter with respect to u and subtract from the former with $dx/d\tau = (dx/du)/(d\tau/du)$ to obtain

$$\frac{1}{\tau}\frac{d\tau}{du} = \frac{u\phi''(u)}{\phi(u) - u\phi'(u)} = -\frac{2(1 + \gamma - 3\gamma u)}{u(1 + \gamma - 2\gamma u)} \tag{6.7.43}$$

The initial condition is given by

$$\tau = \tau_0 \quad \text{at} \quad u = u_1$$

because the shock attains the state u_1 on the right side at the point of confluence B. Thus the integration gives the solution in the following form:

$$\frac{\tau}{\tau_0} = \frac{\phi(u_1) - u_1\phi'(u_1)}{\phi(u) - u\phi'(u)} = \left(\frac{u_1}{u}\right)^2 \frac{1 + \gamma - 2\gamma u_1}{1 + \gamma - 2\gamma u}$$

or, upon substituting Eqs. (6.7.41) and (6.7.39) or (6.7.40),

$$\tau = \frac{u_0}{\phi(u) - u\phi'(u)} = \frac{u_0}{u^2(1 + \gamma - 2\gamma u)} \tag{6.7.44}$$

The interaction will be over when $u = 1$, that is, when

$$\begin{aligned} \tau = \tau_1 &= -\frac{u_0}{\phi'(1)} = \frac{u_0}{1 - \gamma} \\ x = x_1 &= 1 - u_0 \end{aligned} \tag{6.7.45}$$

where the latter follows after the second equation of Eq. (6.7.42). Here the shock becomes stationary. The deposited layer is built up along the characteristic of unit concentration so that the process of sedimentation is complete at $\tau = \tau_1$.

Finally, let us examine the case when u_0 belongs to the upper range, that is, $\alpha \leq u_0 < 1$. The characteristics from the bottom bearing concentrations between u_0 and 1 fan clockwise, and thus a centered simple wave develops as shown in Fig. 6.46. The left end characteristic bearing the concentration u_0 has the equation

$$x - 1 = \lambda(u_0)\tau = \phi'(u_0)\tau = \{1 - 2(1 + \gamma)u_0 + 3\gamma u_0^2\}\tau$$

and this will meet the shock path from the top [cf. Eq. (6.7.36)] at $Q(x_0, \tau_0)$, where

$$\begin{aligned} \tau_0 &= \frac{u_0}{\phi(u_0) - u_0\phi'(u_0)} = \frac{1}{u_0(1 + \gamma - 2\gamma u_0)} \\ x_0 &= \frac{\phi(u_0)}{\phi(u_0) - u_0\phi'(u_0)} = \frac{(1 - u_0)(1 - \gamma u_0)}{u_0(1 + \gamma - 2\gamma u_0)} \end{aligned} \tag{6.7.46}$$

Thereafter, the two waves are involved in an interaction so that the shock path is not linear but satisfies Eq. (6.7.42) or, equivalently, Eq. (6.7.43). Here the appropriate initial condition is

$$\tau = \tau_0 \quad \text{at} \quad u = u_0$$

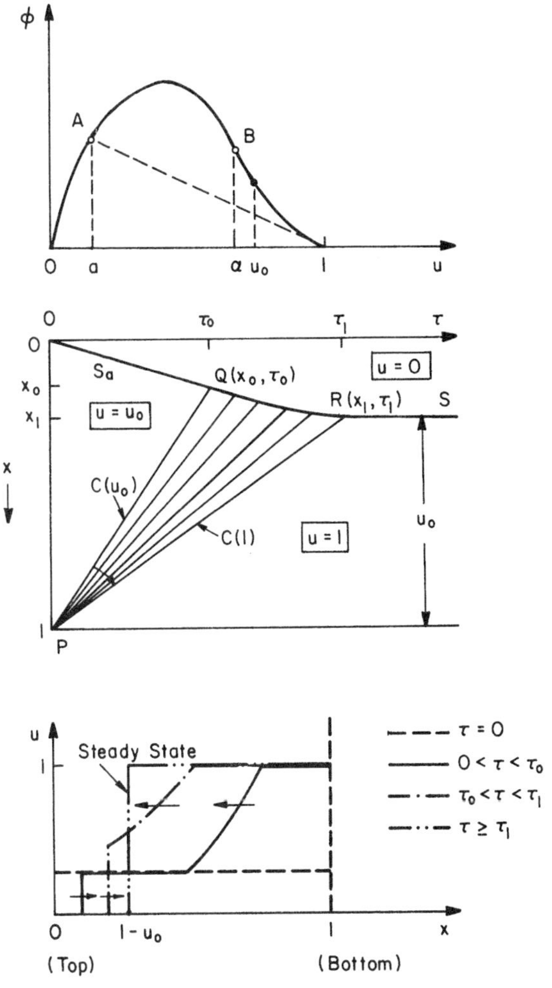

Figure 6.46

Thus Eq. (6.7.43) is integrated from this initial point to give

$$\frac{\tau}{\tau_0} = \frac{\phi(u_0) - u_0\phi'(u_0)}{\phi(u) - u\phi'(u)} = \left(\frac{u_0}{u}\right)^2 \frac{1+\gamma-2\gamma u_0}{1+\gamma-2\gamma u}$$

or $\qquad$ (6.7.47)

$$\tau = \frac{u_0}{\phi(u) - u\phi'(u)} = \frac{u_0}{u^2(1+\gamma-2\gamma u)}$$

as the equation of the curved part QR of the shock path, which is the trajectory of the upper surface of the sediment. We see that $u = 1$ at the point R, so

$$\tau = \frac{u_0}{-\phi'(1)} = \frac{u_0}{1 - \gamma} = \tau_1 \tag{6.7.48}$$

and, from Eq. (6.7.42), $x = 1 - u_0 = x_1$. At this point the shock attains the full strength and becomes stationary. Thus the process of sedimentation is completed.

The behavior with the quartic relationship between ϕ and u has an added feature, which may be illustrated as in Fig. 6.47. Besides the two inflection points at $u = \alpha$ and $u = \beta$ (here labeled C and F, respectively), the chord AEG tangential to the curve at E defines two points of importance. Let B represent a typical point on AC, and draw the tangent from B to the flux curve to locate the point of tangency D. We may let lowercase letters denote the abscissas of these points. If u_0 is in the interval $(0, a]$, the situation of Fig. 6.44 obtains. This figure also applies if $e \leq u_0 < g$, for the relevant part of the flux curve is seen on the left side as we move along the chord EG from E to G. If $u_0 = b$, a typical value in the range (a, c), then a semishock from B to D must be introduced. This is followed by a centered simple wave corresponding to the arc DE and another semishock EG. These two semishocks propagate side by side with the simple wave without intervals, as shown in Fig. 6.48. Finally, if $\alpha = c \leq u_0 < e$, there is no first semishock but a continuous variation from u_0 to e through a simple wave and then a semishock from E to G, as shown in Fig. 6.49.

The situation is summarized in Table 6.2, which gives all the relevant formulas for the quadratic, cubic, and quartic expressions for $\phi(u)$. The

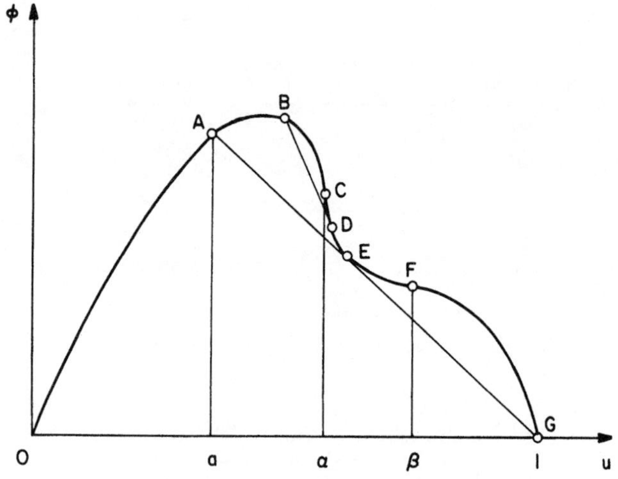

Figure 6.47

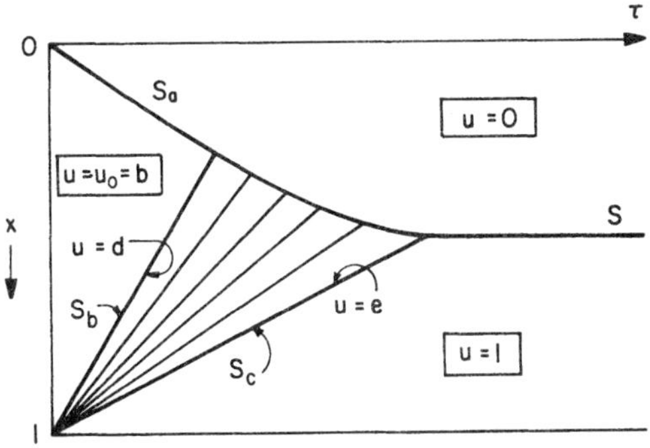

Figure 6.48

obvious abbreviations $\phi(u_0) = \phi_0$, $\phi(u_1) = \phi_1$ are used and $\chi(u) = \phi(u) - u\phi'(u)$.

Throughout the analysis here we have assumed that the deposited layer obtains the maximum possible concentration instantaneously; that is, the sediment is incompressible. Increasing weight of solids per unit area, however, tends to increase the ultimate solids concentration at the bottom of the column, which is reached gradually rather than instantaneously. In this zone of compressible (or compacting) sediment, the assumption that the flux is a function of the concentration alone may not be valid, and thus the present analysis may not be applied to predict the growth of the deposited layer [see, for example, Tory and Shannon (1965) and Scott (1968)]. More recently, some modifications have been proposed by Tiller (1981) and Fitch (1983) to treat the case when a zone of compressible sediment forms at the bottom of the column.

A word might now be said about the theory of the ultracentrifuge, where sedimentation takes place in a force field that varies with position. The

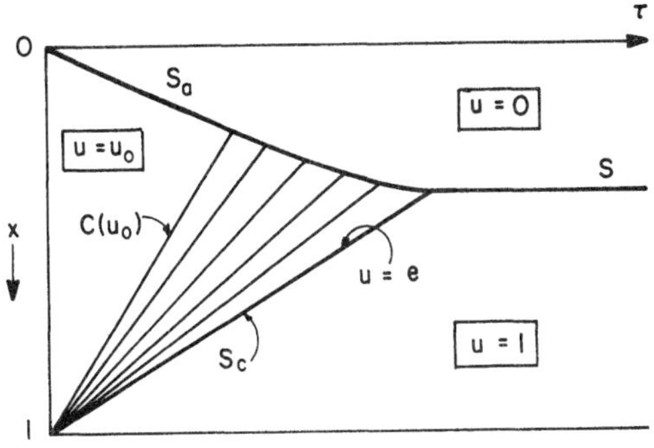

Figure 6.49

TABLE 6.2

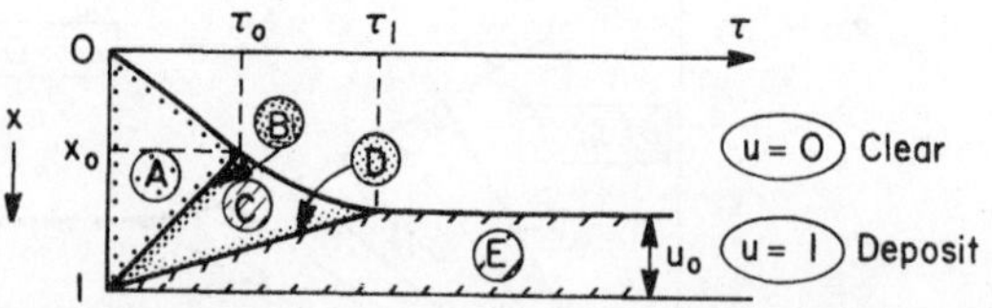

$\chi(u) \equiv \phi(u) - u\phi'(u)$

Case	(A)	$1 - x_0$	τ_0	(B)	(C)	(D)	τ_1	Remarks
I_*	u_0	u_0	τ_1	u_0	—	1	1	Quadratic form of $\phi(u)$. B, C & D collapse.
II_1	u_0	u_0	τ_1	u_0	—	1	$\dfrac{u_0(1-u_0)}{\phi(u_0)}$	B, C & D collapse.
II_2	u_0	$\dfrac{u_0(\phi_0-\phi_1)}{u_1\phi_0-u_0\phi_1}$	$\dfrac{u_0(u_1-u_0)}{u_1\phi_0-u_0\phi_1}$	u_1	$u_1 \to 1$	1	$-u_0/\phi'(1)$	$\phi(u_0)-u_0\phi'(u_1)=\chi(u_1)$
II_3	u_0	$\dfrac{-u_0\phi'(u_0)}{\chi(u_0)}$	$u_0/\chi(u_0)$	u_0	$u_0 \to 1$	1	$-u_0/\phi'(1)$	No semi-shocks.
III_1	u_0	u_0	τ_1	u_0	—	1	$\dfrac{u_0(1-u_0)}{\phi(u_0)}$	B, C & D collapse.
III_2	u_0	$\dfrac{u_0(\phi_0-\phi_1)}{u_1\phi_0-u_0\phi_1}$	$\dfrac{u_0(u_1-u_0)}{u_1\phi_0-u_0\phi_1}$	u_1	$u_1 \to u^*$	u^*	$-u_0/\phi'(1)$	u_0 and u_1 related as in II_2: u^* fixed.
III_3	u_0	$\dfrac{-u_0\phi'(u_0)}{\chi(u_0)}$	$u_0/\chi(u_0)$	u_0	$u_0 \to u^*$	u^*	$-u_0/\phi'(1)$	No semi-shock at AB.
III_4	u_0	u_0	τ_1	u_0	—	1	$\dfrac{u_0(1-u_0)}{\phi(u_0)}$	B, C & D collapse.

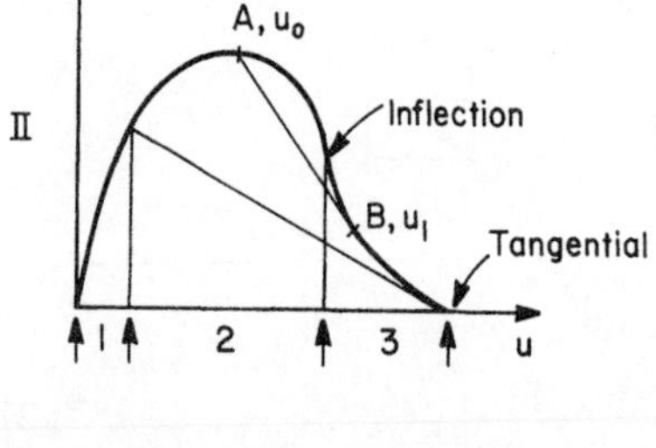

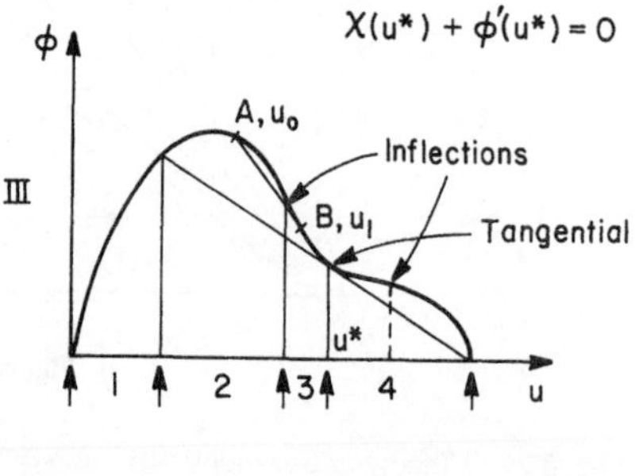

ultracentrifuge cell is a segment of a radial section of a cylinder between radii r_1 and r_2, as shown in Fig. 6.50, and the body force causing sedimentation acts perpendicularly to the axis of rotation. Thus, using a radial coordinate to any cylindrical surfaces, the force per unit mass on the particle is $\omega^2 r$, where ω is the angular velocity of rotation. The sedimentation coefficient, μ, is defined as the ratio of the velocity of sedimentation to this force, so the outward radial flux is

$$f = \mu\omega^2 rc \tag{6.7.49}$$

when the concentration is c. The sedimentation coefficient μ has the dimensions of time and is a function of concentration, pressure, and temperature. In a typical ultracentrifugation, the temperature is held constant and the effect of pressure on the physical properties is ignored. Thus, as before, the flux is primarily a function of concentration. The area over which the flux takes place varies with r so that in setting up the kinematic equation we are led to

$$\frac{\partial c}{\partial r} + \frac{1}{r}\frac{\partial}{\partial r}[\omega^2 r^2 \mu(c)c] = 0 \tag{6.7.50}$$

If $\mu_0 = \mu(0)$, a natural set of variables would be

$$\mu = \mu_0\psi(c), \qquad \tau = \omega^2\mu_0 t, \qquad \rho = \frac{r}{r_1}, \qquad R = \frac{r_2}{r_1} \tag{6.7.51}$$

Then

$$\frac{\partial c}{\partial \tau} + \frac{1}{\rho}\frac{\partial}{\partial \rho}[\rho^2\psi(c)c] = 0$$

or

$$\frac{\partial c}{\partial \tau} + \rho(\psi c)'\frac{\partial c}{\partial \rho} + 2\psi c = 0 \tag{6.7.52}$$

The characteristic differential equation for τ is $d\tau/ds = 1$, where s is the variable along a characteristic; but this can be interpreted rather simply to mean $s = \tau$, and we might just as well take τ as the characteristic parameter.

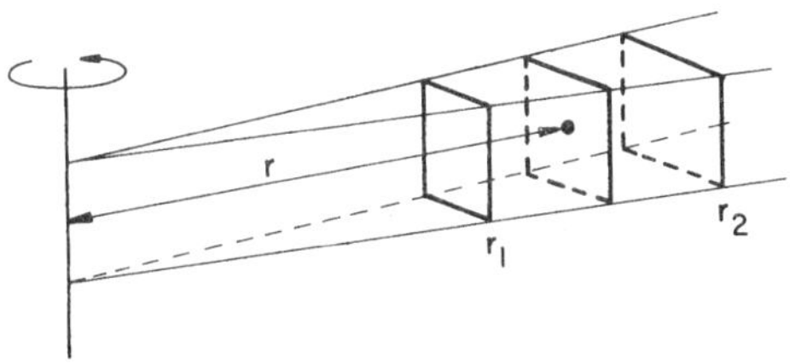

Figure 6.50

Then the characteristic differential equations are given by

$$\frac{d\rho}{d\tau} = \rho(\psi c)'$$
$$\frac{dc}{d\tau} = -2(\psi c) \tag{6.7.53}$$

If the initial concentration at $\tau = 0$ is $c_0(\rho)$, we can put it into the parametric form; that is,

$$\text{at} \quad \tau = 0, \qquad \rho = \xi \quad \text{and} \quad c = c_0(\xi)$$

Then the first equation divided by the second is readily integrated to give

$$\rho^2 c\psi(c) = \xi^2 c_0 \psi(c_0) \tag{6.7.54}$$

and the second equation immediately gives

$$2\tau = \int_c^{c_0} \frac{dc'}{c'\psi(c')} \tag{6.7.55}$$

Again because the cell is limited in extent to the region $1 \le \rho \le R$, there will be a buildup of sediment where c first reaches a maximum concentration. A discontinuity will move at a speed

$$\frac{d\rho}{d\tau} = \rho\frac{c^l\psi(c^l) - c^r\psi(c^r)}{c^l - c^r} = \tilde{\lambda}(c^l, c^r) \tag{6.7.56}$$

so that, for a discontinuity leaving clear liquid to the left ($c^l = 0$),

$$\frac{d\rho}{d\tau} = \rho\psi(c^r) = \tilde{\lambda}(0, c^r) \tag{6.7.57}$$

and for one building up maximum concentration c_m at the right,

$$\frac{d\rho}{d\tau} = -\rho\frac{c^l\psi(c^l)}{c_m - c^l} \tag{6.7.58}$$

If the initial concentration is uniform, that is, c_0 = constant and

$$\psi(c) = 1 - \frac{c}{c_m}$$

we have, from Eq. (6.7.55),

$$c = c_0 e^{-2\tau}\left[1 - \frac{c_0}{c_m}(1 - e^{-2\tau})\right]^{-1} \tag{6.7.59}$$

an expression for the variation of concentration along a characteristic, whose equation is now determined by substituting Eq. (6.7.59) in Eq. (6.7.54) as

$$\rho = \xi e^{\tau}\left[1 - \frac{c_0}{c_m}(1 - e^{-2\tau})\right] \tag{6.7.60}$$

Since the flux curve, $c\psi(c)$ versus c, here is convex, the discontinuities are all shocks. Falling of the top surface is governed by Eq. (6.7.57), and its path is obtained if we integrate this equation with $c^r = c$ and with the initial condition, $\rho = 1$ at $\tau = 0$; thus

$$\rho_{\text{top}} = e^{\tau}\left[1 - \frac{c_0}{c_m}(1 - e^{-2\tau})\right]^{1/2} \tag{6.7.61}$$

Similarly, the bottom layer builds up according to Eq. (6.7.58) with $c^l = c$; that is,

$$\frac{d\rho}{d\tau} = -\rho\frac{c}{c_m}$$

with the initial condition $\rho = R$ at $\tau = 0$. We therefore have

$$\rho_{\text{bottom}} = R\left[1 - \frac{c_0}{c_m}(1 - e^{-2\tau})\right]^{1/2} \tag{6.7.62}$$

These two shocks meet when $\tau = \ln R$ and the process of sedimentation is complete. At this final stage of the process the depth of the sediment is $(c_0/c_m)(R^2 - 1)/R^2 = (c_0/c_m)(r_2^2 - r_1^2)/r_2^2$. Figure 6.51 shows the solution. If the shock fronts are followed by Schlieren photography, the time of completion, t_c, can be determined, and hence the sedimentation coefficient

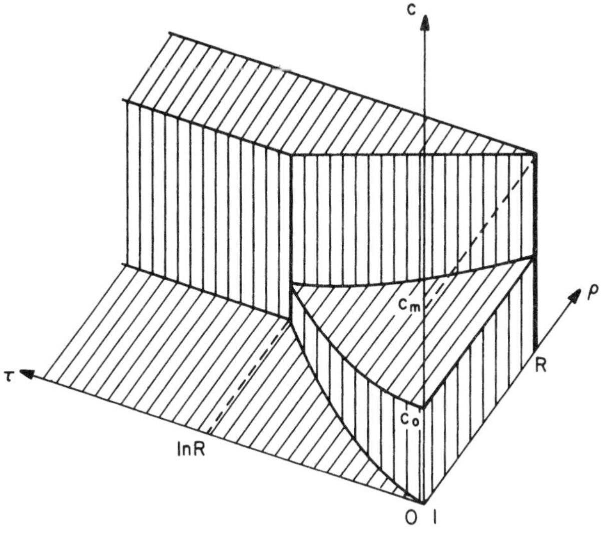

Figure 6.51

can be estimated by

$$\mu_0 = \frac{\ln R}{\omega^2 t_c} \tag{6.7.63}$$

Naturally, there is much more to the theory of the ultracentrifuge than has been indicated here. In particular, diffusion will blur the sharp front, and different solutes will interact with one another.

Exercises

6.7.1. Obtain the equations when $A(z)$, the cross-sectional area of the sedimentation column, varies as a function of height. How would the actual propagation speed be related to $A(z)$? How would the process of sedimentation in general be affected by the variation in the cross-sectional area?

6.7.2. Show that for an arbitrary initial distribution and $\phi(u) = u(1 - u)$ the settling will be complete after a time

$$\bar{\tau} = \frac{1}{u_0^2(\bar{\xi})} \int_{\bar{\xi}}^1 u_0(\xi')\, d\xi'$$

where $\bar{\xi}$ is the solution of

$$[u_0(\bar{\xi})]^2 \int_0^{\bar{\xi}} u_0(\xi')\, d\xi' + [1 - u_0(\bar{\xi})]^2 \int_{\bar{\xi}}^1 u_0(\xi')\, d\xi' = \bar{\xi}[u_0(\bar{\xi})]^2$$

provided that $2u_0'(\xi)\tau + 1$ has no root in $0 \le \tau \le \bar{\tau}$, $1 \ge \xi \ge \bar{\xi}$.

6.7.3. Establish Eq. (6.7.50) and make a suitable reduction to dimensionless variables when ω is a function of time.

6.7.4. If the centrifuge accelerates uniformly to its maximum speed ω_0 in a time $t_a < t_c$, show that the sedimentation coefficient, μ_0, estimated by Eq. (6.7.63) is in error by a factor of $1 - (2t_a/3t_c)$.

6.7.5. Find the characteristics of Eq. (6.7.50) when $\psi(c) = (1 + \gamma c)^{-1}$.

REFERENCES

6.1 through 6.3. The basic references for most of these sections are

R. Courant and K. O. Friedrichs, *Supersonic Flow and Shock Waves*, Wiley-Interscience, New York, 1948.

H.-K. Rhee, "Studies on the Theory of Chromatography: Part I. Chromatography of a Single Solute," Ph.D. Thesis, University of Minnesota, 1968.

G. B. Whitham, *Linear and Nonlinear Waves*, Wiley, New York, 1974.

For the base curve method, see the thesis by Rhee, who has discussed the formation and propagation of shocks in some detail with many examples.

The decay of a single shock and the propagation of an N-wave have been treated by Courant and Friedrichs and by Whitham. See also

P. D. Lax, "Hyperbolic Systems of Conservation Laws II," *Comm. Pure Appl. Math.* **10**, 537–566 (1957).

P. D. Lax, "The Formation and Decay of Shock Waves," *Amer. Math. Monthly* **79**, 227–241 (1972).

P. D. Lax, "Hyperbolic Systems of Conservation Laws and the Mathematical Theory of Shock Waves," *Regional Conference Series in Applied Mathematics*, No. 11, SIAM, Philadelphia, Pa., 1973.

The decay of discontinuous solutions is discussed from a different angle in

J. D. Murray, "Perturbation Effects on the Decay of Discontinuous Solutions of Nonlinear First Order Wave Equations," *SIAM J. Appl. Math.* **19**, 273–298 (1970).

Concerning the first example of Sec. 6.3, Goldstein has demonstrated that the solution can be obtained as the limit of the kinetic theory:

S. Goldstein, "On the Mathematics of Exchange Processes in Fixed Columns. II. The Equilibrium Theory as the Limit of the Kinetic Theory," *Proc. Roy. Soc. Lond.* **A219**, 171–185 (1953).

For quasi-linear equations in more than one space dimension, the formation and decay of shocks is treated in

E. D. Conway, "The Formation and Decay of Shocks for a Conservation Law in Several Dimensions," *Arch. Rat. Mech. Anal.* **64**, 47–57 (1977).

For the impulse response of the chromatographic system, see

J. Loureiro, C. Costa, and A. Rodrigues, "Propagation of Concentration Waves in Fixed-bed Adsorption Reactors," *Chem. Eng. J.* **27**, 135–148 (1983).

6.4. The propagation of pulse-type signals, as well as the simultaneous interaction between a shock and two simple waves, has been treated in the thesis by Rhee (1968) with examples. Examples of the periodic wave are found in the thesis by Rhee (1968) and in the book by Whitham (1974) cited previously. The asymptotic behavior of shocks that originated from pulse-type data, as well as from periodic data, has been discussed in the three papers by Lax (1957, 1972, 1973) and in the book by Whitham (1974) all cited previously.

6.5. This treatment of the countercurrent adsorber is in the line of

H.-K. Rhee and N. R. Amundson, "An Analysis of an Adiabatic Adsorption Column: Part I. Theoretical Development," *Chem. Eng. J.* **1**, 241–254 (1970).

H.-K. Rhee, R. Aris, and N. R. Amundson, "Multicomponent Adsorption in Continuous Countercurrent Exchangers," *Phil. Trans. Roy. Soc. Lond.* **A269**, 187–215 (1971).

6.6. The traffic-flow problem was first treated in this line by

M. J. Lighthill and G. B. Whitham, "On Kinematic Waves: II. A Theory of Traffic Flow on Long Crowded Roads," *Proc. Roy. Soc. Lond.* **A229**, 317–345 (1955).

P. I. Richards, "Shock Waves on the Highway," *Opns. Res.* **4**, 42–51 (1956).

Also see

W. D. Ashton, *The Theory of Road Traffic Flow*, Methuen, London, 1966.

G. B. Whitham, *Linear and Nonlinear Waves*, Wiley, New York, 1974.

Experimental evidence for the quadratic flux relationship is found in

B. D. Greenshields, "A Study of Traffic Capacity," *Proc. Highway Research Board* **14**, 448–474 (1935).

For the logarithmic flux relation, see

H. Greenberg, "An Analysis of Traffic Flow," *Opns. Res.* **7**, 79–85 (1959).

Various models for road traffic flow, including the kinematic model, have been discussed by

W. D. Ashton, "Models for Road Traffic Flow," *Bull. Inst. Maths. Appl.*, 318–322 (Nov. 1972).

6.7. The theory of sedimentation was first given in this form by

G. J. Kynch, "A Theory of Sedimentation," *Trans. Faraday Soc.* **48**, 166–176 (1952).

See also

P. T. Shannon, E. Stroupe, and E. M. Tory, "Batch and Continuous Thickening: Basic Theory, Solids Flux for Rigid Spheres," *Ind. Eng. Chem. Fundam.* **2**, 203–211 (1963).

P. T. Shannon and others, "Batch and Continuous Thickening: Prediction of Batch Settling Behavior from Initial Rate Data with Results for Rigid Spheres," *Ind. Eng. Chem. Fundam.* **3**, 250–260 (1964).

E. M. Tory and P. T. Shannon, "Reappraisal of the Concept of Settling in Compression," *Ind. Eng. Chem. Fundam.* **4**, 194–204 (1965).

K. J. Scott, "Thickening of Calcium Carbonate *Ind. Eng. Chem. Fundam.* **4**, 484–490 (1968).

The last two discuss the subject of compressible sediment.

Excellent surveys are to be found in

P. T. Shannon and E. M. Tory, "Settling of Slurries," *Ind. Eng. Chem.* **57**, 18–25 (1965).

P. Kos, "Fundamentals of Gravity Thickening," *Chem. Eng. Progr.* **73**, 99–105 (Nov. 1977).

B. Fitch, "Sedimentation of Flocculent Suspensions: State of the Art," *AIChE J.* **25**, 913–930 (1979).

The case of compressible sediment is treated in

F. M. Tiller, "Revision of Kynch Sedimentation Theory," *AIChE J.* **27**, 823–829 (1981).

B. Fitch, "Kynch Theory and Compression Zones," *AIChE J.* **29**, 940–947 (1983).

F. M. Tiller and Z. Khatib, "The Theory of Sediment Volumes of Compressible, Particulate Structures," *J. Colloid Interface Sci.* **100**, 55–67 (1984).

For some of the considerations involved in the physics of settling, see

B. Fitch, "A Mechanism of Sedimentation," *Ind. Eng. Chem. Fundam.* **5**, 129–134 (1966).

Sedimentation of a single substance undergoing association or dissociation is treated in

G. A. Gilbert, "Sedimentation and Electrophoresis of Interacting Substances: I. Idealized Boundary Shape for a Single Substance Aggregating Reversibly," *Proc. Roy. Soc. Lond.* **A250**, 377–388 (1959).

In regard to continuous thickening, useful material is found in

K. J. Scott, "Experimental Study of Continuous Thickening of a Flocculated Silica Slurry," *Ind. Eng. Chem. Fundam.* **7**, 582–595 (1968).

M. V. Maljian and J. A. Howell, "Dynamic Response of a Continuous Thickener to Overloading and Underloading," *Trans. Inst. Chem. Engr.* **56**, 56–61 (1978).

The theory of the ultracentrifuge is covered in

H. Fujita, *Mathematical Theory of Sedimentation Analysis*, Academic Press, New York, 1962.

7 Conservation Equations, Weak Solutions, and Shock Layers

Since this chapter has a number of important and interconnected ideas, as well as some key examples that are taken up in various ways, it will be well to describe the warp of ideas and woof of examples in its fabric. If some of the topics are taken up in rather painful detail, this is because the original sources are often marked by the mathematicians' love of terse, comprehensive, and general presentation and are not easy reading. The reader who is impatient of our prolixity may go to "the masters," to whom references are given at the end of the chapter.

As usual, the example of conservation equations that is used in most parts of this chapter is that of chromatography, but the theory of finite-amplitude sound waves is also taken up in the latter part. We want to see how much can be said in general about the solutions of conservation equations, particularly those with discontinuities. The chromatographic equations are revisited with the intention of seeing how the initial conditions may be specified

(Sec. 7.1) and they are compared with other forms of conservation equations (Sec. 7.2; see also the later comparison in Sec. 7.8). The propagation direction of a discontinuity is then formulated from an integral balance of the solute material, and its approximation for weak discontinuities is discussed (Sec. 7.2).

In Sec. 7.3 we pause to look at some properties of convex functions that will be needed later and to introduce the Legendre transformation. This is illustrated with the Langmuir isotherm. We are then in a position to introduce the formal definition of a weak or generalized solution and to show that a discontinuity will propagate in the same direction as was found in Sec. 7.2. In concluding Sec. 7.4, mention is made of some important features of nonlinear equations, including the entropy condition to which repeated reference will be made. In Sec. 7.5, Lax's explicit formula is discussed. The development is first motivated by the notion of maximum operation, an elegant idea in its own right. Then the saturation of an adsorption column is used as an example to show how Lax's formula works. An outline proof of the formula by the method of maximum operation is followed by the derivation given by Lax. The connection between strictly convective conservation equations and parabolic equations with a viscous or diffusion term is discussed next (Sec. 7.6), and it is seen how the smooth solutions are sharpened in the limit of vanishing viscosity. This section also introduces the conditions necessary to ensure the uniqueness of the solution. The scheme of the existence and uniqueness theorem is described, but it is not proved in detail. Some asymptotic properties of weak solutions are proved, and the triangular form of the wave is stated but not proved here; the latter links very closely with the special case treated in the next section.

Returning to physical examples, some aspects of finite-amplitude sound waves are now taken up (Sec. 7.7). Burger's equation can be solved explicitly, and the limiting case of zero diffusivity recrystallizes Lax's formula in a particularly simple form. Because of this simple form, a very elegant construction due to Whitham can be used to plot the shock paths, and this then makes clear how the asymptotic triangular form of the wave arises. This is compared with the exact solution of Burger's equation at the end of this section. In Sec. 7.8, the chromatographic equations are compared in detail to the equations for sound waves and, although the full construction does not go through, the point of inflection on a certain curve does show where the shock will start to form. The implicit form of the solution is related to Lax's formula.

For an arbitrary convex function $f(c)$, the asymptotic behavior of shocks for large time is rigorously formulated in Sec. 7.9. The approach here is motivated by the equal area requirement due to Whitham, and the results are compared to those established by Lax. It is also pointed out that the asymptotic formulas are in full agreement with the previous results determined for specific examples in Secs. 6.3 and 6.6. This section also extends the discussion to the cases of the N-wave and the periodic wave. The final section

(Sec. 7.10) on shock layer takes up the case where the axial dispersion in the fluid phase and the interphase transfer resistance are considered. The existence and uniqueness of the shock layer is proved in a general context by applying perturbation theory, and the concept of shock-layer thickness is introduced to show that the shock layer converges uniformly to the corresponding shock in the limit as both of the added effects tend to zero. Special cases are also discussed and illustrated with explicit formulas for the case of the Langmuir isotherm.

7.1 Chromatographic Equations and Initial Data

Here again we will consider the chromatography of a single solute with the assumption of local equilibrium and look into how the initial data may be specified. Suppose a fluid stream flows with a linear velocity V through a packed bed of void fraction ε. If we let $c(z, t)$ and $n(z, t)$ denote the concentrations at position z from the inlet and time t in the fluid and solid phases, respectively, we have Eq. (1.2.1), from the mass balance of the solute,

$$\varepsilon V\frac{\partial c}{\partial z} + \varepsilon\frac{\partial c}{\partial t} + (1 - \varepsilon)\frac{\partial n}{\partial t} = 0 \qquad (7.1.1)$$

or, in terms of the volume ratio,

$$\nu = \frac{1 - \varepsilon}{\varepsilon} \qquad (7.1.2)$$

and the dimensionless variables

$$x = \frac{z}{Z}, \qquad \tau = \frac{Vt}{Z} \qquad (7.1.3)$$

the following equation

$$\frac{\partial c}{\partial x} + \frac{\partial c}{\partial \tau} + \nu\frac{\partial n}{\partial \tau} = 0 \qquad (7.1.4)$$

In Eq. (7.1.3), Z represents a characteristic length of the packed bed. We will define f† as

$$f = c + \nu n$$

The adsorption isotherm is the relation that obtains between n and c at equilibrium, $n = n(c)$; and if we assume that local equilibrium is established at all times, f becomes a function of c alone:

$$f(c) = c + \nu n(c) \qquad (7.1.5)$$

†Note that here f does not represent the adsorption isotherm. Compare with Eq. (5.3.5).

Bayle and Klinkenberg [*Rec. Trav. Chim. Pays-Bas* **73**, 1037 (1954); **76**, 593 (1957)] suggested the term *equilibrium column isotherm* for $f(c)$, since it involves the constants of the column as well as those of solute and adsorbent. Thus Eq. (7.1.4) is rewritten as

$$\frac{\partial c}{\partial x} + \frac{\partial}{\partial \tau} f(c) = 0 \tag{7.1.6}$$

or

$$\frac{\partial c}{\partial x} + f'(c)\frac{\partial c}{\partial \tau} = 0 \tag{7.1.7}$$

We shall consider only convex isotherms, for which $n'(c) > 0$ and $n''(c) < 0$ when $c > 0$. By Eq. (7.1.5), the column isotherm $f(c)$ is also convex since

$$f'(c) = 1 + \nu n'(c), \qquad f''(c) = \nu n''(c) \tag{7.1.8}$$

A particularly important example is afforded by the Langmuir isotherm,

$$n(c) = \frac{\gamma c}{1 + Kc} \tag{7.1.9}$$

for which

$$f(c) = c + \frac{\nu \gamma c}{1 + Kc} \tag{7.1.10}$$

From Eq. (7.1.7), we have

$$\frac{\partial c}{\partial x} + \left[1 + \frac{\nu \gamma}{(1 + Kc)^2}\right]\frac{\partial c}{\partial \tau} = 0$$

and here it is convenient to take

$$\alpha = \nu \gamma, \qquad u = Kc \tag{7.1.11}$$

so that

$$\frac{\partial u}{\partial x} + \left[1 + \frac{\alpha}{(1 + u)^2}\right]\frac{\partial u}{\partial \tau} = 0 \tag{7.1.12}$$

In regard to the initial data, we assume that the initial distributions of concentrations are in equilibrium, so instead of specifying $c(x, 0)$ and $n(x, 0)$ separately, we will rather specify the combination,

$$c(x, 0) + \nu n(c(x, 0)) = f(c(x, 0)) = F(x) \tag{7.1.13}$$

It is convenient to denote $f(c(x, 0))$ by $\bar{f}(x, 0)$. At the inlet, $x = 0$, we can only specify the concentration of the solute in the fluid phase; that is,

$$c(0, \tau) = G(\tau) \tag{7.1.14}$$

Therefore, the initial curve here consists of the pair of positive axes (see Fig. 5.11). With the Langmuir isotherm, Eq. (7.1.9), $f'(c)$ becomes nowhere zero or infinite, so this initial curve is never tangent to a characteristic direction

$$\frac{d\tau}{dx} = f'(c) \tag{7.1.15}$$

In some problems it is convenient to prescribe c over the whole of just one of the axes. For example, consider the elution problem where $G(\tau) = 0$ and $F(x)$ represents some initial distribution of solute that is to be eluted from the column. In this case we would specify f (and so also c) over the whole x-axis by

$$\bar{f}(x, 0) = f(c(x, 0)) = \begin{cases} 0, & x < 0 \\ F(x), & x \geqslant 0 \end{cases} \tag{7.1.16}$$

for the solution in the second quadrant would be identically zero, and so the value of c on the τ-axis would be zero. It would be slightly more troublesome to specify $c(0, \tau)$ on the whole τ-axis in the general case, for we would have to calculate back and find out what inlet distribution for negative times, $\tau < 0$, would result in the distribution found on the column at time $\tau = 0$. In some cases this is not difficult; if $F(x)$ were constant, $F(x) = f(c_0)$, say, then $c(0, \tau) = c_0$, $\tau < 0$, would ensure that the column was saturated at a total concentration of $f(c_0)$ at time $\tau = 0$. Then we would have the equivalence of

$$\begin{cases} \bar{f}(x, 0) = f(c_0), & x \geqslant 0 \\ c(0, \tau) = 0, & \tau > 0 \end{cases} \quad \text{and} \quad c(0, \tau) = \begin{cases} c_0, & \tau \leqslant 0 \\ 0, & \tau > 0 \end{cases}$$

Similarly, if a clean bed $[F(x) = 0]$ is being fed by a fluid stream with concentration $G(\tau)$, we have the equivalence of

$$\begin{cases} \bar{f}(x, 0) = 0, & x > 0 \\ c(0, \tau) = G(\tau), & \tau \geqslant 0 \end{cases} \quad \text{and} \quad c(0, \tau) = \begin{cases} 0, & \tau < 0 \\ G(\tau), & \tau \geqslant 0 \end{cases}$$

This variation of initial and boundary conditions is introduced because it is often useful to look for the solution in the complete half-plane.

Exercise

7.1.1. Consider the Freundlich isotherm,

$$f(c) = 1 + \nu\alpha c^{\beta}, \qquad 0 < \beta < 1$$

and discuss how one would transform the initial data given by Eqs. (7.1.13) and (7.1.14) to an equivalent one with data prescribed over the whole τ-axis or over the whole x-axis.

7.2 Conservation Equations and the Jump Condition

We have repeatedly stressed that first-order partial differential equations arise from convective conservation laws and that the chromatographic equations are examples of these. There is quite a body of work on the structure of the solutions of conservation equations. Equation (7.1.6) or (7.1.7) is a conservation law of the form

$$\frac{\partial c}{\partial x} + \frac{\partial}{\partial \tau} f(c) = \frac{\partial c}{\partial x} + f'(c)\frac{\partial c}{\partial \tau} = 0 \tag{7.2.1}$$

where $\varepsilon f(c)$ is the total concentration in equilibrium with c and τ $(= Vt/Z)$ has the role of time. But in discussing kinematic waves in general we have seen that they may also take the form†

$$\frac{\partial u}{\partial t} + \frac{\partial}{\partial x} \phi(u) = 0 \tag{7.2.2}$$

where u is a concentration and ϕ a flux. More generally, u and ϕ can be vector-valued functions, and the flux may have components in more than one space dimension. A familiar example is the continuity equation for the density ρ of a nonreacting substance. In a velocity field with components V_1, V_2, V_3 in the direction Ox_1, Ox_2, Ox_3, the continuity equation is

$$\frac{\partial \rho}{\partial t} + \mathbf{V} \cdot \nabla \rho = \frac{\partial \rho}{\partial t} + V_1\frac{\partial \rho}{\partial x_1} + V_2\frac{\partial \rho}{\partial x_2} + V_3\frac{\partial \rho}{\partial x_3} = 0$$

If $V_i \rho = \phi_i$ is the component of the flux in the direction Ox_i, this has the form

$$\frac{\partial \rho}{\partial t} + \sum_{i=1}^{3} \frac{\partial \phi_i}{\partial x_i} = 0 \tag{7.2.3}$$

On the other hand, we might have the simultaneous conservation of several species, the flux of any one of them being a function of all their concentrations, and so

$$\frac{\partial u_i}{\partial t} + \frac{\partial}{\partial x} \phi_i(u_1, u_2, \ldots, u_n) = 0 \tag{7.2.4}$$

The interest in conservation equations arises partly because of their fundamental physical importance and partly because of their interesting mathematical features. In particular, the way in which discontinuities in the solution can arise and are propagated is of cardinal importance and gives rise

†Note that we may regard the roles of x and t here equivalent to those of τ and x, respectively, in Eq. (7.2.1).

to the concept of a weak solution. We will explore this further in a later section (Sec. 7.4); for the moment, we need to establish the conservation law for a discontinuity. Although we have already formulated this in Sec. 5.4, we would like to do it from a different angle.

Let us recall that the characteristic differential equations of Eq. (7.2.1),

$$\frac{dx}{ds} = 1, \qquad \frac{d\tau}{ds} = f'(c), \qquad \frac{dc}{ds} = 0 \tag{7.2.5}$$

imply that c is constant along a characteristic and that the characteristic is therefore a straight line. In the (x, τ)-plane the slope of this line,

$$\frac{d\tau}{dx} = f'(c) \equiv \sigma(c) \tag{7.2.6}$$

is inversely proportional to the speed with which a point of continuity of concentration c will move and might be called the "slowness" of the state of concentration c.

Now we shall consider a discontinuity in the values of c such as is shown in Fig. 7.1(a). Suppose that the discontinuity moves along a smooth curve $x = y(\tau)$ and that, on each side of this curve, Eq. (7.1.7) is satisfied. Let the superscripts l and r denote the states on the left and right sides, respectively, of the discontinuity, and choose a and b in such a way that the curve $x = y(\tau)$ intersects the interval $a \le x \le b$ at time τ [see Fig. 7.1(b)]. If M represents the total number of moles of the solute in the section from a to b of the column, we have

$$M = \int_{x=a}^{x=b} \{\varepsilon c + (1 - \varepsilon)n\}A\, dz = \varepsilon AZ \int_a^b f\, dx \tag{7.2.7}$$

in which A is the cross-sectional area of the column and Z the characteristic length [see Eq. (7.1.3)]. The conservation law asserts that

$$\frac{dM}{dt} = \varepsilon AVc_a - \varepsilon AVc_b$$

or

$$\frac{1}{\varepsilon AZ}\frac{dM}{d\tau} = c_a - c_b \tag{7.2.8}$$

where the subscripts a and b are used to denote the states at $x = a$ and $x = b$, respectively. On the other hand, by differentiating Eq. (7.2.7), we obtain

$$\frac{1}{\varepsilon AZ}\frac{dM}{d\tau} = \frac{d}{d\tau}\left\{\int_a^y f\,dx + \int_y^b f\,dx\right\} = \int_a^y \frac{\partial f}{\partial \tau}dx + f^l\frac{dy}{d\tau} + \int_y^b \frac{\partial f}{\partial \tau}dx - f^r\frac{dy}{d\tau}$$

Here we have used the abbreviations $f(c^l) = f^l$ and $f(c^r) = f^r$. Since on either side of the curve $x = y(\tau)$ Eq. (7.1.6) is satisfied, we may set

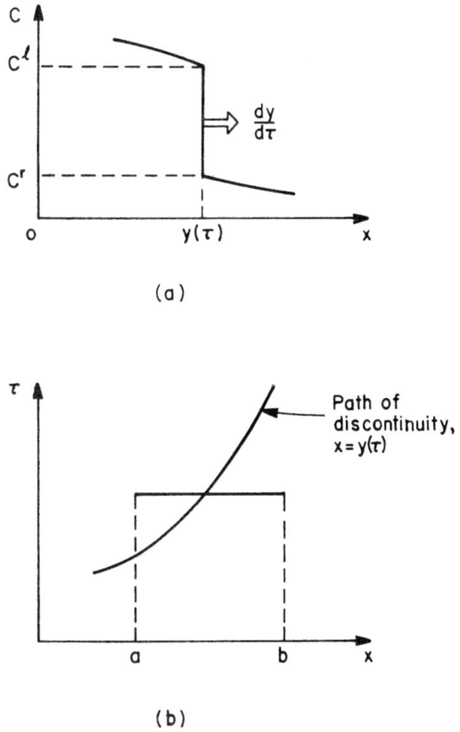

Figure 7.1

$\partial f/\partial\tau = -\partial c/\partial x$ in the integral terms and carry out the integration to have

$$\frac{1}{\varepsilon AZ}\frac{dM}{d\tau} = c_a - c^l - c_b + c^r + (f^l - f^r)\frac{dy}{d\tau} \tag{7.2.9}$$

Combining this with Eq. (7.2.8), we obtain

$$\frac{dy}{d\tau} = \frac{c^l - c^r}{f^l - f^r} = \frac{[c]}{[f]} \equiv \tilde{\lambda}(c^l, c^r) \tag{7.2.10}$$

for the propagation speed of the discontinuity or

$$\frac{d\tau}{dx} = \frac{[f]}{[c]} \equiv \tilde{\sigma}(c^l, c^r) \tag{7.2.11}$$

for the propagation direction of the discontinuity, where the bracket [ ] indicates the jump in the quantity enclosed. This is usually called the jump condition and is essentially the same as Eq. (5.4.2).

Consider a discontinuity with c_1 on one side and c_2 on the other, where

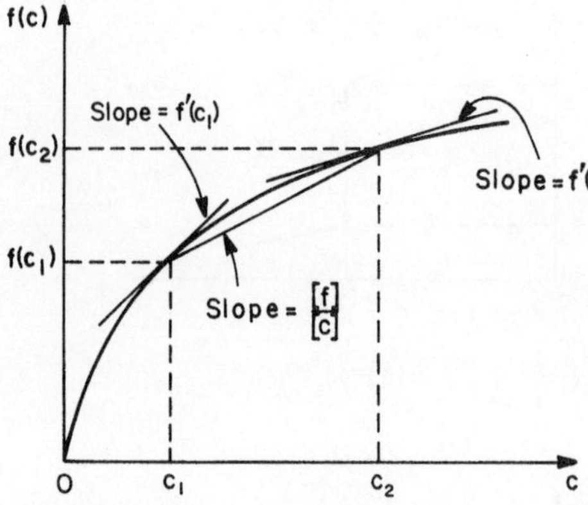

Figure 7.2

$c_1 < c_2$, without prejudice on which is on which side. When we compare Eq. (7.2.11) to Eq. (7.2.6) in the light of the convexity of the curve $f(c)$, as shown in Fig. 7.2, we immediately notice that

$$f'(c_1) \geq \frac{[f]}{[c]} \geq f'(c_2) \tag{7.2.12}$$

in which equality obtains only if $c_1 = c_2$ or the discontinuity vanishes. Thus the speed (or slowness) of a discontinuity always lies between the speeds of continuity of the states on either side of it; in an obvious notation,

$$\sigma(c_1) \geq \tilde{\sigma}(c_1, c_2) \geq \sigma(c_2) \tag{7.2.13}$$

For the Langmuir isotherm, Eq. (7.1.10), we have

$$\sigma(c) = f'(c) = 1 + \frac{\nu\gamma}{(1 + Kc)^2} \tag{7.2.14}$$

and

$$\tilde{\sigma}(c_1, c_2) = \frac{[f]}{[c]} = 1 + \frac{\nu\gamma}{(1 + Kc_1)(1 + Kc_2)} \tag{7.2.15}$$

so that Eq. (7.2.13) immediately follows.

In many situations, the discontinuities are weak in the sense that $(c_2 - c_1)/c_1$ is small. It may be useful to have some approximations for such cases. We note that the direction

$$\tilde{\sigma} = \frac{f(c_2) - f(c_1)}{c_2 - c_1}$$

tends to the characteristic direction

$$\sigma(c) = f'(c)$$

in the limit as the strength of the discontinuity $(c_2 - c_1) \to 0$. For weak discontinuities, we may expand $f(c_2)$ in a Taylor series in $(c_2 - c_1)$ to have

$$f(c_2) = f(c_1) + f'(c_1)(c_2 - c_1) + \frac{1}{2}f''(c_1)(c_2 - c_1)^2 + 0(c_2 - c_1)^3$$

and thus

$$\tilde{\sigma}(c_1, c_2) = f'(c_1) + \frac{1}{2}(c_2 - c_1)f''(c_1) + 0(c_2 - c_1)^2$$

$$= \sigma(c_1) + \frac{1}{2}(c_2 - c_1)\sigma'(c_1) + 0(c_2 - c_1)^2$$

The characteristic direction $\sigma(c_2)$ may also be expanded as

$$\sigma(c_2) = \sigma(c_1) + (c_2 - c_1)\sigma'(c_1) + 0(c_2 - c_1)^2$$

Consequently,

$$\tilde{\sigma}(c_1, c_2) = \frac{1}{2}\{\sigma(c_1) + \sigma(c_2)\} + 0(c_2 - c_1)^2 \tag{7.2.16}$$

and this implies that, in the (x, τ)-plane, the path of a weak discontinuity bisects the angle between the characteristics that meet the discontinuity. The propagation speed is the mean of the speeds of the states on the two sides of it. Clearly, the relation is exact when $f(c)$ is quadratic, as one might have noticed in Sec. 6.6.

Exercises

7.2.1. Put the chromatographic equation into the form of Eq. (7.2.2) and discuss the form of $\phi(u)$ for the Langmuir isotherm.

7.2.2. Find the jump condition for the conservation equation (7.2.2).

7.2.3. Find the conditions under which the equation

$$f(x, t, u)\frac{\partial u}{\partial t} + g(x, t, u)\frac{\partial u}{\partial x} = h(x, t, u)$$

can be reduced to

$$\frac{\partial v}{\partial t} + \frac{\partial}{\partial x}\phi(x, t, v) = 0$$

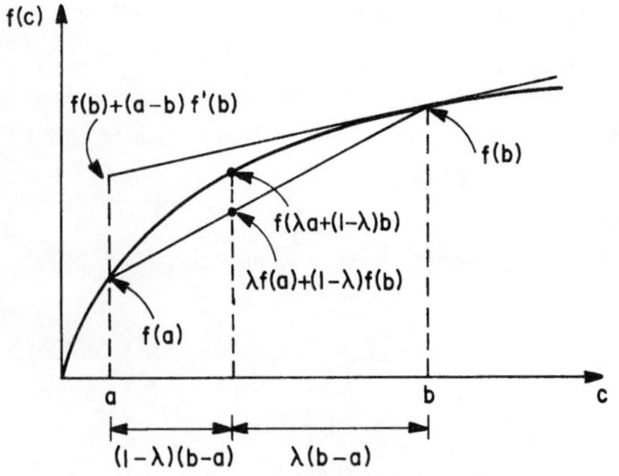

Figure 7.3

7.3 Intermezzo on Convex Function and the Legendre Transformation

If the function $f(c)$ is such that the chord joining the points $c = a$ and $c = b$ lies wholly beneath the curve, that is,

$$\lambda f(a) + (1 - \lambda)f(b) \leq f(\lambda a + (1 - \lambda)b), \qquad 0 \leq \lambda \leq 1 \tag{7.3.1}$$

then the function is convex (see Fig. 7.3). It is strictly convex if the equality only obtains when $\lambda = 0$ or 1. More precisely, we might want to say that such a curve was convex upward or concave downward. We will reserve "convex" for this type of curve and "concave" for its converse, the curve that is concave upward or convex downward or the function for which

$$\lambda f(a) + (1 - \lambda)f(b) \geq f(\lambda a + (1 - \lambda)b), \qquad 0 \leq \lambda \leq 1 \tag{7.3.2}$$

Since we will only be concerned with smooth functions having continuous first derivatives, we will not stop to consider the details that have to be taken care of when there are sharp bends in the curve; but it will be evident that suitable modifications can always be made. As is clear from Fig. 7.3, a convex function has a monotone-decreasing derivative and hence a negative second derivative. Since we are only concerning ourselves with functions having a second derivative, we can take

$$f''(c) < 0 \tag{7.3.3}$$

as the definition of a strictly convex function, and we have seen that this is a general property of the column isotherm, Eq. (7.1.5).

We will need to use later the following elementary property of a strictly convex function:

$$f(c - a) + f(c + a) > f(c - b) + f(c + b), \qquad \text{if} \quad a < b \tag{7.3.4}$$

This says that a shorter chord with the same center-point coordinate as a longer chord must lie above it. For by the mean-value theorem,

$$f(c+b) - f(c+a) = (b-a)f'(c_+)$$

and

$$f(c-a) - f(c-b) = (b-a)f'(c_-)$$

where c_+ is in the interval $(c + a, c + b)$ and c_- is in $(c - b, c - a)$. But the first of these intervals lies wholly to the right of the second, so that $f'(c_+) < f'(c_-)$. The inequality follows from this.

A similar inequality for a concave function, g, is

$$g(c-a) + g(c+a) < g(c-b) + g(c+b), \qquad \text{if} \quad a < b \tag{7.3.5}$$

With the convex function $f(c)$, we can associate a conjugate concave function $g(s)$ defined by

$$g(s) = \max_c [f(c) - sc] \tag{7.3.6}$$

This is shown in Fig. 7.4. It is evident that $g(s)$ is the greatest distance of the curve $f(c)$ above the straight line of slope s issuing from the origin. We also see that $g(s)$ is only defined between s^* and s_*, the greatest and the least values of $f'(c)$. For $s < s_*$, $g(s)$ is infinite, and for $s > s^*$ we can think of $g(s)$ being constant and equal to zero. For $s_* < s < s^*$, $g(s)$ is monotonically

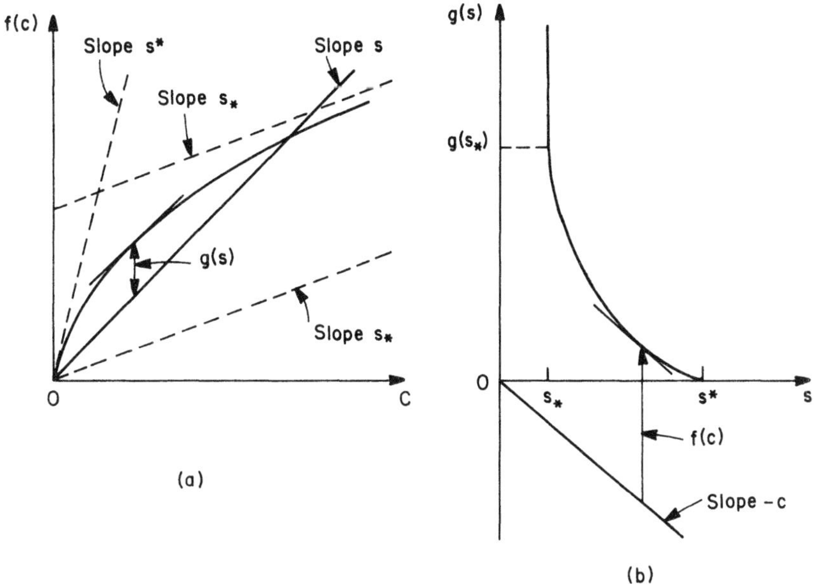

Figure 7.4

decreasing. Clearly, the curve $f(c)$ could be reconstructed from $g(s)$, since the line $\phi = sc + g(s)$ is tangent to the curve $\phi = f(c)$, and so $f(c)$ is the envelope of the family of such lines.

To find the value of c at which $f(c) - sc$ is greatest, we differentiate it with respect to c:

$$f'(c) - s = 0 \tag{7.3.7}$$

For clarity, let us denote $f'(c)$ by $\psi(c)$; that is,

$$\psi(c) = s \tag{7.3.8}$$

and let $\chi(s)$ be its inverse function,

$$\chi(s) = c \tag{7.3.9}$$

We may also write

$$\chi(\psi(c)) = c, \qquad \psi(\chi(s)) = s \tag{7.3.10}$$

Then Eq. (7.3.9) provides the solution of Eq. (7.3.6) as

$$g(s) = f(\chi(s)) - s\chi(s) \tag{7.3.11}$$

Now

$$g'(s) = \{f'(\chi(s)) - s\}\chi'(s) - \chi(s) = -\chi(s) \tag{7.3.12}$$

as a result of Eq. (7.3.7). Hence $-g'(s)$ is the inverse function of $f'(c)$. But if $f'(c) = \psi(c)$ is monotone decreasing, so is the inverse function $\chi(s)$. Since $g'(s) = -\chi(s)$, it is monotone increasing and hence $g(s)$ is a concave function.

Moreover, the definition of $g(s)$ can be expressed as

$$g(s) = f(c) - cs + r$$

where r is a positive quantity that is only zero if the correct value of c is chosen to give the maximum; this choice is $c = \chi(s)$. Rearranging this equation gives

$$f(c) = g(s) + cs - r$$

so $f(c)$ is always less than $g(s) + cs$, except when s is chosen to make this minimum, $s = \psi(c)$. Thus

$$f(c) = \min_s [g(s) + cs] = g(\psi(c)) + c\psi(c) \tag{7.3.13}$$

is the inverse relation. This is schematically shown in Fig. 7.4(b), in which the tangent to the curve $g(s)$ also has a slope of $-c$.

The functions $f(c)$ and $g(s)$ are said to be *conjugate*. The formulas (7.3.6) and (7.3.13) are also sometimes said to define the *maximum transform* and its inverse.

To illustrate this, we will consider the Langmuir isotherm for which we have

$$f(c) = c + \frac{\alpha c}{1 + Kc} \tag{7.3.14}$$

with $\alpha = \nu\gamma$ and thus

$$f'(c) = 1 + \frac{\alpha}{(1 + Kc)^2} = \psi(c) = s \tag{7.3.15}$$

by Eqs. (7.3.7) and (7.3.8). From this equation we find that $s_* = 1$ and $s^* = 1 + \alpha$. Now the inverse function can be expressed as

$$c = \frac{1}{K}\left(\sqrt{\frac{\alpha}{s - 1}} - 1\right) = \chi(s) = -g'(s) \tag{7.3.16}$$

by Eqs. (7.3.9) and (7.3.12). The function $g(s)$ is then determined by substituting Eq. (7.3.16) in Eq. (7.3.11) or by integrating Eq. (7.3.16) with the condition $g = 0$ at $s = s^* = 1 + \alpha$. This gives

$$g(s) = \frac{1}{K}(\sqrt{s - 1} - \sqrt{\alpha})^2 \tag{7.3.17}$$

If Eq. (7.3.15) is substituted in Eq. (7.3.13), we recover the function $f(c)$ as in Eq. (7.3.14).

A closely allied property of a convex function is the fact that its tangents always lie above it. Thus $f(b) + (c - b)f'(b)$, which corresponds to the point of abscissa c on the tangent to the curve at b, is above the point $f(c)$ on the curve, and the two only coincide when $b = c$ if $f(c)$ is strictly convex (see Fig. 7.3). Another way of putting this is to say that

$$f(c) = \min_b [f(b) + (c - b)f'(b)] \tag{7.3.18}$$

the minimum being attained when $b = c$.

Now this may be used to linearize certain processes by the method of quasi-linearization or maximum operation that Bellman and Kalaba have developed. For example, if $M(c) = 0$ is a nonlinear equation involving $f(c)$ as its only nonlinearity, we can replace $f(c)$ by $cf'(b) + [f(b) - bf'(b)]$, which is linear in c, and obtain a linear equation $L(c; b) = 0$. If L is a positive operator in the sense that $L(c; b) > 0$ implies $c > 0$, then we can recover the solution of $M(c) = 0$ by solving the linear equation $L(c; b) = 0$. This will give $c = C(b)$ as a function of b, but it can be shown that the original nonlinear equation is solved by

$$c = \max_b C(b)$$

To show this, we write

$$f(c) = f(b) - bf'(b) + cf'(b) - r$$

where $r > 0$. Then

$$M(c) = L(c; b) - r$$

But if we solve $L(c; b) = 0$ instead of $L(c; b) = r$, we will get a smaller value c than we would if r were not zero because L is a positive operator. Hence $c \geq C(b)$, and we can find c by finding the maximum of $C(b)$. For a concave function, the operations of maximizing and minimizing are reversed.

The transformation from the pair of variables c, $f(c)$ to the pair s, $g(s)$ is also known as the *Legendre transformation*. It is an involutory transformation being, as we have seen, its own inverse. It can be generalized to any number of variables. Let $f(c_1, c_2, \ldots, c_n)$ be a function whose Hessian determinant $|\partial^2 f/\partial c_i \partial c_j|$ does not vanish, and let

$$s_i = \frac{\partial f}{\partial c_i} \tag{7.3.19}$$

We define

$$g(s_1, \ldots, s_n) = \operatorname*{ext}_{[c_i]} \left[f(c_1, \ldots, c_n) - \sum_{j=1}^{n} s_j c_j \right] \tag{7.3.20}$$

by analogy to Eq. (7.3.6), using ext to denote the appropriate extremum (maximum for a convex and minimum for a concave function). The conditions under which g would attain its extremum are given by Eq. (7.3.9). Then we have to be able to solve the equations in Eq. (7.3.19) for the c_i as functions of s_i and substitute in Eq. (7.3.20). But the definiteness of the Hessian determinant is just the condition that allows this to be done. We observe by differentiating Eq. (7.3.20) that

$$\frac{\partial g}{\partial s_i} = -c_i + \sum_{j=1}^{n} \left(\frac{\partial f}{\partial c_j} - s_j \right) \frac{\partial c_j}{\partial s_i} = -c_i \tag{7.3.21}$$

Thus, by an analogy to Eqs. (7.3.12) and (7.3.13), we have

$$f(c_1, \ldots, c_n) = \operatorname*{ext}_{[s_i]} \left[g(s_1, \ldots, s_n) + \sum_{j=1}^{n} c_j s_j \right] \tag{7.3.22}$$

Exercises

7.3.1. From the inequality (7.3.1), show that $f''(c) \leq 0$.

7.3.2. For the Freundlich isotherm,

$$f(c) = c + \alpha c^{\beta}, \qquad 0 < \beta < 1$$

show that in $1 \leq s \leq \infty$,

$$-g'(s) = \left(\frac{s-1}{\alpha\beta}\right)^{1/(\beta-1)}$$

and

$$g(s) = (1-\beta)\alpha^{1/1-\beta}\left(\frac{s-1}{\beta}\right)^{\beta/1-\beta}$$

7.3.3. Discuss the form of the conjugate function $g(s)$ if $f(c)$ is a continuous convex function with a sharp corner.

7.3.4. Show that for conjugate functions

$$g''(s) = -\frac{1}{f''(c)}$$

7.3.5. Let $f_1(c)$ and $f_2(c)$ be two convex functions and $g_1(s)$ and $g_2(s)$ their conjugates. If

$$F(c) = \max_{0 \leq b \leq c} [f_1(c-b) + f_2(b)]$$

show that $G(s)$, the conjugate of $F(c)$, is given by

$$G(s) = g_1(s) + g_2(s)$$

7.3.6. Define

$$F(z) = \max_{x \geq 0} [e^{-xz} f(x)]$$

for any positive monotone-increasing function $f(x)$ on $x \geq 0$. What restriction must be placed on $f(x)$ if $F(z)$ is to exist for $z \geq 0$? How can $f(x)$ be expressed in terms of $F(z)$? Find $F(z)$ if

$$f(x) = \max_{0 \leq y \leq x} [f_1(x-y) + f_2(y)]$$

7.3.7. If the Hessian determinant of $f(c_1, \ldots, c_n)$ is positive definite and $g(s_1, \ldots, s_n)$ is the Legendre transform or conjugate function, show that, for any $c_1, \ldots, c_n$ and $s_1, \ldots, s_n$,

$$\sum_{j=1}^{n} c_j s_j \leq f(c_1, \ldots, c_n) - g(s_1, \ldots, s_n)$$

7.4 Weak Solutions and the Entropy Condition

In Chapters 5 and 6 we encountered solutions with discontinuities in both the value of the dependent variable and its derivatives, and we would like to define them more carefully. The former have previously been referred to as "discontinuous" solutions, but, following Lax, the term *weak*† will be used for all solutions that are not continuously differentiable. The definition follows the type of definition that must be given for generalized functions.

For example, although it is highly suspect to say that the Dirac delta function $\delta(t)$ is everywhere zero except at the origin where it is infinite, this function can be properly defined by saying that it is such that, for any test function $w(t)$ that is continuous at t,

$$\int_0^\infty \delta(t - \eta)w(\eta)\, d\eta - w(t) = 0 \tag{7.4.1}$$

The property of continuity that the delta function lacks is transferred to the class of test functions.

Similarly, we can set up a class of test functions and define a weak solution of a partial differential equation as one that makes certain integrals vanish. Let us consider the chromatographic problem

$$\frac{\partial c}{\partial x} + \frac{\partial}{\partial \tau} f(c) = 0 \tag{7.4.2}$$

with

$$c(0, \tau) = \Gamma(\tau), \qquad -\infty < \tau < \infty \tag{7.4.3}$$

Here it is to be understood that the initial data have been transformed in such a way that the data be specified over the whole of the τ-axis (see Sec. 7.1). If we have discontinuities in the solution, $c(x, \tau)$, the jump condition must be satisfied at every point of discontinuity, and so we have, from Eq. (7.2.11),

$$\frac{d\tau}{dx} = \frac{[f]}{[c]} = \tilde{\sigma}(c^l, c^r) \tag{7.4.4}$$

Now we set up the class of smooth test functions $w(x, \tau)$ that possesses continuous first partial derivatives and vanishes for $x + |\tau|$ large enough. (The latter condition is often referred to as having *compact support*.) Then $c(x, \tau)$ is called a weak solution of Eqs. (7.4.2) and (7.4.3) if c and $f(c)$ are integrable over every bounded set of the half-plane $x \geq 0$, and for all such test functions $w(x, \tau)$,

$$\int_0^\infty dx \int_{-\infty}^\infty d\tau \left\{\frac{\partial w}{\partial x} c + \frac{\partial w}{\partial \tau} f(c)\right\} + \int_{-\infty}^\infty w(0, \tau)\Gamma(\tau)\, d\tau = 0 \tag{7.4.5}$$

To see that this ties back into our understanding of the solution of the partial differential equations, we will show, first, that wherever a weak solution is smooth it satisfies the differential equation (7.4.2) and, second, that a discontinuity in a weak solution satisfies the jump condition, Eq. (7.4.4).

If $c(x, \tau)$ is smooth (i.e., it has continuous first derivatives) in any region of the (x, τ)-plane and satisfies Eq. (7.4.5), then we may integrate by parts

†Oleinik used the expression "generalized solutions" in her series of celebrated articles.

and obtain

$$\left[\int_{-\infty}^{\infty} wc\,d\tau\right]_{x=0}^{x=\infty} + \left[\int_{0}^{\infty} wf(c)\,dx\right]_{\tau=-\infty}^{\tau=+\infty}$$
$$- \int_0^\infty dx \int_{-\infty}^{\infty} d\tau \left\{ w\frac{\partial c}{\partial x} + w\frac{\partial f}{\partial \tau} \right\} + \int_{-\infty}^{\infty} w(0, \tau)\Gamma(\tau)\,d\tau = 0$$

Now the integrated parts for $x = \infty$ and $\tau = \pm\infty$ vanish by the vanishing of w and that for $x = 0$ cancels with the boundary integral (i.e., the last term). Hence

$$\int_0^\infty dx \int_{-\infty}^{\infty} w\left\{\frac{\partial c}{\partial x} + \frac{\partial}{\partial \tau}f(c)\right\} d\tau = 0 \tag{7.4.6}$$

Since this is true for any smooth function $w(x, \tau)$, the factor inside the bracket of the integrand must vanish, so $c(x, \tau)$ satisfies Eq. (7.4.2). If the solution $c(x, \tau)$ is continuous and piecewise smooth, it satisfies the differential equation in each subregion where it is smooth; moreover, any "edge" must be a characteristic, for c is constant along the characteristic, and a discontinuity in derivative can only occur in a direction normal to it.

If, however, $c(x, \tau)$ is discontinuous, then we must show that Eq. (7.4.4) is satisfied. We can suppose without loss of generality that there is a single curve of discontinuity, S, that crosses the half-plane $x \geq 0$. If there is more than one, we have to repeat the argument; if it does not cross the whole region, we have only to put the jumps in c and f equal to zero. Let S_+ and S_- denote the upper and lower sides of S, D_+, and D_-, the domains above and below S, and let $(B_\pm + I_\pm + S_\pm)$ be the boundaries of D_+ and D_-, as shown in Fig. 7.5. By Green's theorem, the double integral

$$\iint \left\{\frac{\partial}{\partial x}(wc) + \frac{\partial}{\partial \tau}(wf)\right\} dx\,d\tau = \oint w(c\,d\tau - f\,dx) \tag{7.4.7}$$

where the single integral is taken around the boundary of the region. Applying this to the domain $D_+ + D_-$, we have

$$\iint_{D_+ + D_-} w\left(\frac{\partial c}{\partial x} + \frac{\partial f}{\partial \tau}\right) dx\,d\tau + \iint_{D_+ + D_-} \left(\frac{\partial w}{\partial x}c + \frac{\partial w}{\partial \tau}f\right) dx\,d\tau =$$
$$\int_{S_+ + S_-} w(\tilde{\sigma}c - f)\,dx + \int_{I_+ + I_-} w(c\,d\tau - f\,dx) + \int_{B_+ + B_-} w(c\,d\tau - f\,dx)$$

in which we put $\tilde{\sigma} = d\tau/dx$ along the curve S. Now let us make the domain $D_+ + D_-$ large enough to occupy the whole of the half-plane $x \geq 0$. On the left side, the first integral vanishes because Eq. (7.4.2) holds in the two separate domains D_+ and D_-. The second double integral is equal to $-\int_{-\infty}^{\infty} w(0, \tau)\Gamma(\tau)\,d\tau$, since c is a weak solution and satisfies Eq. (7.4.5). But the integral over $I_+ + I_-$ is also equal to $-\int_{-\infty}^{\infty} w(0, \tau)\Gamma(\tau)\,d\tau$, for $dx = 0$

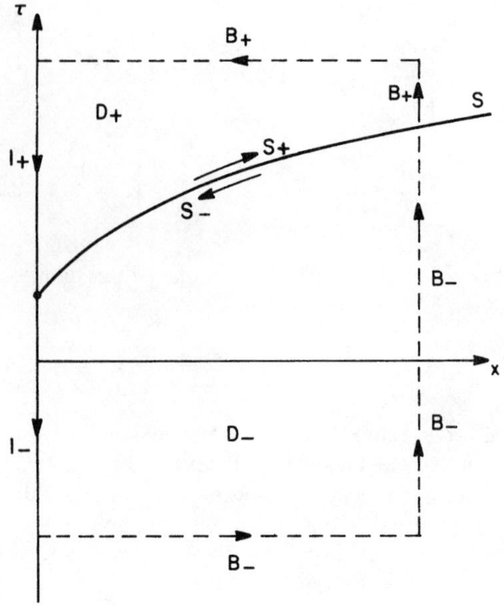

Figure 7.5

on this path, which is traversed in the negative direction, and $c(0, \tau) = \Gamma(\tau)$. The integral over $B_+ + B_-$ vanishes by the vanishing of w, and we are left with

$$\int_{S_+} w(\tilde{\sigma} c_+ - f_+)\, dx + \int_{S_-} w(\tilde{\sigma} c_- - f_-)\, dx = 0$$

But since S_- is traversed in the opposite direction to S_+, this gives

$$\int_S w(\tilde{\sigma}[c] - [f])dx = 0$$

where $[c] = c_+ - c_-$ and $[f] = f_+ - f_-$. Arguing again from the arbitrariness of w, we see that

$$\tilde{\sigma}[c] = [f]$$

on any line of discontinuity.

Some important features of weak solutions of quasi-linear equations are worth emphasizing. First, a weak solution is not uniquely determined by the initial data. This is demonstrated by the following simple example. (Also, see Exercises 7.4.1 and 7.4.2.)

Consider the conservation equation

$$\frac{\partial u}{\partial x} + \frac{\partial}{\partial \tau}\left(\frac{1}{2}u^2\right) = 0 \tag{7.4.8}$$

with the initial data

$$u(0, \tau) = \Gamma(\tau) = \begin{cases} 1, & \tau > 0 \\ 0, & \tau < 0 \end{cases} \tag{7.4.9}$$

The jump condition is

$$\frac{d\tau}{dx} = \frac{1}{2}(u^l + u^r) \tag{7.4.10}$$

Then the function

$$u_1(x, \tau) = \begin{cases} 1, & x < \tau \\ \dfrac{\tau}{x}, & 0 < \tau < x \\ 0, & \tau \leq 0 \end{cases} \tag{7.4.11}$$

which is continuous and piecewise differentiable, satisfies the differential equation (7.4.8), as well as the initial condition (7.4.9). Note, however, that the discontinuous function

$$u_2(x, \tau) = \begin{cases} 1, & \tau > \dfrac{x}{2} \\ 0, & \tau < \dfrac{x}{2} \end{cases} \tag{7.4.12}$$

is also a solution since it satisfies the jump condition (7.4.10) and takes on the same prescribed initial data. The layouts of the characteristics are shown for these solutions in Fig. 7.6. Since u_1 is continuous, it is the preferred

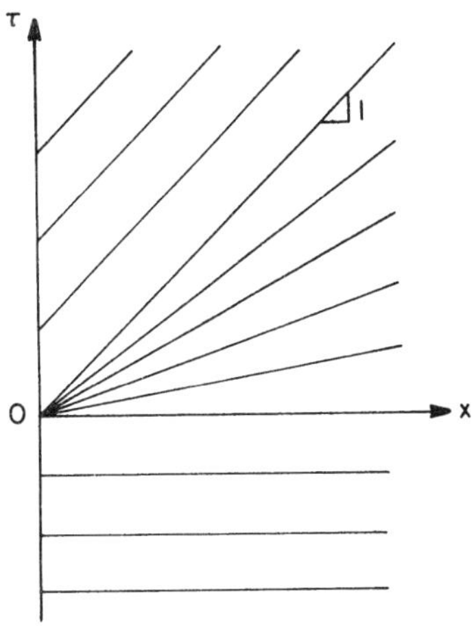

(a) u_1, centered simple wave

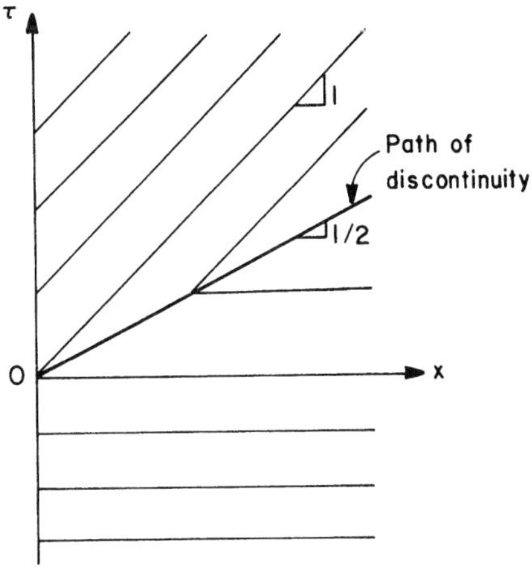

(b) u_2, inadmissible solution

Figure 7.6

solution, whereas u_2 is not desired. The question then is how to eliminate in a systematic way such an undesirable discontinuous solution.

A clue to the answer comes from those situations in which the characteristics intersect each other when continued in the direction of positive x (see Sec. 5.3). Indeed, among all solutions we are only interested in the one that prevents the crossing of characteristics and so gives the one-valuedness that is necessary on physical grounds; this is called the *admissible* (or *acceptable*) *solution*. Another way of putting this criterion is as follows: In an admissible solution, the characteristics starting on either side of the path of a discontinuity when continued in the positive x-direction intersect the path of the discontinuity. Clearly, this is not the case with u_2, as shown in Fig. 7.6(b), and so it is inadmissible and rejected. To see an example of an admissible solution, let us again consider Eq. (7.4.8) with the initial condition

$$u(0, \tau) = \Gamma(\tau) = \begin{cases} 0, & \tau > 0 \\ 1, & \tau < 0 \end{cases} \tag{7.4.13}$$

By comparing this with Eq. (7.4.9), we immediately find a discontinuous solution in the form

$$u(x, \tau) = \begin{cases} 0, & \tau > \dfrac{x}{2} \\ 1, & \tau < \dfrac{x}{2} \end{cases} \tag{7.4.14}$$

As shown in Fig. 7.7, this solution satisfies the preceding criterion and thus it is admissible.

It is now evident that, for every discontinuity in an admissible solution, we must have

$$\sigma(c^l) < \tilde{\sigma}(c^l, c^r) < \sigma(c^r) \tag{7.4.15}$$

But this implies that

$$f'(c^l) < \frac{[f]}{[c]} < f'(c^r) \tag{7.4.16}$$

and, from Fig. 7.2, we observe that $c^l > c^r$ for a convex isotherm $f(c)$, that is, for $f''(c) < 0$. For a strictly concave function, the opposite ($c^l < c^r$) would be true.

A discontinuity, if it satisfies the inequality condition in Eq. (7.4.15), is known as a *shock*, and it is in this sense that Eq. (7.4.15) is referred to as the *shock condition* (see Sec. 5.4). In fluid mechanics, the corresponding condition is easily interpreted to mean that the entropy of the fluid must increase as it passes through the shock, and thus it is often called the *entropy condition*. The strict convexity or concavity of $f(c)$ ensures that the conservation equation (7.4.2) is genuinely nonlinear, for at no value of c is $f'(c)$

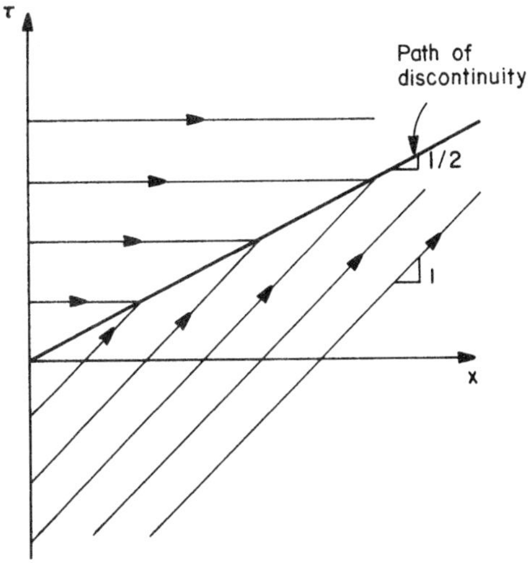

Figure 7.7

locally constant [i.e., $f''(c) = 0$]. It then follows that every discontinuity in the solution of a genuinely nonlinear conservation equation is a shock.

As we have already noticed in Sec. 5.4, the condition (7.4.15) is not sufficient to guarantee the unique determination of solutions by the initial data if the sign of $f'(c)$ is not uniform. A replacement for this condition has been found by Oleinik in the following form: At every point of discontinuity in the solution to a conservation equation, we must have

$$\tilde{\sigma}(c^l, c^r) \geq \tilde{\sigma}(c^l, c) \tag{7.4.17}$$

for every c between c^l and c^r. We now recall from Sec. 5.4 that this condition can be rephrased as in the following rule:

> RULE: If two states c^l and c^r are to be connected by a discontinuity, identify the chord connecting the two points L and R on the curve $f(c)$ for which $c = c^l$ and $c = c^r$, respectively. Then, for the conservation equation (7.4.2), we must see the curve $f(c)$ on the right side as we move along the chord from L to R.

Lax put this rule in another way and suggested inequality conditions:

(i) If $c^l > c^r$, then the graph of $f(c)$ over (c^l, c^r) lies above the chord [see Fig. 7.8(a)]:

$$f(\alpha c^l + (1 - \alpha)c^r) \geq \alpha f(c^l) + (1 - \alpha)f(c^r) \qquad \text{for} \qquad 0 \leq \alpha \leq 1 \tag{7.4.18}$$

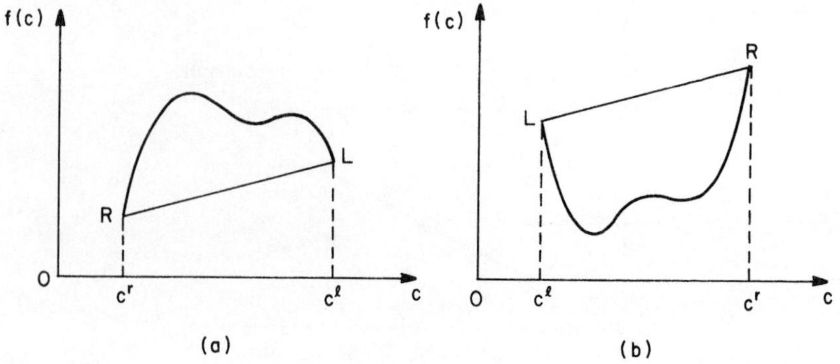

Figure 7.8

(ii) If $c^l < c^r$, the graph of $f(c)$ over (c^l, c^r) lies below the chord [see Fig. 7.8(b)]:

$$f(\alpha c^l + (1-\alpha)c^r) \leq \alpha f(c^l) + (1-\alpha) f(c^r) \qquad \text{for} \qquad 0 \leq \alpha \leq 1 \tag{7.4.19}$$

However, we observe that these three expressions are really equivalent to one another: any of these may be called the *generalized entropy condition*.

Two more features are to be mentioned before concluding this section. First, in contrast to the situation with linear equations, where discontinuities can only arise from discontinuities in the initial data, the smoothest initial data (analytic) may lead to the development of shocks, and these are not propagated in a characteristic direction (see Sec. 6.1). Second, for a linear equation, discontinuous solutions may be obtained as the limit of continuous ones and may be so defined. This is not the case for quasi-linear conservation laws, although their weak solutions may sometimes be obtained as the degenerated case of higher-order equations when a viscositylike parameter tends to zero. This will be discussed in Sec. 7.6.

Exercises

7.4.1. Show that if

$$\frac{\partial u}{\partial x} - \frac{\partial}{\partial \tau}\left(\frac{1}{3}u^3\right) = 0$$

with

$$u(0, \tau) = \begin{cases} 0, & \tau > 0 \\ 1, & \tau < 0 \end{cases}$$

then the solutions

$$u_1(x, \tau) = \begin{cases} 0, & \tau > \dfrac{-x}{3} \\ 1, & \tau < \dfrac{-x}{3} \end{cases}$$

and

$$u_2(x, \tau) = \begin{cases} 0, & \tau \geq 0 \\ \left(-\dfrac{\tau}{x}\right)^{1/2}, & 0 < -\tau < x \\ 1, & \tau < -x \end{cases}$$

both satisfy the equation. Draw the layouts of the characteristics and argue that the discontinuous solution u_1 is not admissible

7.4.2. Show that

$$\frac{\partial u}{\partial t} + u\frac{\partial u}{\partial x} = 0$$

$$u(x, 0) = -\operatorname{sgn} x$$

can have any number of solutions with discontinuities on the lines $t = \alpha x$, $x = 0$, and $t = -\alpha x$. If the further condition that $[u(x_1, t) - u(x_2, t)]/(x_1 - x_2) \leq K$ for any pair of x_1 and x_2 is imposed, show that only one of the possible solutions survives.

7.4.3. Obtain a suitable definition of a weak solution for the problem

$$\frac{\partial u}{\partial t} + \frac{\partial}{\partial x}f(u) = 0$$

subject to $u(x, 0) = U(x)$.

7.4.4. Frame a suitable definition for a weak solution of

$$\frac{\partial c}{\partial x} + \frac{\partial}{\partial \tau}f(c) = 0$$

subject to the conditions

$$c(x, 0) = F(x)$$

$$c(0, \tau) = G(\tau)$$

7.4.5. Obtain the jump condition (7.4.4) by considering the integral (7.4.7) over a very narrow rectangle sitting over a portion of the path S of the discontinuity.

7.4.6. If

$$\frac{\partial}{\partial x}f(x, t, u) + \frac{\partial}{\partial t}g(x, t, u) = h(x, t, u)$$

show that a discontinuity between u_1 and u_2 moves with speed

$$\frac{dx}{dt} = \frac{f(x, t, u_1) - f(x, t, u_2)}{g(x, t, u_1) - g(x, t, u_2)}$$

7.5 Lax's Solution for the Quasi-linear Conservation Law

Lax has given a very elegant and interesting formula for the admissible solution of a conservation equation. We will motivate it by considering Bellman and Kalaba's notion of maximum operation, but go on to give an outline of Lax's proof of it.

Consider the problem

$$\frac{\partial c}{\partial x} + f'(c)\frac{\partial c}{\partial \tau} = 0$$
$$c(0, \tau) = \Gamma(\tau), \qquad -\infty < \tau < \infty \tag{7.5.1}$$

where $f''(c) < 0$, so that the equation is genuinely nonlinear. Let

$$\phi(x, \tau) = \int_{-\infty}^{\tau} c(x, \tau')\, d\tau' \tag{7.5.2}$$

and

$$\phi(0, \tau) = \int_{-\infty}^{\tau} \Gamma(\tau')\, d\tau' = \Phi(\tau) \tag{7.5.3}$$

Since

$$c(x, \tau) = \frac{\partial \phi}{\partial \tau}$$

we have, from Eq. (7.5.1),

$$\frac{\partial^2 \phi}{\partial x\, \partial \tau} + f'\left(\frac{\partial \phi}{\partial \tau}\right)\frac{\partial^2 \phi}{\partial \tau^2} = 0 \tag{7.5.4}$$

Integrating this once with respect to τ,

$$\frac{\partial \phi}{\partial x} + f\left(\frac{\partial \phi}{\partial \tau}\right) = 0 \tag{7.5.5}$$

Now the function f is convex, so from Eq. (7.3.18) with $c = \partial\phi/\partial\tau$ and $b = u$ we have

$$f\left(\frac{\partial \phi}{\partial \tau}\right) = \min_u \left[f(u) + \left(\frac{\partial \phi}{\partial \tau} - u\right) f'(u) \right]$$

$$= f'(u)\frac{\partial \phi}{\partial \tau} + f(u) - uf'(u) - r \tag{7.5.6}$$

where $r \geq 0$ and indeed is zero only when $u = \partial\phi/\partial\tau$. In this expression, u is an arbitrary function of x and τ and, although r depends on $\partial\phi/\partial\tau$ and u in an unknown way, we know that it cannot be negative, and this is all that is needed. For any given $u(x, \tau)$, consider the linear equation

$$\frac{\partial v}{\partial x} + f'(u)\frac{\partial v}{\partial \tau} = uf'(u) - f(u) \tag{7.5.7}$$

with

$$v(0, \tau) = \Phi(\tau)$$

Since this is a strictly linear equation and $f'(u(x, \tau))$ [> 0] is uniquely defined for any given piecewise continuous u, the characteristics of Eq. (7.5.7) are given by

$$\frac{d\tau}{dx} = f'(u(x, \tau)) \tag{7.5.8}$$

and the value of v by

$$\begin{aligned} \frac{dv}{dx} &= uf'(u) - f(u) \\ v &= \Phi(\tau) \qquad \text{at } x = 0 \end{aligned} \tag{7.5.9}$$

But the last is a positive operator in the sense defined previously (see Sec. 7.3). Substituting for $f(\partial\phi/\partial\tau)$ from Eq. (7.5.6) into Eq. (7.5.5) shows that ϕ satisfies

$$\frac{\partial \phi}{\partial x} + f'(u)\frac{\partial \phi}{\partial \tau} = uf'(u) - f(u) + r \tag{7.5.10}$$

and, by comparison with Eq. (7.5.7), we have

$$\phi(x, \tau) > v(x, \tau)$$

The difference will only be zero if u is so chosen that $r = 0$; hence we may write

$$\phi(x, \tau) = \max_u[v(x, \tau)] \tag{7.5.11}$$

where $v(x, \tau)$ is the solution of the linear equation, Eq. (7.5.7).

Now let us take Lax's formula, illustrate it by a simple example, and link it up with this motivation. After this we will prove the formula in the way Lax does.

We recall from Sec. 7.3 that with the convex function $f(c)$ we can associate a conjugate concave function $g(s)$. With Φ defined by Eq. (7.5.3), consider the function

$$F(\eta) = \Phi(\eta) - xg\left(\frac{\tau - \eta}{x}\right) \tag{7.5.12}$$

Note that here the argument of g is replaced by $(\tau - \eta)/x$ and that g is defined between the lower limit s_* and the upper limit s^*. For fixed x and τ, the function F certainly approaches $-\infty$ as $(\tau - \eta)/x$ approaches s_*. If $(\tau - \eta)/x$ is greater than s^* (i.e., $\eta < \tau - s^*x$), then g becomes zero and, as $\eta \to -\infty$, $\Phi(\eta)$ approaches some finite value, perhaps zero. Thus $F(\eta)$ will have some maximum value for $\eta = \eta_0(x, \tau)$, say. The value of η_0 will satisfy

$$F'(\eta) = \Phi'(\eta) + g'\left(\frac{\tau - \eta}{x}\right) = 0$$

and so we have

$$-g'\left(\frac{\tau - \eta_0}{x}\right) = \Gamma(\eta_0) \tag{7.5.13}$$

But, from Eqs. (7.3.9) and (7.3.12), $-g'(s)$ is identical to the function $\chi(s)$, which is the inverse of $f'(c)$. It then follows that the solution of Eq. (7.5.1) is given by

$$c(x, \tau) = \chi\left(\frac{\tau - \eta_0}{x}\right) = -g'\left(\frac{\tau - \eta_0}{x}\right) = \Gamma(\eta_0) \tag{7.5.14}$$

and this is exactly what Lax's theorem asserts. This equation may be rewritten to give

$$\frac{\tau - \eta_0}{x} = f'(\Gamma(\eta_0)) \tag{7.5.15}$$

Equations (7.5.14) and (7.5.15) express the fact that c remains constant along straight lines and that the slope of the line emanating from the point $(0, \eta_0)$ is $f'(\Gamma(\eta_0))$. Clearly, this is the structure of the solution that we have already observed by the method of characteristics.

To illustrate this, we will take

$$\Gamma(\tau) = \begin{cases} 0, & \tau < 0 \\ c_0, & \tau > 0 \end{cases} \tag{7.5.16}$$

and

$$f(c) = c + \frac{\alpha c}{1 + Kc} \tag{7.5.17}$$

for which we know the solution to be

$$c(x, \tau) = \begin{cases} 0, & \tau < x\left(1 + \dfrac{\alpha}{1 + Kc_0}\right) \\[2ex] c_0, & \tau > x\left(1 + \dfrac{\alpha}{1 + Kc_0}\right) \end{cases} \tag{7.5.18}$$

Now

$$\Phi(\tau) = \begin{cases} 0, & \tau < 0 \\ c_0\tau, & \tau > 0 \end{cases} \tag{7.5.19}$$

and, from Eq. (7.3.17), we have

$$g(s) = \frac{1}{K}(\sqrt{s-1} - \sqrt{\alpha})^2, \qquad 1 \le s \le 1 + \alpha \tag{7.5.20}$$

Let us write $\eta = \tau - sx$ so that

$$F(s) = \Phi(\tau - sx) - xg(s) = c_0(\tau - sx) - \frac{x}{K}(\sqrt{s-1} - \sqrt{\alpha})^2 \tag{7.5.21}$$

and $\eta_0 = \tau - s_0 x$, where s_0 maximizes the function F. Three possibilities for the configuration arise, as shown in the three parts of Fig. 7.9. These figures represent the situations for three different values of τ with x fixed. In each of them the term $\Phi(\tau - sx)$ gives the ramp of slope $-c_0$ coming into a corner at $s = \tau/x$, to the right of which Φ remains fixed at zero. The curve $xg(s)$ has been drawn, and the function F comes up from below the s-axis. In the lower part of the figure, the derivative curves are shown, and the crossing of $-g'(s)$ and $\Gamma(\tau - sx)$ corresponds to Eq. (7.5.13).

In the configuration of Fig. 7.9(a), the ramp Φ lies wholly underneath the curve $xg(s)$; thus, although F may have a local maximum to the left of s^*, this maximum is negative, so the maximum value of $F(s)$ is zero and occurs at $s_0 = s^*$. Here $g'(s)$ is zero, so $c(x, \tau) = -g'(s_0) = 0$. In the configuration of Fig. 7.9(c), however, the ramp Φ crosses the curve $xg(s)$, so F has a unique positive maximum at P, $s = s_0(x, \tau)$. Sincc Φ and xg must be parallel at such a point, $g'(s) = -c_0$ and thus $c(x, \tau) = -g'(s_0) = c_0$. In the lower part of these figures, we see that, although the derivative curves may intersect, the triangular areas A and B are not equal, corresponding to the fact that there is a unique maximum value of F.

In the configuration of Fig. 7.9(b), however, the ramp just touches the curve, and F has two maxima of the same value, one at the point Q, where both slopes are $-c_0$ and one at $s = s^*$, where both slopes are zero. The values of c associated with these two values of $s_0(x, \tau)$ are thus c_0 and zero, and these are the values on either side of the discontinuity. The condition for the discontinuity is therefore that the ramp Φ should just be tangent to the curve $xg(s)$ or, equivalently, that the tangent to the curve of slope $-c_0$ should meet the s-axis at τ/x. This gives two conditions:

$$g'(s) = -c_0 = \frac{g(s)}{s - \tau/x} \tag{7.5.22}$$

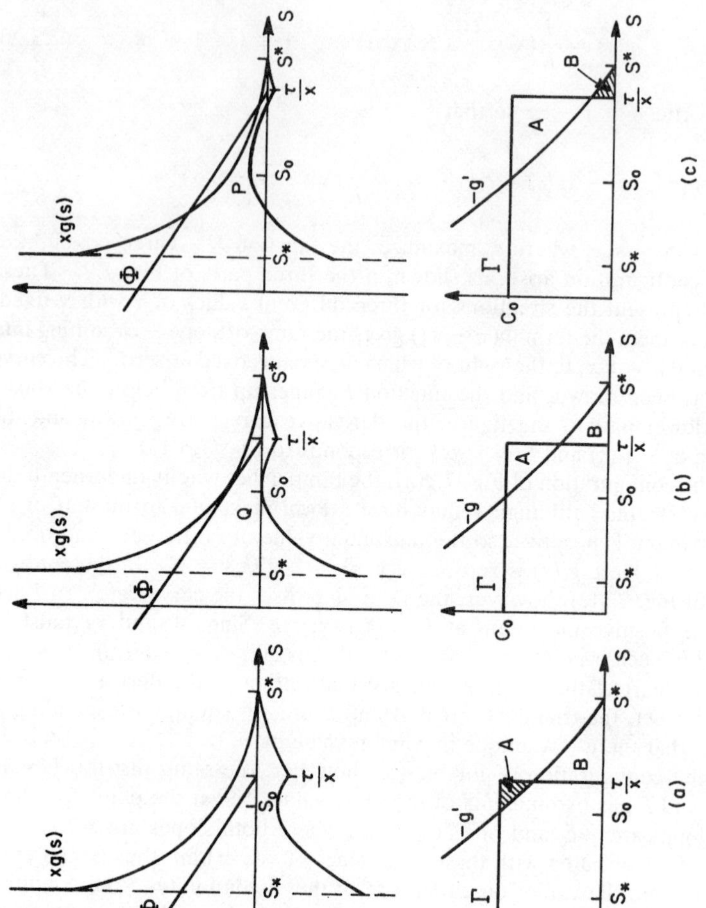

Figure 7.9

But, from Eq. (7.3.16),

$$g'(s) = \frac{1}{K}\left(\sqrt{\frac{\alpha}{s - 1}} - 1\right) \tag{7.5.23}$$

giving†

$$s = 1 + \frac{\alpha}{(1 + Kc_0)^2}$$

and, by using Eq. (7.5.20) for $g(s)$,

$$\frac{\tau}{x} = 1 + \frac{g(s)}{c_0} = 1 + \frac{\alpha}{1 + Kc_0} \tag{7.5.24}$$

This is the path of the discontinuity. If $\tau/x < 1 + \alpha/(1 + Kc_0)$, we have the situation of Fig. 7.9(a), so $c = 0$ and, with the inequality sign reversed, Fig. 7.9(c) will be the case to give $c = c_0$. This is the solution given by Eq. (7.5.18), with which we are familiar.

It is interesting to notice that we have two maximization problems associated with the solution $c(x, \tau)$: one for the function $F(\eta)$ and another for $v(x, \tau)$ of Eq. (7.5.11). It ought therefore to be possible to see that these two problems are related in some way. For this, we will consider Eq. (7.5.7) for any given $u(x, \tau)$ with the initial data $v(0, \tau) = \Phi(\tau)$. The characteristic issuing from the point $(0, \eta)$ is determined by integrating Eq. (7.5.8) in the form

$$\frac{\tau - \eta}{x} = f'(u) \tag{7.5.25}$$

Taking x as the characteristic parameter, we then integrate Eq. (7.5.9) along the characteristic line from the point $(0, \eta)$ to an arbitrary point (x, τ) to obtain

$$v(x, \tau) - v(0, \eta) = x\{uf'(u) - f(u)\} \tag{7.5.26}$$

On the other hand, we recall that $\psi(c) = f'(c)$ in Eq. (7.3.13) and write it for $c = u$:

$$f(u) = g(f'(u)) + uf'(u)$$

or

$$uf'(u) - f(u) = -g(f'(u)) = -g\left(\frac{\tau - \eta}{x}\right) \tag{7.5.27}$$

Substituting this into Eq. (7.5.26), we get

$$v(x, \tau) = \Phi(\eta) - xg\left(\frac{\tau - \eta}{x}\right) = F(\eta) \tag{7.5.28}$$

†Note that s is equivalent to the characteristic direction σ. This is expected from Eq. (7.3.7).

which implies that the function $F(\eta)$ is identical to $v(x, \tau)$. The coordinates x and τ are fixed, and u may be manipulated in any way that we please to maximize v. We have now shown that the maximization of $v(x, \tau)$ with respect to u is equivalent to the maximization of $F(\eta)$ with respect to η. Since the maximum of $v(x, \tau)$ is given by $\phi(x, \tau)$, we can write

$$F(\eta_0(x, \tau)) = \phi(x, \tau) = \int_{-\infty}^{\tau} c(x, \tau')\, d\tau' \tag{7.5.29}$$

Let us now prove the Lax formula (7.5.14) by Lax's method. To do this we need the following lemma.

LEMMA: Let η_1 and η_2 be the values of η at which $F(\eta)$ is greatest for (x, τ_1) and (x, τ_2), respectively. Then if $\tau_1 < \tau_2$, $\eta_1 \geq \eta_2$.

Proof: Since $F_1(\eta) = \Phi(\eta) - xg(\tau_1 - \eta)/x$ is greatest when $\eta = \eta_1$, $F_1(\eta_1) > F_1(\eta_2)$. Thus

$$\Phi(\eta_1) - xg\left(\frac{\tau_1 - \eta_1}{x}\right) > \Phi(\eta_2) - xg\left(\frac{\tau_1 - \eta_2}{x}\right)$$

and similarly

$$\Phi(\eta_2) - xg\left(\frac{\tau_2 - \eta_2}{x}\right) > \Phi(\eta_1) - xg\left(\frac{\tau_2 - \eta_1}{x}\right)$$

By addition and division by $-x$,

$$g\left(\frac{\tau_1 - \eta_1}{x}\right) + g\left(\frac{\tau_2 - \eta_2}{x}\right) < g\left(\frac{\tau_1 - \eta_2}{x}\right) + g\left(\frac{\tau_2 - \eta_1}{x}\right)$$

But by putting

$$\tau_1 + \tau_2 - \eta_1 - \eta_2 = 2cx$$

$$\tau_1 - \tau_2 - \eta_1 + \eta_2 = 2ax$$

$$\tau_1 - \tau_2 + \eta_1 - \eta_2 = 2bx$$

we have

$$g(c + a) + g(c - a) < g(c + b) + g(c - b)$$

and by Eq. (7.3.5) this implies that $a < b$ or $\eta_1 > \eta_2$.

This shows that $\eta_0(x, \tau)$ is a monotone-nonincreasing function of τ for any x. Suppose that η_0^* and η_{0*} are the largest and smallest values of η_0 for which F is maximum; that is, $F(\eta_0^*(x, \tau)) = F(\eta_{0*}(x, \tau))$, $\eta_{0*}(x, \tau) < \eta_0(x, \tau) < \eta_0^*(x, \tau)$. Then $\eta_{0*}(x, \tau) < \eta_0^*(x, \tau)$ by definition, and $\eta_{0*}(x, \tau_1) > \eta_0^*(x, \tau_2)$, if $\tau_1 < \tau_2$, by the lemma. Thus η_{0*} and η_0^* can differ at points of discontinuity, and since both are monotone nonincreasing, this can only happen at a denumerable number of points.

To show that†

$$c(x,\tau) = \Gamma(\eta_0(x,\tau)) = \chi\left(\frac{\tau - \eta_0(x,\tau)}{x}\right) = -g'\left(\frac{\tau - \eta_0(x,\tau)}{x}\right) \tag{7.5.30}$$

is a weak solution of the conservation equation (7.5.1), we should do three things. First, we show that when Γ is smooth $c(x, \tau)$ coincides with the smooth solution just as long as it exists. Second, we will show that $c(x, \tau)$ and $f(x, \tau)$ are the limits of sequences of functions $c_N(x, \tau)$ and $f_N(x, \tau)$ that satisfy the same conservation equation, and so $c(x, \tau)$ is a weak solution. Third, we wish to show continuous dependence on the initial data.

1. If $\Gamma(\tau)$ is a smooth function, then there is a neighborhood of the initial line in which the solution is smooth. This is covered by the nonintersecting family of characteristics, for which we can write the equation

$$\tau = \eta + f'(c)x \tag{7.5.31}$$

where $c = \Gamma(\eta)$. Since χ is the inverse function of f', this can be written

$$c(x, \tau) = \Gamma(\eta) = \chi\left(\frac{\tau - \eta}{x}\right) = -g'\left(\frac{\tau - \eta}{x}\right) \tag{7.5.32}$$

But when $\Gamma(\eta)$ is differentiable, this is precisely the condition for it to be maximum [cf. Eq. (7.5.13)]. Hence smooth solutions are certainly given by Eq. (7.5.30).

2. Consider the expression

$$v_N(\eta) = \frac{\exp NF(\eta)}{\int_{-\infty}^{\infty} \exp NF(\eta)\, d\eta} \tag{7.5.33}$$

where $F(\eta)$ is given by Eq. (7.5.12); that is,

$$F(\eta) = \Phi(\eta) - xg\left(\frac{\tau - \eta}{x}\right) \tag{7.5.34}$$

so that x and τ only appear as parameters in $v_N(\eta)$. As $N \to \infty$, this function behaves like a delta function around $\eta = \eta_0(x, \tau)$, for $\int_{-\infty}^{\infty} v_N(\eta)\, d\eta = 1$, and at any other point than η_0, $v_N \to 0$. Thus, by defining c_N and f_N as

$$c_N(x, \tau) = \int_{-\infty}^{\infty} -g'\left(\frac{\tau - \eta}{x}\right) v_N(\eta)\, d\eta \tag{7.5.35}$$

$$f_N(x, \tau) = \int_{-\infty}^{\infty} f\left(-g'\left(\frac{\tau - \eta}{x}\right)\right) v_N(\eta)\, d\eta \tag{7.5.36}$$

†Note that χ is the inverse function of f' and identical to $-g'$ [cf. Eq. (7.3.12)].

we can write

$$c(x, \tau) = \lim_{N\to\infty} c_N(x, \tau) \tag{7.5.37}$$

$$f(c(x, \tau)) = \lim_{N\to\infty} f_N(x, \tau) \tag{7.5.38}$$

If we let

$$V_N = \int_{-\infty}^{\infty} \exp NF(\eta)\, d\eta$$

we have

$$\frac{\partial V_N}{\partial \tau} = N \int_{-\infty}^{\infty} \frac{\partial F}{\partial \tau} \exp NF(\eta)\, d\eta$$

$$= N \int_{-\infty}^{\infty} -g'\left(\frac{\tau - \eta}{x}\right) \exp NF(\eta)\, d\eta = NV_N c_N$$

or

$$c_N(x, \tau) = \frac{\partial}{\partial \tau}\left(\frac{1}{N} \ln V_N\right)$$

Similarly,

$$\frac{\partial V_N}{\partial x} = N \int_{-\infty}^{\infty} \frac{\partial F}{\partial x} \exp NF(\eta)\, d\eta$$

$$= -N \int_{-\infty}^{\infty} \left\{ g\left(\frac{\tau - \eta}{x}\right) - \frac{\tau - \eta}{x} g'\left(\frac{\tau - \eta}{x}\right) \right\} \exp NF(\eta)\, d\eta$$

Recalling Eq. (7.3.13) with $s = \psi(c)$ and so $c = \chi(s) = -g'(s)$, we can write

$$f(-g'(s)) = g(s) - sg'(s)$$

and thus

$$\frac{\partial V_N}{\partial x} = -NV_N \int_{-\infty}^{\infty} f\left(-g'\left(\frac{\tau - \eta}{x}\right)\right) v_N(\eta)\, d\eta = NV_N f_N$$

or

$$f_N(x, \tau) = -\frac{1}{NV_N} \frac{\partial V_N}{\partial x} = -\frac{\partial}{\partial x}\left(\frac{1}{N} \ln V_N\right)$$

It then follows that

$$\frac{\partial c_N}{\partial x} + \frac{\partial f_N}{\partial \tau} = \frac{\partial^2}{\partial x\, \partial \tau}\left(\frac{1}{N} \ln V_N\right) - \frac{\partial^2}{\partial \tau\, \partial x}\left(\frac{1}{N} \ln V_N\right) = 0$$

and so

$$\int_0^{\infty} dx \int_{-\infty}^{\infty} d\tau \left(\frac{\partial w}{\partial x} c_N + \frac{\partial w}{\partial \tau} f_N \right) + \int_{-\infty}^{\infty} w(0, \tau)\Gamma(\tau)\, d\tau = 0$$

for all smooth functions $w(x, \tau)$ that vanish for large $x + |\tau|$. Consequently, c and $f(c)$ defined by Eqs. (7.5.37) and (7.5.38) form a weak solution.

Also†

$$\lim_{N\to\infty} \left(\frac{1}{N} \ln V_N \right) = \lim_{N\to\infty} (\ln V_N^{1/N}) = \max_{\eta} F(\eta)$$

Hence

$$\int_{-\infty}^{\infty} c(x, \tau')\, d\tau' = F(\eta_0(x, \tau)) = \lim_{N\to\infty} \left(\frac{1}{N} \ln V_N \right) \tag{7.5.39}$$

3. If $\Gamma_n(\tau)$ is a sequence of initial data converging weakly to $\Gamma(\tau)$, and $c_n(x, \tau)$ is the solution given by Eq. (7.5.30),

$$c_n(x, \tau) = \Gamma_n(\eta_{0n}(x, \tau))$$

then $c_n(x, \tau) \to c(x, \tau)$ at all points of continuity. For let $\Phi_n(\tau)$ and $\Phi(\tau)$ be the integrals of $\Gamma_n(\tau)$ and $\Gamma(\tau)$, respectively, as before. If (x, τ) is a point where $F(\eta)$ has a unique maximum at $\eta_0(x, \tau)$, then it is a point of continuity of the solution. But

$$F_n(\eta) = \Phi_n(\eta) - xg\left(\frac{\tau - \eta}{x}\right)$$

will have a maximum at a point $\eta_{0n}(x, \tau)$, which tends to $\eta_0(x, \tau)$ as $n \to \infty$. Hence, by the explicit formula,

$$c_n(x, \tau) = \Gamma_n(\eta_{0n}(x, \tau)) \to \Gamma(\eta_0(x, \tau)) = c(x, \tau)$$

We will return to Lax's formula again in discussing the general properties of the solution in Secs. 7.6 and 7.8.

Exercises

7.5.1. Obtain the solution given in Sec. 6.3 for the initial data

$$\Gamma(\tau) = \begin{cases} 0, & \tau < 0 \\ c_1, & 0 < \tau < \alpha \\ 0, & \tau > \alpha \end{cases}$$

by means of Lax's formula.

†See, for example, J. Korevaar, *Mathematical Methods: Vol. 1. Linear Algebra/Normed Spaces/Distributions/Integration*, Academic Press, New York, 1968, pp. 234–243.

7.5.2. Show that the maximum value of $F(\eta)$ is the integral of $c(x, \tau)$ with respect to τ by using Eq. (7.5.10) in the form

$$\frac{\partial \phi}{\partial x} + f'(u)\frac{\partial \phi}{\partial \tau} \geq uf'(u) - f(u)$$

7.5.3. Reformulate the explicit solution for the conservation equation

$$\frac{\partial c}{\partial x} + \frac{\partial}{\partial \tau} f(c) = 0$$

when the initial data are given on the whole x-axis.

7.5.4. Show that the solution given by Lax exhibits the semigroup property with respect to x; that is, if $c(x_1, \tau)$ is the solution with $c(0, \tau) = \Gamma(\tau)$, then $c(x_2, \tau) = C(x_2 - x_1, \tau)$, $x_2 > x_1$, where $C(x, \tau)$ is the solution with $C(0, \tau) = c(x_1, \tau)$.

7.6 Some Additional Properties of Weak Solutions

We remarked in Sec. 7.4 and saw in exercises that weak solutions of a conservation equation are not uniquely defined without an additional condition being imposed. Although this additional condition, known as the entropy condition, has been formulated for an arbitrary form of the curve $f(c)$, we will confine our discussion here to the case of a convex isotherm so that $f''(c) < 0$. Then we have the shock condition (7.4.15),

$$\sigma(c^l) < \tilde{\sigma}(c^l, c^r) < \sigma(c^r) \tag{7.6.1}$$

and this implies that a shock must propagate faster than the state on its right and with lesser speed than the state on its left. In other words, a shock tends to overtake the state on the right side and to be overtaken by the state on the left side, so it remains sharp.

The shock condition has been expressed in two other ways. Both Lax and Oleinik have written it in the form

$$\frac{c(x, \tau_1) - c(x, \tau_2)}{\tau_1 - \tau_2} \geq K(x) \tag{7.6.2}$$

for any two values τ_1 and τ_2 and some continuous function $K(x)$. The lower bound may tend to $-\infty$ as x tends to zero. This is simply to say that the slope of the chord joining any two points at position x is bounded below so that a discontinuity with c decreasing in the positive τ-direction (or with c increasing in the positive x-direction) is prohibited (see Exercise 7.4.2). Another form introduced by Oleinik is the requirement that, for any smooth nonnegative function $W(x, \tau)$ that vanishes as $\tau \to \pm\infty$ for all x, there should be a continuous function $K(x)$ such that

$$\int_{-\infty}^{\infty} \left\{ \frac{\partial W}{\partial \tau} c(x, \tau) + W(x, \tau) K(x) \right\} d\tau \leq 0 \tag{7.6.3}$$

(see Exercise 7.6.1).

One method of deriving the condition that can be generalized to more variables is to discuss the stability of a given solution. Suppose that $c(x, \tau)$ is a solution of the conservation equation (7.5.1) subject to the initial condition $c(0, \tau) = \Gamma(\tau)$. For definiteness, we will suppose that there is just one shock path S along which

$$\frac{d\tau}{dx} = \frac{f(c^l) - f(c^r)}{c^l - c^r} = \tilde{\sigma}(c^l, c^r) \tag{7.6.4}$$

In fact, $c^l(x, \tau)$ and $c^r(x, \tau)$ can denote the solutions to the left and right of S, respectively. If a small perturbation $\delta\Gamma$ is made in the initial data and propagates as a small deviation $\delta c(x, \tau)$, it obeys the linear differential equation

$$\frac{\partial \delta c}{\partial x} + f'(c) \frac{\partial \delta c}{\partial \tau} = 0 \tag{7.6.5}$$

with a piecewise continuous coefficient $f'(c)$, which is a known function of x and τ. Thus δc is propagated unchanged along a characteristic

$$\frac{d\tau}{dx} = f'(c) = \sigma(c) \tag{7.6.6}$$

It must be possible to find the perturbations δc^l and δc^r at corresponding sides of the shock, and this means that the characteristics must be "incoming" from both sides, as shown in Fig. 7.10. But the slopes of the characteristics are, respectively, $\sigma(c^l)$ and $\sigma(c^r)$, and thus we get Eq. (7.6.1).

A most instructive way to look at the weak solution is as the limit of solutions of a higher-order equation. For example, we might consider the equation

$$\frac{\partial c}{\partial x} + \frac{\partial}{\partial \tau} f(c) = \mu \frac{\partial^2 c}{\partial x^2} \tag{7.6.7}$$

which returns to our equation in the limit as $\mu \to 0$. The added second-order derivative term represents longitudinal dispersion in the fluid phase, which will smear out any sharp discontinuity, and we might hope to reach the discontinuous solutions as this dispersion becomes completely negligible. In fluid mechanics, such a term corresponds to a viscous stress, and the parameter μ is often referred to as the viscosity. We may denote the solution of the parabolic equation (7.6.7) by $c(x, \tau; \mu)$ and suppose that it satisfies

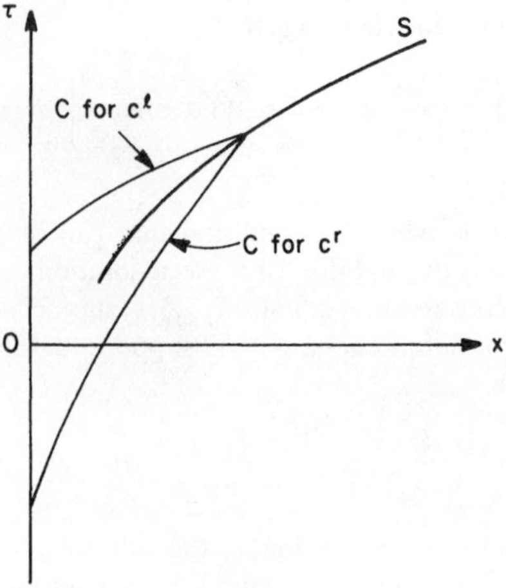

Figure 7.10

the condition

$$c(0, \tau) - \mu\frac{\partial c}{\partial x}(0, \tau) = \Gamma(\tau), \qquad \text{on} \quad x = 0 \tag{7.6.8}$$

and that $c(x, \tau)$ remains finite as $x + |\tau| \to \infty$. Then this condition also reduces to the previous ones when $\mu \to 0$, and the finiteness of $c(x, \tau)$ implies that the derivative of c at any point in any direction tends to zero as that point goes to infinity.

Let us consider a point (x, τ) on a shock path in the solution of the equation with $\mu = 0$, and let $\tilde{\sigma}$ be the slope of the path there. If we define a new coordinate

$$\xi = \tilde{\sigma}x - \tau \tag{7.6.9}$$

the tangent to the shock path at this point is represented by ξ = constant. Since we anticipate that the solution of the equation with $\mu \neq 0$ will approximate the discontinuous one, and since it is the local behavior that will be important, we may look for solutions that are functions of ξ only. Then $\partial/\partial x = \tilde{\sigma}(d/d\xi)$ and $\partial/\partial\tau = -d/d\xi$, so we have

$$\mu\tilde{\sigma}^2\frac{d^2c}{d\xi^2} - \tilde{\sigma}\frac{dc}{d\xi} + \frac{df}{d\xi} = 0 \tag{7.6.10}$$

Since $\tilde{\sigma}$ is positive, ξ is negative to the left of the shock and positive to the right. By looking at solutions $c(\xi; \mu)$ of this form, we are essentially looking at them along a line perpendicular to the shock path.

Moreover, if we replace ξ by ξ/μ, the parameter μ disappears, and we see

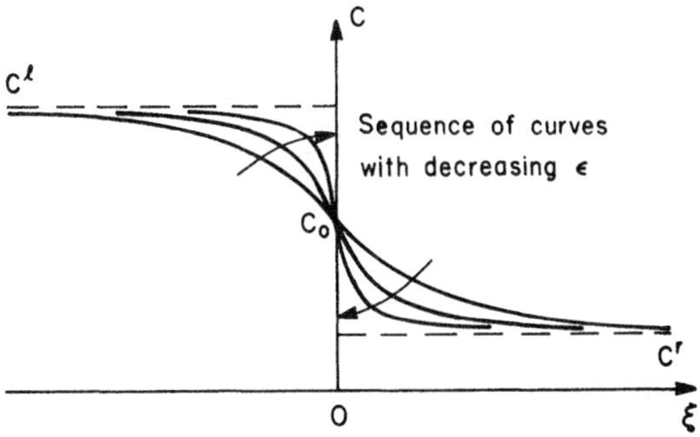

Figure 7.11

that

$$c(\xi;\ \mu) = c\left(\frac{\xi}{\mu};\ 1\right) \tag{7.6.11}$$

Thus letting $\mu \to 0$ pulls the distant parts of the solution close in to $\xi = 0$, as illustrated in Fig. 7.11. Consider now a solution of Eq. (7.6.7) with

$$c(\xi;\ \mu) \to c^l \qquad \text{as} \quad \xi \to -\infty$$

and

$$c(\xi;\ \mu) \to c^r \qquad \text{as} \quad \xi \to +\infty$$

Then, by Eq. (7.6.11), this will tend to a discontinuous solution with

$$c(\xi) = c^l \qquad \text{for} \quad \xi < 0$$

$$c(\xi) = c^r \qquad \text{for} \quad \xi > 0$$

as $\mu \to 0$. Eq. (7.6.10) can be integrated once with respect to ξ to give

$$\mu\tilde{\sigma}^2 \left[\frac{dc}{d\xi}\right]_{-\infty}^{\infty} - \tilde{\sigma}(c^r - c^l) + f(c^r) - f(c^l) = 0$$

Since $dc/d\xi$ vanishes at both limits, the first term disappears and

$$\tilde{\sigma} = \frac{f(c^l) - f(c^r)}{c^l - c^r}$$

so we recover the jump condition.

By integrating Eq. (7.6.10) from $\xi = -\infty$ to $\xi = \xi$, we have

$$\mu\tilde{\sigma}^2 \frac{dc}{d\xi} = \tilde{\sigma}c - f(c) - \tilde{\sigma}c^l + f(c^l) \equiv F(c) \tag{7.6.12}$$

an ordinary differential equation for $c(\xi)$ that may be solved by quadratures:

$$\xi = \mu\tilde{\sigma}^2 \int_{c_0}^{c} \frac{dc'}{F(c')} \tag{7.6.13}$$

Now if ξ is to go from $-\infty$ to ∞ as c goes from c^l to c^r, the integrand must diverge to $-\infty$ and $+\infty$ as $c \to c^l$ and $c \to c^r$, respectively, and must be bounded between these limits. We know that c^l and c^r are zeros by the definition of $F(c)$ in Eq. (7.6.12). The boundedness of the integrand requires that c^l and c^r should be adjacent zeros. Moreover, we see that $F''(c) = -f''(c)$, and so the graph of $F(c)$ will be as shown in Fig. 7.12. If c is to pass from c^l to c^r as ξ increases, $dc/d\xi$ [and so $F(c)$] must have the same sign as $(c^r - c^l)$. This implies that $F'(c^l) > 0$ and $F'(c^r) < 0$, or

$$f'(c^l) < \tilde{\sigma} < f'(c^r)$$

which is the shock condition itself. Hence a shock can be obtained as the limit of smooth solutions of the parabolic equation (7.6.7).

From this derivation, we have the added fact that c^l and c^r must be adjacent zeros of

$$f(c) = \tilde{\sigma}c - \tilde{\sigma}c^l + f(c^l) \tag{7.6.14}$$

If $f(c)$ is convex, this equation has only two zeros, but if $f(c)$ has a point of inflection, this added condition becomes important, as we have seen in Secs. 5.5

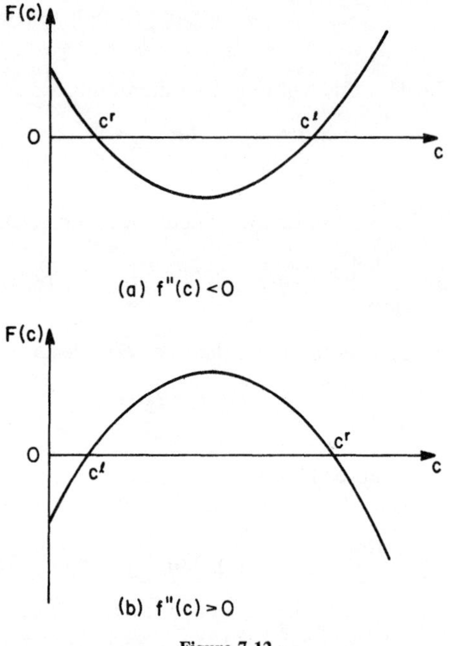

Figure 7.12

and 6.7 in connection with the chromatography with the BET isotherm (see Exercise 7.6.2) and sedimentation, respectively. It is this condition that is so difficult to generalize to systems of equations.

We can now state the existence and uniqueness theorem.

THEOREM: If $\Gamma(\tau)$ is a bounded, measurable function and the inequality (7.6.2) is imposed, the weak solution of

$$\frac{\partial c}{\partial x} + \frac{\partial}{\partial \tau} f(c) = 0, \qquad f''(c) \neq 0$$

$$c(0, \tau) = \Gamma(\tau)$$

exists and is unique.

We will not give the details of the proof, which are to be found in Oleinik's papers. She divides the proof into three parts, establishing first the uniqueness and then the existence for $\Gamma(\tau)$ when this is a bounded function with bounded derivative. Finally, it is shown that a sequence of solutions with the smoother initial conditions converges to the solution when $\Gamma(\tau)$ is bounded and measurable. One method of proof is by the limiting case of the viscous solution. Another method, used by Lax, Glim, and others, is based on the convergence of a finite-difference scheme.

Because the chromatographic equations more naturally give rise to a nonlinear function in the derivative with respect to the timelike variable τ, it should be mentioned that the mathematical literature more often has the nonlinearity in the derivative with respect to the spacelike variable x; for example,

$$\frac{\partial u}{\partial t} + \frac{\partial}{\partial x} \phi(u) = 0 \tag{7.6.15}$$

with initial data on the line $t = 0$. Douglis (1959) has introduced an ordering principle to study weak solutions of Eq. (7.6.15) under the condition that $\phi(u)$ is strongly convex. Given a bounded weak solution $u(x, t)$, let us consider the triangular region T given by

$$x_0 + At \leq x \leq x_1 - At, \qquad 0 \leq t \leq \frac{x_1 - x_0}{2A}$$

where $x_0 < x_1$ and A is constant. We will call T a triangle of determinacy for u if

$$A \geq \max_x \mid \phi'(u(x, 0)) \mid$$

and the triangle's intersection with the x-axis is called its base. We say the ordering principle holds within a class Ω of bounded weak solutions of Eq. (7.6.15) if the members of Ω, in a certain sense, are partially ordered like their initial data. Let u and v be two bounded weak solutions belonging to Ω, and let T be a common triangle of determinacy. The ordering principle requires that, if $u \le v$ at almost every point of the base of T, then $u \le v$ at every point of the interior of T. Douglis has used this principle to prove the following uniqueness theorem: If $\phi(u)$ is strongly convex, two bounded, normalized weak solutions of Eq. (7.6.15), which coincide at almost every point of the base of a common triangle of determinacy T, also coincide at all points of the interior of T as well. He has also developed elementary means to construct weak solutions and examined the properties of weak solutions. This approach has been extended by Wu (1963) and Ballou (1970) to establish the existence and uniqueness of weak solutions for cases without convexity conditions, that is, when $\phi''(u)$ changes sign.

Some care has to be exercised in the transformation of variables, for as Lax and others have pointed out, the class of weak solutions depends on the form in which the equation is written. To see this, we may quote an example of Gelfand's. The equation

$$\frac{\partial u}{\partial t} + u\frac{\partial u}{\partial x} = 0$$

might be written as

$$\frac{\partial u}{\partial t} + \frac{\partial}{\partial x}\left(\frac{1}{2}u^2\right) = 0$$

or as

$$\frac{\partial}{\partial t}\left(\frac{1}{2}u^2\right) + \frac{\partial}{\partial x}\left(\frac{1}{3}u^3\right) = 0$$

If in the second equation we write $2v = u^2$, it becomes

$$\frac{\partial v}{\partial t} + \frac{\partial}{\partial x}\left\{\frac{1}{3}(2v)^{3/2}\right\} = 0$$

The characteristic directions are equivalent, for

$$\frac{dx}{dt} = (2v)^{1/2} = u$$

But for the speed of a discontinuity, we have in the one case

$$\frac{dx}{dt} = \frac{1}{2}\frac{(u^l)^2 - (u^r)^2}{u^l - u^r} = \frac{1}{2}(u^l + u^r)$$

and in the other

$$\frac{dx}{dt} = \frac{2^{3/2}}{3}\frac{(v^l)^{3/2} - (v^r)^{3/2}}{v^l - v^r} = \frac{2}{3}\frac{(u^l)^3 - (u^r)^3}{(u^l)^2 - (u^r)^2} = \frac{2}{3}\frac{(u^l)^2 + u^l u^r + (u^r)^2}{u^l + u^r}$$

which are not the same.

The asymptotic properties of the solution of the conservation equation at large distances or after a long time are also of importance. Consider our standard equation

$$\begin{aligned} \frac{\partial c}{\partial x} + \frac{\partial}{\partial \tau} f(c) &= 0 \\ c(0, \tau) &= \Gamma(\tau) \end{aligned} \tag{7.6.16}$$

We will say that $c(x, \tau)$ has the mean value $m(x)$ if the limit

$$\lim_{L \to \infty} \frac{1}{L} \int_a^{a+L} c(x, \tau)\, d\tau = m(x) \tag{7.6.17}$$

exists uniformly with respect to a. For example, if $c(x, \tau)$ is zero outside of a finite band, then its mean value is zero. We can also show that, for a bounded weak solution of Eq. (7.6.16), the mean value is independent of x. To see this, we consider a plane vector field with x and τ components:

$$w(\tau)c(x, \tau) \quad \text{and} \quad \int_{-\infty}^{\tau} w(\tau') \frac{\partial f}{\partial \tau'}(x, \tau')\, d\tau' = \phi(x, \tau)$$

respectively, where $w(\tau)$ is any smooth function. The divergence of this vector field is

$$\frac{\partial}{\partial x}(wc) + \frac{\partial \phi}{\partial \tau} = w\left(\frac{\partial c}{\partial x} + \frac{\partial f}{\partial \tau}\right)$$

which vanishes everywhere (except on a set of measure zero). Hence we have

$$\oint (wc\, d\tau - \phi dx) = 0$$

when the integral is taken around any closed curve. If we take the contour to be a long stretch of the τ-axis and a parallel interval on the line x = constant, the ends being joined by lines parallel to the x-axis, and let these ends go to $+\infty$ and $-\infty$, respectively, then we have

$$\int_{-\infty}^{\infty} w(\tau)c(0, \tau)\, d\tau + \int_{\infty}^{-\infty} w(\tau)c(x, \tau)\, d\tau = 0$$

or

$$\int_{-\infty}^{\infty} w(\tau)c(x, \tau) d\tau = \int_{-\infty}^{\infty} w(\tau)c(0, \tau)\, d\tau = \int_{-\infty}^{\infty} w(\tau)\Gamma(\tau)\, d\tau$$

Now, if $w(\tau)$ is a smooth function, zero outside the interval $(a, a + L)$ and equal to 1 within the interval $(a + \varepsilon, a + L - \varepsilon)$, where ε is arbitrarily small, we see that, by letting $L \to \infty$,

$$m(x) = \operatorname*{Lim}_{L \to \infty} \frac{1}{L} \int_a^{a+L} \Gamma(\tau)\, d\tau = m = \text{constant} \tag{7.6.18}$$

Furthermore, it can be shown that not only is m constant but that the solution tends to m uniformly as $x \to \infty$. Without loss of generality, we can take $m = 0$, since otherwise we could modify the problem by replacing $\Gamma(\tau)$ by $\Gamma(\tau) - m$. Let us denote $f'(0)$ by σ_0 and then $g'(\sigma_0) = 0$, since $-g'(s)$ is the inverse function of $f'(c)$. Thus σ_0 corresponds to the least value of $g(s)$. Since we have

$$c(x, \tau) = -g'\left(\frac{\tau - \eta_0(x, \tau)}{x}\right) \tag{7.6.19}$$

by Lax's formula, we will succeed in showing that $c \to 0$ if we can show that $(\tau - \eta_0)/x \to \sigma_0$, as $x \to \infty$. Let us put

$$\xi = \tau - \sigma_0 x$$

so that we need to show that $\xi \to \eta_0$ as $x \to \infty$. We will do so by proving that otherwise the function

$$F(\eta) = \Phi(\eta) - xg\left(\frac{\tau - \eta}{x}\right)$$

would have a greater value at $\eta = \xi$ than at $\eta = \eta_0$, which is contrary to the definition of η_0.

Consider the difference

$$F(\eta_0) - F(\xi) = \{\Phi(\eta_0) - \Phi(\xi)\} - x\left\{g\left(\sigma_0 + \frac{\xi - \eta_0}{x}\right) - g(\sigma_0)\right\} \tag{7.6.20}$$

Since $g(s)$ is strictly concave and has its minimum at $s = \sigma_0$,

$$g\left(\sigma_0 + \frac{\xi - \eta_0}{x}\right) - g(\sigma_0) \geq \varepsilon\left|\frac{\xi - \eta_0}{x}\right|$$

provided that $|(\xi - \eta_0)/x| > \delta$. Since the mean value of $\Gamma(\tau)$ is zero, we can find an L such that

$$\Phi(\eta_0) - \Phi(\xi) = \int_{\xi}^{\eta_0} \Gamma(\tau)\, d\tau < \varepsilon\,|\eta_0 - \xi|$$

provided that $|\eta_0 - \xi| > L$. Then if $|\eta_0 - \xi|$ is both greater than L and $x\delta$, the second term of Eq. (7.6.20) is greater in magnitude than the first, and this implies that $F(\xi) > F(\eta_0)$, which is contradictory. Thus we cannot have both $|\eta_0 - \xi| > L$ and $|(\eta_0 - \xi)/x| > \delta$, so that if $x > L/\delta$, then $|(\eta_0 - \xi)/x| < \delta$ and $\xi \to \eta_0$ as $x \to \infty$.

An even more remarkable result on the asymptotic behavior applies to the finite chromatogram. Suppose that $\Gamma(\tau)$ is positive and vanishes outside a finite interval and that

$$M = \int_{-\infty}^{\infty} \Gamma(\tau)\, d\tau \tag{7.6.21}$$

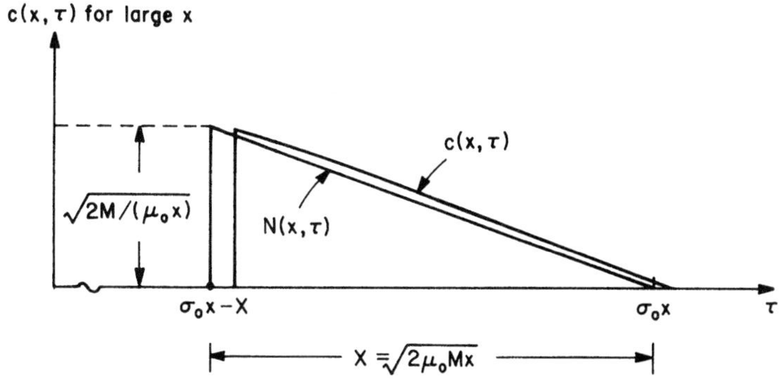

Figure 7.13

Let

$$\sigma_0 = f'(0), \qquad \mu_0 = -f''(0) \tag{7.6.22}$$

Then the asymptotic shape of $c(x, \tau)$ for large x is

$$c(x, \tau) \sim N(x, \tau) = \begin{cases} \dfrac{1}{\mu_0}\left(\sigma_0 - \dfrac{\tau}{x}\right), & \sigma_0 x - X < \tau < \sigma_0 x \\ \\ 0, & \text{elsewhere} \end{cases} \tag{7.6.23}$$

where

$$X = \sqrt{2\mu_0 M x} \tag{7.6.24}$$

This is to say that for large x every point of the graph of $c(x, \tau)$ is within a distance $0(\sqrt{x})$ of the graph of $N(x, \tau)$, and so the shape of $c(x, \tau)$ approaches that of the *N-wave*†, as shown in Fig. 7.13. It would be instructive to recall the discussion of Sec. 6.3 and compare with the observation here. We will not give the complete proof here, as a very similar phenomenon will be discussed in the next section and the formulation will be presented in Sec. 7.9.

For conservation laws in more than one space dimension, Smoller and Conway have established existence and uniqueness theorems. The conditions that have to be imposed are natural generalizations of Lax's Eq. (7.6.2). They also prove that the solution has the same upper and lower bounds as the initial data and that monotonicity is preserved (see Exercise 7.6.4). An independent study has been given by Vol'pert (1967), who has proven the existence and uniqueness of the solution, as well as its stability with respect to the initial data. He has also showed that the solution is the limit of solutions of parabolic equations into which a small "viscosity" is introduced. Kruzkov

†Because of the shape of the curve (7.6.23), it is known as the *N*-wave.

(1969, 1970) has relaxed some of the conditions and established uniqueness and stability theorems by applying Lebesque's theorem. To prove the existence theorem, Kruzkov has applied the vanishing viscosity method. The uniqueness and stability have also been proved by B. Keyfitz Quinn (1971) by using the equivalence of the generalized entropy condition to L_1-contractibility.

More recently, the theory of singularities has been introduced to investigate the *generic* properties of weak solutions. Schaeffer (1973) has proved that generically the weak solution of the Cauchy problem for Eq. (7.6.15) is piecewise smooth, having jump discontinuities along a finite number of smooth shock paths if $\phi(u)$ is convex. This proof has been extended by Guckenheimer (1975) to the case without the convexity condition. Golubitsky and Schaeffer (1975) have also extended the result of Schaeffer to show that generically the topological and differentiable structure of the shock set is unaffected by small perturbations of the initial data.

Exercises

7.6.1. Show that the conditions (7.6.2) and (7.6.3) are equivalent.

7.6.2. Consider the function $F(c)$ defined by Eq. (7.6.12) with the BET isotherm [cf. Eq. (5.3.22)]. Sketch the graph of F versus c and thereby formulate the entropy condition.

7.6.3. Compare Eqs. (7.6.23) and (7.6.24) with the results of Sec. 6.3 and discuss.

7.6.4. Show that if $\Gamma(\tau)$ is bounded below and above by m and M, respectively, then for the solution $c(x, \tau)$ of

$$\frac{\partial c}{\partial x} + \frac{\partial}{\partial \tau} f(c) = 0$$

$$c(0, \tau) = \Gamma(\tau)$$

it is also true that

$$m < c(x, \tau) < M$$

Show that if $\Gamma(\tau)$ is monotonic then $c(x, \tau)$ has the same monotonicity.

7.7 Sound Waves of Finite Amplitude

A topic that ties in closely with this chapter is the consideration of finite-amplitude sound waves. We will not attempt to give the full derivation of the equations, for this is treated by Lighthill in an essay in which physical and mathematical considerations are beautifully balanced. Let it suffice to say that, if u denotes the excess velocity above the undisturbed speed of sound a_0, t is the time, and $X = x - a_0 t$, the distance from a moving origin, then

$$\frac{\partial u}{\partial t} + u\frac{\partial u}{\partial X} = \frac{1}{2}\delta\frac{\partial^2 u}{\partial X^2} \tag{7.7.1}$$

Here δ is the diffusivity of sound, a linear combination of the shear and bulk kinematic viscosities and the thermal diffusivity. The right side is clearly analogous to the viscosity term we introduced previously in examining the structure of the weak solution. The left side is the continuity equation. It is the simplest form of an equation showing the competition between the convective and diffusive effects. It is sometimes known as Burger's equation, since he used it to give the simplest possible basis for examining turbulence theory.

If we write Eq. (7.7.1) in the form

$$\frac{\partial u}{\partial t} + \frac{\partial}{\partial X}\left(\frac{1}{2}u^2 - \frac{1}{2}\delta\frac{\partial u}{\partial X}\right) = 0$$

we see that there exists a potential ϕ such that

$$u = -\frac{\partial \phi}{\partial X}, \qquad \frac{1}{2}u^2 - \frac{1}{2}\delta\frac{\partial u}{\partial X} = \frac{\partial \phi}{\partial t}$$

Setting

$$\phi = \int_X^\infty u(x, t)\, dx \tag{7.7.2}$$

and integrating Eq. (7.7.1) from X to infinity gives

$$\frac{\partial \phi}{\partial t} = \frac{1}{2}\left(\frac{\partial \phi}{\partial X}\right)^2 + \frac{1}{2}\delta\frac{\partial^2 \phi}{\partial X^2} \tag{7.7.3}$$

If we put

$$\psi = \exp\left(\frac{\phi}{\delta}\right) = \exp\left\{\frac{1}{\delta}\int_X^\infty u(x, t)\, dx\right\} \tag{7.7.4}$$

then

$$\frac{\partial \psi}{\partial t} = \frac{\psi}{\delta}\frac{\partial \phi}{\partial t}, \qquad \frac{\partial \psi}{\partial X} = \frac{\psi}{\delta}\frac{\partial \phi}{\partial X}$$

$$\frac{\partial^2 \psi}{\partial X^2} = \frac{\psi}{\delta^2}\left(\frac{\partial \phi}{\partial X}\right)^2 + \frac{\psi}{\delta}\frac{\partial^2 \phi}{\partial X^2}$$

and thus the nonlinear term can be eliminated from Eq. (7.7.3) to give

$$\frac{\partial \psi}{\partial t} = \frac{1}{2}\delta\frac{\partial^2 \psi}{\partial X^2} \tag{7.7.5}$$

Given the wave form $u(X, 0)$ at time $t = 0$, we can calculate $\psi(X, 0)$ by Eq. (7.7.4) and determine the solution of Eq. (7.7.5) in the form

$$\psi(X, t) = (2\pi\delta t)^{-1/2}\int_{-\infty}^{\infty} \psi(Y, 0)\exp\left[-\frac{(X - Y)^2}{2\delta t}\right] dY \tag{7.7.6}$$

Note that here and in the following Y takes the role of a parameter running along the initial line $t = 0$. But since $u = -\partial\phi/\partial X = -\delta(\partial\psi/\partial X)/\psi$, this gives

$$u(X,t) = \frac{\displaystyle\int_{-\infty}^{\infty} \frac{X-Y}{t} \exp\left\{\frac{1}{\delta}\left[\int_{Y}^{\infty} u(Y',0)\,dY' - \frac{(X-Y)^2}{2t}\right]\right\} dY}{\displaystyle\int_{-\infty}^{\infty} \exp\left\{\frac{1}{\delta}\left[\int_{Y}^{\infty} u(Y',0)\,dY' - \frac{(X-Y)^2}{2t}\right]\right\} dY} \tag{7.7.7}$$

We might write this as

$$u(X, t) = \int_{-\infty}^{\infty} \frac{X - Y}{t} D(X, Y, t; \delta)\, dY \tag{7.7.8}$$

where

$$D(X, Y, t; \delta) = \frac{\exp\,[F(X, Y, t)/\delta]}{\displaystyle\int_{-\infty}^{\infty} \exp\,[F(X, Y, t)/\delta]\, dY} \tag{7.7.9}$$

and

$$F(X, Y, t) = \int_{Y}^{\infty} u(Y', 0)\, dY' - \frac{(X - Y)^2}{2t} \tag{7.7.10}$$

Clearly, D has the property that

$$\int_{-\infty}^{\infty} D(X, Y, t; \delta)\, dY = 1$$

for all δ, and as $\delta \to 0$ the character of D is shaped entirely by the behavior of F near its maximum value. Thus, supposing that F takes its maximum value F_M at $Y = Y_M$ and writing $F = F_M - (F_M - F)$, we see that

$$D = \frac{\exp\,[-(F_M - F)/\delta]}{\displaystyle\int_{-\infty}^{\infty} \exp\,[-(F_M - F)/\delta]\, dY}$$

so that $D \to 0$ as $\delta \to 0$ whenever $Y \neq Y_M$. Thus $D(X, Y, t; \delta)$ acts as a Dirac delta function in the limit $\delta = 0$. It then follows from Eq. (7.7.8) that in the limit $\delta = 0$ we have

$$u(X, t) = \frac{X - Y_M(X, t)}{t} \tag{7.7.11}$$

as the solution of Eq. (7.7.1) without the second-order derivative term.

Now F is a continuously differentiable function of Y, so $dF/dY = 0$ at $Y = Y_M$. In other words, Y_M is determined by differentiating Eq. (7.7.10) with respect to Y and is the root of

$$\frac{X - Y}{t} = u(Y, 0) \tag{7.7.12}$$

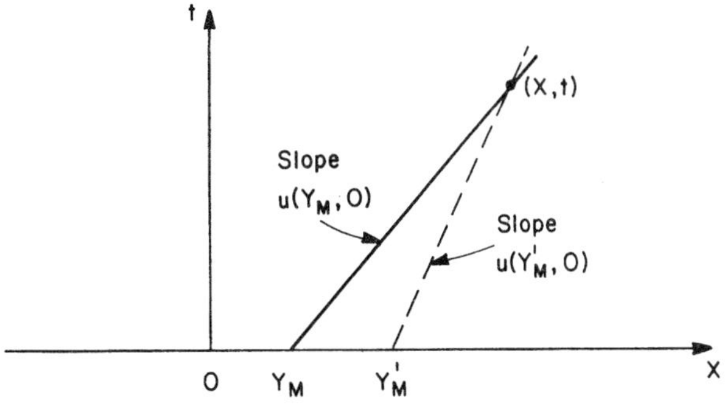

Figure 7.14

If the root is unique, we have

$$u(X, t) = u(Y_M(X, t), 0) \tag{7.7.13}$$

where

$$X = Y_M + tu(Y_M, 0) \tag{7.7.14}$$

But this is just to say that Y_M is the point on the initial line from which the characteristic comes that passes through the point (X, t), as shown in Fig. 7.14. If there were two roots Y_M and Y'_M of Eq. (7.7.12) that give the same maximum value of F, then, as shown by the dashed line in Fig. 7.14, we could claim two values of u at the same point. The only way of reconciling this with physical common sense is by introducing a discontinuity that would be a shock, because here we have $f(u) = \frac{1}{2}u^2$. The connection to the Lax method is now quite evident, for shocks there arose in exactly the same way, and we are obtaining his formulas in the special case of $f(u) = \frac{1}{2}u^2$.

A very neat method of obtaining the shock paths comes from the graphical interpretation of Eq. (7.7.12). In Fig. 7.15, an arbitrary initial distribution of velocity is the curve shown. This is the right side of Eq. (7.7.12), the left side of which can be represented as the straight line of intercept X and slope $-1/t$. If $t \to 0$, this line becomes vertical, and we have $u(X, t) = u(X, 0)$, as we should expect (see PQ on Fig. 7.15 drawn for very small t). If X has a value corresponding to the point A and t a value giving such a slope as that of the line $ABCD$, we see that there can be three solutions to Eq. (7.7.12). Since the second-order derivative of F is

$$\frac{d^2F}{dY^2} = -u'(Y, 0) - \frac{1}{t} \tag{7.7.15}$$

we only have a maximum of F if $u'(Y, 0)$, the slope of the curve, is greater than $-1/t$, the slope of the line. Thus B and D correspond to maxima, and E is a minimum between them. Let Y_1, Y_2, and Y_3 denote the abscissas of the points D, C, and B, respectively. Then the maximum value of $F(X, Y,$

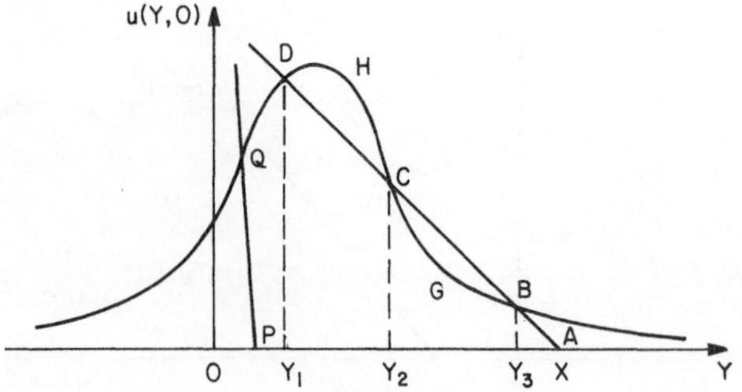

Figure 7.15

t) at D minus the minimum value at C is

$$F(X, Y_1, t) - F(X, Y_2, t) = \int_{Y_1}^{Y_2} u(Y, 0)\, dY - \frac{(X - Y_1)^2}{2t} + \frac{(X - Y_2)^2}{2t}$$

$$= \int_{Y_1}^{Y_2} u(Y, 0)\, dY - \frac{Y_2 - Y_1}{2}\{u(Y_1, 0) + u(Y_2, 0)\}$$

since $u(Y_i, 0) = (X - Y_i)/t$. Similarly, we can write

$$F(X, Y_3, t) - F(X, Y_2, t) = \int_{Y_3}^{Y_2} u(Y, 0)\, dY - \frac{(X - Y_3)^2}{2t} + \frac{(X - Y_2)^2}{2t}$$

$$= \frac{Y_3 - Y_2}{2}\{u(Y_2, 0) + u(Y_3, 0)\} - \int_{Y_2}^{Y_3} u(Y, 0)\, dY$$

Geometrically, the last expressions in these two equations represent the areas of the lobes *DHC* and *CGB*, respectively, so the greater maximum will be associated with the larger lobe (D in Fig. 7.15) and will give the corresponding u.

On the other hand, a shock is required when the two maxima are equal (i.e., the two lobes are equal in area). The path of the shock is easy to plot in the (X, t)-plane if we can find the family of chords of the initial distribution curve that have negative slope and cut off lobes of equal area. As shown in Fig. 7.16(a), these will clearly start with the tangent at the point of inflection and form a one-parameter family. The position of the shock at time t is the intercept X of the chord of slope, $-1/t$ and the values of u at either side are the values of $u(Y, 0)$ at the ends of the chord.

If the form of the curve is as shown in Fig. 7.16(b), where $u = 0$ at point A and the curve has the greatest negative slope there, then the shock will start at the front of the initial distribution at a time given by the reciprocal of the slope there.

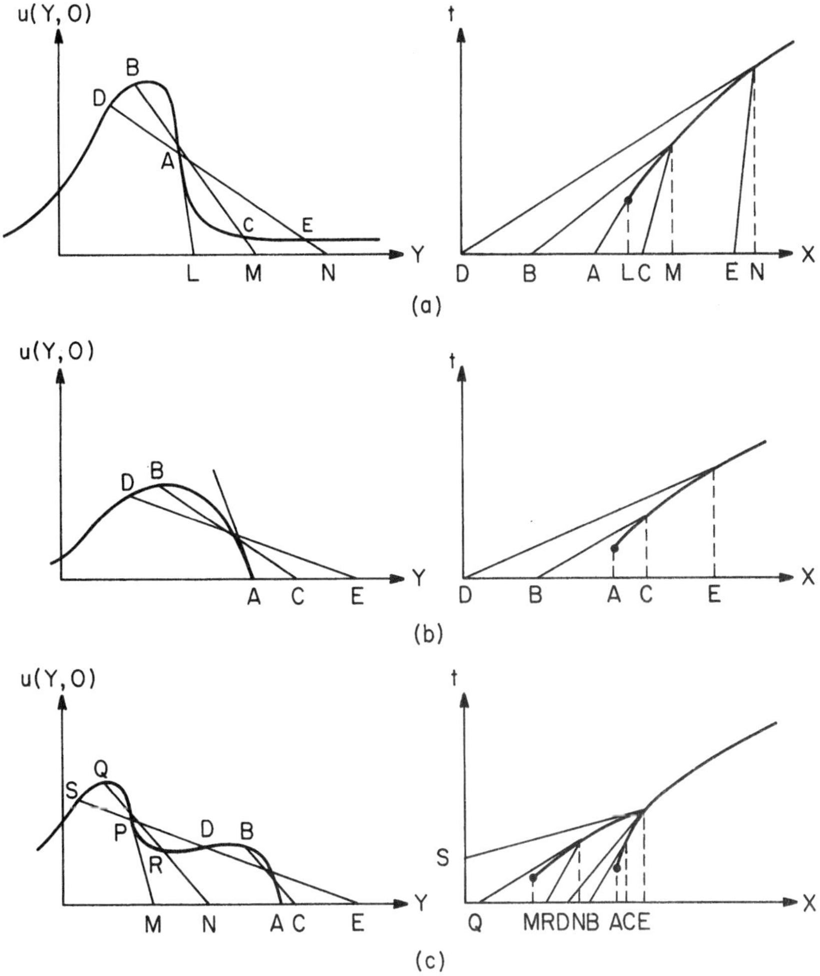

Figure 7.16

In the case shown in Fig. 7.16(c), there is the possibility of two shocks being initiated, one at M at a time corresponding to the reciprocal of the slope at the point of inflection P and another at A, where the slope is greatest for the lower part of the curve. At a slightly later time corresponding to the reciprocal of the common slope of the chords BC and QRN, the two shocks have progressed to the points C and N. As these two chords rotate counterclockwise keeping the same slope, the one initiated at M, which at first lies to the left and below the other, will eventually come into coincidence

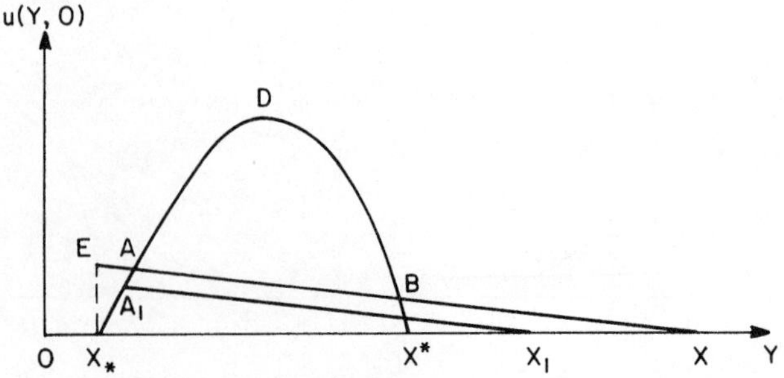

Figure 7.17

with it. This is shown as the chord SDE, and it corresponds to the two shocks meeting at E. Thereafter there is only one shock corresponding to a chord of even smaller slope than that of SDE, such that the net area between the curve and chord (i.e., counting the area of the one or two lobes above as positive and those below as negative) is zero.

We can now see the reason why a shock always forms at the head of a finite pulse and understand what lies behind the asymptotic behavior mentioned in the last section. Let us suppose for definiteness that the initial data $u(Y, 0)$ are always positive and are zero outside the finite interval $X_* \leq Y \leq X^*$, as shown in Fig. 7.17, with area under the curve

$$\int_{-\infty}^{\infty} u(Y, 0)\, dY = M \tag{7.7.16}$$

When t becomes very large, the chord is very nearly parallel to the Y-axis, and the area ABD above it must be equal to the narrow triangular area X^*BX. If we add the area ABX^*X_* to each and neglect the small triangular area EAX_*, we have in the one case the area under the initial curve and in the other the area of the triangle XX_*E. Thus we have

$$\frac{1}{2}(X - X_*)\frac{1}{t} = M$$

or

$$X = X_* + \sqrt{2Mt} \tag{7.7.17}$$

which gives the position of the shock at time t. Its magnitude is the difference between the ordinates of A and B. For very large t, however, this difference may be approximated by the ordinate of E,

$$\frac{X - X_*}{t} = \sqrt{\frac{2M}{t}} \tag{7.7.18}$$

in which Eq. (7.7.17) is substituted. Given the time t, values of $u(X, t)$ for points between X_* and $X_* + \sqrt{2Mt}$ are obtained by drawing lines parallel to AX. For example, at X_1, $u(X_1, t)$ has the value of $u(Y, 0)$ at A_1. Thus if the initial distribution curve has a nonzero slope at $X = X_*$, we can approximate X_*A by a straight line and the distribution $u(x, t)$ is evidently linear in X. This gives the N-wave mentioned in the last section. The structure of the shock can be examined in more detail by using the steepest-descent approximation to the integrals in Eq. (7.7.7), and Lighthill gives some elegant results on the velocity distribution during the confluence of two shock waves, as well as during the formation and decay of shock waves.

This treatment of the limiting case (i.e., $\delta = 0$) suggests that we might look for solutions of Burger's equation that preserve their shape but change in amplitude and span. Thus, if we seek a solution of the form

$$u(X, t) = t^m f(Xt^n) = t^m f(\zeta)$$

we see from Eq. (7.7.1) that f must satisfy

$$mt^{m-1}f + nXt^{m+n-1}f' + t^{2m+n}ff' = \frac{1}{2}\delta t^{m+2n}f''$$

A factor of a power of t can be canceled only if $m = n = -\frac{1}{2}$, and then

$$\delta f'' = 2ff' - \zeta f' - f \tag{7.7.19}$$

If

$$R = \frac{M}{\delta} = \frac{1}{\delta}\int_{-\infty}^{\infty} u(X, 0)\, dX$$

is the Reynolds number of the pulse, it is invariant with time, and the potential ψ defined by Eq. (7.7.4) must tend to 1 as $X \to \infty$ and e^R as $X \to -\infty$. The solution of Eq. (7.7.5) that has these limiting values and is a function of $X/\sqrt{t}$ is

$$\psi(x, t) = 1 + \frac{1}{2}(e^R - 1)\operatorname{erfc}\left(\frac{X}{\sqrt{2\delta t}}\right) \tag{7.7.20}$$

where

$$\operatorname{erfc} x = \frac{2}{\sqrt{\pi}}\int_x^{\infty} e^{-\xi^2}\, d\xi \tag{7.7.21}$$

Since $u = -\delta(\partial\psi/\partial X)/\psi$, this gives

$$u(X,t) = (e^R - 1)\sqrt{\frac{\delta}{2\pi t}}e^{-X^2/2\delta t}\left\{1 + \frac{1}{2}(e^R - 1)\operatorname{erfc}\frac{X}{\sqrt{2\delta t}}\right\}^{-1} \tag{7.7.22}$$

When R is small (i.e., for a weak pulse), then

$$u(X, t) \simeq R\sqrt{\frac{\delta}{2\pi t}}e^{-X^2/2\delta t} + O(R^2) \tag{7.7.23}$$

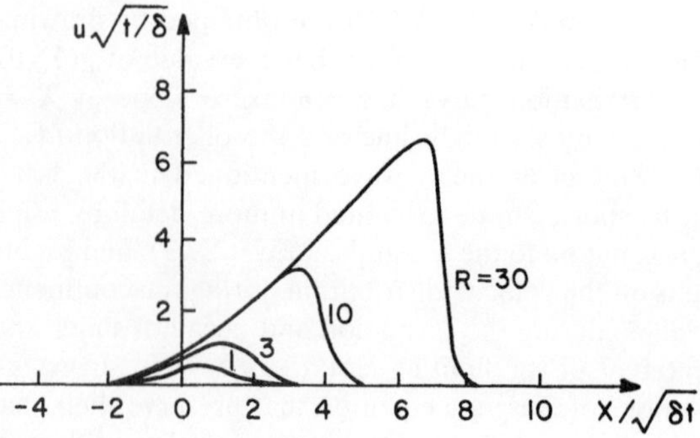

Figure 7.18

since the complementary error function defined by Eq. (7.7.21) never exceeds the value 2. Thus a weak pulse is asymptotically Gaussian. If R is large, however, the Gaussian character is lost for positive values of $X/\sqrt{2\delta t}$, which are of any magnitude. For $X/\sqrt{2\delta t} > 2$, the complementary error function is quite well approximated by

$$\frac{1}{2}\operatorname{erfc}\frac{X}{\sqrt{2\delta t}} \simeq \sqrt{\frac{\delta t}{2\pi}}\frac{1}{X}\exp\left(-\frac{X^2}{2\delta t}\right)$$

so

$$u(X, t) \simeq \sqrt{\frac{\delta}{2\pi t}}e^{R - X^2/2\delta t}\left\{1 + \sqrt{\frac{\delta}{2\pi t}}\frac{t}{X}e^{R - X^2/2\delta t}\right\}^{-1} \tag{7.7.24}$$

Thus, if $4 < X^2/2\delta t << R$, the exponential is quite large and

$$u(X, t) \simeq \frac{X}{t} \tag{7.7.25}$$

However, when $X^2/2\delta t$ becomes comparable to R, the dominant term in the denominator is the constant 1, and we have a Gaussian fall-off in the neighborhood of $X^2/2\delta t = R$. This analysis, which follows Lighthill's very closely, fully accounts for the softened form of N-wave that is given by Eq. (7.7.22) and shown in Fig. 7.18.

Exercises

7.7.1. If $u(X, 0)$ is zero outside of the interval (X_*, X^*) and $u'(X_*, 0) = 0$ but $u''(X_*, 0) = U > 0$, then show that the asymptotic form of the wave is given by

$$u(X, t) = \{1 + Ut(X - X_*) - [1 + 2Ut(X - X_*)]^{1/2}\}Ut^2$$

within the interval $X_* \leq X \leq X_* + (2Mt)^{1/2} + O(t^{-1/4})$.

7.7.2. By using the approximation

$$\int_{-\infty}^{\infty} G(Y)e^{F(Y)/\delta}\, dY \simeq \sum_M G(Y_M) \sqrt{\frac{2\pi\delta}{-F''(Y_M)}}\, e^{F(Y_M)/\delta}$$

where the summation is over the maxima of $F(Y)$, show that in a shock wave between u^l and u^r

$$\frac{u^l - u(X,t)}{u(X,t) - u^r} \cong \left\{\frac{1 + t(\partial/\partial X)\, u(Y^l, O)}{1 + t(\partial/\partial X)\, u(Y^r, O)}\right\}^{1/2} exp \left\{\frac{[u](X - X_s)}{\delta}\right\}$$

where $[u] = u^l - u^r$, and X_s denotes the position of the shock wave at time t.

7.8 Some General Properties of Chromatograms

We have been able to treat the equations of finite amplitude sound waves with particular completeness because of the very simple form $f(u) = \frac{1}{2}u^2$ in the equation

$$\frac{\partial u}{\partial t} + \frac{\partial}{\partial X} f(u) = 0 \tag{7.8.1}$$

The chromatographic equation

$$\frac{\partial c}{\partial x} + \frac{\partial}{\partial \tau} f(c) = 0 \tag{7.8.2}$$

even with the Langmuir isotherm, is not quite so simple. Nevertheless, we can make some progress in generalizing the results for sound waves and learn more about the origin and path of shocks. Let us first compare the variables and equations as listed in Table 7.1, in which the equation numbers (7.8.3) through (7.8.12) are assigned.

We can now see how the graphical construction of the last section can be modified. Since $-f'(c)$ is the inverse function of $g'(s)$, the differential condition (7.8.11) for the maximum of F may be written

$$\frac{\tau - \eta_0}{x} = f'(\Gamma(\eta_0)) = \sigma(\Gamma(\eta_0)) \tag{7.8.13}$$

Thus, in place of $u(Y, 0)$, a graph of $f'(\Gamma(\eta))$ may be drawn as in Fig. 7.19. The distribution $\Gamma(\eta)$ is inverted, and its proportions are changed by the transformation $f'(\Gamma(x))$, but, however great its range, it is now in the strip between σ_0 and zero. The point $\eta_0(x, \tau)$ is the intersection of the curve σ with the line of slope $-1/x$ and intercept τ. Again we see that there may be three intersections if $\tau > \tau_{\text{inf}}$ and $x > x_{\text{inf}}$, where these are defined by the intercept and slope of the tangent at the inflection point. It should be remembered that this is at the front of the wave when we express the initial

TABLE 7.1

	Chromatography	Sound waves	Equation Number
Equation	$\frac{\partial c}{\partial x} + f'(c)\frac{\partial c}{\partial \tau} = 0$	$\frac{\partial u}{\partial \tau} + u\frac{\partial u}{\partial X} = 0$	(7.8.3)
Initial data	$c(0, \tau) = \Gamma(\tau)$	$u(X, 0)$ given	(7.8.4)
Analogous variables	c x τ	u t X	(7.8.5)
Convex nonlinearity	$f(c)$	$\frac{1}{2}u^2$	(7.8.6)
Conjugate function	$g(s)$	$-\frac{1}{2}s^2$	
Derivative of nonlinearity Inverse function	$f'(c)$ $-g'(s)$	u s	(7.8.7)
Explicit solution	$c(x, \tau) = -g'\left(\frac{\tau - \eta_0}{x}\right)$	$u(X, t) = \frac{X - Y_M}{t}$	(7.8.8)
Maximizing function	$\eta_0(x, \tau)$	$Y_M(X, t)$	(7.8.9)
Function to be maximized	$F(\eta) = \int_{-\infty}^{\eta} \Gamma(\tau)\, d\tau - xg\left(\frac{\tau - \eta}{x}\right)$	$F(Y) = \int_{Y}^{\infty} u(X, 0)\, dX - \frac{(X - Y)^2}{2t}$	(7.8.10)
Differential condition for maximum	$\Gamma(\eta_0) + g'\left(\frac{\tau - \eta_0}{x}\right) = 0$	$u(Y_M, 0) = \frac{X - Y_M}{t}$	(7.8.11)
Condition for shock	Two values of η_0	Y_M give the same maximum value of F	(7.8.12)

data in the form $c(0, \tau) = \Gamma(\tau)$, for the concentration $\Gamma(\tau_1)$ has been put into the bed before the concentration $\Gamma(\tau_2)$ whenever $\tau_1 < \tau_2$.

The inflection points of the σ versus η curve where the tangent has negative slope are thus the starting points of the shocks. But the simple graphical construction by chords cutting off lobes of equal area cannot be carried out, since the distortion of scale in Fig. 7.19 destroys the equality of area. The

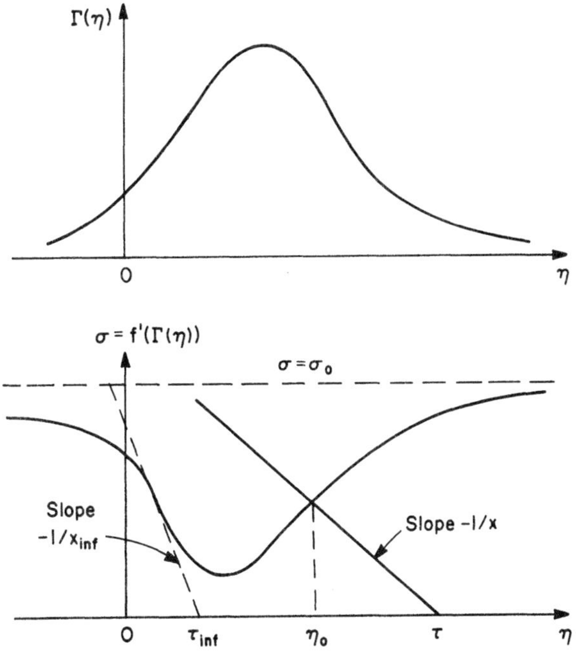

Figure 7.19

kind of diagram in which equality of area is maintained is like that in the lower part of Fig. 7.9, but here of course the intersecting line is the descending curve and not so easily drawn. Lighthill and Whitham in their paper on waves suggest that it can be used in the limiting cases of shocks just developing or finally dying. The general treatment will be discussed in the next section.

Although the simple construction cannot be carried through, quite a bit can be said about the inception of the shock. To see this, let us take the implicit solution of Eqs. (7.8.3) and (7.8.4) listed in Table 7.1,

$$c(x, \tau) = \Gamma(\tau - xf'(c(x, \tau))) \tag{7.8.14}$$

Clearly, this satisfies the initial condition, and by differentiating we have

$$\frac{\partial c}{\partial x} = \Gamma'(\tau - xf'(c))\left\{-f'(c) - xf''(c)\frac{\partial c}{\partial x}\right\}$$

or

$$\frac{\partial c}{\partial x} = -\frac{f'(c)\Gamma'(\tau - xf'(c))}{1 + xf''(c)\Gamma'(\tau - xf'(c))} \tag{7.8.15}$$

Similarly,

$$\frac{\partial c}{\partial \tau} = \frac{\Gamma'(\tau - xf'(c))}{1 + xf''(c)\Gamma'(\tau - xf'(c))} \tag{7.8.16}$$

which shows that the differential equation is satisfied. We notice that this implicit solution agrees with Lax's formula whenever it defines a unique value of c. For if

$$\eta = \tau - f'(c)x \tag{7.8.17}$$

then

$$c(x, \tau) = \Gamma(\eta)$$

But using the inverse function, Eq. (7.8.17) can be solved to give

$$c(x, \tau) = -g'\left(\frac{\tau - \eta}{x}\right) = \Gamma(\eta) \tag{7.8.18}$$

which is the same as Lax's formula (7.5.14). What Lax's formula does is to tell us just how to choose η when this inversion is not unique.

Obviously, the trouble all starts when the denominators in Eqs. (7.8.15) and (7.8.16) become zero, for then the solution has a vertical tangent plane and afterward must become multivalued. Thus the condition for the inception of a shock is

$$x\Gamma'(\tau - xf'(c))f''(c) = -1$$

Since the value $c = \Gamma(\eta)$ is borne on the characteristic $\tau = \eta + xf'(c)$, this condition is

$$x\Gamma'(\eta)f''(\Gamma(\eta)) = -1 \tag{7.8.19}$$

For a convex isotherm with $f''(c) < 0$, a shock can only form on a characteristic emanating from a region where $\Gamma'(\eta) > 0$. Since

$$x = -\{\Gamma'(\eta)f''(\Gamma(\eta))\}^{-1} = -\left\{\frac{df'(\Gamma(\eta))}{d\eta}\right\}^{-1} = -\left(\frac{d\sigma}{d\eta}\right)^{-1} \tag{7.8.20}$$

at the point of inception, this must correspond to a point of inflection on the σ versus η curve, as we have seen in Sec. 6.1.

7.9 Asymptotic Behavior

We have seen in Sec. 7.7 the ingenious procedure that enables one to find the shock path by application of the equal-area requirement to the initial data curve. However, the actual procedure is not straightforward unless the nonlinearity is quadratic, as we discussed in the previous section. Here we would like to extend the argument to the case of an arbitrary convex function (i.e., a genuinely nonlinear equation) and apply it to predict the asymptotic behavior of shocks for large time. In this regard, we will closely follow the elegant formulation laid down by Whitham.

Let us once again consider the chromatographic equation

$$\frac{\partial c}{\partial x} + \frac{\partial}{\partial \tau} f(c) = 0 \tag{7.9.1}$$

with the initial condition

$$c(0, \tau) = \Gamma(\tau) \tag{7.9.2}$$

and recall that, if the concentration profile contains a multivalued portion, it must be replaced by a discontinuity, which is a shock, as shown in Fig. 7.20. We notice that the profiles are drawn for the case $f''(c) < 0$, and so $c^l > c^r$; but the formulation in the following does hold for either case, $f''(c) < 0$ or $f''(c) > 0$, if the roles of c^l and c^r are interchanged (see Exercise 7.9.2). Clearly, the multivalued curve as well as the discontinuous one must satisfy the conservation law, and thus the area under each curve should be the same. It then follows that the two lobes cut off by the shock must be equal in area, and this argument is valid for both profiles in Fig. 7.20. Here we prefer to work with the c versus τ profile because of the way the initial data are prescribed.

For the continuous solution, we can write

$$\begin{aligned} c(x, \tau) &= \Gamma(\eta) \\ \tau &= \eta + \sigma(c)x \end{aligned} \tag{7.9.3}$$

If we let the inverse of the first equation be $\eta = \eta(c)$, the second equation may be put in the form

$$\tau(c, x) = \eta(c) + \sigma(c)x \tag{7.9.4}$$

and interpreted as the inverse of the function $c = c(x, \tau)$ in the multivalued solution. This means that we focus our attention on a particular state c and observe when it is introduced to the system, $\eta(c)$, and when it appears at a fixed position x, $\tau(c, x)$. With reference to Fig. 7.20(b), we may express the equal-area requirement as

$$(c^l - c^r)\tau = \int_{c^r}^{c^l} \tau(c, x)\, dc = \int_{c^r}^{c^l} \{\eta(c) + \sigma(c)\, x\}\, dc$$

But we see that $\sigma(c) = f'(c)$, and thus

$$(c^l - c^r)\tau - \{f(c^l) - f(c^r)\}x = \int_{c^r}^{c^l} \eta(c)\, dc \tag{7.9.5}$$

Now suppose we have a situation as shown in Fig. 7.21. At an arbitrary point $P(x, \tau)$ on the shock path where the states on the left and right sides are c^l and c^r, respectively, we can locate two characteristics with intercepts

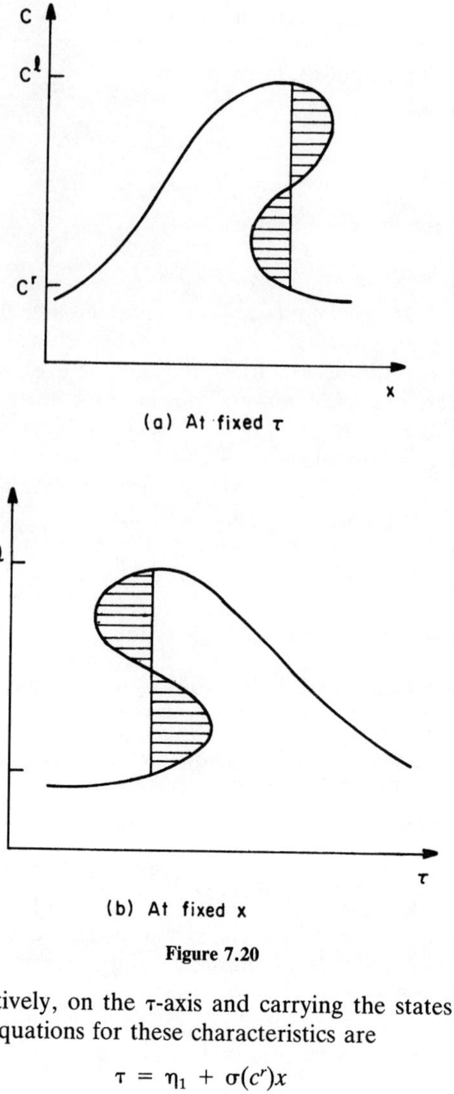

Figure 7.20

η_2 and η_1, respectively, on the τ-axis and carrying the states c^l and c^r, respectively. The equations for these characteristics are

$$\begin{aligned} \tau &= \eta_1 + \sigma(c^r)x \\ \tau &= \eta_2 + \sigma(c^l)x \end{aligned} \tag{7.9.6}$$

so that we have

$$x = -\frac{\eta_2 - \eta_1}{\sigma(c^l) - \sigma(c^r)} \tag{7.9.7}$$

and

$$(c^l - c^r)\tau = \{c^l\sigma(c^l) - c^r\sigma(c^r)\}x + \eta_2 c^l - \eta_1 c^r \tag{7.9.8}$$

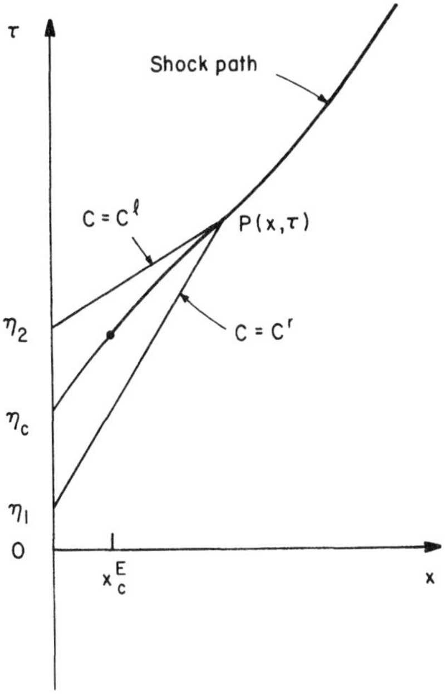

Figure 7.21

On the other hand, the right side of Eq. (7.9.5) can be integrated by parts to give

$$\eta_2 c^l - \eta_1 c^r - \int_{\eta_1}^{\eta_2} c(\eta)\, d\eta \tag{7.9.9}$$

Substituting these into Eq. (7.9.5) and replacing $c(\eta)$ with the initial data $\Gamma(\eta)$, we obtain

$$\{(c^l\sigma^l - c^r\sigma^r) - (f^l - f^r)\}\frac{\eta_2 - \eta_1}{\sigma^l - \sigma^r} = \int_{\eta_1}^{\eta_2} \Gamma(\eta)\, d\eta \tag{7.9.10}$$

in which abbreviated symbols are used for f and σ; for example, $f^l = f(c^l)$ and $\sigma^r = \sigma(c^r)$.

Given the initial data, we may be able to solve the three equations in Eqs. (7.9.6) and (7.9.10) to obtain τ, η_1, and η_2 as functions of x. The shock path is given by $\tau(x)$, whereas $\eta_1(x)$ and $\eta_2(x)$ will determine the states c^r and c^l, respectively, by Eq. (7.9.6). We notice that the shock behavior can be completely determined without being involved with differential equations as in Chapter 6. Incidentally, at the point where the shock is first formed, the shock strength is equal to zero, and the two characteristics merge to a single characteristic as noted in Fig. 7.21. Thus, from Eq. (7.9.7), we get

$$x^E = -\left\{\frac{d\sigma}{d\eta}\right\}^{-1} = -\left\{\frac{d\sigma}{dc}\frac{d\Gamma}{d\eta}\right\}^{-1} \tag{7.9.11}$$

which is indeed the equation of the envelope of characteristics [cf. Eq. (6.1.5)], and obviously the shock starts to form at the least value of x^E. Moreover, it may be proved by differentiation that Eqs. (7.9.6) and (7.9.10) satisfy the jump condition

$$\frac{d\tau}{dx} = \frac{[f]}{[c]} = \frac{f(c^l) - f(c^r)}{c^l - c^r} \tag{7.9.12}$$

[see Exercise 7.9.4].

To examine the asymptotic behavior of a shock, we will take an example where the initial data contain a pulse-type signal $G(\tau)$ above a uniform level of concentration c_0; that is,

$$\Gamma(\eta) = \begin{cases} c_0, & \eta < 0 \\ c_0 + G(\eta), & 0 < \eta < 1 \\ c_0, & \eta > 1 \end{cases} \tag{7.9.13}$$

where $G(\eta) \geq 0$. Then Eq. (7.9.10) gives

$$-\int_{\eta_1}^{\eta_2} G(\eta)\, d\eta = x\{c^l\sigma^l - c^r\sigma^r) - (f^l - f^r)\} + c_0(\eta_2 - \eta_1)$$

As the shock propagates after its inception, the state c^r, for example, decreases and eventually becomes equal to c_0. At this stage, $\eta_1 < 0$ and c^r is fixed at c_0 so that $\sigma^r = \sigma(c^r) = \sigma(c_0) = \sigma_0$. Furthermore, the factor $(\eta_2 - \eta_1)$ can be eliminated by Eq. (7.9.7) and $G(\eta) = 0$ for $\eta < 0$. Therefore,

$$-\int_0^{\eta_2} G(\eta)\, d\eta = x\{\sigma^l(c^l - c_0) - (f^l - f_0)\} \tag{7.9.14}$$

and since $c^l = \Gamma(\eta_2) = c_0 + G(\eta_2)$, this equation can be solved for η_2. The shock path is then given by the equation

$$\tau = \eta_2 + \sigma(c^l)x = \eta_2 + \sigma(\Gamma(\eta_2))x \tag{7.9.15}$$

In the limit as $t \to \infty$, we see $\eta_2 \to 1$, $\sigma^l \to \sigma_0$, and

$$f^l - f_0 = f(c^l) - f(c_0) \cong f'(c_0)(c^l - c_0) + \frac{1}{2}f''(c_0)(c^l - c_0)^2$$

$$= \sigma_0(c^l - c_0) + \frac{1}{2}f''(c_0)(c^l - c_0)^2$$

so that Eq. (7.9.14) gives

$$-2\int_0^1 G(\eta)\, d\eta = f''(c_0)(c^l - c_0)^2 x \tag{7.9.16}$$

Now letting M denote the strength of the initial signal, that is,

$$M = \int_0^1 G(\eta)\, d\eta \tag{7.9.17}$$

and putting

$$\mu_0 = -f''(c_0) = -\sigma'(c_0) > 0 \tag{7.9.18}$$

we finally obtain the asymptotic formula for the shock strength as

$$c^l - c_0 \sim \sqrt{\frac{2M}{\mu_0 x}} \tag{7.9.19}$$

At this stage the shock becomes sufficiently weak, and we can write

$$\sigma(c^l) - \sigma_0 \cong -\mu_0(c^l - c_0) \sim -\sqrt{\frac{2\mu_0 M}{x}} \tag{7.9.20}$$

The asymptotic equation for the shock path is then determined from Eq. (7.9.15) as

$$\tau \sim 1 + \sigma_0 x - \sqrt{2\mu_0 M x} \sim \sigma_0 x - \sqrt{2\mu_0 M x} \tag{7.9.21}$$

For large time, therefore, the shock decays like $1/\sqrt{x}$ and the shock path is asymptotically parabolic as illustrated in Fig. 7.22. We notice that this is the generalization of what we have already observed in a specific example with the Langmuir isotherm in Sec. 6.3 (see Exercise 7.9.1).

The simple wave solution to the left of the shock path is given by Eq. (7.9.15) with c^l replaced by c and η_2 by η, where η spans from η_2 to $\eta = 1$. Thus

$$\sigma(c) = \frac{\tau - \eta}{x} \sim \frac{\tau}{x}, \qquad \sigma_0 x - \sqrt{2\mu_0 M x} < \tau < \sigma_0 x$$

and, by virtue of Eq. (7.9.20), we obtain

$$c - c_0 \sim \frac{1}{\mu_0}\left(\sigma_0 - \frac{\tau}{x}\right), \qquad \sigma_0 x - \sqrt{2\mu_0 M x} < \tau < \sigma_0 x \tag{7.9.22}$$

which is identical to Eq. (7.6.23). It is interesting that the shape of the initial signal $G(\eta)$ is completely lost and that the asymptotic behavior depends only on the strength M of the initial signal and the nonlinearity, $\mu_0 = -f''(c_0)$, of the differential equation. Indeed, these asymptotic formulas, (7.9.19), (7.9.21), and (7.9.22), are equivalent to those proved by Lax (1957, 1972, 1973) for the conservation equation

$$\frac{\partial u}{\partial t} + \frac{\partial}{\partial x} f(u) = 0$$

with $f''(u) > 0$. Thus the roles of x and τ are equivalent to those of t and x, respectively.

At the ultimate stage, the shock path tends to be linear with the slope approaching σ_0 asymptotically. Hence we may replace x by τ/σ_0 in Eq.

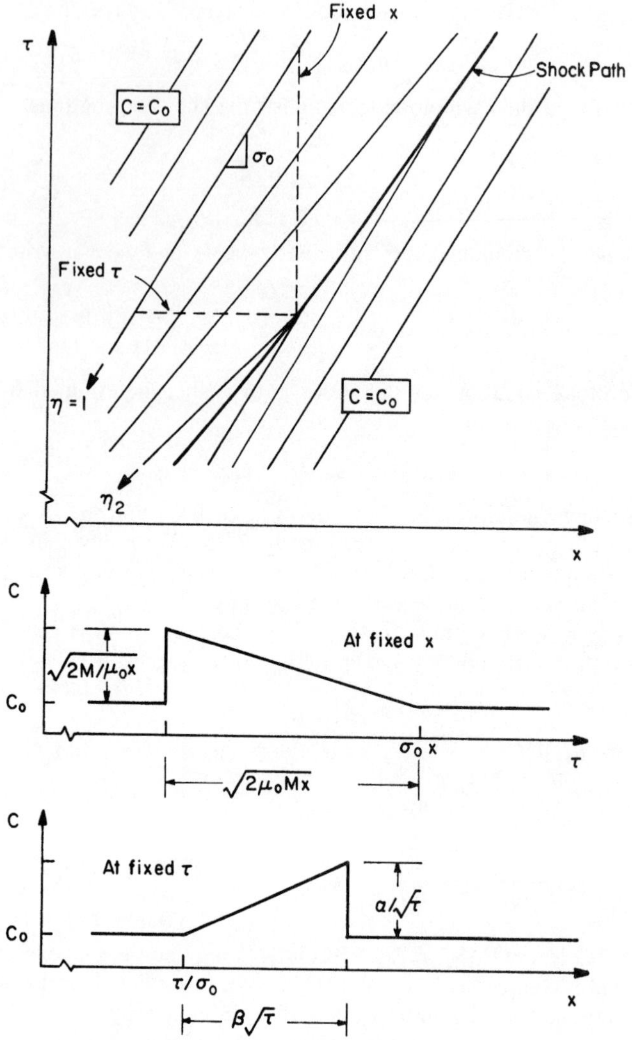

Figure 7.22

(7.9.19) to express the asymptotic formula in terms of τ, and then we have

$$c^l - c_0 \sim \frac{\alpha}{\sqrt{\tau}} \tag{7.9.23}$$

where

$$\alpha = \sqrt{\frac{2\sigma_0 M}{\mu_0}} \tag{7.9.24}$$

For the shock path, we put Eq. (7.9.15) in the form $x \sim \tau/\sigma(c')$ and then use Eqs. (7.9.20) and (7.9.23) to obtain

$$x \sim \frac{\tau}{\sigma_0} + \beta\sqrt{\tau} \tag{7.9.25}$$

in which

$$\beta = \sqrt{\frac{2\mu_0 M}{\sigma_0^3}} \tag{7.9.26}$$

With respect to Eq. (7.9.22), we would observe the profile at fixed x and the parameter η varies from η_2 to 1; but now we view the profile at fixed τ so that η runs in the reverse direction (see Fig. 7.22). Hence the asymptotic formula corresponding to Eq. (7.9.22) is

$$c - c_0 \sim \frac{\sigma_0^2}{\mu_0}\left(\frac{x}{\tau} - \frac{1}{\sigma_0}\right), \qquad \frac{\tau}{\sigma_0} < x < \frac{\tau}{\sigma_0} + \beta\sqrt{\tau} \tag{7.9.27}$$

and this is depicted in Fig. 7.22.

If the initial signal $G(\eta)$ has a shape similar to one cycle of a sine wave, a positive phase being followed by a negative phase about a uniform level of concentration c_0, we expect to have two compression waves, so two separate shocks will appear: one in the front side and another at the rear side. For large time τ, therefore, the concentration profile will take the form of an N-wave, as shown in Fig. 7.23. For the shock in the front side, we have

$$c_+ - c_0 \sim \frac{\alpha_+}{\sqrt{\tau}}$$

$$x_+ \sim \frac{\tau}{\sigma_0} + \beta_+\sqrt{\tau}$$

where α_+ and β_+ are given by Eqs. (7.9.24) and (7.9.26), respectively, with M denoting the area of the curve $G(\eta)$ above $c = c_0$. Similarly, we can write

$$c_- - c_0 \sim \frac{\alpha_-}{\sqrt{\tau}}$$

$$x_- \sim \frac{\tau}{\sigma_0} - \beta_-\sqrt{\tau}$$

for the shock in the rear side, with α_- and β_- associated with the area of $G(\eta)$ below $c = c_0$. Between these two shocks, we have a simple wave, which may be expressed by the asymptotic formula

$$c - c_0 \sim \frac{x}{\tau}, \qquad \frac{\tau}{\sigma_0} - \beta_-\sqrt{\tau} < x < \frac{\tau}{\sigma_0} + \beta_+\sqrt{\tau}$$

because the slope at fixed τ is obviously equal to $1/\tau$.

Suppose now the order of the positive and negative phases of the curve

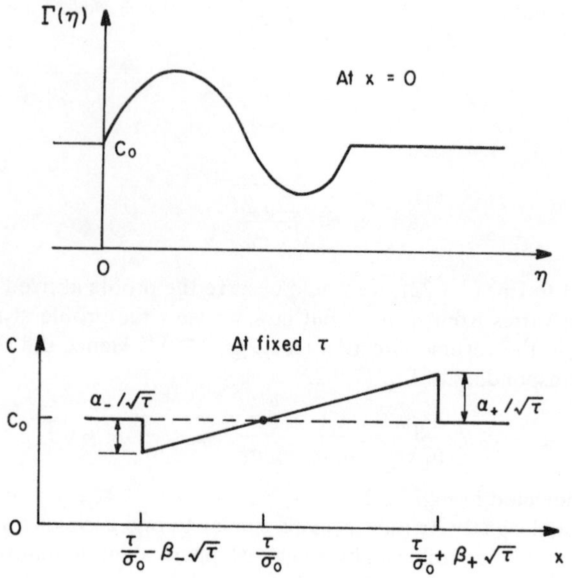

Figure 7.23

$G(\eta)$ is reversed. We then have a compression wave in the middle so that only one shock appears with a simple wave on each side, as shown in Fig. 7.24. If we consider the two simple waves together, the region is bounded by two parallel characteristics of slope σ_0, and yet the asymptotic profile may be comparable to the N-wave with the negative phase translated to the right so that the two shocks are united. Therefore, it is expected that the shock in this case will decay faster than in the previous case. Now, since the initial signal $G(\eta)$ is introduced over an interval 1, the distance between the pair of characteristics is always equal to $1/\sigma_0$. Then, in order for the asymptotic profile to be linear in x with slope $1/\tau$, the shock strength must be asymptotically equal to $1/(\sigma_0\tau)$, and so the shock must decay like $1/\tau$.

This argument may be more convincing if we consider a periodic wave. Clearly, the portrait in the (x, τ)-plane, as well as the asymptotic profile, will be given by a train of each of the corresponding units in Fig. 7.24. The asymptotic profile over an interval of two periods is given in Fig. 7.25. (See also Figs. 6.20 and 6.21.) Looking at the two shocks and the simple wave in between, we may regard the periodic wave as approaching to a train of N-waves. Since the period here in the x-direction is $p = 1/\sigma_0$, the strength of the shock is asymptotically equal to p/τ. This nature of the periodic wave has been proved in general by Lax (1957, 1972) and illustrated in detail for a quadratic function $f(c)$ by Whitham (1974). For conservation laws without convexity conditions, Dafermos (1972) has treated the asymptotic behavior of solutions under periodic initial conditions by using Liapunov functionals

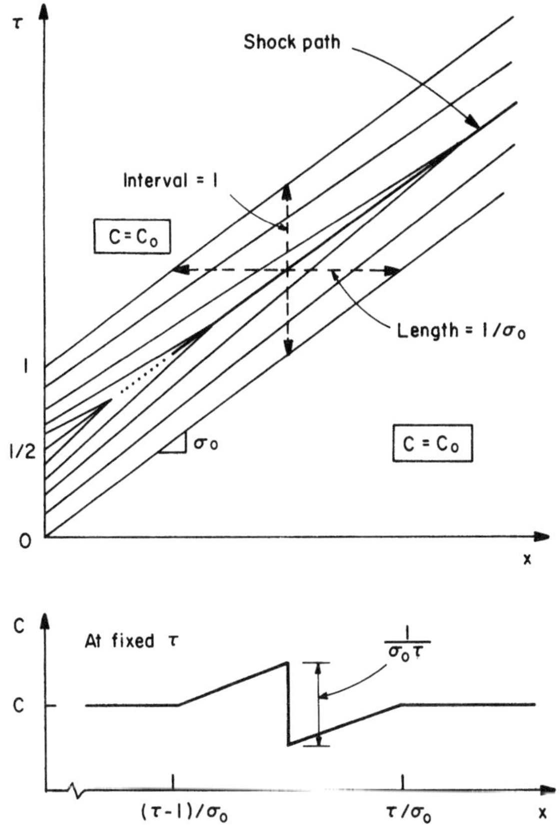

Figure 7.24

and established a similar result to that of Lax. A more elaborate result has been given by Greenberg and Tong (1973): for convex functions $f(c)$ such that $f''(c) \sim -bc^{2k}$ as $c \to 0$, where $k \geq 0$ is an integer and b a positive constant, and with periodic initial data of period p whose integral over one period vanishes, the solution satisfies

$$| c(x, \tau) | \leq \frac{C(k, b, p)}{\tau^{1/(2k+1)}}$$

in which the constant C may be chosen independently of the particular initial data. When $k = 0$, this estimate coincides with that of Lax.

For conservation laws in more than one space dimension, Conway (1977) has treated the formation and decay of shocks and proved that, if the flux function is genuinely nonlinear, a solution with compact support for fixed time must decay like $1/\sqrt{\tau}$ for large time.

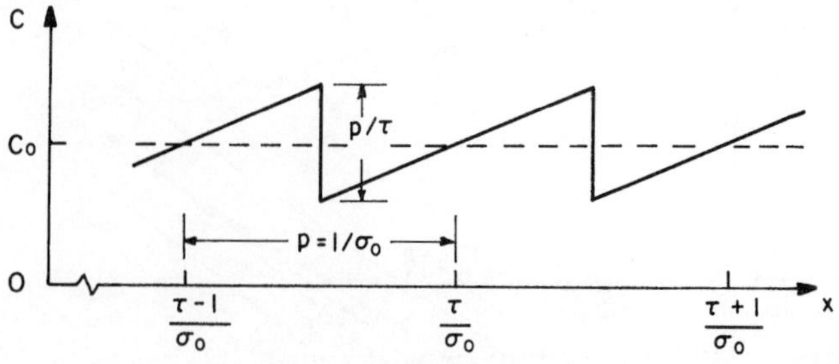

Figure 7.25

Exercises

7.9.1. For the Langmuir isotherm

$$f(c) = c + \frac{\nu\gamma c}{1 + Kc}$$

with the initial condition

$$c(0, \tau) = \begin{cases} 0, & \tau < 0 \\ c_1, & 0 < \tau < \alpha \\ 0, & \tau > \alpha \end{cases}$$

show that Eqs. (7.9.19) and (7.9.21) give the asymptotic forms of Eqs. (6.3.11) and (6.3.10), respectively.

7.9.2 Repeat the formulation of this section for the case $f''(c) > 0$, and confirm that the asymptotic formula for the shock strength is given by Eq. (7.9.19) with c^r in place of c^l and $\mu_0 = f''(c_0)$. Also, find the formulas corresponding to Eqs. (7.9.21) and (7.9.22).

7.9.3. Consider the conservation equation

$$\frac{\partial \rho}{\partial t} + \frac{\partial}{\partial x}\phi(\rho) = 0$$

subject to the initial condition

$$\rho(x, 0) = \begin{cases} \rho_0, & x < 0 \\ \rho_0 + F(\rho), & 0 < x < 1 \\ \rho_0, & x > 1 \end{cases}$$

where $F(\rho)$ is a positive, pulselike signal. For the case $\phi''(\rho) < 0$, apply the procedure of this section to find the asymptotic formulas describing the shock behavior. Also, show that for the special case of $\phi(\rho) = \rho(1 - \rho)$ the formulas are reduced to the asymptotic forms of Eqs. (6.6.37), (6.6.38), and (6.6.40) if

$$\int_0^1 F(\rho)\, d\rho = \rho_1 - \rho_0$$

7.9.4. Show by differentiation that Eqs. (7.9.5) and (7.9.6) satisfy the jump condition (7.9.12). [*Hint:* Differentiate Eq. (7.9.5) with respect to x and then use Eq. (7.9.6).]

7.10 Shock-layer Analysis

In the previous two chapters we have repeatedly encountered discontinuous solutions, especially shocks, and we have established in this chapter the mathematical basis for discontinuous solutions. In the practical sense, however, discontinuous solutions are rather unrealistic since we naturally expect that the profiles of physical quantities would be smooth. On the other hand, we recall from Sec. 4.2 that the concentration profile gets steeper as phase equilibrium is approached between two phases and also from Sec. 7.6 that the profile approaches the discontinuous solution as the longitudinal diffusion becomes negligible. It is thus expected that axial dispersion in the fluid phase, as well as interphase transfer resistance, will smear out any sharp discontinuity. The question is then to what extent these factors affect the structure of the profile. According to experimental observation, the profile tends to approach a constant pattern that moves with a constant speed without changing its shape. This implies that the profile may become diffuse only to a certain degree and no more. The same phenomenon was recognized even earlier in compressible fluid flow problems in which the term "shock layer" was coined to represent the solution of constant pattern, because, as we will see soon, it corresponds to the shock in the limit as the diffusive effects become completely negligible. [See, for example, von Mises (1950) and Gilbarg (1951).] For this reason, we will use the term *shock layer* in the following.

To demonstrate this, as well as to investigate the influence of axial dispersion and interphase transfer resistance, we will take the chromatographic equation from Eq. (7.1.4) with the axial dispersion term included,

$$\frac{1}{Pe}\frac{\partial^2 c}{\partial x^2} = \frac{\partial c}{\partial x} + \frac{\partial c}{\partial \tau} + \nu\frac{\partial n}{\partial \tau} \tag{7.10.1}$$

and the rate expression motivated from Eq. (1.2.6),

$$\frac{\partial x}{\partial \tau} = St(f(c) - n) \tag{7.10.2}$$

in which we have

$$Pe = \frac{VZ}{E_a} \tag{7.10.3}$$

$$St = \frac{kZ}{V} \tag{7.10.4}$$

and $f(c)$ is used to represent the adsorption equilibrium isotherm. Here E_a denotes the effective axial dispersion coefficient, k the overall mass-transfer coefficient based on the solid phase,† and ν, V, and Z have their usual meanings [cf. Egs. (7.1.2) and (7.1.3)]. To exclude any end effects, the domain will be assumed to be infinitely large so that the appropriate boundary conditions are

$$\begin{aligned} c &= c^l(\text{constant}) \qquad \text{at } x = -\infty \\ c &= c^r(\text{constant}) \qquad \text{at } x = +\infty \end{aligned} \tag{7.10.5}$$

Now we search for a moving coordinate ξ:

$$\xi = x - \tilde{\lambda}\tau \tag{7.10.6}$$

in which the solution to Eqs. (7.10.1) and (7.10.2) can be expressed in the form

$$\begin{aligned} c(x, \tau) &= c(\xi) \\ n(x, \tau) &= n(\xi) \end{aligned} \tag{7.10.7}$$

and satisfies the following conditions:

$$\begin{aligned} c = c^l, \qquad \frac{dc}{d\xi} = \frac{dn}{d\xi} = 0 \qquad \text{at } \xi = -\infty \\ c = c^r, \qquad \frac{dc}{d\xi} = \frac{dn}{d\xi} = 0 \qquad \text{at } \xi = +\infty \end{aligned} \tag{7.10.8}$$

If there exists such a real value $\tilde{\lambda}$, Eq. (7.10.7) represents the shock layer of the system described by Eqs. (7.10.1) and (7.10.2), and $\tilde{\lambda}$ is the constant speed of the shock layer.

Suppose there exists a shock layer so that we can write

$$\frac{\partial c}{\partial x} = \frac{dc}{d\xi}\frac{\partial \xi}{\partial x} = \frac{dc}{d\xi}, \qquad \frac{\partial^2 c}{\partial x^2} = \frac{d^2 c}{d\xi^2}$$

$$\frac{\partial c}{\partial \tau} = \frac{dc}{d\xi}\frac{\partial \xi}{\partial \tau} = -\tilde{\lambda}\frac{dc}{d\xi}$$

and substituting these into Eqs. (7.10.1) and (7.10.2) to have

$$\begin{aligned} \frac{1}{Pe}\frac{d^2c}{d\xi^2} - (1 - \tilde{\lambda})\frac{dc}{d\xi} + \nu\tilde{\lambda}\frac{dn}{d\xi} = 0; \\ -\tilde{\lambda}\frac{dn}{d\xi} = St(f(c) - n) \end{aligned} \tag{7.10.9}$$

†Based on eq. (1.2.6), k may be regarded as the symbol representing the desorption rate constant.

When compared with Eq. (7.10.8), these equations suggest the auxiliary conditions

$$\begin{aligned} \frac{d^2c}{d\xi^2} &= 0 \qquad n = f(c^l), \qquad \text{at } x = -\infty \\ \frac{d^2c}{d\xi^2} &= 0 \qquad n = f(c^r), \qquad \text{at } x = +\infty \end{aligned} \tag{7.10.10}$$

Eliminating n and $dn/d\xi$ between the two equations in Eq. (7.10.9), we obtain

$$\frac{\tilde{\lambda}}{PeSt}\frac{d^3c}{d\xi^3} - \left\{\frac{1}{Pe} + \frac{\tilde{\lambda}(1-\tilde{\lambda})}{St}\right\}\frac{d^2c}{d\xi^2} + (1-\tilde{\lambda})\frac{dc}{d\xi} - \nu\tilde{\lambda}\frac{df}{d\xi} = 0$$

and, by integrating this from $\xi = -\infty$, where both $dc/d\xi$ and $d^2c/d\xi^2$ vanish, to ξ, we have

$$-\frac{\tilde{\lambda}}{PeSt}\frac{d^2c}{d\xi^2} + \left\{\frac{1}{Pe} + \frac{\tilde{\lambda}(1-\tilde{\lambda})}{St}\right\}\frac{dc}{d\xi} = (1-\tilde{\lambda})(c - c^l) - \nu\tilde{\lambda}\{f(c) - f(c^l)\} \tag{7.10.11}$$

In the limit as $\xi \to +\infty$, the left side becomes equal to zero, and thus

$$\frac{1}{\tilde{\lambda}} = 1 + \nu\frac{f(c^l) - f(c^r)}{c^l - c^r} = 1 + \nu\frac{[f]}{[c]} \tag{7.10.12}$$

which is the very jump condition to be satisfied by the shock with the states c^l and c^r on either side. Equation (7.10.11) is then rearranged in the form

$$-\frac{\tilde{\lambda}}{PeSt}\frac{d^2c}{d\xi^2} + \left\{\frac{1}{Pe} + \frac{\tilde{\lambda}(1-\tilde{\lambda})}{St}\right\}\frac{dc}{d\xi} = \nu\tilde{\lambda}F(c;\, c^l, c^r) \tag{7.10.13}$$

where

$$F(c;\, c^l, c^r) = \frac{[f]}{[c]}(c - c^l) - f(c) + f(c^l) \tag{7.10.14}$$

Consequently, if there exists a shock layer, it satisfies Eq. (7.10.13), and its propagation speed $\tilde{\lambda}$ is uniquely determined by the equilibrium relation and the end states. This speed is independent of Pe and St and is the same as that of the corresponding shock. It is interesting to notice that the dispersion (Pe^{-1}) and the mass-transfer resistance (St^{-1}) assume an equivalent role in Eq. (7.10.13) and that, in addition to the individual contribution that appears as the first-order derivative term, there exists a coupled effect that is given by the second-order derivative term.

For a given pair of states c^l and c^r, we now ask if Eq. (7.10.13) with the predetermined value of $\tilde{\lambda}$ has a solution that satisfies the conditions (7.10.8) and further if it is unique. We will establish the proof for the case $f''(c) < 0$ so that $F''(c;\, c^l, c^r) > 0$ and F has only two zeros, c^l and c^r. Let us rearrange Eq. (7.10.13) in the form

$$\begin{aligned} \frac{dc}{d\xi} &= q \equiv G(c, q) \\ \frac{dq}{d\xi} &= \left\{(1-\tilde{\lambda})Pe + \frac{St}{\tilde{\lambda}}\right\}q - \nu PeStF(c; c^l, c^r) \equiv H(c, q) \end{aligned} \tag{7.10.15}$$

and consider the portrait in the (c, q)-plane. Since F is a single-valued function of c and has exclusive singular points at c^l and c^r, the curves $G = 0$ and $H = 0$ separate the half-plane $(c \geq 0)$ into five distinct regions, as shown in Fig. 7.26, and each region is simply connected. Both $G(c, q)$ and $H(c, q)$ assume a uniform sign in each region, and so all the integral curves of Eq. (7.10.15) must satisfy these conditions. In particular, the integral curve is allowed to cross the curve $G = 0$ only with an infinite slope and the curve $H = 0$ only with a zero slope.

Let A and B denote the singular points as shown in Fig. 7.26 and note that, if $c = c^r$ at A, then $c = c^l$ at B, or vice versa. We will now examine the nature of the direction field in the neighborhood of these points. The characteristic equation of the system (7.10.15) is

$$\begin{vmatrix} \dfrac{\partial G}{\partial c} - \gamma & \dfrac{\partial G}{\partial q} \\ \\ \dfrac{\partial H}{\partial c} & \dfrac{\partial H}{\partial q} - \gamma \end{vmatrix} = 0$$

or

$$\gamma^2 - \left\{(1-\tilde{\lambda})Pe + \frac{St}{\tilde{\lambda}}\right\}\gamma + \nu PeStF'(c; c^l, c^r) = 0 \tag{7.10.16}$$

The two roots are determined as

$$\begin{aligned} \gamma \pm &= (1-\tilde{\lambda})\,Pe + \frac{St}{\tilde{\lambda}} \\ &\quad \pm \sqrt{\left\{(1-\tilde{\lambda})Pe + \frac{St}{\tilde{\lambda}^2}\right\}^2 - 4\nu PeStF'(c; c^l, c^r)} \\ &= (1-\tilde{\lambda})Pe + \frac{St}{\tilde{\lambda}} \\ &\quad \pm \sqrt{\left\{(1-\tilde{\lambda})Pe - \frac{St}{\tilde{\lambda}}\right\}^2 + 4\nu PeStf'(c)} \end{aligned} \tag{7.10.17}$$

in which Eqs. (7.10.12) and (7.10.14) have been introduced. Since the isotherm $f(c)$ is a monotone-increasing function, $f'(c) > 0$ so that both γ_+

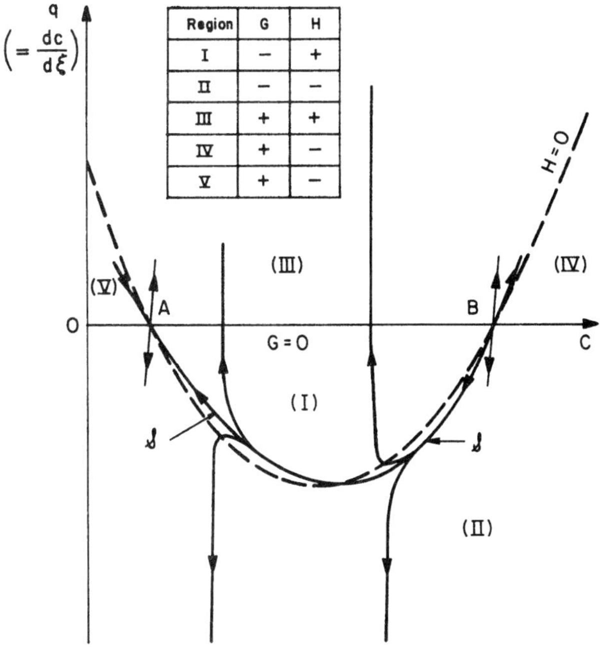

Figure 7.26

and γ_- are real. At point A, $F' < 0$, and so $\gamma_- < 0 < \gamma_+$, whereas at B, $F' > 0$ so that $0 < \gamma_- < \gamma_+$. Hence A is a saddle point and B is an unstable node. According to perturbation theory, there are two integral curves of Eq. (7.10.15) that approach the saddle point A in the direction $(1, \gamma_-)$ as $\xi \rightarrow +\infty$ and two that approach A in the direction $(1, \gamma_+)$ as $\xi \rightarrow -\infty$. Comparing with the sign of the ratio H/G, we notice that one of the integral curves converging to A as $\xi \rightarrow +\infty$ approaches it from region I. Let this integral curve be designated by $\mathscr{S}(\xi)$.

Clearly, $\mathscr{S}(\xi)$ is montonic with negative slope while in region I, and yet it must approach the node B as $\xi \rightarrow -\infty$ because there are no other singular points. All the integral curves of Eq. (7.10.15), however, can approach B only with positive slopes as $\xi \rightarrow -\infty$ and, thus, if we follow backward along $\mathscr{S}(\xi)$, we must cross the curve $H = 0$ with zero slope toward region II. This is permitted because $dc/d\xi = G < 0$ in regions I and II. Now $\mathscr{S}(\xi)$ remains monotonic with positive slope in region II, so there is only one crossover between $\mathscr{S}(\xi)$ and the curve $H = 0$. Once we cross the curve $H = 0$, therefore, we must approach B from region II as $\xi \rightarrow -\infty$. With this integral curve, we have $c = c^r$ at A and $c = c^l$ *at* B, or $c^r < c^l$. It is a monotone-decreasing function of ξ and has an inflection point given by the intersection between $\mathscr{S}(\xi)$ and the curve $H = 0$, as illustrated in Fig. 7.27. Such an integral curve $\mathscr{S}(\xi)$ is certainly a solution of Eq. (7.10.15), and this proves

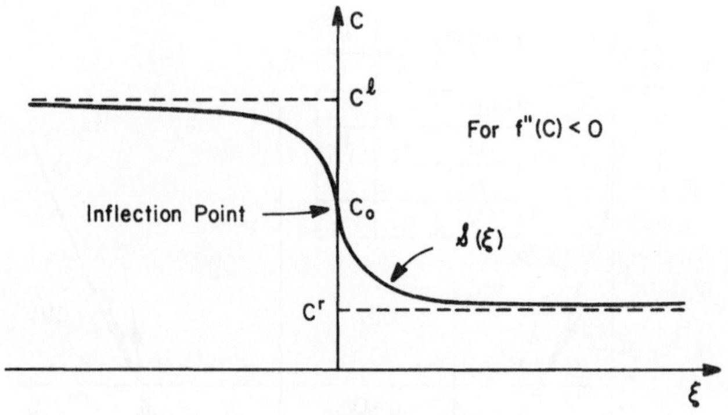

Figure 7.27

that there exists a shock layer of the system (7.10.1) and (7.10.2) if $c^r < c^l$. We notice that the condition $c^r < c^l$ is equivalent to the shock condition.

To prove the uniqueness of the shock layer, we will first examine the situation in the neighborhood of point A carefully. Since point B (unstable node) can be approached by the integral curve only as $\xi \to -\infty$, point A should be approached as $\xi \to +\infty$. Point A being a saddle point, we see that there are exactly two integral curves of Eq. (7.10.15) that approach A as $\xi \to +\infty$, one from region I and the other from region V. While the former is identical to $\mathcal{S}(\xi)$, the latter is the only integral curve other than $\mathcal{S}(\xi)$ that can connect points A and B. We will call this curve $\overline{\mathcal{S}}(\xi)$. If $\overline{\mathcal{S}}(\xi)$ is also a shock layer, then $\mathcal{S}(\xi)$ and $\overline{\mathcal{S}}(\xi)$ form a closed curve enclosing a simply connected region D in the (c, q)-plane. On the other hand, we have already observed that there are two integral curves of Eq. (7.10.15) that approach A with positive slope γ_+ as $\xi \to -\infty$. One of these curves enters the region D as ξ increases. Since A and B are exclusive singular points, it cannot terminate or approach a limit cycle in region D. Therefore, the integral curve must intersect $\mathcal{S}(\xi)$ or $\overline{\mathcal{S}}(\xi)$. This is a contradiction to the uniqueness of the integral curves and so the uniqueness of the shock layer is established.

If $f''(c) > 0$, point B becomes a saddle point and point A is an unstable node. By applying the same procedure, we can show that there exists a unique shock layer if $c^l < c^r$. The case in which $f''(c)$ changes sign is more involved, and as a typical example we will take the BET isotherm, which is characterized by the conditions $f''(c) < 0$ for $c < c_I$, $f''(c_I) = 0$, and $f''(c) > 0$ for $c > c_I$, where c_I corresponds to the inflection point of the isotherm. Then it can be argued that there exists a unique shock layer if the end states c^l and c^r satisfy one of the conditions

$$c^r < c^l \leq c_t \qquad \text{for } c^r < c_I$$

or (7.10.18)

$$c_t \leq c^l < c^r \qquad \text{for } c^r > c_I$$

where c_t corresponds to the tangential point between the isotherm and the straight line from the point of c^r. These conditions are equivalent to the shock condition (see Exercises 7.6.2 and 7.10.3).

Now that we have proved the existence and uniqueness of the shock layer, we will examine the limit of the shock layer as $Pe \to \infty$ and $St \to \infty$. Suppose that the axial dispersion coefficients E_a and the mass-transfer resistance $1/k$ can be independently made arbitrarily small while the end states c^l and c^r remain fixed. We can thus take the limit one by one, for example, first the limit as $1/k \to 0$ and then the limit as $E_a \to 0$. The first limit directly eliminates the second-order derivative term from Eq. (7.10.13) because $(\tilde{\lambda}/Pe)d^2c/d\xi^2$ remains finite while $St \to \infty$, and thus we have the equation

$$\frac{1}{\nu\tilde{\lambda}Pe}\frac{dc}{d\xi} = F(c; c^l, c^r) \tag{7.10.19}$$

Although we have already argued in Sec. 7.6 that the solution of this equation will tend to the corresponding shock as $Pe \to \infty$ (see Fig. 7.28), we would like to prove it here with a different motivation.

Since the shock layer has only one inflection point, we can make use of it as a fixed point. Let c_0 be the c value at the inflection point and then $F'(c_0) = 0$ or, from Eq. (7.10.14), we have

$$f'(c_0) = \frac{[f]}{[c]} = \frac{f(c^l) - f(c^r)}{c^l - c^r} \tag{7.10.20}$$

Now we will assign the value $c = c_0$ at $\xi = 0$ so that Eq. (7.10.19) may be solved by quadratures:

$$\nu\tilde{\lambda}Pe\xi(c) = \int_{c_0}^{c}\frac{dc'}{F(c'; c^l, c^r)} \tag{7.10.21}$$

which is equivalent to Eq. (7.6.13). But the integral on the right side diverges for both $c = c^l$ and $c = c^r$, so we will introduce the concept of the *shock*

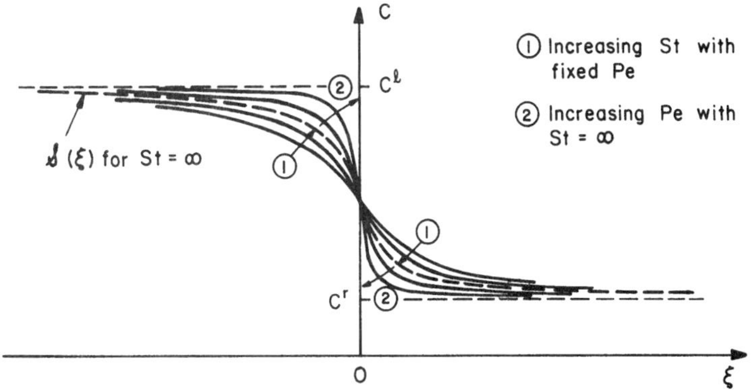

Figure 7.28

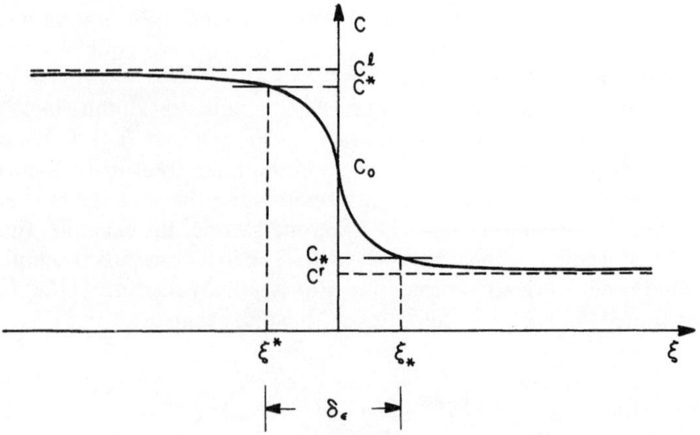

Figure 7.29

layer thickness. Let us consider two concentration bounds, c^* and c_*, associated with a parameter ε, as follows:

$$\frac{c^l - c^*}{c^l - c^r} = \frac{c_* - c^r}{c^l - c^r} = \varepsilon\left(< \frac{1}{2}\right) \tag{7.10.22}$$

and define the shock layer thickness δ_ε by Eq. (7.10.21) as

$$\delta_\varepsilon = \xi_* - \xi^* = \frac{1}{\nu\tilde{\lambda}Pe}\int_{c^*}^{c_*} \frac{dc'}{F(c'; c^l, c^r)} \tag{7.10.23}$$

This is illustrated in Fig. 7.29. If we specify $\varepsilon = 0.005$, then δ_ε represents the width over which 99 percent of the total variation is accomplished.

From the mathematical point of view, this concept gives rise to the parametrization of the shock layer and enables us to investigate the limit as $Pe \rightarrow \infty$. For a fixed value of ε, the integral of Eq. (7.10.23) yields a constant, and so δ_ε is inversely proportional to Pe (or linearly proportional to E_a). Consequently, by making Pe sufficiently large, the shock-layer thickness can be made arbitrarily small for all values of ε; that is,

$$\lim_{Pe\rightarrow\infty} \delta_\varepsilon = 0$$

This implies that a shock layer approaches the corresponding shock in the limit as $Pe \rightarrow \infty$, as shown in Fig. 7.28, and consequently we have

$$\lim_{\substack{St\rightarrow\infty \\ Pe\rightarrow\infty}} c(\xi; Pe, St) = \begin{cases} c^l, & \text{for } -\infty < \xi < \xi^s \\ c^r, & \text{for } \xi^s < \xi < +\infty \end{cases} \tag{7.10.24}$$

the convergence being uniform in every closed interval not containing $\xi = \xi^s$, where $\xi^* < \xi_s < \xi_*$.

In summary, we see that for an arbitrary isotherm $f(c)$ there exists a unique shock layer of Eqs. (7.10.1) and (7.10.2) if and only if the end states c^l and c^r satisfy the shock condition. The shock layer propagates with the same speed as that of the corresponding shock and converges uniformly to the corresponding shock in the limit as $Pe \to \infty$ and $St \to \infty$ independently. A physical argument may be worth adding at this point. Without axial dispersion and mass-transfer resistance, we would have a shock as long as the shock condition is satisfied. The shock has a self-sharpening tendency because the state c^l tends to overtake the state c^r, and the shock just keeps this from taking place. On the other hand, the dispersion, as well as the mass-transfer resistance, would make the solution more and more diffuse as time increases. It is then evident that the two effects tend to be balanced against each other as time goes on and ultimately give rise to the shock layer. If the shock condition is not satisfied, the solution would be given either by an expansive simple wave or by a combined wave. Now both dispersion and mass-transfer resistance will make the expansion wave all the more diffuse, and in the case of the combined wave their effect on the simple wave part will destroy the self-sharpening tendency of the shock part, so it is not expected that a shock layer forms in either case.

The profile of the shock layer can be constructed by integrating the nonlinear differential equation (7.10.13) or the system (7.10.15). Since the domain is infinitely long and we have a singularity at either end, the numerical integration does not seem straightforward. However, the nature of the direction field in the (c, q)-plane reveals that the backward integration would give a fast approach to the shock layer, whereas the forward integration always becomes unstable. Thus it is suggested to start the numerical integration from $\xi = -\infty$ by using the conditions $c = c^l$ and $dc/d\xi = \kappa$, where κ denotes a very small number of the order of, say, 10^{-10} and $\kappa < 0$ if $f''(c) < 0$ and $\kappa > 0$ if $f''(c) > 0$.

In the practice of chromatography, however, both Pe and St are usually very large, and so in Eq. (7.10.13) the second-order derivative term would be very small in comparison to the first-order derivative term throughout the entire range of c values of interest; that is, as an approximation, the coupled effect may be neglected in comparison to the sum of the individual effects of dispersion and mass-transfer resistance.† If this is the case, the shock layer and its thickness are approximated by Eqs. (7.10.21) and (7.10.23), respectively, with $[1/Pe + \tilde{\lambda}(1 - \tilde{\lambda})/St]$ in place of $1/Pe$. In the limiting case, when axial dispersion becomes completely negligible (i.e., $Pe = \infty$), we have exact expressions given by Eqs. (7.10.21) and (7.10.23), with $\tilde{\lambda}(1 - \tilde{\lambda})/St$ instead of $1/Pe$. Thus the shock-layer thickness is now inversely proportional to the mass-transfer coefficient k.

†In a specific example with the Langmuir isotherm, the ratio of the two remains lower than 0.02 except for some extreme cases.

As an illustration, we will consider the Langmuir isotherm

$$f(c) = \frac{\gamma c}{1 + Kc} \tag{7.10.25}$$

for which we have $f''(c) < 0$, and so only the case of $c^l > c^r$ is of our concern. If we let

$$\beta^l = 1 + Kc^l$$

$$\beta^r = 1 + Kc^r$$

we have, from Eqs. (7.10.12) and (7.10.14),

$$\frac{1}{\tilde{\lambda}} = 1 + \frac{\nu\gamma}{\beta^l\beta^r} \tag{7.10.26}$$

$$F(c; c^l, c^r) = \frac{\gamma K}{\beta^l\beta^r}\frac{(c - c^l)(c - c^r)}{1 + Kc} \tag{7.10.27}$$

In the limiting case, when $St \to \infty$, Eq. (7.10.20) gives

$$Kc_0 = -1 + \sqrt{\beta^l\beta^r} < \frac{1}{2}K(c^l + c^r) \tag{7.10.28}$$

for the c value at the inflection point, and this shows that the shock layer has no point of symmetry. Equation (7.10.21) is readily integrated to give the shock layer

$$\left(\frac{c^l - c}{c^l - c_0}\right)^{\beta^l}\left(\frac{c_0 - c^r}{c - c^r}\right)^{\beta^r} = \exp\left\{\frac{\nu\gamma(\beta^l - \beta^r)}{\nu\gamma + \beta^l\beta^r}\mathrm{Pe}\xi\right\} \tag{7.10.29}$$

From Eq. (7.10.23), we also obtain an explicit expression for the shock-layer thickness as

$$Pe\delta_\varepsilon = \left(1 + \frac{\beta^l\beta^r}{\nu\gamma}\right)\frac{\beta^l + \beta^r}{\beta^l - \beta^r}\ln\frac{1 - \varepsilon}{\varepsilon} \tag{7.10.30}$$

If $\nu = 1.5$, $\gamma = 1$, $K = 5$ l/mol, $c^l = 0.07$ mol/l, and $c^r = 0.02$ mol/l, we obtain $\tilde{\lambda} = 0.1654$, $c_0 = 0.0437$ mol/l, and

$$Pe\delta_\varepsilon = 11.74\ln\frac{1 - \varepsilon}{\varepsilon}$$

For $Pe = 80$, then 99 percent ($\varepsilon = 0.005$) of the total saturation will be accomplished in a zone of thickness 0.777, and every fluid element will pass this zone in a fractional time 0.932. The accomplishment of 99.9 percent saturation ($\varepsilon = 5 \times 10^{-4}$) will require a zone thickness of 1.115.

Grad (1952) suggested another way of defining the shock-layer thickness, which may be expressed as

$$\delta = \frac{c^l - c^r}{|\,dc/d\xi\,|_{\max}} \tag{7.10.31}$$

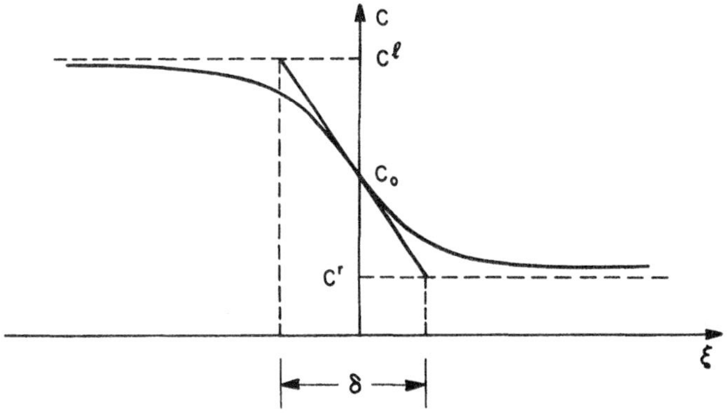

Figure 7.30

where $|dc/d\xi|_{max}$ is the maximum slope of the shock-layer profile.† In the present case, the maximum slope is obtained at the inflection point, as depicted in Fig. 7.30, and thus, by substituting Eqs. (7.10.26), (7.10.27), and (7.10.28) into Eq. (7.10.19), we have

$$\frac{1}{Pe}\left|\frac{dc}{d\xi}\right|_{max} = \nu\tilde{\lambda}F(c_0; c^l, c^r) = \frac{\nu\gamma/K}{\nu\gamma + \beta^l\beta^r}(\sqrt{\beta^l} - \sqrt{\beta^r})^2 \qquad (7.10.32)$$

Hence the shock layer thickness δ is

$$Pe\delta = \left(1 + \frac{\beta^l\beta^r}{\nu\gamma}\right)\frac{\sqrt{\beta^l} + \sqrt{\beta^r}}{\sqrt{\beta^l} - \sqrt{\beta^r}} \qquad (7.10.33)$$

For the numerical data given previously, this equation gives $\delta = 0.5$.

There is experimental evidence for the validity of Eq. (7.10.29). Also, it has been demonstrated by numerical integration that the transient solution of Eqs. (7.10.1) and (7.10.2) rapidly converges to the shock layer as time increases. According to the experimental observation with a packing of good quality, the shock-layer thickness is in the range of a few particle diameters, and this is in analogy with the shock layer in compressible fluid flow. (See the references given at the end of this chapter.)

When the solid phase moves countercurrently against the fluid phase, some interesting features are noted in regard to the shock layer. Here the flow-rate ratio takes an important role, and the shock layer exists only for the flow-rate ratios in certain ranges. In particular, it is possible to have a stationary shock layer and to have the solid phase concentration higher than the equilibrium value (see Exercise 7.10.5).

†According to this definition, δ is a purely local property of the shock layer and becomes meaningless if the shock layer has a vertical tangent.

Exercises

7.10.1. Consider the Langmuir isotherm

$$f(c) = \frac{\gamma c}{1 + Kc}$$

with $\gamma = 1$ and $K = 5$ l/mol. If $\nu = 1$, $c^l = 0.07$ mol/l and $c^r = 0.02$ mol/l, determine the integral curves of Eq. (7.10.15), starting from various points (c, q) on the phase plane, and make a plot as illustrated in Fig. 7.26. Assume that $Pe = 80$ and $St = 20$.

7.10.2. With the data given in Exercise 7.10.1, determine the shock layers of Eqs. (7.10.1) and (7.10.2) for various sets of values of Pe and St. Plot the shock-layer profiles together by using the point with $c = (c^l + c^r)/2$, say, as a fixed point and discuss. Also, compare the shock layers by plotting them on the same phase plane of c and q $(= dc/d\xi)$ and examine the effects of Pe and/or St on the shock layer. Assume that $Pe = 80$ and $St = 20$.

7.10.3. For the BET-type isotherm, sketch the isotherm $f(c)$ to locate the point for c_t and the curve $F(c; c^l, c^r)$ to locate the points of zeros. Then discuss the situation in the $(c, dc/d\xi)$-plane to derive the conditions given by Eq. (7.10.18).

7.10.4. By using Eqs. (7.10.9) and (7.10.13), show that the departure from equilibrium may be expressed as

$$f(c) - n = \frac{1}{\nu\tilde{\lambda}Pe}\frac{dc}{d\xi} - F(c; c^l, c^r)$$

and further that, if we neglect the coupled effect of the axial dispersion and the mass-transfer resistance, the preceding expression can be approximated by

$$f(c) - n \cong -\frac{F(c; c^l, c^r)}{1 + (St/Pe)/[\tilde{\lambda}(1 - \tilde{\lambda})]}$$

Thus the departure depends on the product (kE_a). Discuss the departure in the limiting cases as $Pe \to \infty$ and as $St \to \infty$, respectively.

7.10.5. Consider the countercurrent moving-bed adsorber with axial dispersion in the fluid phase and mass-transfer resistance (see Sec. 6.5) and derive the ordinary differential equation describing the shock layer. For the case $f''(c) < 0$, formulate the conditions for the existence of the shock layer. Can you argue that the departure from equilibrium, $f(c) - n$, may become negative in certain cases? Can you also argue that in some extreme case the shock layer exists if $c^l < c^r$? Finally, show that mass-transfer resistance has no influence over the stationary shock layer $(\tilde{\lambda} = 0)$.

REFERENCES

The basic references for much of this chapter are

P. D. Lax, "Weak Solutions of Nonlinear Hyperbolic Equations and Their Numerical Computation," *Comm. Pure Appl. Math.* **7**, 159–193 (1954).

P. D. Lax, "Hyperbolic Systems of Conservation Laws II," *Comm. Pure Appl. Math.* **10**, 537–566 (1957).

O. A. Oleinik, "Discontinuous Solutions of Nonlinear Differential Equations," *Uspekhi Mat. Nauk* **12**, 3–73 (1957); English translation in *Amer. Math. Soc. Transl.*, Ser. 2, **26**, 95–192 (1963).

O. A. Oleinik, "Uniqueness and Stability of the Generalized Solution of the Cauchy Problem for a Quasi-linear Equation," *Uspekhi Mat. Nauk* **14**, 165–176 (1959); English translation in *Amer. Math. Soc. Transl.*, Ser. 2, **33**, 285–290 (1963).

O. A. Oleinik, "Discontinuous Solutions of Nonlinear Differential Equations," Seminari dell'Instituto Nazionalle di Alta Matematica, 1962–1963.

P. D. Lax, "The Formation and Decay of Shock Waves," *Amer. Math. Monthly* **79**, 227–241 (1972).

P. D. Lax, "Hyberbolic Systems of Conservation Laws and the Mathematical Theory of Shock Waves," *Regional Conference Series in Applied Mathematics*, No. 11, SIAM, Philadelphia, Pa., 1973.

Excellent, although somewhat outdated, surveys are

I. M. Gelfand, "Some Problems in the Theory of Quasi-linear Equations," *Uspekhi Mat. Nauk* **14**, 87–158 (1959); English translation in *Amer. Math. Soc. Transl.*, Ser. 2, **29**, 295–381 (1963).

P. D. Lax, "Nonlinear Hyperbolic Systems of Conservation Laws," in *Nonlinear Problems*, R. Langer (ed.), University of Wisconsin Press, Madison, Wisc., 1963, pp. 1–12.

See also

G. B. Whitham, *Linear and Nonlinear Waves*, Wiley, New York, 1974.

J. Smoller, *Shock Waves and Reaction-Diffusion Equations*, Springer-Verlag, New York, 1983.

7.2. Conditions under which an arbitrary system of quasi-linear equations can be reduced to the conservation form (7.2.4) are discussed by

B. L. Rozdestvenskii, "Systems of Quasilinear Equations," *Dokl. Akad. Nauk SSSR* **115**, 454–457 (1957): English translation in *Amer. Math. Soc. Transl.*, Ser. 2, **42**, 13–17 (1964).

7.3. For the maximum transform, see

R. Bellman, "Functional Equations in the Theory of Dynamic Programming. V. Positivity and Quasilinearity," *Proc. Nat. Acad. Sci. USA*, **41**, 743–746 (1955).

R. Bellman, *Dynamic Programming*, Princeton University Press, Princeton, N.J. 1957.

R. Kalaba, "On Nonlinear Differential Equations, the Maximum Operation, and Monotone Convergence," *J. Math. Mech.* **8**, 519–574 (1959).

R. Bellman and W. Karush, "On a New Functional Transform in Analysis: The Maximum Transform," *Bull. Amer. Math. Soc.* **67**, 501–503 (1961).

See also

R. Bellman and S. E. Dreyfus, *Applied Dynamic Programming*, Princeton University Press, Princeton, N.J., 1962, Appendix IV.

There is nothing particularly new about this transform since it is none other than the Legendre transform. The convolution properties established by Bellman and mentioned in Exercises 7.3.5 and 7.3.6 are probably new.

7.4. The definition of the weak solution was first given in Lax's 1954 paper cited previously. For the generalized entropy condition, see Oleinik's 1959 paper and Lax's 1973 paper cited previously, and also

E. Hopf, "On the Right Weak Solution of the Cauchy Problem for a Quasilinear Equation of First Order," *J. Math. Mech.* **19**, 483–487 (1969).

The generalized entropy condition has been proved equivalent to the L_1-contractiveness of the solution operator by

B. Keyfitz Quinn, "Solutions with Shocks: An Example of an L_1 Contractive Semigroup," *Comm. Pure Appl. Math.* **24**, 125–132 (1971).

7.5. For the method used to motivate Lax's solution, see the papers by Bellman (1955), Kalaba (1959), and Bellman and Karush (1961) all cited previously for Sec. 7.3.

The proof of the explicit solution is in Lax's 1957 paper, cited previously. For the continuous dependence on the initial data, see also

O. A. Oleinik, "The Cauchy Problem for Nonlinear Equations in a Class of Discontinuous Functions," *Dokl. Akad. Nauk SSSR* **115**, 451–454 (1954): English Translation in *Amer. Math. Soc. Transl.*, Ser. 2, **42**, 7–12 (1964).

A. Douglis, "The Continuous Dependence of Generalized Solutions of Non-linear Partial Differential Equations upon Initial Data," *Comm. Pure Appl. Math.* **14**, 267–284 (1961).

7.6. For the use of vanishing viscosity, see

E. Hopf, "The Partial Differential Equation $u_t + uu_x = \mu u_{xx}$," *Comm. Pure Appl. Math.* **3**, 201–230 (1950).

O. A. Oleinik, "Construction of a Generalized Solution of the Cauchy Problem for a Quasi-linear Equation of First Order by the Introduction of Vanishing Viscosity," *Uspekhi Mat. Nauk* **14**, 159–164 (1959); English translation in *Amer. Math. Soc. Transl.*, Ser. 2, **33**, 277–283 (1963).

H.-K. Rhee, B. F. Bodin, and N. R. Amundson, "A Study of the Shock Layer in Equilibrium Exchange Systems," *Chem. Eng. Sci.* **26**, 1571–1580 (1971).

See also Gelfand's 1959 paper and Oleinik's 1962 paper, cited previously. The existence theorem based on the convergence of a finite-difference scheme is treated in Lax's 1954 paper, Oleinik's 1957 paper, and

J. Glim "Solutions in the Large for Nonlinear Hyperbolic Systems of Equations," *Comm. Pure Appl. Math.* **18**, 697–715 (1965).

For the ordering principle and its application, see

A. Douglis, "An Ordering Principle and Generalized Solutions of Certain Quasi-linear Partial Differential Equations," *Comm. Pure Appl. Math.* **12**, 87–112 (1959).

The existence and uniqueness theorem for cases without convexity conditions are treated in

Z.-Q. Wu, "On the Existence and Uniqueness of the Generalized Solutions of the Cauchy Problem for Quasilinear Equations of First Order without Convexity Conditions," *Acta Math. Sinica* **13**, 513–530 (1963); English translation in *Chinese Math.* **4**, 561–577 (1964).

D. P. Ballou, "Solutions to Nonlinear Hyperbolic Cauchy Problems without Convexity Conditions," *Trans. Amer. Math. Soc.* **152**, 441–460 (1970).

For the asymptotic behavior, see Lax's 1957 paper, cited previously, and those given for Sec. 7.9.

Results for equations with many spacelike variables are to be found in

E. Conway and J. Smoller, "Global Solutions of the Cauchy Problem for Quasilinear First-order Equations in Several Space Variables," *Comm. Pure Appl. Math.* **19**, 95–105 (1966).

E. Conway and J. Smoller, "Uniqueness and Stability Theorem for the Generalized Solution of the Initial-value Problem for a Class of Quasi-linear Equations in Several Space Variables," *Arch. Rat. Mech. Anal.* **23**, 399–408 (1967).

A. I. Vol'pert, "The Spaces BV and Quasilinear Equations," *Mat. Sbornik* **73** (**115**), 255–302 (1967); English translation in *Math. USSR Sbornik* **2**, 225–267 (1967).

S. N. Kruzkov, "Generalized Solutions of the Cauchy Problem in the Large for First Order Nonlinear Equations," *Dokl. Akad. Nauk SSSR* **187**, 29–32 (1969); English Translation in *Soviet Math. Dokl.* **10**, 785–788 (1969).

S. N. Kruzkov "First Order Quasilinear Equations in Several Independent Variables," *Mat. Sbornik* **81** (**123**), 228–255 (1970); English translation in *Math. USSR Sbornik* **10**, 217–243 (1970).

See also B. Keyfitz Quinn's paper (1971), cited previously.

The generic properties of weak solutions are discussed in

D. G. Schaeffer, "A regularity theorem for Conservation Laws," *Adv. Math.* **11**, 368–386 (1973).

J. Guckenheimer, "Solving a Single Conservation Law," in *Dynamical Systems*, A. D. Dold and B. Eckmann (eds.), *Lecture Notes Series in Mathematics*, No. 468, Springer-Verlag, Berlin, 1975, pp. 108–134.

M. Golubitsky and D. G. Schaeffer, "Stability of Shock Waves for a Single Conservation Law," *Adv. Math.* **15**, 65–71 (1975).

in which the theory of singularities has been introduced.

The structure of weak solutions has been investigated by various authors. See, for example,

D. P. Ballou, "Weak Solutions with a Dense Set of Discontinuities," *J. Diff. Eqns.* **10**, 270–280 (1971).

C. M. Dafermos, "Generalized Characteristics and the Structure of Solutions of Hyperbolic Conservation Laws," *Indiana U. Math. J.* **26**, 1097–1119 (1977).

7.7. A beautiful exposition of the theory of sound waves of finite amplitude is to be found in

M. J. Lighthill, "Viscosity Effects in Sound Waves of Finite Amplitude," in *Surveys in Mechanics: The G. I. Taylor 70th Anniversary Volume*, G. K. Batchelor and R. M. Davis (eds.), Cambridge University Press, New York, 1956.

Among other things, the formation of N-waves is fully discussed here. We have drawn on it for Exercises 7.7.1 and 7.7.2.

Burger's work on Eq. (7.7.1) is summed up in

J. M. Burger, "A Mathematical Model Illustrating the Theory of Turbulence," *Adv. Appl. Mech.* **1**, 171–199 (1948).

The graphical method of finding shock paths goes back to

G. B. Whitham, "The Flow Pattern of a Supersonic Projectile," *Comm. Pure Appl. Math.* **5**, 301–348 (1952).

An excellent expository article on shocks is

F. Y. Thomas, "Mathematical Foundations of the Theory of Shocks in Gases," *J. Math. Anal. Appl.* **26**, 595–629 (1969).

Some of the more recent results on waves are collected in

M. Froissart (ed.), *Hyperbolic Equations and Waves*, Springer-Verlag, Berlin-Heidelberg, 1970.

S. Leibovich and A. R. Seebass, *Nonlinear Waves*, Cornell University Press, Ithaca, N.Y., 1974.

See also

T. Taniuti and K. Nishihara, *Nonlinear Waves*, Iwanami Shoten, Tokyo, 1977; English translation by Pitman Books Limited, London, 1983.

P. L. Bhatnagar, *Nonlinear Waves in One-dimensional Dispersive Systems*, Oxford University Press, New York, 1979.

7.9. The formulation of asymptotic behavior follows somewhat on the line laid down in

G. B. Whitham, *Linear and Nonlinear Waves*, Wiley, New York, 1974, pp. 42–55.

See also Lax's 1957, 1972, and 1973 papers, cited previously, and

C. M. Dafermos, "Applications of the Invariance Principle for Compact Processes II. Asymptotic Behavior of Solutions of a Hyperbolic Conservation Law," *J. Diff. Eqns.* **11**, 416–424 (1972).

J. M. Greenberg and D. D. M. Tong, "Decay of Periodic Solutions of $\partial u/\partial t + \partial f(u)/\partial x = 0$," *J. Math. Anal. Appl.* **43**, 56–71 (1973).

The decay of discontinuous solutions is also treated from singular perturbation solutions in

J. D. Murray, "Singular Perturbations of a Class of Nonlinear Hyperbolic and Parabolic Equations," *J. Math. and Phys.* **47**, 111–133 (1968).

J. D. Murray, "Perturbation Effects on the Decay of Discontinuous Solutions of Nonlinear First-order Wave Equations," *SIAM J. Appl. Math.* **19**, 273–298 (1970).

Discussions for equations in more than one space dimension are given by

E. D. Conway, "The Formation and Decay of Shocks for a Conservation Law in Several Dimensions," *Arch. Rat. Mech. Anal.* **64**, 47–57 (1977).

7.10. The development here is taken from

H.-K. Rhee, B. F. Bodin, and N. R. Amundson, "A Study of the Shock Layer in Equilibrium Exchange Systems," *Chem. Eng. Sci.* **26**, 1571–1580 (1971).

H.-K. Rhee and N. R. Amundson, "A Study of the Shock Layer in Nonequilibrium Exchange Systems," *Chem. Eng. Sci.* **27**, 199–211 (1972).

Although they are concerned with a pair of equations, the following articles are found instructive:

R. von Mises, "On the Thickness of a Steady Shock Wave," *J. Aeronaut. Sci.* **17**, 551–594 (1950).

D. Gilbarg, "The Existence and Limit of the One-dimensional Shock Layer," *Amer. J. Math.* **73**, 256–274 (1951).

H. Grad, "The Profile of a Steady Plane Shock Wave," *Comm. Pure Appl. Math.* **5**, 257–300 (1952).

D. Gilbarg and D. Paolucci, "The Structure of Shock Waves in the Continuum Theory of Fluids," *J. Rat. Mech. Anal.* **2**, 617–642 (1953).

In the practice of adsorption processes, the concentration profile of constant pattern has been recognized since as early as 1920. The existence theorem has been established by

D. O. Cooney and E. N. Lightfoot, "Existence of Asymptotic Solutions to Fixed-bed Separations and Exchange Equations," *Ind. Eng. Chem. Fundam.* **4**, 233–236 (1965).

Some of the more recent works in this direction are to be found in

K. R. Hall, L. C. Eagleton, A. Acrivos, and T. Vermeulen, "Pore- and Solid-diffusion Kinetics in Fixed-bed Adsorption under Constant-pattern Conditions," *Ind. Eng. Chem. Fundam.* **5**, 212–223 (1966).

R. D. Fleck, Jr., D. J. Kirwan, and K. R. Hall, "Mixed-resistance Diffusion Kinetics in Fixed-bed Adsorption under Constant-pattern Conditions," *Ind. Eng. Chem. Fundam.* **12**, 95–99 (1973).

A. P. Coppola and M. D. Levan, "Adsorption with Axial Diffusion in Deep Beds," *Chem. Eng. Sci.* **36**, 967–971 (1981).

References to earlier works are given in these papers.

See also

A. Acrivos, "On the Combined Effect of Longitudinal Diffusion and External Mass Transfer Resistance in Fixed-bed Operations," *Chem. Eng. Sci.* **13**, 1–6 (1960).

Experimental evidence for the shock layer is to be found in

E. Glueckauf and J. I. Coates, "Theory of Chromatography: Part IV. The Influence of Incomplete Equilibrium on the Front Boundary of Chromatograms and on the Effectiveness of separation," *J. Chem. Soc.* **1947**, 1315–1321 (1947).

W. J. Thomas and J. L. Lombardi, "Binary Adsorption of Benzene–Toluene Mixtures," *Trans. Instn. Chem. Engrs.* **49**, 240–250 (1971)

The convergence of the transient profile to the shock layer is illustrated in the first two papers cited for this section. The formal proof, although given for the equilibrium model ($St \to \infty$), is established in

A. M. Il'in and O. A. Oleinik, "Behavior of the Solutions of the Cauchy Problem for Certain Quasilinear Equations for Unbounded Increase of the Time," *Dokl. Akad. Nauk SSSR* **120**, 25–28 (1958); English translation in *Amer. Math. Soc. Transl.*, Ser. 2, **42**, 19–23 (1964).

For the shock layer in a countercurrent moving-bed adsorber, see

H.-K. Rhee and N. R. Amundson, "Asymptotic Solution to Moving-bed Exchange Equations," *Chem. Eng. Sci.* **28**, 55–62 (1973).

Nonhomogeneous Quasi-linear Equations

8

Although the mathematical theory for nonhomogeneous quasi-linear equations is essentially the same as that for homogeneous equations, actual construction of the solution can be quite complicated because the characteristics are all curved due to the presence of the nonhomogeneous term. In this chapter, application of the theoretical developments is discussed with illustrative examples. The physical examples used are the fixed-bed adsorber with reaction, the countercurrent adsorber with reaction, and transient volumetric pool boiling.

A summary of the theoretical features for nonhomogeneous equations is given in the first section, which then takes up the case of a fixed-bed adsorber with a first-order reaction taking place in the fluid phase. For this simple example of the chromatographic reactor, both a step input and a square-wave input are considered. In Sec. 6.2, transient volumetric pool boiling is treated, which presents the unique feature that the characteristic direction, as well as the direction of the shock path, is dependent not only on the local state but also on the integral of the source function. Although the analysis is restricted to the case of convex flux relations, variations expected with nonconvex flux curves are also discussed.

The concept of the black-box steady state, clearly distinguishable from the true steady state, is introduced in Sec. 8.3 and elucidated with a simple example in which the characteristic direction changes its sign from positive to negative as the state varies along the characteristic. In such a case it is demonstrated that the black-box steady state is attained in a finite time (or even instantaneously), whereas the true steady state is achieved at a later time or only approached asymptotically. The countercurrent adsorber with a reaction taking place on the solid phase presents itself as a physical example of this kind and is treated in detail. The last two sections (Secs. 8.4 and 8.5) are concerned with the countercurrent adsorber without and with reaction, respectively, starting from the nonequilibrium condition. The analysis of the countercurrent adsorber without reaction has the merit of being so simple that it can be done completely and affords a comprehensive picture of the system. Although the emphasis falls on steady-state solutions, it is essential to look at the development of the steady state and the insight that comes by maneuvering between the transient model and the steady-state model and/or the nonequilibrium model and the equilibrium model. The full picture of the steady-state solution is displayed in tables for both cases, without and with reaction.

8.1 Nonhomogeneous Equations with Two Independent Variables

If a system involves the generation of species and/or energy, an additional term (i.e., the source term) must be included in its mathematical description. Thus the conservation equation, for example, describing fixed-bed adsorption with a chemical reaction may take the form

$$\frac{\partial c}{\partial x} + \frac{\partial}{\partial \tau} f(c) = g(c) \tag{8.1.1}$$

and the initial condition is

$$c(0, \tau) = \Gamma(\tau) \tag{8.1.2}$$

According to Lax (1954), $c(x, \tau)$ is called a weak solution of Eq. (8.1.1), with the initial value $\Gamma(\tau)$, if the integral relation

$$\int_0^\infty dx \int_{-\infty}^\infty d\tau \left\{ \frac{\partial w}{\partial x} c + \frac{\partial w}{\partial \tau} f(c) + w g(c) \right\} + \int_{-\infty}^\infty w(0, \tau) \Gamma(\tau)\, d\tau = 0 \tag{8.1.3}$$

holds for every test function $w(x, \tau)$ that has continuous first partial derivatives and that vanishes as $x + |\tau|$ becomes sufficiently large. This equation is to be compared with Eq. (7.4.5). Since the weak solution has been treated in detail with respect to homogeneous equations in Chapter 7, we will briefly discuss only the salient features concerning the nonhomogeneous equations and then quickly move on to treat physical examples.

First we will see that a weak solution, if continuous, satisfies the partial differential equation (8.1.1) and that Eq. (8.1.3), if applied to a discontinuity in a weak solution, reduces to the jump condition.

Suppose $c(x, \tau)$ is smooth in a region of the (x, τ)-plane and satisfies Eq. (8.1.3). We can then integrate this equation by parts to obtain

$$\left[\int_{-\infty}^{\infty} wc\, d\tau\right]_{x=0}^{x=\infty} - \int_0^{\infty} dx \int_{-\infty}^{\infty} d\tau w\frac{\partial c}{\partial x} + \left[\int_0^{\infty} wf(c)dx\right]_{\tau=-\infty}^{\tau=+\infty}$$
$$- \int_0^{\infty} dx \int_{-\infty}^{\infty} d\tau w\frac{\partial}{\partial \tau}f(c) + \int_0^{\infty} dx \int_{-\infty}^{\infty} d\tau wg(c) + \int_{-\infty}^{\infty} w(0, \tau)\Gamma(\tau)d\tau = 0$$

In this equation the integrated parts for $x = \infty$ and $\tau = \pm\infty$ vanish due to the vanishing of w and that for $x = 0$ just cancels with the last term. Thus we have

$$\int_0^{\infty} dx \int_{-\infty}^{\infty} d\tau w\left\{\frac{\partial c}{\partial x} + \frac{\partial}{\partial \tau}f(c) - g(c)\right\} = 0 \tag{8.1.4}$$

This is true for every test function $w(x, \tau)$ defined previously, and so the expression inside the parentheses must be equal to zero. This implies that $c(x, \tau)$ satisfies Eq. (8.1.1). If $c(x, \tau)$ is discontinuous, we refer to Fig. 7.5 and make use of Green's theorem (7.4.7). When applied to the region $D_+ + D_-$, this gives

$$\iint_{D_+ + D_-} w\left\{\frac{\partial c}{\partial x} + \frac{\partial}{\partial \tau}f(c) - g(c)\right\} dx\, d\tau$$
$$+ \iint_{D_+ D_-} \left\{\frac{\partial w}{\partial x}c + \frac{\partial w}{\partial \tau}f(c) + wg(c)\right\} dx\, d\tau = \int_{G_+ G_-} w(\tilde{\sigma}c - f)\, dx$$
$$+ \int_{I_+ + I_-} w(c\, d\tau - f\, dx) + \int_{C_+ + C_-} w(c\, d\tau - f\, dx)$$

in which the term $\iint_{D_+ + D_-} wg(c)\, dx\, d\tau$ has been subtracted and then added again. On the left side, the first integral vanishes because Eq. (8.1.1) holds in each of D_+ and D_-. Since $c(x, \tau)$ is a weak solution, it satisfies Eq. (8.1.3), and so the second integral is equal to $-\int_{-\infty}^{\infty} w(0, \tau)\Gamma(\tau)\, d\tau$. The integral terms on the right side are identical to those for the homogeneous equation, which we discussed in Sec. 7.4. Therefore, the same argument can be applied here to give

$$\tilde{\sigma}(c^l, c^r) = \frac{d\tau}{dx} = \frac{[f]}{[c]} = \frac{f(c^l) - f(c^r)}{c^l - c^r} \tag{8.1.5}$$

which is the jump condition to be satisfied by the discontinuity (c^l, c^r). The fact that the source term $g(c)$ does not appear explicitly in Eq. (8.1.5) is not surprising, since from the physical point of view no reaction can take place at the sharp discontinuity.

Next we note that the Oleinik's proof of the existence and uniqueness theorem was established, in fact, for the nonhomogeneous equation (8.1.1) with the entropy condition

$$\frac{c(x, \tau_1) - c(x, \tau_2)}{\tau_1 - \tau_2} \geq K(x, \tau_1, \tau_2) \tag{8.1.6}$$

where $f''(c) < 0$. It has been extended to the case of an arbitrary function $f(c)$ with the generalized entropy condition. Here the entropy condition (8.1.6) is identical to that for the homogeneous equation, and this is also true for the generalized entropy condition (see Secs. 5.4 and 7.4).

Now it would be natural to think that the fundamental character of the solution to the nonhomogeneous equation is essentially the same as that for the homogeneous equation. We recall, however, that with the homogeneous equation the dependent variable c remains unchanged along each characteristic, and thus the characteristic curves are all straight and parallel to the (x, τ)-plane. It is this fact that makes the construction of the solution so simple. Obviously, this is not the case if the equation is nonhomogeneous. The characteristics are necessarily curved, and so, although the structure of the solution may be similar, actual construction of the solution appears to be much more difficult.

To see this in a general context, we will consider the equation

$$P(x, y, z)\frac{\partial z}{\partial x} + Q(x, y, z)\frac{\partial z}{\partial y} = R(x, y, z) \tag{8.1.7}$$

which, in principle, may be put into the form of a conservation law (see Exercise 7.2.3). As we have seen in Sec. 2.5, the solution may be constructed from characteristic curves that satisfy the characteristic differential equations

$$\begin{aligned} \frac{dx}{ds} &= P(x, y, z) \\ \frac{dy}{ds} &= Q(x, y, z) \\ \frac{dz}{ds} &= R(x, y, z) \end{aligned} \tag{8.1.8}$$

where s is the parameter running along the characteristic curves. In particular, if we require the solution surface to pass through the initial curve

$$x = X(\xi), \qquad y = Y(\xi), \qquad z = Z(\xi) \tag{8.1.9}$$

we may solve the characteristic differential equations with these initial values for x, y, and z when $s = 0$. As shown in Fig. 2.2, we are then just picking out the characteristic curves that pass through the given initial curve and so obtaining the solution surface in parametric form as

$$x = x(s, \xi), \qquad y = y(s, \xi), \qquad z = z(s, \xi) \tag{8.1.10}$$

These curves generate the solution surface $z = z(x, y)$ if we can express s and ξ as functions of x and y. This can be done as long as the Jacobian $\partial(x, y)/\partial(s, \xi)$ does not vanish along the initial curve (8.1.9).

Notice that, in contrast to the linear case, it is not generally possible to solve the first two equations of Eq. (8.1.8) independently of the third and so lay down the characteristics (or the characteristic ground curves) once and for all. Neither is it allowed to integrate the first two equations from each point along the initial curve with the value of z fixed, as was done for the homogeneous quasi-linear equations. Here the projection of the characteristic curves onto the (x, y)-plane will give a family of curves, but these are not ground curves since they vary from solution to solution not only depending on the specified initial data but also due to the nonhomogeneous term.

In many cases of physical interest, P, Q, and R are all dependent on z only. Then the construction of the solution may be somewhat simpler, because the third equation of Eq. (8.1.8) can be solved independently of the first two to give $z = z(s, \xi)$, and the characteristic direction in the (x, y)-plane is given by

$$\frac{dy}{dx} = \frac{Q(z)}{P(z)} = \sigma(z) \tag{8.1.11}$$

If, for example, $P \neq 0$, we can divide Eq. (8.1.7) through by P so that the first equation of Eq. (8.1.8) reads $dx/ds = 1$. It is then possible to identify the characteristic parameter s with x up to an additive constant. Thus Eq. (8.1.11) can be solved by quadratures to give the characteristics in the (x, y)-plane along each of which we have already determined the solution z. We find a typical example of this case in Eq. (8.1.1).

If the solution contains a discontinuity, we will have to determine its propagation path in such a way that the jump condition is satisfied. We recall from Sec. 7.6 that it is meaningful to discuss the jump condition only after the equation is put into the form of a conservation law, and so we will take Eq. (8.1.1) here. The jump condition is then given by Eq. (8.1.5), for which both c^l and c^r are to be found from the pair of characteristics that meet the path of the discontinuity at the same point. Therefore, both c^l and c^r vary along the path, so the discontinuity may never obtain a constant speed unless it becomes stationary. Of course, we have to impose the entropy condition wherever the solution becomes discontinuous. For genuinely nonlinear equations [i.e., $f''(c) \neq 0$], we have the condition (8.1.6), but this is indeed equivalent to the shock condition

$$\sigma(c^l) < \tilde{\sigma}(c^l, c^r) < \sigma(c^r) \tag{8.1.12}$$

where the characteristic direction $\sigma(c)$ is given by

$$\frac{d\tau}{dx} = f'(c) \equiv \sigma(c) \tag{8.1.13}$$

So the discontinuity in genuinely nonlinear systems will be called a shock. Otherwise, we impose the generalized entropy condition, which reads

$$\tilde{\sigma}(c^l, c^r) \geq \tilde{\sigma}(c^l, c) \tag{8.1.14}$$

for every c between c^l and c^r.

For the purpose of illustration, we will consider the fixed-bed adsorption of a single solute when the adsorption follows the Langmuir isotherm and the solute undergoes a first-order, irreversible reaction in the fluid phase. We will assume that the product material does not adsorb, and so the solute species is not aware of the presence of the product. In terms of the usual notation, the mass balance for the solute gives the equation

$$\varepsilon V\frac{\partial c}{\partial z} + \frac{\partial}{\partial t}\{\varepsilon c + (1 - \varepsilon)n\} = -\varepsilon k_r c \tag{8.1.15}$$

in which k_r denotes the reaction rate constant. Let us define the dimensionless parameters

$$\nu = \frac{1 - \varepsilon}{\varepsilon} \tag{8.1.16}$$

$$\beta = \frac{k_r Z}{V} \tag{8.1.17}$$

and the dimensionless variables

$$x = \frac{z}{Z}, \qquad \tau = \frac{Vt}{Z} \tag{8.1.18}$$

where Z is the characteristic length of the system, so that Eq. (8.1.15) can be rewritten as

$$\frac{\partial c}{\partial x} + \frac{\partial}{\partial \tau}\{c + \nu n(c)\} = -\beta c \tag{8.1.19}$$

in which $n(c)$ represents the adsorption isotherm. With the Langmuir isotherm, we have

$$n(c) = \frac{\gamma c}{1 + Kc} \tag{8.1.20}$$

and since $n''(c) < 0$, Eq. (8.1.19) is genuinely nonlinear. We note that this is a special case of Eq. (8.1.1) with $f(c) = c + \nu n(c)$ and $g(c) = -\beta c$.

The characteristic differential equations are

$$\frac{dx}{ds} = 1$$

$$\frac{d\tau}{ds} = 1 + \nu n'(c) = 1 + \frac{\nu\gamma}{(1 + Kc)^2} \tag{8.1.21}$$

$$\frac{dc}{ds} = -\beta c$$

and since $dx = ds$ along the characteristic curves, we have

$$\frac{d\tau}{dx} = 1 + \nu n'(c) = 1 + \frac{\nu\gamma}{(1 + Kc)^2} \equiv \sigma(c) \tag{8.1.22}$$

for the characteristic direction, and

$$\frac{dc}{dx} = -\beta c \tag{8.1.23}$$

for the variation of c in the characteristic direction. Given the initial data, we can solve Eq. (8.1.23) to obtain $c = c(x, \xi)$, where ξ is the x-intercept of the characteristic, and this will be used in Eq. (8.1.22) to locate the characteristics in the (x, τ)-plane. The jump condition for a discontinuity in c is given from Eq. (8.1.5) as

$$\frac{d\tau}{dx} = 1 + \nu\frac{[n]}{[c]} = 1 + \frac{\nu\gamma}{(1 + Kc^l)(1 + Kc^r)} \equiv \tilde{\sigma}(c^l, c^r) \tag{8.1.24}$$

It is readily seen that the shock condition (8.1.12) is satisfied, and so discontinuities that we may encounter here are all shocks.

Let us now suppose that the solute is being fed with constant concentration c_0 into an initially fresh bed, and thus the initial data are prescribed as

$$\begin{aligned} &\text{at } \tau = 0, \qquad c = 0 \\ &\text{at } x = 0, \qquad c = c_0 \end{aligned} \tag{8.1.25}$$

For the characteristics emanating from each point along the x-axis, we have $c(x, \xi) = c_I(\xi) \exp[-\beta(x - \xi)]$ from Eq. (8.1.23); but $c_I(\xi) = 0$, and so $c(x, \xi) = 0$ along each characteristic. Since the characteristic direction is now constant (i.e., $\sigma = 1 + \nu\gamma$), the characteristics become straight, and these have the equation

$$\tau = (1 + \nu\gamma)(x - \xi) \tag{8.1.26}$$

Hence the constant state of zero concentration develops in the region adjacent to the x-axis, as shown in Fig. 8.1. For each point along the τ-axis, we will assign $\tau = \eta$, and further we have $x = 0$ and $c = c_0$ from the initial data. Thus, from Eq. (8.1.23), we obtain

$$c = c_0 \exp(-\beta x) \tag{8.1.27}$$

and substituting this into Eq. (8.1.22) gives

$$\frac{d\tau}{dx} = 1 + \frac{\nu\gamma}{[1 + Kc_0 \exp(-\beta x)]^2} \tag{8.1.28}$$

for the characteristics emanating from the τ-axis. Integrating subject to the initial condition $\tau = \eta$ at $x = 0$, we get the equation of these characteristics as follows:

$$\begin{aligned}\tau - \eta &= \int_0^x \left\{1 + \frac{\nu\gamma}{[1 + Kc_0 \exp(-\beta x')]^2}\right\} dx' \\ &= (1 + \nu\gamma)x + \frac{\nu\gamma}{\beta}\left\{\frac{1}{1 + Kc_0} - \frac{1}{1 + Kc_0 \exp(-\beta x)}\right. \\ &\quad \left. + \ln \frac{1 + Kc_0 \exp(-\beta x)}{1 + Kc_0}\right\}\end{aligned} \tag{8.1.29}$$

These characteristics are curved and concave upward since $d\tau/dx$ is monotone increasing with x. But this family of curves is generated by simply translating one curve in the τ-direction, and thus they are all vertically parallel, as depicted in Fig. 8.1. Furthermore, since c depends on x only, the line of constant c is parallel to the τ-axis. Hence the concentration profile (c versus x), once established, will remain unchanged as time goes on.

Turning our attention to the situation at the origin, we see that there is an initial discontinuity, and this is bound to propagate through the domain as a shock since the shock condition is satisfied. At an arbitrary point $D(x, \tau)$ on the shock path, we can write

$$\frac{d\tau}{dx} = 1 + \frac{\nu\gamma}{1 + Kc^l}$$

because $c^r = 0$, but c^l is given by Eq. (8.1.27), so we have

$$\frac{d\tau}{dx} = 1 + \frac{\nu\gamma}{1 + Kc_0 \exp(-\beta x)} \tag{8.1.30}$$

This is now integrated from the origin to give the equation

$$\tau = (1 + \nu\gamma)x + \frac{\nu\gamma}{\beta} \ln \frac{1 + Kc_0 \exp(-\beta x)}{1 + Kc_0} \equiv \theta(x) \tag{8.1.31}$$

for the shock path. Clearly, the shock becomes weak in strength as it propagates, and so it is decelerated continuously. Since the concentration c drops

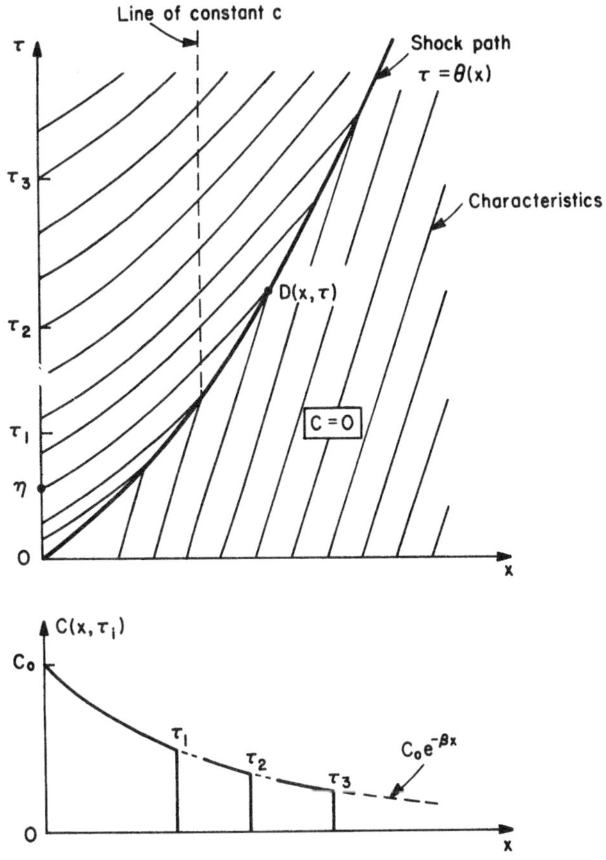

Figure 8.1

to zero across the shock, Eq. (8.1.27) may be rewritten as

$$c(x, \tau) = c_0[\exp(-\beta x)]H(\tau - \theta(x)) \tag{8.1.32}$$

where H represents the Heaviside step function and $\theta(x)$ is defined in Eq. (8.1.31). This is illustrated in Fig. 8.1.

In the present case, the concentration of the product can be readily determined by using the preceding results. If the reaction is $A \rightarrow B$, the equation for c_B, the concentration of the product B, is given by

$$\frac{\partial c_B}{\partial x} + \frac{\partial c_B}{\partial \tau} = \beta c \tag{8.1.33}$$

which is strictly linear. The characteristic direction is $d\tau/dx = 1$, and this is valid in either side of the shock path. In the region to the left of the shock path, we have

$$\frac{dc_B}{dx} = \beta c \tag{8.1.34}$$

along the characteristic so that, if the feed stream contains no B, by using Eq. (8.1.27) we obtain

$$c_B = c_0[1 - \exp(-\beta x)] \tag{8.1.35}$$

Here again the line of constant c_B is parallel to the τ-axis. Since there is no solute A on the right side of the shock path, we have $dc_B/dx = 0$, and this implies that c_B remains constant along the characteristics. Hence the line of constant c_B consists of two branches with the corner on the shock path, as denoted by the dotted lines in Fig. 8.2. It is now straightforward to draw the profiles of c_B at various times based on the information from the (x, τ)-diagram. This is illustrated in Fig. 8.2. To find an expression for c_B, however, let us consider the two points $D(x, \tau)$ and $E(\tilde{x}, \tilde{\tau})$ as marked in Fig. 8.2. At both D and E, we have the same value of c_B, which is equal to $c_0[1 - \exp(-\beta\tilde{x})]$ from Eq. (8.1.35), where $\tilde{x}$ satisfies Eq. (8.1.31) for given $\tilde{\tau}$ [i.e., $\tilde{\tau} = \theta(\tilde{x})$]. But we note that $\tilde{\tau} = \tau - (x - \tilde{x})$ because the line DE has a slope 1, and thus we have $\tau - (x - \tilde{x}) = \theta(\tilde{x})$. Solving this equation for $\tilde{x}$ and substituting in $c_B = c_0[1 - \exp(-\beta\tilde{x})]$, we obtain

$$c_B = c_0 \frac{\exp\left\{\dfrac{\beta(\tau - x)}{\nu\gamma}\right\} - 1}{\exp\left\{\dfrac{\beta(\tau - x)}{\nu\gamma}\right\} - \dfrac{Kc_0}{1 + Kc_0}}$$

This equation holds only in the region between the shock path and the characteristic issuing from the origin. Below this characteristic, we have $c_B = 0$ because the initial bed would be naturally free of B, and the B produced cannot move faster than the fluid, the speed of which is just equal to 1. Consequently, we can write

$$c_B(x, \tau) = c_0\{1 - \exp(-\beta x)\}H(\tau - \theta(x))$$
$$+ c_0 \frac{\exp\left\{\dfrac{\beta(\tau - x)}{\nu\gamma}\right\} - 1}{\exp\left\{\dfrac{\beta(\tau - x)}{\nu\gamma}\right\} - \dfrac{Kc_0}{1 + Kc_0}}\{H(\tau - x) - H(\tau - \theta(x))\} \tag{8.1.36}$$

It is to be noted that, while the reactant A tends to be retained in the reactor due to adsorption, the product B passes through along with the carrier fluid. Therefore, not only the conversion is enhanced but also the partial separation is realized at the same time. If the product B also adsorbs, then Eq. (8.1.33)

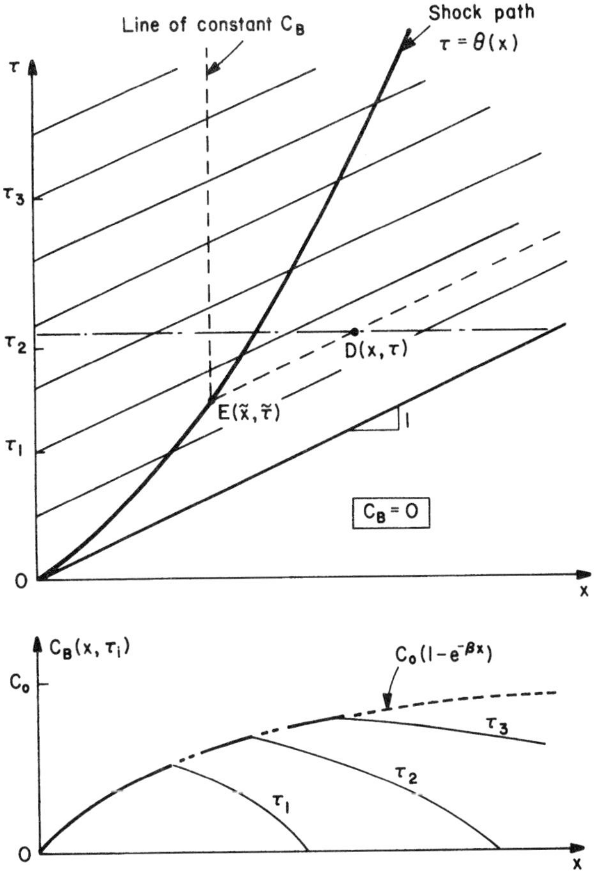

Figure 8.2

becomes quasi-linear and coupled with Eq. (8.1.19) through the isotherm expression so that we have a pair of quasi-linear equations. This is the subject to be treated in the second volume.

Next we will consider the same system with the square-wave input of the solute. Thus the initial data are

$$
\begin{aligned}
&\text{at} \quad \tau = 0, \qquad c = 0 \\
&\text{at} \quad x = 0, \qquad c = c_0 \qquad \text{for } 0 \le \tau \le a \\
&\qquad\qquad\qquad\quad c = 0 \qquad \text{for } \tau > a
\end{aligned}
\tag{8.1.37}
$$

In the neighborhood of the origin, we expect the solution to be the same as in the previous case. As shown in Fig. 8.3, however, there is a fan of characteristics all emanating from the point $D(0, a)$. Each of these curves

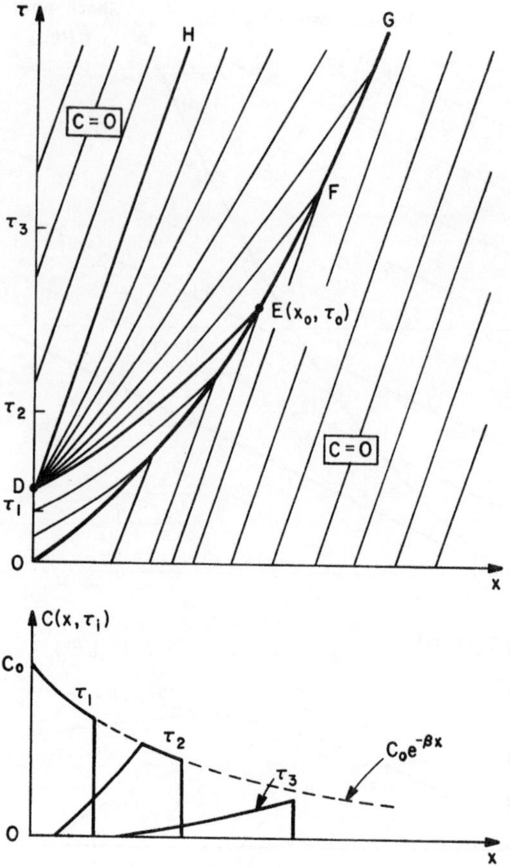

Figure 8.3

starts with a particular concentration $\bar{c}$, which ranges from c_0 for DE to $\bar{c} = 0$ for the uppermost characteristic DH. Hence the solution in the region ODE is given by Eqs. (8.1.29), (8.1.31), and (8.1.32), whereas we have $c = 0$ to the left of the shock path as well as above the characteristic DH. The point E is determined by solving Eq. (8.1.29) with $\eta = a$ and Eq. (8.1.31), and so it has the coordinates

$$\begin{aligned} x = x_0 &= \frac{1}{\beta} \ln \frac{\nu\gamma + a\beta(1 + Kc_0)}{\nu\gamma - a\beta(1 + 1/Kc_0)} \\ \tau = \tau_0 &= (1 + \nu\gamma)x_0 + \frac{\nu\gamma}{\beta} \ln \frac{1 + Kc_0 \exp(-\beta x_0)}{1 + Kc_0} \end{aligned} \tag{8.1.38}$$

Inside the region $HDEFG$, we have Eq. (8.1.23), and integration of this with the initial condition, $c = \bar{c}$ at $x = 0$, gives

$$c = \bar{c}\exp(-\beta x), \qquad 0 \le \bar{c} \le c_0 \tag{8.1.39}$$

The equation of the characteristic is given by Eq. (8.1.29) with a for η and $\bar{c}$ for c_0; that is,

$$\tau - a = (1 + \nu\gamma)x + \frac{\nu\gamma}{\beta}\left\{\frac{1}{1 + K\bar{c}} - \frac{1}{1 + K\bar{c}\exp(-\beta x)} + \ln\frac{1 + K\bar{c}\exp(-\beta x)}{1 + K\bar{c}}\right\} \tag{8.1.40}$$

At an arbitrary point $F(x, \tau)$ on the shock path EG, we can write the jump condition, which now reads

$$\frac{d\tau}{dx} = 1 + \frac{\nu\gamma}{1 + Kc} \tag{8.1.41}$$

where c represents the state on the left side of the shock. But from Eq. (8.1.23), we have

$$\frac{dx}{dc} = -\frac{1}{\beta c} \tag{8.1.42}$$

along each of the characteristics intersecting the shock path so that, regarding c as a parameter running along the shock path, we can put $d\tau/dx = (d\tau/dc)/(dx/dc)$ and eliminate dx/dc between Eqs. (8.1.41) and (8.1.42) to obtain the equation

$$\frac{d\tau}{dc} = -\frac{1 + \nu\gamma}{\beta}\frac{1}{c} + \frac{\nu\gamma}{\beta}\frac{K}{1 + Kc} \tag{8.1.43}$$

with the initial condition, $\tau = \tau_0$ when $c = c_0\exp(-\beta x_0)$, which corresponds to the point E. Direct integration gives

$$\tau = \tau_0 + \ln\left\{\frac{c_0\exp(-\beta x_0)}{c}\right\}^{(1+\nu\gamma)/\beta}\left\{\frac{1 + Kc}{1 + Kc_0\exp(-\beta x_0)}\right\}^{\nu\gamma/\beta}$$

or, by substituting Eq. (8.1.39), we get the equation

$$\tau = \tau_0 + \ln\left\{\frac{c_0\exp(-\beta x_0)}{\bar{c}\exp(-\beta x)}\right\}^{(1+\nu\gamma)/\beta}\left\{\frac{1 + K\bar{c}\exp(-\beta x)}{1 + Kc_0\exp(-\beta x_0)}\right\}^{\nu\gamma/\beta} \tag{8.1.44}$$

for the shock path. We note that $\bar{c}$ is a parameter ranging from c_0 to zero as τ increases, and thus Eq. (8.1.44) cannot determine the shock path by itself. In fact, we combine Eq. (8.1.44) with Eq. (8.1.40), and for various values of $\bar{c}$ we solve the pair of equations for x and τ to locate the shock path. At this stage (EFG), the shock will decelerate further because there is the added effect on the concentration on its left side due to the fall of the input concentration. The concentration profiles can be constructed from the (x, τ)-diagram as illustrated in Fig. 8.3.

The product concentration c_B can be determined in the same manner as in the previous case. In the angular region DEO, we have Eq. (8.1.35) for c_B, and in the region to the right of the shock path, c_B remains constant along the characteristics all having slope 1. The values of c_B along the shock path are determined by integrating Eq. (8.1.34) in the characteristic direction (i.e., along the lines of slope 1). To the left of the characteristic DH, it is obvious that c_B vanishes. Along the shock path c_B increases first to reach the highest value, very likely at point E of Fig. 8.3, and then decreases. The highest value is carried by the characteristic passing through the point so that the profiles of c_B will show a peak at the position of this characteristic.

Exercises

8.1.1. Recall Fig. 7.1 in Sec. 7.2 and suppose that the solute undergoes a chemical reaction in the fluid phase. If the reaction rate is $r(c)$, apply the argument of Sec. 7.2 to this system and show that the jump condition is unaffected by the reaction term and is given by Eq. (7.2.11) or Eq. (8.1.5).

8.1.2. If a compressible fluid is used in the heat exchanger immersed in a bath of temperature T_a (see Secs. 1.6 and 3.2), the velocity V becomes a function of temperature T, and, since the heat-transfer coefficient is a function of V, this also would be a function of T. Derive a differential equation for T at a distance x from the inlet at time t. Suppose T_a is constant and the initial data are specified as

$$\text{at} \quad s = 0, \qquad x = x_0(\xi), \qquad t = t_0(\xi), \qquad T = T_0(\xi)$$

where s and ξ are parameters along the characteristic and the initial curve, respectively. If no discontinuity is to be formed how would you formulate the solution?

8.1.3. A river broadens out in such a way that the *stage* h of a flood is governed by

$$\frac{\partial h}{\partial t} + \frac{8}{3}a(z)h^{2/3}\frac{\partial h}{\partial z} + a'(z)h^{5/3} = 0$$

(see Exercise 1.11.2). Discuss the subsidence of a flood for which $h(z, 0) = \beta z^{\mu}$ when $a(z) = \alpha z^{\gamma}$. If $\gamma = 1$ and $\mu = 0$, show that h remains independent of z.

8.1.4. If the adsorption follows the Freundlich isotherm, $n(c) = Kc^m$, solve the chromatographic equation (8.1.19) subject to the initial condition (8.1.25).

8.1.5. Consider the nth order, irreversible reaction in Eq. (8.1.19) with the Langmuir isotherm (8.1.20). For the initial data given by Eq. (8.1.25), construct the solution in the (x, τ)-plane with as many analytical expressions as possible. What kind of difficulty would you expect if $n < 1$?

8.1.6. For the Freundlich isotherm, $n(c) = Kc^m$, solve the chromatographic equation (8.1.19) subject to the initial condition (8.1.37).

8.1.7. Substitute Eq. (8.1.39) into Eq. (8.1.41) and regard $\bar{c}$ as a parameter along the shock path. (This is allowed by virtue of the connection by the characteristics.) By eliminating $d\tau/d\bar{c}$ between this equation and Eq. (8.1.40), show that the differential equation

$$\beta\frac{dx}{d(K\bar{c})} = \frac{[1 - \exp(-\beta x)]\{2 + K\bar{c}[1 + \exp(-\beta x)]\}}{(1 + K\bar{c})^2\{2 + K\bar{c}\exp(-\beta x)\}}, \qquad 0 < \bar{c} < c_0$$

holds along the shock path *EFG* in Fig. 8.3 with the initial condition, $x = x_0$ when $\bar{c} = c_0$, where x_0 is given by Eq. (8.1.38).

8.2 Analysis of Transient Volumetric Pool Boiling

In this section we would like to discuss the transient volumetric boiling that can be described by a nonhomogeneous quasi-linear equation. Since this has been treated in detail by Condiff and Epstein (1976), we will present only the summary of the analysis in connection with the mathematical developments of this chapter.

Let us consider a pool of liquid with unconfined free surface at the top and with a uniform depth. We will assume that the vapor void fraction is distributed homogeneously and uniformly across any horizontal section, so the two-phase vapor–liquid flow is one dimensional. Thus, if we define

α = vapor void fraction
u_v, u_l = vapor and liquid velocities, respectively
ρ_v, ρ_l = constant vapor and liquid densities, respectively
$\dot{m}$ = local vapor-mass production rate per unit volume of the mixture

and make the z-axis vertical with the origin at the bottom of the pool, the vapor and liquid phase continuity equations at time t are

$$\frac{\partial \alpha}{\partial t} + \frac{\partial}{\partial z}(\alpha u_v) = \frac{\dot{m}}{\rho_v} \tag{8.2.1}$$

$$\frac{\partial}{\partial t}(1 - \alpha) + \frac{\partial}{\partial z}[(1 - \alpha)u_l] = -\frac{\dot{m}}{\rho_l} \tag{8.2.2}$$

The vapor-mass production rate $\dot{m}$ is related to the volumetric heat-generation rate $\dot{Q}$ (power output) by

$$\dot{m} = \frac{\dot{Q}\omega(\alpha)}{\Delta H} \tag{8.2.3}$$

in which ΔH denotes the latent heat of vaporization and $\omega(\alpha)$ is the source function, which represents the variation of power input with increase of voids.

A convenient choice would be

$$\omega(\alpha) = (1 - \alpha)^m \tag{8.2.4}$$

where m is nonnegative because we expect the source function to be monotone decreasing as voidage increases. The volume average velocity,

$$u = \alpha u_v + (1 - \alpha)u_l \tag{8.2.5}$$

is determined by adding Eqs. (8.2.1) and (8.2.2) with the approximation $\rho_v^{-1} - \rho_l^{-1} \cong \rho_v^{-1}$ and substituting Eq. (8.2.3). Thus we have

$$\frac{\partial u}{\partial z} = \frac{\dot{Q}\omega(\alpha)}{\rho_v \Delta H}$$

and this is integrated at fixed time to give

$$u = u_0 + \frac{\dot{Q}}{\rho_v\, \Delta H}\int_0^z \omega(\bar{\alpha})d\bar{z} \tag{8.2.6}$$

in which u_0 is the volume average velocity at the pool bottom to be determined by the boundary condition, and the overbar is used in the integral term to distinguish the integration variable from the upper bound. Now the *drift fluxes* are defined as

$$j_{vl} \equiv \alpha(u_v - u) = -(1 - \alpha)[u_l - u) \equiv -j_{lv} \tag{8.2.7}$$

and if we assume that the relative motion between liquid and vapor is dependent only on the void fraction, we can write

$$j_{vl} = v_\infty \phi(\alpha), \qquad \phi(0) = \phi(1) = 0 \tag{8.2.8}$$

This is indeed the definition of the flux function $\phi(\alpha)$, where v_∞ is the terminal rise velocity of a single bubble of average size at a convenient void fraction. An empirical expression for this flux function is

$$\phi(\alpha) = \alpha(1 - \alpha)^n \tag{8.2.9}$$

where the exponent n is to be determined by experiment. Combining Eqs. (8.2.6), (8.2.7), and (8.2.8), we obtain

$$\alpha u_v = \alpha u_0 + \frac{\alpha \dot{Q}}{\rho_v\, \Delta H}\int_0^z \omega(\bar{\alpha})\, d\bar{z} + v_\infty \phi(\alpha) \tag{8.2.10}$$

We now return to Eq. (8.2.1) with Eqs. (8.2.3) and (8.2.10), and find it convenient to make the independent variables dimensionless:

$$\tau = \frac{\dot{Q}t}{\rho_v\, \Delta H}, \qquad x = \frac{\dot{Q}z}{\rho_v v_\infty\, \Delta H} \tag{8.2.11}$$

Then we have

$$\frac{\partial \alpha}{\partial \tau} + \frac{\partial}{\partial x}\left\{v_0\alpha + \alpha\int_0^x \omega(\bar{\alpha})\, d\bar{x} + \phi(\alpha)\right\} = \omega(\alpha) \tag{8.2.12}$$

where $v_0 = u_0/v_\infty$ is the dimensionless volume average velocity at the pool bottom. This is in the form of a conservation law, and it can be rewritten as

$$\frac{\partial \alpha}{\partial \tau} + \left\{ v_0 + \int_0^x \omega(\bar{\alpha})\, d\bar{x} + \phi'(\alpha) \right\} \frac{\partial \alpha}{\partial x} = (1 - \alpha)\omega(\alpha) \qquad (8.2.13)$$

The characteristic differential equations are

$$\frac{d\tau}{ds} = 1$$

$$\frac{dx}{ds} = v_0 + \int_0^x \omega(\bar{\alpha})\, d\bar{x} + \phi'(\alpha) \qquad (8.2.14)$$

$$\frac{d\alpha}{ds} = (1 - \alpha)\omega(\alpha)$$

But since $d\tau = ds$ along the characteristic curves, we may take τ as the characteristic parameter and write

$$\frac{dx}{d\tau} = v_0 + \int_0^x \omega(\bar{\alpha})\, d\bar{x} + \phi'(\alpha) \equiv \lambda(\alpha) \qquad (8.2.15)$$

for the reciprocal of the characteristic direction or the propagation speed of the state α, and

$$\frac{d\alpha}{d\tau} = (1 - \alpha)\omega(\alpha) \qquad (8.2.16)$$

for the variation of α along the characteristics.

If there is a discontinuity in α, it must satisfy the jump condition, which can be read directly from Eq. (8.1.5) after making comparison between Eqs. (8.1.1) and (8.2.12). Thus, noting that the roles of x and τ are interchanged here, we can write

$$\frac{dx}{d\tau} = \frac{\left[v_0\alpha + \alpha \int_0^x \omega(\bar{\alpha})\, d\bar{x} + \phi(\alpha)\right]}{[\alpha]}$$

$$= v_0 + \int_0^x \omega(\bar{\alpha})\, d\bar{x} + \frac{[\phi(\alpha)]}{[\alpha]} \equiv \tilde{\lambda}(\alpha^l, \alpha^r) \qquad (8.2.17)$$

for the propagation speed of the discontinuity or the reciprocal direction of the propagation path. As usual, the square brackets [ ] represent the jump of the quantity enclosed across the discontinuity. Of course, one can set up the material balances for the vapor and liquid phases, respectively, around a sharp discontinuity and formulate the jump condition to obtain Eq. (8.2.17) (see Exercise 8.2.1). By comparing Eq. (8.2.17) to Eq. (8.2.15), we notice that the entropy condition is to be given in terms of the flux function $\phi(\alpha)$

alone. As long as the entropy condition is concerned, the situation here is indeed the same as that in water flooding (Sec. 5.6), traffic flow (Sec. 6.6), or sedimentation (Sec. 6.7). Hence the generalized entropy condition should be the same as the rule of Sec. 5.6, which is rephrased here:

> RULE: If two states α^l and α^r are to be connected by a discontinuity, identify the chord connecting the two points L and R on the flux curve for which $\alpha = \alpha^l$ and $\alpha = \alpha^r$, respectively. Then, as we move along the chord from L to R, we must see the flux curve on the left side.

Note that the superscripts l and r represent the left and right sides of the discontinuity as we look at it at fixed time. If $n > 1$ in Eq. (8.2.9), the flux curve has an inflection point at $\alpha = 2/(n + 1)$, and so we expect to see both shocks and semishocks (see Secs. 5.6 and 6.7). On the other hand, if $0 < n \leq 1$, the flux curve is purely convex and the preceding rule is satisfied whenever $\alpha^l < \alpha^r$, as shown in Fig. 8.4. Clearly, we see

$$\phi'(\alpha^l) > \frac{[\phi]}{[\alpha]} > \phi'(\alpha^r)$$

and thus the shock condition

$$\lambda(\alpha^l) > \tilde{\lambda}(\alpha^l, \alpha^r) > \lambda(\alpha^r) \tag{8.2.18}$$

is satisfied. Since we will confine our discussion to the case of a convex flux curve, the entropy condition is $\alpha^l < \alpha^r$ and the discontinuities we may encounter here are all shocks.

Suppose now we have an initially saturated pool of liquid of depth d_0 open to the atmosphere, and the volumetric heat-generation rate is increased from zero to $\dot{Q}$ so that bulk boiling begins. At the bottom of the pool there is neither heat flux nor vapor-mass flux, so $v_0 = 0$ and the vanishing of vapor-mass flux implies $\alpha = 0$ at the bottom. Thus the appropriate initial and

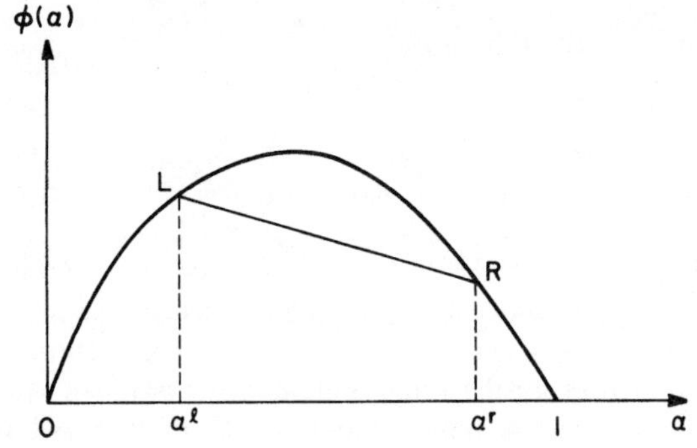

Figure 8.4

boundary conditions are

$$\begin{aligned} &\text{at} \quad x = 0, &\alpha = 0 & \\ &\text{at} \quad \tau = 0, &\alpha = 0 &\quad \text{for } 0 < x < \delta_0 \\ & &\alpha = 1 &\quad \text{for } x > \delta_0 \end{aligned} \tag{8.2.19}$$

in which δ_0 is the dimensionless initial pool depth defined by Eq. (8.2.11); that is,

$$\delta_0 \equiv \frac{\dot{Q} d_0}{\rho_v v_\infty \, \Delta H} \tag{8.2.20}$$

First we will parameterize the τ-axis with the parameter η so that we have the condition

$$x = 0, \qquad \tau = \eta, \qquad \alpha = 0$$

along the τ-axis. For the characteristics emanating from each point of the τ-axis, Eq. (8.2.16) can be solved by quadratures

$$\tau - \eta = \int_0^\alpha \frac{d\zeta}{(1 - \zeta)\omega(\zeta)} \equiv F(\alpha) \tag{8.2.21}$$

Eliminating τ between Eqs. (8.2.15) and (8.2.16) with $v_0 = 0$, we obtain

$$(1 - \alpha)\omega(\alpha)\frac{dx}{d\alpha} - \int_0^x \omega(\bar{\alpha}) \, d\bar{x} = \phi'(\alpha) \tag{8.2.22}$$

or

$$\frac{d}{dx}\left\{(1 - \alpha) \int_0^x \omega(\bar{\alpha}) \, d\bar{x}\right\} - \phi'(\alpha)$$

which upon integration becomes

$$\int_0^x \omega(\bar{\alpha}) \, d\bar{x} = \frac{\phi(\alpha)}{1 - \alpha} \tag{8.2.23}$$

with the constant of integration vanishing because at $x = 0$, $\alpha = \phi(\alpha) = 0$. Differentiating this equation with respect to α, dividing through by $\omega(\alpha)$, and then integrating from $x = 0$, where $\alpha = 0$, we find the equation

$$x = \int_0^\alpha \frac{d}{d\zeta}\left\{\frac{\phi(\zeta)}{1 - \zeta}\right\}\frac{d\zeta}{\omega(\zeta)} \equiv G(\alpha) \tag{8.2.24}$$

The two equations (8.2.21) and (8.2.24) define the relationship among x, τ, and α along the characteristic emanating from the point $\tau = \eta$ on the τ-axis. According to Eq. (8.2.24), the state α is dependent on x only, and so the lines of constant α are parallel to the τ-axis. On the other hand, Eq. (8.2.21) implies that α depends only on $(\tau - \eta)$. It then follows that the

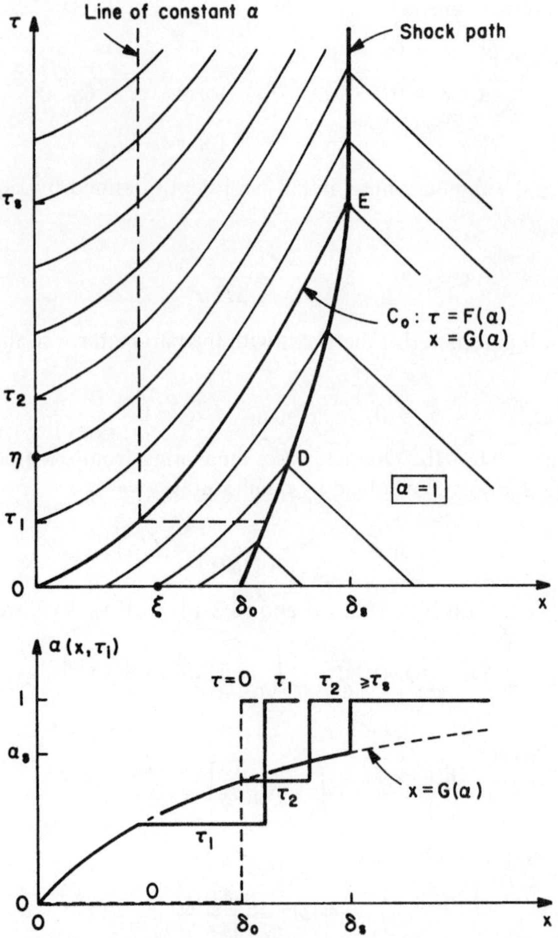

Figure 8.5

characteristics based on the τ-axis are all vertically parallel, as shown in Fig. 8.5. Among these, we will pick the one emanating from the origin and denote it by C_0, for which we have $\tau = F(\alpha)$ and $x = G(\alpha)$. The region bounded between C_0 and the τ-axis is the range of influence of the data specified on the τ-axis, and the situation in this region is called the *steady boiling regime*. Clearly, $F'(\alpha) = 1/\{(1 - \alpha)\omega(\alpha)\} > 0$ and $F'(\alpha) \to \infty$ as $\alpha \to 1$. Also, from Eq. (8.2.24) we have

$$G'(\alpha) = \frac{d}{d\alpha}\left\{\frac{\phi(\alpha)}{1 - \alpha}\right\}\frac{1}{\omega(\alpha)} = \frac{1}{(1 - \alpha)\omega(\alpha)}\left\{\phi'(\alpha) + \frac{\phi(\alpha)}{1 - \alpha}\right\} \tag{8.2.25}$$

and for convex flux functions we see that $\phi'(\alpha) + \phi(\alpha)/(1 - \alpha) > 0$ (see Fig. 8.4), and it can be shown that $G'(\alpha) \to \infty$ as $\alpha \to 1$. Therefore, both $F(\alpha)$ and $G(\alpha)$ are monotone increasing with α and tend to diverge as $\alpha \to 1$. This then implies that the characteristics are monotone-increasing curves. They may have inflection points depending on the form of the flux function, since

$$\frac{dx}{d\tau} = \frac{G'(\alpha)}{F'(\alpha)} = \phi'(\alpha) + \frac{\phi(\alpha)}{1 - \alpha}$$

and so

$$\frac{d^2x}{d\tau^2} = \frac{d}{d\alpha}\left(\frac{dx}{d\tau}\right)\frac{d\alpha}{d\tau} = (1 - \alpha)\omega(\alpha)\left\{\phi''(\alpha) + \frac{\phi'(\alpha)}{1 - \alpha} + \frac{\phi(\alpha)}{(1 - \alpha)^2}\right\}$$

which can change its sign from minus for small α to plus for higher α. For the flux function (8.2.9) with $n = 1$, however, we can show that $d^2x/d\tau^2 = -(1 - \alpha)\omega(\alpha) < 0$, and the characteristics in Fig. 8.5 are shown for this case or under the premise that the inflection point is not reached in the region of interest here.

Now let us introduce the parameter ξ along the x-axis so that the initial condition may read

$$x = \xi, \qquad \tau = 0, \qquad \alpha = 0 \qquad \text{for } 0 < \xi < \delta_0$$

Along the characteristics emanting from each point on the interval $(0, \delta_0)$ of the x-axis, we have, from Eqs. (8.2.16) and (8.2.21), the equation

$$\tau = F(\alpha) \tag{8.2.26}$$

This implies that the line of constant α is now parallel to the x-axis. For the relationship between x and α, we have Eq. (8.2.22), but the integral term can be rewritten here as

$$\int_0^x \omega(\bar{\alpha})\, d\bar{x} = \int_0^{G(\alpha)} \omega(\bar{\alpha})\, d\bar{x} + \int_{G(\alpha)}^x \omega(\bar{\alpha})\, d\bar{x}$$

$$= \int_0^{G(\alpha)} \omega(\bar{\alpha})\, d\bar{x} + \omega(\alpha)\{x - G(\alpha)\}$$

for the integration is made at constant time and α remains constant in this direction to the right of the curve C_0 on which we have $x = G(\alpha)$. Thus Eq. (8.2.22) gives

$$(1 - \alpha)\omega(\alpha)\frac{dx}{d\alpha} - \omega(\alpha)\{x - G(\alpha)\} = \int_0^{G(\alpha)} \omega(\bar{\alpha})\, d\bar{x} + \phi'(\alpha)$$

But recalling that $x = G(\alpha)$ is the solution of Eq. (8.2.22) in the steady boiling regime, we immediately notice that the right side of the preceding equation is equal to $(1 - \alpha)\omega(\alpha)\, dG(\alpha)/d\alpha$. It then follows that

$$(1 - \alpha)\frac{d}{d\alpha}\{x - G(\alpha)\} = x - G(\alpha)$$

and integrating with the condition $x - G(\alpha) = \xi$ at $\alpha = 0$, we obtain

$$x = G(\alpha) + \frac{\xi}{1 - \alpha} \tag{8.2.27}$$

This equation, together with Eq. (8.2.26), gives the relationship among x, τ, and α along the characteristics based on the interval $(0, \delta_0)$ of the x-axis. These characteristics are not parallel but tend to fan clockwise due to the second term on the right side of Eq. (8.2.27), as depicted in Fig. 8.5. Thus the solution is determined in the region between the characteristic C_0 and the pool top. The situation in this region is known as the *transient boiling regime*.

At the pool top, we see a discontinuity in the initial data, and it satisfies the entropy condition with $\alpha^l = 0$ and $\alpha^r = 1$ so that it starts to propagate as a shock in the positive direction. This implies that the pool top begins to rise upward. As the shock propagates, the state on its right remains fixed at $\alpha = 1$, whereas on the left side the void fraction α tends to increase because the characteristics impinging on the shock path bear successively higher void fractions. If we let α stand for the void fraction on the left side of the shock and δ be the pool-top position at time τ, we can write at an arbitrary point $D(\phi, \tau)$ on the shock path the equation

$$\frac{d\delta}{d\tau} = \int_0^{\delta} \omega(\bar{\alpha})\, d\bar{x} + \frac{\phi(\alpha)}{\alpha - 1} \tag{8.2.28}$$

from Eq. (8.2.17), with $v_0 = 0$, $\alpha^r = 1$, and $\alpha^l = \alpha$. But Eq. (8.2.16) also holds, so Eq. (8.2.28) can be rewritten as

$$(1 - \alpha)\omega(\alpha)\frac{d\delta}{d\alpha} = \int_{G(\alpha)}^{\delta} \omega(\bar{\alpha})\, d\bar{x} + \int_0^{G(\alpha)} \omega(\bar{\alpha})\, d\bar{x} - \frac{\phi(\alpha)}{1 - \alpha}$$

On the right side, the first term is equal to $\omega(\alpha)[\delta - G(\alpha)]$ since α remains constant in the x-direction while the last two terms cancel each other because of Eq. (8.2.23). Thus we have

$$\frac{d}{d\alpha}\{(1 - \alpha)\delta\} = -G(\alpha) \tag{8.2.29}$$

with the initial condition $\delta = \delta_0$ at $\alpha = 0$. This can be integrated directly, but the result will be expressd in the form of a double integral [see Eq. (8.2.24) for the definition of $G(\alpha)$]. To avoid this, we simply add $(1 - \alpha)G'(\alpha)$ on both sides of Eq. (8.2.29) and obtain the equation

$$\frac{d}{d\alpha}\{(1 - \alpha)(\delta - G(\alpha))\} = (1 - \alpha)G'(\alpha)$$

which upon integration gives

$$\delta = G(\alpha) + \frac{\delta_0}{1 - \alpha} + \frac{1}{1 - \alpha} \int_0^\alpha (1 - \zeta) G'(\zeta)\, d\zeta$$

since $G(0) = 0$. Consequently, by putting

$$\psi(\alpha) = \int_0^\alpha (1 - \zeta) G'(\zeta)\, d\zeta = \int_0^\alpha \frac{d}{d\zeta}\left\{\frac{\phi(\zeta)}{1 - \zeta}\right\} \frac{1 - \zeta}{\omega(\zeta)}\, d\zeta \tag{8.2.30}$$

we have the equation

$$\delta = G(\alpha) + \frac{\delta_0 - \psi(\alpha)}{1 - \alpha} \tag{8.2.31}$$

for the position of the shock that represents the pool top. The time required to reach this position is given by Eq. (8.2.26) [i.e., $\tau = F(\alpha)$]. Since $\delta \geq G(\alpha)$, it is obvious that we must have $\delta_0 \geq \psi(\alpha)$. Now we see

$$\frac{d\delta}{d\alpha} = G'(\alpha) - \frac{\psi'(\alpha)}{1 - \alpha} + \frac{\delta_0 - \psi(\alpha)}{(1 - \alpha)^2} = \frac{\delta_0 - \psi(\alpha)}{(1 - \alpha)^2} \geq 0 \tag{8.2.32}$$

because $(1 - \alpha)G'(\alpha) = \psi'(\alpha)$ by definition, and so δ is monotone increasing with α, as shown in Fig. 8.5. Also, we have

$$\psi'(\alpha) = \frac{d}{d\alpha}\left\{\frac{\phi(\alpha)}{1 - \alpha}\right\} \frac{1 - \alpha}{\omega(\alpha)} = \frac{1}{\omega(\alpha)}\left\{\phi'(\alpha) + \frac{\phi(\alpha)}{1 - \alpha}\right\}$$

and this is positive for convex flux functions [cf. Eq. (8.2.25)], so $\psi(\alpha)$ is a monotone-increasing function.

The distance between the shock path and the characteristic C_0 [i.e., $\delta - G(\alpha)$] decreases as the shock propagates, and the two will intersect each other when the condition

$$\delta_0 = \psi(\alpha) \tag{8.2.33}$$

is satisfied. The function $\psi(\alpha)$ being monotone increasing, Eq. (8.2.33) has either no root or only one root, and the latter will be denoted by α_s. Since $\psi(\alpha)$ is defined in the range $0 \leq \alpha \leq 1$, it certainly assumes a maximum at $\alpha = 1$ so that, if $\delta_0 \leq \psi(1)$, the shock path necessarily meets the characteristic C_0 when the void fraction on its left side becomes equal to α_s. This is marked by the point $E(\delta_s, \tau_s)$ in Fig. 8.5, the coordinates of which are given by

$$\delta_s = G(\alpha_s), \qquad \tau_s = F(\alpha_s) \tag{8.2.34}$$

At this point, $d\delta/d\alpha = 0$ from Eq. (8.2.32), or $d\delta/d\tau = 0$ from Eq. (8.2.23) and (8.2.28), and hence the shock becomes stationary. This implies that the boiling has reached a steady state, with the steady boiling regime prevailing throughout the pool. At this stage the void fraction α gradually increases from zero at the pool bottom to α_s just beneath the pool top and then abruptly rises to 1 across the pool surface. The profiles of α at successive times are

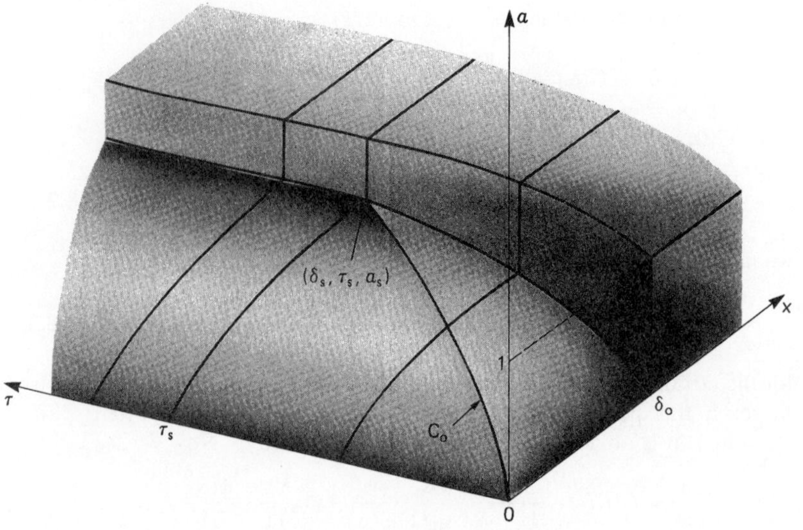

Figure 8.6

illustrated in the lower part of Fig. 8.5, whereas a three-dimensional view of the solution is given in Fig. 8.6.

On the other hand, if $\delta_0 > \psi(1)$, the shock never meets the characteristic C_0 so that both the steady boiling regime and the transient boiling regime tend to expand with the pool top rising continuously. The system never reaches the steady state but will be saturated physically and eventually brings about some physical change.

To illustrate this, we will consider Eqs. (8.2.4) and (8.2.9) for the source function $\omega(\alpha)$ and the flux function $\phi(\alpha)$, respectively; but since we are concerned here with convex flux functions, the index n will be restricted to $0 < n \leq 1$. Then, from Eqs. (8.2.21) and (8.2.24), we have

$$F(\alpha) = \frac{1}{m}\{(1 - \alpha)^{-m} - 1\} \tag{8.2.35}$$

and

$$\begin{aligned} G(\alpha; m, n) &= \frac{m}{(n - m)(n - m - 1)}\{(1 - \alpha)^{n-m} - 1\} \\ &\quad + \frac{n - 1}{n - m - 1}\alpha(1 - \alpha)^{n-m-1} \\ &= \frac{n - 1}{n - m - 1}\{(1 - \alpha)^{n-m-1} - 1\} \\ &\quad - \frac{n}{n - m}\{(1 - \alpha)^{n-m} - 1\} \end{aligned} \tag{8.2.36}$$

Based on Eq. (8.2.30), we can immediately write

$$\psi(\alpha; m, n) = G(\alpha; m - 1, n) \tag{8.2.37}$$

For $\alpha = 1$, therefore, we find that

$$\psi(1; m, n) = \begin{cases} \dfrac{1 - m}{(n - m)(n - m + 1)}, & \text{for} \quad n > m \quad \text{or} \quad n = 1 \\ \infty \quad , & \text{for} \quad n < m \quad \text{and} \quad n \neq 1 \end{cases} \tag{8.2.38}$$

This implies that, if $n < m$ and $n \neq 1$, pool boiling necessarily reaches a steady state after finite time, whereas for $n > m$ or $n = 1$, no steady state will be established unless $\delta_0 < \psi(1; m, n)$. In the special case when $m = n = 1$, the first equation of Eq. (8.2.38) gives $\psi(1; 1, 1) = 1$ and the boiling reaches a steady state only if $\delta_0 < 1$.

Along the characteristic C_0, we have $\tau = F(\alpha)$ and $x = G(\alpha; m, n)$ so that Eq. (8.2.35) gives

$$1 - \alpha = (1 + m\tau)^{-1/m} \tag{8.2.39}$$

and substituting this in Eq. (8.2.36) gives

$$x = x_0(\tau) = \frac{n - 1}{n - m - 1}(1 + m\tau)^{(m-n+1)/m} - \frac{n}{n - m}(1 + m\tau)^{(m-n)/m} - \frac{n - 1}{n - m - 1} + \frac{n}{n - m} \tag{8.2.40}$$

Here we note that $d^2x/d\tau^2 = 0$ for $\alpha = \alpha_{\text{inf}} = (2n - 1)/n^2$ or at $\tau = \tau_{\text{inf}} = (1/m)\{[n/(1 - n)]^{2m} - 1\}$, and obviously $0 < \alpha_{\text{inf}} < 1$ and $\tau_{\text{inf}} > 0$ if $\frac{1}{2} < n < 1$. With such flux functions, therefore, the characteristic C_0 will change its convexity unless $\alpha_s < \alpha_{\text{inf}}$ or $\tau_s < \tau_{\text{inf}}$. In the limit as $m \to 0$, we have $(1 + m\tau)^{1/m} \to \exp \tau$, and hence

$$\begin{aligned} \alpha &= 1 - \exp(-\tau) \\ x &= x_0(\tau) = \exp[(1 - n)\tau] - \exp(-n\tau) \end{aligned} \tag{8.2.41}$$

For the path of the shock at the pool top, Eq. (8.2.39) holds, and thus from Eqs. (8.2.31), (8.2.36) and (8.2.37) we obtain

$$\delta = x_0(\tau) + \left\{\delta_0 + \frac{n}{n - m + 1} - \frac{n - 1}{n - m}\right\}(1 + m\tau)^{1/m} + \frac{n}{n - m + 1}(1 + m\tau)^{(m-n)/m} - \frac{n - 1}{n - m}(1 + m\tau)^{(m-n+1)/m} \tag{8.2.42}$$

In the limit as $m \to 0$, this takes the form

$$\delta = \left\{\delta_0 - \frac{1}{n(n+1)}\right\}\exp \tau + \frac{1}{n}\exp[(1-n)\tau] - \frac{1}{n+1}\exp(-n\tau) \qquad (8.2.43)$$

so that the pool top position is an exponential function of time if the vapor production rate is uniform throughout the pool. The time τ_s required to reach the steady state is readily determined by letting $\delta = x_0(\tau_s)$ in Eq. (8.2.42).

If the flux curve has an inflection point, we see from Eq. (8.2.25) that $G'(\alpha)$ vanishes when the condition $\phi'(\alpha) = \phi(\alpha)/(\alpha - 1)$ is satisfied. When such a value of α is reached, the characteristics emanating from the τ-axis will turn backward and thus cross among themselves. Furthermore, $\psi'(\alpha)$ also vanishes for the same value of α, and this implies that the shock path representing the pool top may also turn backward. These are the major features that make the problem much more involved when the flux curve is nonconvex. This has been treated by Condiff and Epstein (1976) in their second paper and indeed their analysis has revealed various phenomena in problems of boilup with foam formation (see Exercises 8.2.3 and 8.2.4).

Exercises

8.2.1. Consider a jump discontinuity in the vapor void fraction α and set up the material balances for both the vapor and liquid phases, respectively. Combining these equations with Eqs. (8.2.5) through (8.2.8), formulate the jump condition (8.2.17).

8.2.2. If the source function is given by Eq. (8.2.4) and the flux function by Eq. (8.2.9), find an equation that determines the time τ_s required to reach the steady state. In particular, show that

$$(n^2 - 1)[\exp(-n\tau_s)] - n^2 \exp[-(n+1)\tau_s] = n(n+1)\delta_0 - 1$$

for $m = 0$, and that

$$\frac{\tau_s}{(1+\tau_s)^n} = \delta_0$$

for $m = 1$.

8.2.3. Consider the flux curve with one inflection point and define α_1 to be the void fraction at the point where the tangent to the flux curve at $\alpha = 1$ intersects the flux curve. Can you argue that if $\delta_0 < \psi(\alpha_1)$, the solution is given by the same equations as presented in this section? What variation would you expect to take place if $\delta_0 > \psi(\alpha_1)$?

8.2.4. In case the flux curve has one inflection point, define α_2 to be the void fraction at the point where the straight line passing through the point, $\alpha = 1$ and $\phi = 0$, becomes tangent to the flux curve, and show that $G'(\alpha_2) = 0$ and $\psi'(\alpha_2) =$

0. In particular, if $\phi(\alpha) = \alpha(1 - \alpha)^n$ with $n > 1$, show that $\alpha_2 = 1/n$. If $\delta_0 > \psi(\alpha_2)$, the characteristics emanating from the τ-axis turn backward at a distance $x = G(\alpha_2)$ from the pool bottom and thus are enveloped by a line of constant $x = G(\alpha_2)$. How would you proceed from here to complete the solution?

8.3 Black-box Steady State

If a process unit is viewed as a black box with certain inputs and outputs, it may be regarded as in a steady state when, for constant inputs, the outputs have become constant. Such a steady state is called a *black-box steady state* (BBSS) or an *ostensible steady state*, and it need not imply that the internal state of the system has reached a steady state. The latter condition, when every variable involved in a full description of the unit is constant in time, is a *true steady state* (TSS).

Clearly, the existence of a TSS implies that the outputs will be constant and hence assures the existence of a BBSS. In many cases the converse is true, and the observation of a BBSS implies the existence of the TSS. These may be reached simultaneously in a finite time (as in the ideal plug-flow reactor, when, after a single holding time, constant inputs will give a steady state) or approached asymptotically (as in a continuous stirred tank reactor). This is our usual experience with linear systems, but in some of the quasi-linear systems the BBSS can be attained in a finite time when the TSS is not attained or may only be approached asymptotically, and under certain circumstances the BBSS may be attained immediately. In fact, we have already seen such a case in Fig. 6.31 of Sec. 6.5. Here we have two shocks starting from each end, and they come together after a finite time τ_s to give a single shock that remains stationary inside the adsorption unit. At both boundaries we have adsorption equilibrium established, so the BBSS is immediately attained, whereas the TSS is reached at $\tau = \tau_s$.

To illustrate this in further detail, we will consider the system governed by the quasi-linear equation

$$\frac{\partial u}{\partial t} + (u - 1)\frac{\partial u}{\partial x} = -u \tag{8.3.1}$$

with the initial state

$$u(x, 0) = u_0 \tag{8.3.2}$$

and the input data

$$u(0, t) = u_1 \tag{8.3.3}$$

The output is then defined by

$$u_L = u(L, t) \tag{8.3.4}$$

The characteristic differential equations are

$$\frac{dt}{ds} = 1$$

$$\frac{dx}{ds} = u - 1$$

$$\frac{du}{ds} = -u$$

and by using $ds = dt$ from the first equation we have

$$\frac{dx}{dt} = u - 1 \equiv \lambda(u) \tag{8.3.5}$$

for the reciprocal direction of the characteristic curve and

$$\frac{du}{dt} = -u \tag{8.3.6}$$

for the variation of u in the characteristic direction. From these equations it is evident that, while u decreases monotonically along a characteristic, its slope in the (x, t)-plane becomes negative when u drops below 1. To see this more clearly, let us consider a characteristic curve passing through the point $x = t = 0$ and $u = \overline{u}$, in the three dimensional space of x, t, and u. Then Eqs. (8.3.5) and (8.3.6) are integrated to give

$$\begin{aligned} u &= \overline{u}e^{-t} \\ x &= \overline{u}(1 - e^{-t}) - t \end{aligned} \tag{8.3.7}$$

and, when the second equation is plotted for various values of $\overline{u}$, we obtain a family of characteristics as shown in Fig. 8.7. Here we note that the curves starting with $\overline{u}$ less than 1 are directed backward from the beginning and those starting with $\overline{u}$ greater than 1 turn backward later when u becomes less than 1.

Since from Eq. (8.3.1) we can think of the flux function in the form $\phi(u) = \frac{1}{2}u^2 - u$, the jump condition gives

$$\frac{dx}{dt} = \frac{[\phi]}{[u]} = \tfrac{1}{2}(u^l + u^r) - 1 \equiv \tilde{\lambda}(u^l, u^r) \tag{8.3.8}$$

for the propagation speed of a discontinuity. Here the system is genuinely nonlinear [i.e., $\phi''(u) \neq 0$], and so every discontinuity is a shock. The flux curve being convex downward, the entropy condition simply reads that we must have $u^l > u^r$ at every point of discontinuity (see the rule of the previous section).

Let us first suppose that the input and initial data are such that $u_1 > u_0 > 1$. The characteristics emanating from the t-axis are all vertically parallel, and thus the line of constant u is vertical, as shown in Fig. 8.8. If we let $t = \eta$ along the t-axis, we have the data $x = 0$ and $u = u_1$ at $t = \eta$ for Eqs.

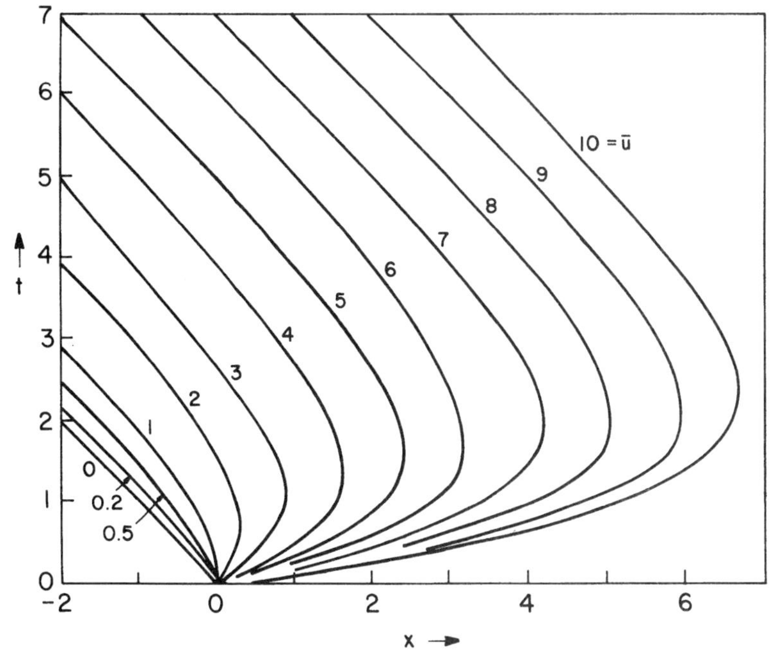

Figure 8.7

(8.3.5) and (8.3.6) so that integration gives

$$u = u_1 e^{-(t-\eta)} \tag{8.3.9}$$

$$x = u_1\{1 - e^{-(t-\eta)}\} - (t - \eta) = u_1 - u - \ln\frac{u_1}{u} \tag{8.3.10}$$

On the other hand, the characteristics starting from the x-axis are all horizontally parallel so that the line of constant u is parallel to the x-axis. By putting $x = \xi$ along the x-axis, we obtain the data $x = \xi$ and $u = u_0$ at $t = 0$, and hence

$$u = u_0 e^{-t} \tag{8.3.11}$$

$$x - \xi = u_0(1 - e^{-t}) - t = u_0 - u - \ln\frac{u_0}{u} \tag{8.3.12}$$

But the slope of these characteristics is greater than that of those emanating from the t-axis and the shock condition is satisfied (i.e., $u^l = u_1 > u_0 = u^r$). Hence the discontinuity at the origin starts to propagate as a shock, as depicted in Fig. 8.8. The shock path is determined by the jump condition (8.3.8), where u^l may be expressed in terms of x only [cf. Eq. (8.3.10)] and u^r in terms of t alone [cf. Eq. (8.3.11)]. Thus we have

$$\frac{dx}{du^l} = -1 + \frac{1}{u^l}, \qquad \frac{du^r}{dt} = -u^r$$

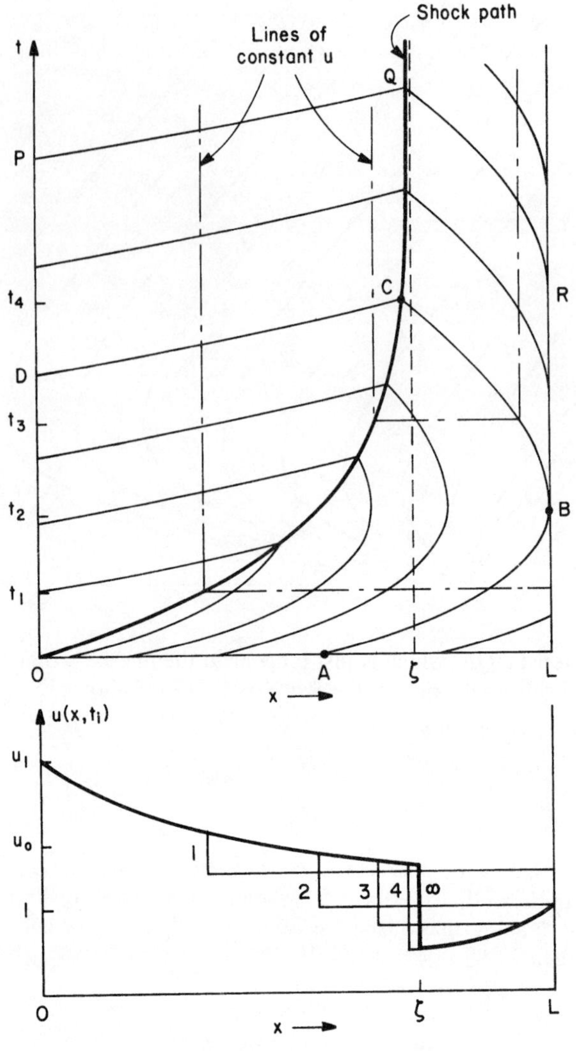

Figure 8.8

and, by combining these with Eq. (8.3.8), we obtain

$$\frac{du^l}{du^r} = \frac{u^r(u^l + u^r - 2)}{2u^l(u^l - 1)} \tag{8.3.13}$$

which is to be integrated with the initial condition, $u^l = u_1$ when $u^r = u_0$, to give the pair (u^l, u^r) along the shock path. The shock path is then determined by Eqs. (8.3.10) and (8.3.11).

As time goes on, however, the characteristics starting from those points of small ξ on the x-axis will turn backward if the length L is sufficiently large, that is, if

$$L > u_1 - 1 - \ln u_1 \tag{8.3.14}$$

[cf Eq. (8.3.10) with $u = 1$]. In particular, there exists a characteristic that just touches the exit boundary ($x = L$), as illustrated by the curve ABC in Fig. 8.8. At the point $B(L, t_B)$, we have $u = 1$ and $t_B = \ln u_0$. We note that the range of influence of the initial data u_0 is bounded by the curve BC, and so Eq. (8.3.13) is valid only for the portion OC of the shock path. At the point C, we have Eq. (8.3.10) with $u = u^l$ along the characteristic DC, but u^r is given by the characteristic BC for which we find

$$\begin{aligned} u^r &= e^{-t} \\ x - L &= 1 - u^r + \ln u^r \end{aligned} \tag{8.3.15}$$

from Eqs. (8.3.5) and (8.3.6) with the condition $u = 1$ at $x = L$. Hence, eliminating x by Eq. (8.3.10) gives

$$(u^l - \ln u^l) - (u^r - \ln u^r) = (u_1 - \ln u_1) - (1 + L) \tag{8.3.16}$$

Along the shock path beyond the point C, u^l and u^r must satisfy Eq. (8.3.16) since at a point Q on this path u^l is given by the characteristic PQ and u^r by the characteristic QR, which is just tangent to the line $x = L$ at R with $u = 1$. In the region on the right of the shock path CQ and above the characteristic CB, the line of constant u is vertical, and thus the profiles of u will take the form shown in the lower part of Fig. 8.8. Here the numbers represent the sequence of times indicated in the upper diagram, where $t_1 < t_B$, $t_2 = t_B$, $t_B < t_3 < t_C$, and $t_4 = t_C$.

A special relation among L, u_0, and u_1 is required for the pair (u^l, u^r) to remain fixed beyond the point C so that the shock path CQ becomes vertical. If this special relation happens to be satisfied, then the steady state is achieved at time t_c corresponding to the point C, and the shock remains stationary at this position. In all other cases, the shock path asymptotically approaches its final position as $t \to \infty$, as shown in Fig. 8.8, and the pair (u^l, u^r) tends to approach the steady-state values. However, $u_L = u(L, t)$ is constant and equal to 1 at all points above B. Thus the BBSS is achieved in time $t_B = \ln u_0$ even though, apart from exceptional circumstances, the TSS is only approached asymptotically.

The TSS here consists of a continuous segment in $0 < x < \zeta$ wherein $u(x)$ decreases from u_1 to u_s^l, according to the equations

$$\begin{aligned} x &= u_1 - u(x) - \ln \frac{u_1}{u(x)} \\ \zeta &= u_1 - u_s^l - \ln \frac{u_1}{u_s^l} \end{aligned} \tag{8.3.17}$$

At $x = \zeta$, there is a discontinuity from u_s^l to u_s^r, where

$$u_s^l + u_s^r = 2 \tag{8.3.18}$$

and the solution then rises continuously in $\zeta < x \leq L$ according to the equations

$$\begin{aligned} x &= \zeta + u_s^r - u(x) - \ln \frac{u_s^r}{u(x)} \\ L &= \zeta + u_s^r - 1 - \ln u_s^r \end{aligned} \tag{8.3.19}$$

Under the assumption of Eq. (8.3.14), the output is independent of u_1 and L, being $u_L = 1$. This steady state is shown by the heavier curve (labeled ∞) in the lower part of Fig. 8.8.

If $u_1 > 1 > u_0 \geq 0$, we again have a shock propagating from the origin, but the characteristics emanating from the x-axis have negative slope, as depicted in Fig. 8.9. The shock path may be treated using the same equations as before up to the point A on the characteristic BA through the point B, with $u = u_0 < 1$. Between BA and BC there is a centered expansion wave region covered by characteristics isuing from B, with values of u at B ranging from u_0 to 1. Thus at a point D on the shock path AC we require the equation

$$\begin{aligned} u^l &= u_1 e^{-(t-\eta)}, \qquad & x &= u_1 - u^l - \ln \frac{u_1}{u^l} \\ u^r &= \bar{u} e^{-t}, \qquad & L - x &= u^r - \bar{u} + \ln \frac{\bar{u}}{u^r} \end{aligned} \tag{8.3.20}$$

to be satisfied, where $\bar{u}$ is the value of u on BD at the point B. The two equations for x may be added to give an equation for $\bar{u}$ as a function of u^l and u^r; that is,

$$L = u_1 - u^l + u^r - \bar{u} + \ln\left(\frac{\bar{u} u^l}{u_1 u^r}\right) \tag{8.3.21}$$

These equations are then combined with the jump condition (8.3.8) to determine u^l, u^r, x, and τ along the shock path in terms of $\bar{u}$. It is in fact convenient to regard $\bar{u}$ as the parameter along the part AC of the shock path and obtain

$$\frac{du^l}{d\bar{u}} = \frac{u^l(\bar{u} - u^r)(u_a - 1)}{\bar{u}(u_a - u^r)(1 - u^l)}, \qquad \frac{du^r}{d\bar{u}} = \frac{u^r(u_a - \bar{u})}{\bar{u}(u_a - u^r)} \tag{8.3.22}$$

$$\frac{dx}{d\bar{u}} = \frac{(\bar{u} - u^r)(u_a - 1)}{\bar{u}(u_a - u^r)}, \qquad \frac{dt}{d\bar{u}} = \frac{\bar{u} - u^r}{\bar{u}(u_a - u^r)} \tag{8.3.23}$$

in which

$$u_a = \frac{u^l + u^r}{2} = \tilde{\lambda}(u^l, u^r) + 1$$

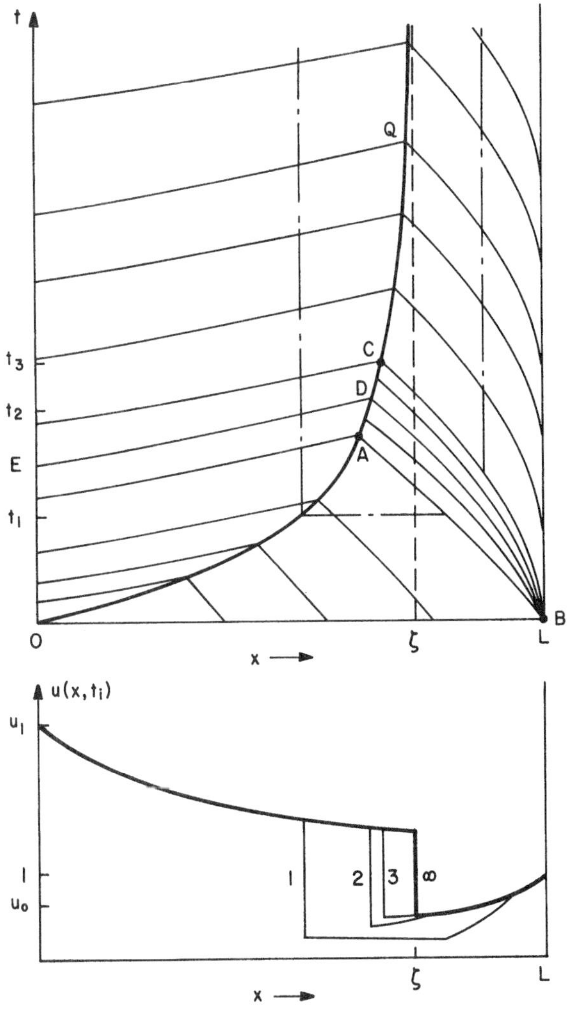

Figure 8.9

The first pair of equations (8.3.22) must be integrated simultaneously, but the second pair (8.3.23) can then be integrated by quadratures. This integration continues until $\bar{u} = 1$ when the point C is reached. Beyond this point (on CQ), the shock path approaches its steady-state position asymptotically in just the same fashion as before. There will again be a particular condition on L, u_1, and u_0 that makes $u^l + u^r = 2$ at C, and so the shock becomes stationary at this position; in this case the TSS is achieved in finite time t_C corresponding to the point C. However, we note that $u_L = u(L, t) = 1$ for all $t > 0$. Thus the BBSS is attained instantaneously even though the TSS is (exceptional circumstances apart) only approached asymptotically. The TSS is expressed by Eqs. (8.3.17), (8.3.18), and (8.3.19), and its profile is drawn by the heavier line (labeled ∞) along with the transient profiles in the lower part of Fig. 8.9.

In summary, if $u_1 > 1$ and Eq. (8.3.14) is satisfied, we see that the BBSS is established in a finite time

$$t = \begin{cases} \ln u_0, & u_0 > 1 \\ 0, & u_0 \leq 1 \end{cases} \tag{8.3.24}$$

This depends only on the initial state u_0, which we take to be a constant. Similar results can be obtained for arbitrary initial conditions. By contrast, the TSS is only approached asymptotically as $t \to \infty$, although there is a particular combination of L, u_1, and u_0 that allows it to be reached in a finite time

$$t = \begin{cases} ln\left(\dfrac{u_0}{u_s^r}\right), & u_0 > 1 \\ \\ \ln\left(\dfrac{1}{u_s^r}\right), & u_0 \leq 1 \end{cases} \tag{8.3.25}$$

As a physical example, let us consider the countercurrent adsorber with reaction (see Sec. 6.5 for the case without reaction). The solute is the reactant and forms products at a rate proportional to n, the solute concentration on the solid phase. The conservation equation can be written from Eq. (1.5.7) as

$$\varepsilon V_f \frac{\partial c}{\partial z} - (1 - \varepsilon) V_s \frac{\partial n}{\partial z} + \varepsilon \frac{\partial c}{\partial t} + (1 - \varepsilon) \frac{\partial n}{\partial t} = -k_r n \tag{8.3.26}$$

where the symbols have their usual meaning. Defining the dimensionless variables

$$x = \frac{z}{L}, \qquad \tau = \frac{V_f t}{L} \tag{8.3.27}$$

in which L denotes the total length of the column, and the dimensionless parameters

$$\nu = \frac{1 - \varepsilon}{\varepsilon} = \text{volume ratio} \tag{8.3.28}$$

$$\mu = \frac{(1 - \varepsilon) V_s}{\varepsilon V_f} = \text{flow rate ratio} \tag{8.3.29}$$

$$\beta = \frac{k_r L}{\varepsilon V_f} = \text{dimensionless reaction rate constant} \tag{8.3.30}$$

we can rewrite Eq. (8.3.26) in the form

$$\frac{\partial c}{\partial x} - \mu\frac{\partial n}{\partial x} + \frac{\partial c}{\partial \tau} + \nu\frac{\partial n}{\partial \tau} = -\beta n \tag{8.3.31}$$

If we assume that local equilibrium is established and the equilibrium relationship is given by the Langmuir isotherm

$$n = n(c) = \frac{\gamma c}{1 + Kc} \tag{8.3.32}$$

we obtain

$$\left\{1 - \frac{\mu\gamma}{(1 + Kc)^2}\right\}\frac{\partial c}{\partial x} + \left\{1 + \frac{\nu\gamma}{(1 + Kc)^2}\right\}\frac{\partial c}{\partial \tau} = -\frac{\beta\gamma c}{1 + Kc} \tag{8.3.33}$$

This equation must be solved subject to an initial condition

$$c(x, 0) = c_0(x) \tag{8.3.34}$$

and the inlet conditions that describe the feed concentrations in the fluid stream at $x = 0$ [a specified function $c_a(t)$] and in the solid stream at $x = 1$ [$n_b(t)$]. Since there may be a discontinuity in the concentrations at one or both ends, boundary conditions must be obtained by the condition that no matter accumulates in the planes $x = 0$ or $x = 1$. If the points just inside the column are denoted by $x = 0+$ and $x = 1-$ (see Fig. 6.23), the boundary conditions are

$$\text{at } x = 0, \qquad c_a - c_{0+} = \mu(n_a - n_{0+}) \tag{8.3.35}$$

$$\text{at } x = 1, \qquad c_{1-} - c_b = \mu(n_{1-} - n_b) \tag{8.3.36}$$

where the concentrations n and c just within the column are in equilibrium according to Eq. (8.3.32). If there is no discontinuity at $x = 0$, then $c_{0+} = c_a(t)$, which is specified, and $n_a(t) = n_{0+}$ is in equilibrium with c_{0+} and hence with $c_a(t)$. If there is a discontinuity, then n_a is determined from Eq. (8.3.35). Similar considerations apply at $x = 1$ with respect to $c_b(t)$. The functions $n_a(t)$ and $c_b(t)$ are outputs, and when they bcome constant for constant c_a and n_b, the BBSS is attained.

Now the characteristic differential equations are

$$\frac{dx}{ds} = 1 - \frac{\mu\gamma}{(1 + Kc)^2} \tag{8.3.37}$$

$$\frac{d\tau}{ds} = 1 + \frac{\nu\gamma}{(1 + Kc)^2} \tag{8.3.38}$$

$$\frac{dc}{ds} = -\frac{\beta\gamma c}{1 + Kc} \tag{8.3.39}$$

According to the last equation (8.3.39), c decreases monotonically along the characteristic so that we can use c as a parameter running along the characteristic and obtain an implicit form of the characteristic curve. For the characteristic curve that passes through the point (x_i, τ_i) in the (x, τ)-plane and has the value c_i there, we eliminate ds between Eqs. (8.3.37) and (8.3.39) and integrate the resulting equation to obtain

$$x - x_i = \Phi(c; c_i) \tag{8.3.40}$$

where

$$\Phi(c; c_i) = \frac{1}{\beta\gamma}\left\{K(c_i - c) + \ln\frac{c_i}{c} + \mu\gamma \ln\frac{c(1 + Kc_i)}{c_i(1 + Kc)}\right\} \tag{8.3.41}$$

and, likewise from Eqs. (8.3.38) and (8.3.39),

$$\tau - \tau_i = \Psi(c; c_i) \tag{8.3.42}$$

where

$$\Psi(c; c_i) = \frac{1}{\beta\gamma}\left\{K(c_i - c) + \ln\frac{c_i}{c} + \nu\gamma \ln\frac{c_i(1 + Kc)}{c(1 + Kc_i)}\right\} \tag{8.3.43}$$

On the other hand, it is convenient to have the characteristic direction from Eqs. (8.3.37) and (8.3.38); that is,

$$\frac{d\tau}{dx} = \frac{(1 + Kc)^2 + \nu\gamma}{(1 + Kc)^2 - \mu\gamma} \equiv \sigma(c) \tag{8.3.44}$$

We notice that the denominator of Eq. (8.3.44) vanishes for

$$c = \frac{1}{K}(\sqrt{\mu\gamma} - 1) \equiv c^* \tag{8.3.45}$$

and thus, if $\mu\gamma > 1$ and $c_i > c^*$, the characteristic will first be directed forward, then become vertical for $c = c^*$, and finally be directed backward for $c < c^*$.

If there is a discontinuity in c, it must satisfy the jump condition that reads, from Eq. (6.5.10),

$$\frac{d\tau}{dx} = \frac{1 + \nu\dfrac{[n]}{[c]}}{1 - \mu\dfrac{[n]}{[c]}}$$

or, by substituting Eq. (8.3.32),

$$\frac{d\tau}{dx} = \frac{(1 + Kc^l)(1 + Kc^r) + \nu\gamma}{(1 + Kc^l)(1 + Kc^r) - \mu\gamma} \equiv \tilde{\sigma}(c^l, c^r) \tag{8.3.46}$$

Furthermore, since the isotherm is convex [i.e., $n''(c) < 0$], the entropy condition is identical to the shock condition, and so every discontinuity arising here is a shock (see Sec. 6.5). In terms of the concentration, the shock condition is expressed as $c^l > c^r$.

We will consider only Riemann problems for which $c_a(t)$, $n_b(t)$, and $c_0(x)$ are identically constant at c_a, n_b, and c_0, respectively. As we have seen in Sec. 6.5, a variety of solutions is expected depending on the three numbers c_a, n_b, and c_0, and so we naturally expect to have steady states of various kinds. Here we will confine our attention to the case $c_a > c_0$ so that the entropy condition is satisfied at the origin, and thus a shock starts to propagate from there.

If $\mu\gamma < 1$, all the characteristics are necessarily directed forward and so is the shock path. Thus the structure of the solution will be similar to that shown in Fig. 8.1. Compare this with the case when μ belongs to the lower range in Sec. 6.5. For the characteristics emanating from the τ-axis, we have $x_i = 0$ and $c_i = c_a$ in Eq. (8.3.40) so that the line of constant c is vertical. When the shock reaches the boundary at $x = 1$, the system attains a steady state with $c_s(x)$ given by $\Phi(c_s; c_a) = x$ and c_{1-} by $\Phi(c_{1-}; c_a) = 1$. The boundary at $x = 1$ is to bear a discontinuity so that c_b is determined by Eq. (8.3.36), but $n_a = n(c_a)$, both being constant. Therefore, the BBSS and the TSS are reached simultaneously in a finite time (see Exercise 8.3.2).

Let us now suppose that $\mu\gamma > 1$ and consider two cases: (1) $c^* < c_0 < c_a$ and (2) $0 \le c_0 \le c^* < c_a$. The solutions are presented on the left and right of Fig. 8.10, in which the upper diagrams show the portraits of the solution in the (x, τ)-plane and the lower parts a sequence of concentration profiles for values of τ indicated in the upper diagram. The two steady states, shown with a heavier line and labeled ∞ in the lower graphs, are the same.

In both cases, a shock starts to propagate from the origin, as shown in Fig. 8.10. To the left of this shock path the solution is given by the characteristics emanating from the τ-axis. These curves are directed forward and have the equations

$$\begin{aligned} x &= \Phi(c; c_a) \\ \tau - \eta &= \Psi(c; c_a) \end{aligned} \tag{8.3.47}$$

where η is the intercept on the τ-axis. These are all parallel in a vertical sense, and so the line of constant c is parallel to the τ-axis.

In case (1) the characteristics based on the x-axis are represented by the equations

$$\begin{aligned} x - \xi &= \Phi(c; c_0) \\ \tau &= \Psi(c; c_0) \end{aligned} \tag{8.3.48}$$

in which ξ denotes the intercept on the x-axis. Since these curves are horizontally parallel, the line of constant c is parallel to the x-axis. As c decreases along these characteristics, they start to turn backward when c drops to c^*, as shown on the left of Fig. 8.10. In particular, there is one characteristic, ABC, that just touches the line $x = 1$ at B before bending back and meeting

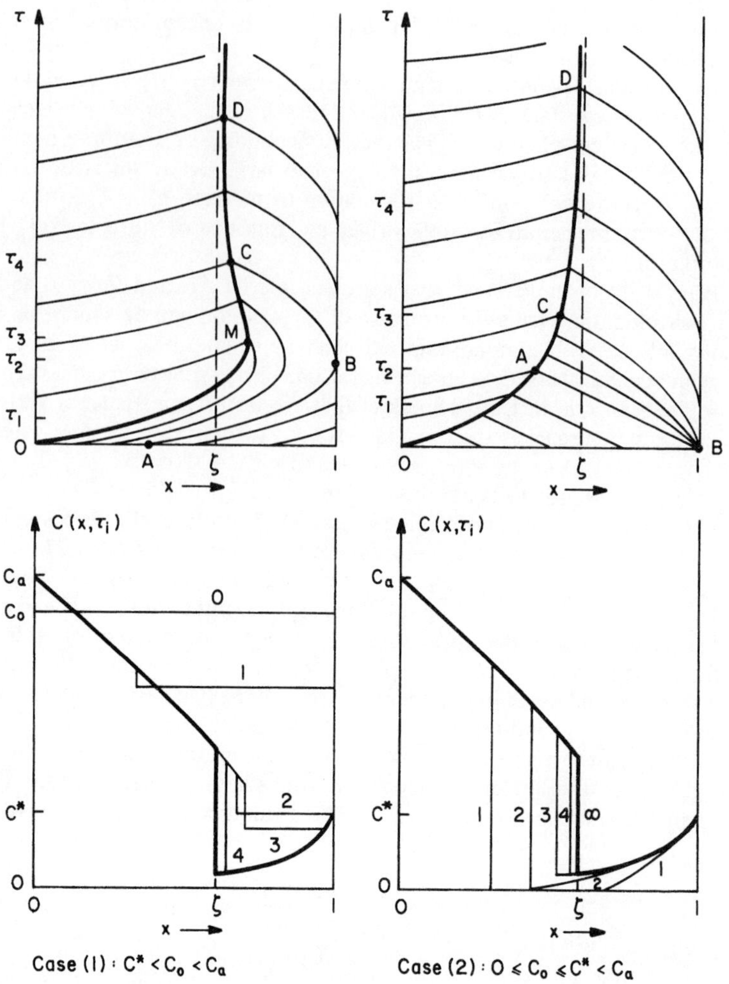

Figure 8.10

the shock path at C. Above the curve BC, the characteristics emanate from the line $x = 1$ where $c = c^*$ so that they are all vertically parallel and tangent to the line $x = 1$. The equations for these characteristics are

$$x - 1 = \Phi(c; c^*) \qquad\qquad (8.3.49)$$
$$\tau = \Psi(c; c^*)$$

The time at B, τ_B, can be calculated from Eq. (8.3.48) with $c = c^*$; that is,

$$\tau_B = \Psi(c^*; c_0) \qquad\qquad (8.3.50)$$

Beyond the point B, c_{1-} is kept constant at c^*, and c_b given by Eq. (8.3.36) also remains fixed, while $n_a = n(c_a) =$ constant. This implies that the BBSS is attained at $\tau = \tau_B$.

The shock path is to be determined by integrating the differential equation (8.3.46) with c^l and c^r given by the characteristics on either side. It is convenient to find dx/dc^l and $d\tau/dc^r$ first from Eqs. (8.3.47) and (8.3.48), respectively, and then to combine these with Eq. (8.3.46) to obtain

$$\frac{dc^r}{dc^l} = \frac{c^r(1 + Kc^r)\{(1 + Kc^l)^2 - \mu\gamma\}\{(1 + Kc^l)(1 + Kc^r) + \nu\gamma\}}{c^l(1 + Kc^l)\{(1 + Kc^r)^2 + \nu\gamma\}\{(1 + Kc^l)(1 + Kc^r) - \mu\gamma\}} \tag{8.3.51}$$

Integrating this with the initial condition $c^r = c_0$ when $c^l = c_a$, we match the pair (c^l, c^r) along the shock path and use this result in Eqs. (8.3.47) and (8.3.48) to locate the shock path. This is a tedious task, but it can be shown that for c_0 near c_a the shock overshoots its final position $x = \zeta$, reaching the point M and gradually retreating to ζ. At the point M, we have

$$(1 + Kc^l)(1 + Kc^r) = \mu\gamma \tag{8.3.52}$$

Beyond the point C, we have Eq. (8.3.49) instead of Eq. (8.3.48) on the right side of the shock, but Eq. (8.3.51) still holds. At C, we have the relation

$$\Phi(c^l; c_a) - \Phi(c^r; c^*) = 1 \tag{8.3.53}$$

from Eqs. (8.3.47) and (8.3.49), and this must be satisfied along the shock path beyond the point C.

As time goes on, the shock tends to approach its final position $x = \zeta$, where it will stop ultimately. Thus the TSS is only approached asymptotically. On the other hand, for c_0 near c^* the shock path is monotonic in the fashion of the diagram on the right of Fig. 8.10. It follows that there is some intermediate value of c_0 for which the point M is the final position of the shock, and then the TSS is reached in a finite time τ_M. Note that $\tau_M \geq \tau_B$.

In the lower part on the left of Fig. 8.10, the concentration profile for $\tau_1 < \tau_B$ is shown by the curve labeled 1. The steady-state profile has been achieved up to the shock, and the concentration is constant thereafter. At $\tau_2 = \tau_B$, the shock has advanced, and the plateau concentration is equal to c^*. Thereafter the concentration c_{1-} remains constant at c^*, and so the exit concentration c_b is fixed. Hence the BBSS has been reached even though the shock continues to move. At $\tau_3 = \tau_M$, it has advanced as far as it will go, and the plateau of constant c meets a short section of the steady-state profile at the very end of the column. The shock recedes further until at $\tau_4 = \tau_C$ the plateau has disappeared, but it only approaches its asymptotic position $x = \zeta$ as $\tau \to \infty$. Thus, although the TSS is not reached in any finite time (exceptional case apart), the BBSS is attained at time τ_B. Figure 8.11 shows an isometric view of the development of the steady state.

The true steady-state solution may be described as follows. The solution $c_s(x)$ has a discontinuity (c_s^l, c_s^r) at $x = \zeta$. To the left of the discontinuity,

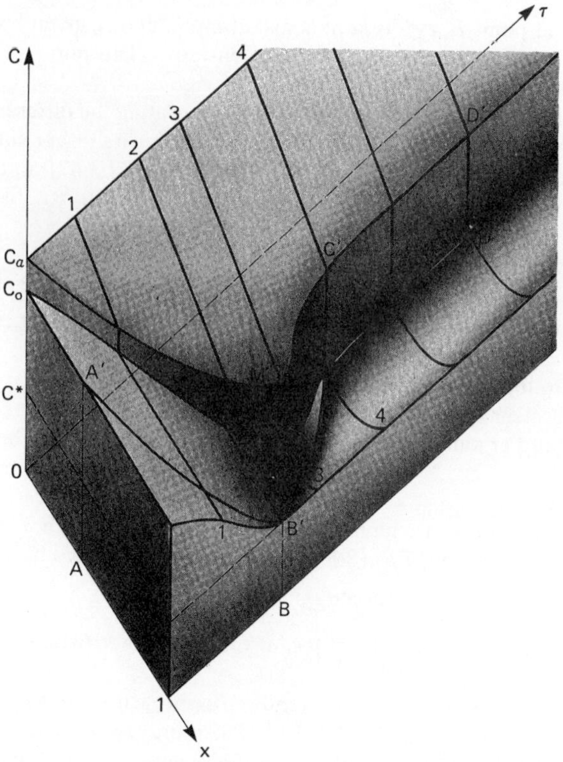

Figure 8.11

the value of c_s is given implicitly by

$$x = \Phi(c_s; c_a), \qquad 0 \le x < \zeta \tag{8.3.54}$$

To the right of the discontinuity,

$$x - 1 = \Phi(c_s; c^*) \tag{8.3.55}$$

and the discontinuity itself is determined by the three equations

$$\zeta = \Phi(c_s^l; c_a) = 1 + \Phi(c_s^r; c^*) \tag{8.3.56}$$

and

$$(1 + Kc_s^l)(1 + Kc_s^r) = \mu\gamma \tag{8.3.57}$$

The value of c_s at $x = 1-$ is equal to c^*, and so

$$c_{bs} = c^* + \mu\{n_b - n(c^*)\} \tag{8.3.58}$$

For case (2) the right side of Fig. 8.10 shows the solution when $c_0 = 0$, which is entirely typical of the behavior for $c_0 \le c^*$. In this case, the characteristics emanating from the x-axis all have negative slope; indeed, for c_0

$= 0$ they are all straight lines of slope $-(\nu\gamma + 1)/(\mu\gamma - 1)$. Thus there is a fan of characteristics in the region ABC emanating from the point B corresponding to concentration $\bar{c}$ in the range from c_0 to c^*. Above BC and to the right of the shock path, the characteristics are all vertically parallel. As in the previous case, the determination of the shock path requires the solution of Eq. (8.3.46), but it can be shown in this case that the shock path approaches the line $x = \zeta$ from the left. Four profiles for $\tau_1 < \tau_A$, $\tau_2 = \tau_A$, $\tau_3 = \tau_C$, and $\tau_4 > \tau_C$ are shown in the lower part on the right of Fig. 8.10, and a three-dimensional view of the solution surface is given in Fig. 8.12. Clearly, the BBSS is attained immediately, whereas the TSS given by Eqs. (8.3.54) through (8.3.58) is only approached asymptotically, and this feature remains the same for any value of c_0 less than or equal to c^*.

If the flow rate ratio μ becomes very large so that $c_0 < c_a < c^*$, not only the characteristics but also the shock path will be directed backward. Then the roles of the two boundaries, $x = 0$ and $x = 1$, are interchanged, and the data we specify at $x = 1$ (i.e., n_b) will pass through the column to reach the opposite end at $x = 0$. The situation is reversed from that for the case $\mu\gamma < 1$, and there is a discontinuity at $x = 0$ at the steady state. Compare this with the case when μ belongs to the upper range in Sec. 6.5. Here again the BBSS and the TSS are attained simultaneously in a finite time (see Exercise 8.3.3).

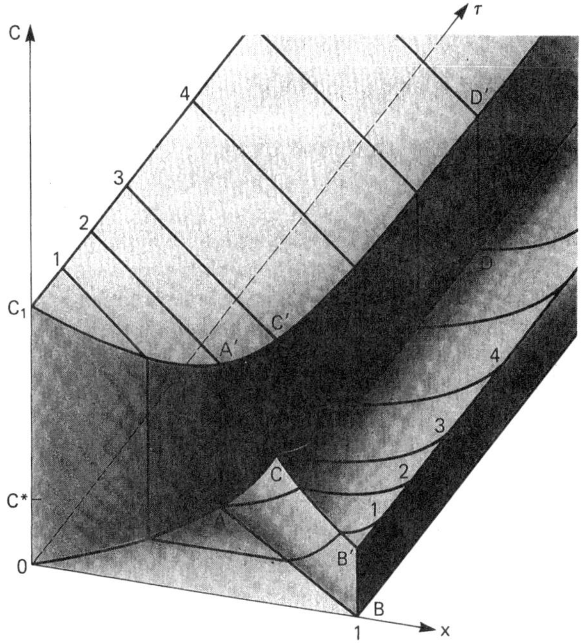

Figure 8.12

The examples treated in this section show that the BBSS is a concept distinguishable from that of the TSS. It raises some interesting questions of control, for though it may be a good objective to stabilize the BBSS, the behavior of the internal state may need to be considered as a constraint.

Exercises

8.3.1. Derive Eqs. (8.3.22) and (8.3.23).

8.3.2. Consider Eqs. (8.3.31) and (8.3.32). If $\mu\gamma < 1$ and $c_0 = 0$, discuss the solution for constant input data c_a and n_b. In particular, show that the time τ_s required to reach the steady state is given by the equation

$$\tau_s = \frac{1}{\beta\gamma}\int_{c_{1-}}^{c_a} \frac{(1 + Kc + \nu\gamma)\{(1 + Kc)^2 - \mu\gamma\}}{c(1 + Kc)\{(1 + Kc) - \mu\gamma\}}\, dc$$

where $\Phi(c_{1-}; c_a) = 1$. [*Hint:* Below the shock path, $c = 0$ so that c^r is fixed at zero.]

8.3.3. For the system of Exercise 8.3.2, consider the case when $\mu\gamma > 1$, $0 < c_0 < c_a < (\sqrt{\mu\gamma} - 1)/K$ and $n_b = 0$. Discuss the solution in detail for constant c_0 and c_a, and show that the shock path is given by the equations

$$\frac{dx}{dc} = \frac{\{\mu\gamma - (1 + Kc)\}\{(1 + Kc)^2 + \nu\gamma\}}{\beta\gamma c(1 + Kc)(1 + Kc + \nu\gamma)}$$

$$\tau = \Phi(c; c_0)$$

where c is the concentration on the left side of the shock. Solve the differential equation by quadratures subject to the initial condition $x = 1$ when $c = c_0$, and find an expression for the time τ_s required to reach the steady state.

8.4 Countercurrent Adsorber under Nonequilibrium Conditions

While we have given a full discussion on the analysis of a countercurrent adsorber in Sec. 6.5, we will revisit the subject here under nonequilibrium conditions. Thus some of the material may seem overlapping, but this section as a whole will serve as a lead to the next section, where a chemical reaction is to be included. We will be more interested in the steady-state analysis, and naturally one may presume that this is a simpler problem. In this system, however, there exists a variety of steady-state solutions, as we have seen in Sec. 6.5, and it is often necessary to examine the development of the steady state. Hence we begin with the unsteady-state equations.

For a single solute we can write the mass-conservation equations on the two phases from Eqs. (1.5.1) and (1.5.3) in the form

$$\begin{aligned}\varepsilon V_f\frac{\partial c}{\partial z} + \varepsilon\frac{\partial c}{\partial t} &= -k_1 c(N - n) + k_2 n \\ &= (1 - \varepsilon)V_s\frac{\partial n}{\partial z} - (1 - \varepsilon)\frac{\partial n}{\partial t}\end{aligned} \tag{8.4.1}$$

where a_v of Eq. (1.5.3) has been absorbed into k_1 and k_2 (see Fig. 1.2 or 6.23 for the notations). Equation (8.4.1) has a unique feature: it may be regarded as a chromatographic equation with a rate expression or as a pair of semilinear equations. In the limit as $k_2 \to \infty$, the rate expression in the middle gives the Langmuir isotherm

$$n = \frac{\gamma c}{1 + Kc} \tag{8.4.2}$$

in which

$$K = \frac{k_1}{k_2}, \qquad \gamma = NK = N\frac{k_1}{k_2} \tag{8.4.3}$$

Using the dimensionless variables

$$u = Kc, \qquad v = \frac{n}{N} \tag{8.4.4}$$

$$x = \frac{z}{Z}, \qquad \tau = \frac{V_f t}{Z} \tag{8.4.5}$$

and the dimensionless parameters

$$\nu = \frac{1 - \varepsilon}{\varepsilon}\gamma, \qquad \mu = \frac{(1 - \varepsilon)V_s}{\varepsilon V_f}\gamma, \qquad \alpha = \frac{Nk_1 Z}{\varepsilon V_f} \tag{8.4.6}$$

we have

$$\frac{\partial u}{\partial x} + \frac{\partial u}{\partial \tau} = -\alpha\{u(1 - v) - v\} = \mu\frac{\partial v}{\partial x} - \nu\frac{\partial v}{\partial \tau} \tag{8.4.7}$$

Note that the characteristic length Z is used in Eq. (8.4.5) so that the total length of the adsorber is not equal to 1 but is given by

$$X = \frac{L}{Z} \tag{8.4.8}$$

and the parameters ν and μ are defined with the factor γ included [cf. Eqs. (6.5.2), (8.3.28), and (8.3.29)]. The boundary conditions are taken from Eqs. (8.3.35) and (8.3.36), which now read

$$\text{at } x = 0, \qquad u_a - u_{0+} = \mu(v_a - v_{0+}) \tag{8.4.9}$$

$$\text{at } x = X, \qquad u_{X-} - u_b = \mu(v_{X-} - v_b) \tag{8.4.10}$$

where the subscripts $0+$ and $X-$ denote the points just inside the column (see Fig. 1.2 or 6.23). The boundary conditions can be satisfied identically if u and v are continuous across either boundary. If not satisfied identically, they serve to determine the outputs n_a and c_b by

$$\begin{aligned} v_a &= v_{0+} + \frac{(u_a - u_{0+})}{\mu} \\ u_b &= u_{X-} + \mu(v_b - v_{X-}) \end{aligned} \tag{8.4.11}$$

This implies that either boundary can bear a discontinuity in u, which is called a *boundary discontinuity* (B.D.) and distinguished from the one appearing inside the domain (see Sec. 6.5). We also have initial conditions

$$u(x, 0) = u_0(x), \qquad v(x, 0) = v_0(x) \tag{8.4.12}$$

Consider first the steady state that is governed by

$$\frac{du}{dx} = \mu\frac{dv}{dx} = -\alpha\{u(1 - v) - v\}. \tag{8.4.13}$$

It follows that for all cases

$$u - \mu v = w \quad \text{(a constant)} \tag{8.4.14}$$

and hence

$$\alpha x = \int_{v(x)}^{v_{0+}} \frac{\mu\, dv}{(w + \mu v)(1 - v) - v} \tag{8.4.15}$$

where

$$w = u_{0+} - \mu v_{0+} = u_{X-} - \mu v_{X-} \tag{8.4.16}$$

Since this model is not patient of discontinuities, Eqs. (8.4.9) and (8.4.10) are satisfied identically.

We will return to this integral later, but suppose for the moment that we have adsorption equilibrium (i.e., $\alpha \to \infty$). Then Eq. (8.4.13) gives

$$v = \frac{u}{1 + u} \tag{8.4.17}$$

which is the Langmuir isotherm (8.4.2) in dimensionless form. Thus the path in the (u, v)-plane is the hyperbola of the Langmuir isotherm. At a point (u, v) not on the isotherm, the derivatives of u and v become infinite, and this means that changes in u and v take place in a vanishingly short distance. Off the isotherm, the paths satisfy Eq. (8.4.14), and so the image of a discontinuity is a segment of a line $u = \mu v + w$. Below the isotherm, $\partial u/\partial x < 0$ so that the isocline of slope $1/\mu$ below the curve should have an arrow downward, as shown in Fig. 8.13; similarly the isoclines above the curve point

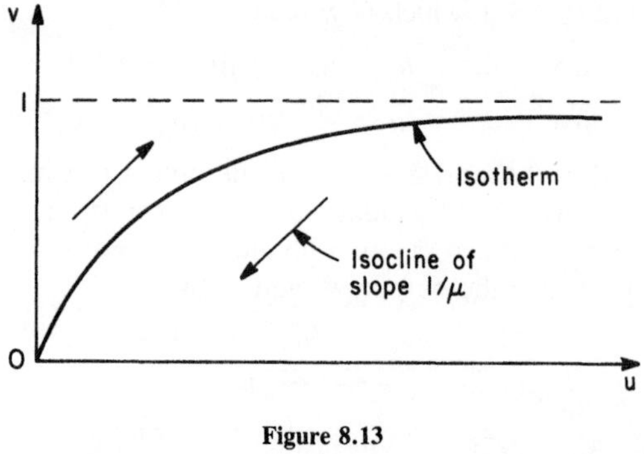

Figure 8.13

upward. But the isotherm is a line of critical points, $\partial u/\partial x = 0$ and $\partial v/\partial x = 0$, for all finite α, so that off the isotherm u and v change infinitely fast and on it they change infinitely slowly. This seems to make no sense, so we had better back off and look at the development of the steady state. Thus let $\alpha \to \infty$ in Eq. (8.4.7) and substitute Eq. (8.4.17) to have

$$\left\{1 - \frac{\mu}{(1+u)^2}\right\}\frac{\partial u}{\partial x} + \left\{1 + \frac{\nu}{(1+u)^2}\right\}\frac{\partial u}{\partial \tau} = 0 \qquad (8.4.18)$$

[cf. with Eq. (6.5.6)]. The characteristic differential equations are

$$\frac{dx}{ds} = 1 - \frac{\mu}{(1+u)^2}$$
$$\frac{d\tau}{ds} = 1 - \frac{\nu}{(1+u)^2} \qquad (8.4.19)$$
$$\frac{du}{ds} = 0$$

so that u is constant along characteristics, and this makes them straight lines of slope

$$\frac{d\tau}{dx} = \frac{(1+u)^2 + \nu}{(1+u)^2 - \mu} \equiv \sigma(u) \qquad (8.4.20)$$

The relation between σ and u is shown in Fig. 8.14, where we observe how those curves differ for $\mu < 1$, $\mu = 1$, and $\mu > 1$, respectively.

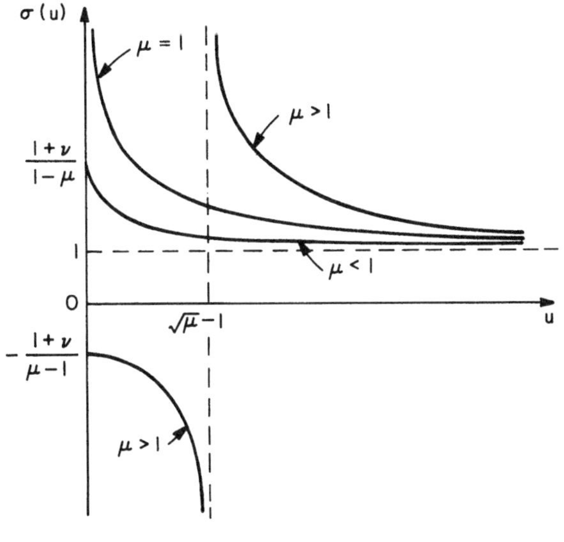

Figure 8.14

Although we are seeking a comprehensive picture, we cannot, at this point, allow everything to be arbitrary and will at least take u_a, v_b, u_0, and v_0 to be constant; indeed, we will take $u_0 = v_0 = 0$ for simplicity and return to the case of nonvanishing u_0 and v_0 later. Because of the convexity of the Langmuir isotherm, a discontinuity in u that may appear inside the domain is a shock, and it propagates in the direction

$$\frac{d\tau}{dx} = \frac{(1 + u^l)(1 + u^r) + \nu}{(1 + u^l)(1 + u^r) - \mu} \equiv \tilde{\sigma}\,(u^l, u^r) \tag{8.4.21}$$

which is taken from Eq. (8.3.46). A shock will be stationary if

$$(1 + u^l)(1 + u^r) = \mu \tag{8.4.22}$$

We now have four different models related as follows (equilibrium refers, of course, to adsorption equilibrium):

I. Transient, nonequilibrium model: Eqs. 7,9, 10, 12 of this section

II. Transient, equilibrium model: Eqs. 18, 9, 10, 12, 21 of this section

III. Steady-state, nonequilibrium model: Eqs. 13, 9, 10, 12 of this section

IV. Steady-state, equilibrium model: Eqs. 14, 15, 16, 22 of this section

For the nonequilibrium models, the boundary conditions (8.4.9) and (8.4.10) are satisfied trivially by the continuity of conditions in the inputs. At first blush it would appear that the fourth model is the simplest, but, as we have said, it seems to be either in shock or completely steady, and we will find that the most insight into the structure of the solutions comes with simultaneous consideration of the models.

Let us first discuss the case of the fixed bed, $\mu = 0$. We know what happens with model II from Secs. 5.3 and 5.5. The bed becomes saturated in a time $\tau_s = X\{1 + \nu/(1 + u_a)\}$ as the shock with $u^l = u_a$ and $u^r = u_0 = 0$ passes through the bed. Thereafter $u_b = u_a$ and the feed stream passes through unchanged. Since there is no movement of the solid, it does not matter what n_b may be. Model IV thus has the solution $u(x) = u_a$, $v(x) = u_a/(1 + u_a)$. Trivial though this case may appear, we will find that it has affinities with the situation when $\mu < 1$. Let us also note that if u_0 had not been taken to zero we would have had $\tau_s = X\{1 + \nu/[(1 + u_a)(1 + u_0)]\}$.

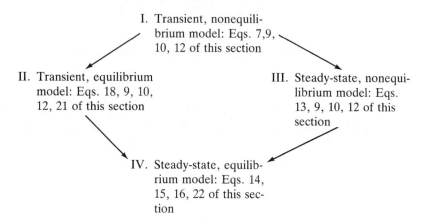

When $0 < \mu < 1$, the carrying capacity of the solid phase is less than that of the fluid phase, and we feel intuitively that the boundary at $x = X$ will be less influential than that at $x = 0$. Since $(1 + u)$ is always greater than 1 and μ less than 1, the characteristics all have a finite positive slope greater than 1. Since the slope of the Langmuir isotherm is less than 1 and the B.D. lines, which represent the images of boundary discontinuities in the (u, v)-plane, have slope $\mu^{-1} > 1$, there is no tangency between the equilibrium curve and the B.D. lines.

In model IV, we have to connect the condition at $x = 0$ with that at $x = X$, bearing in mind that there may be a boundary discontinuity either at $x = 0$ or at $x = X$. But it is clear that wherever the point (u_a, v_b) may lie in the (u, v)-plane (see Fig. 8.15), the only way to get from the vertical $u = u_a$ to the horizontal $v = v_b$ is to start from the point $(u_a, u_a/(1 + u_a))$ on the isotherm and take a discontinuous jump to the point $(u_a - \mu(\{u_a/(1 + u_a)\} - v_b), v_b)$. Thus the concentration will be constant throughout the column with

$$u(x) = u_a, \qquad v(x) = u_a/(1 + u_a) \tag{8.4.23}$$

and outputs will be

$$v_a = \frac{u_a}{1 + u_a}, \qquad u_b = u_a + \mu v_b - \frac{\mu u_a}{1 + u_a} \tag{8.4.24}$$

Because of the assumption of instantaneous adsorption, all these results are independent of the length of the column.

We can now understand how the integral (8.4.15) is to be calculated for given u_a and v_b. Because the rate of adsorption is finite, there can be no discontinuities, although there may be steep changes in $u(x)$ and $v(x)$. Thus we have $u_a = u_{0+}$ and $v_b = v_{X-}$, but $v_{0+} = v_a$ and $u_{X-} = u_b$, which we do not know except in the trivial case of $X = 0$ when $v_a = v_b$ and $u_b = u_a$ because nothing can have happened. Suppose in the first instance that the point (u_a, v_b) lies below the equilibrium isotherm, as shown in Fig. 8.16.

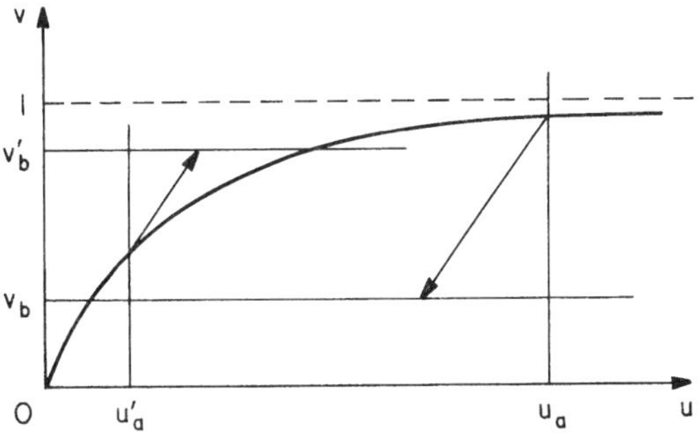

Figure 8.15

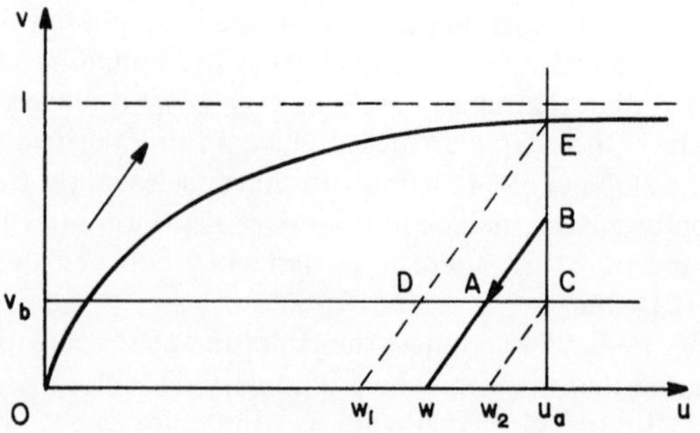

Figure 8.16

Then to each w in the interval (w_1, w_2) there corresponds a path going, for example, from $B(u_a, (u_a - w)/\mu)$ to $A(w + \mu v_b, v_b)$ and the value of X given by

$$\alpha X = \int_{v_b}^{(u_a - w)/\mu} \frac{\mu dv}{(w + \mu v)(1 - v) - v} \tag{8.4.25}$$

As the diagonal moves from C to DE, X increases from zero to infinity, for at E, the intersection of the diagonal and the isotherm, the denominator of the integrand is zero. We see that for a long column (i.e., a path near DE) the system is close to equilibrium for the greater part of the column and makes a more abrupt transition near the end of the column, as illustrated in Fig. 8.17. This again confirms that in the limiting case ($\alpha \to \infty$) the boundary discontinuity is at $x = X$.

The bounds of w corresponding to any (u_a, v_b) are

$$w_1 = u_a - \mu \frac{u_a}{1 + u_a}, \qquad w_2 = u_a - \mu v_b \tag{8.4.26}$$

Moreover, we could compute once and for all the integral

$$I(v; w, \mu) = \int_0^v \frac{\mu dv'}{(w + \mu v')(1 - v') - v'} \tag{8.4.27}$$

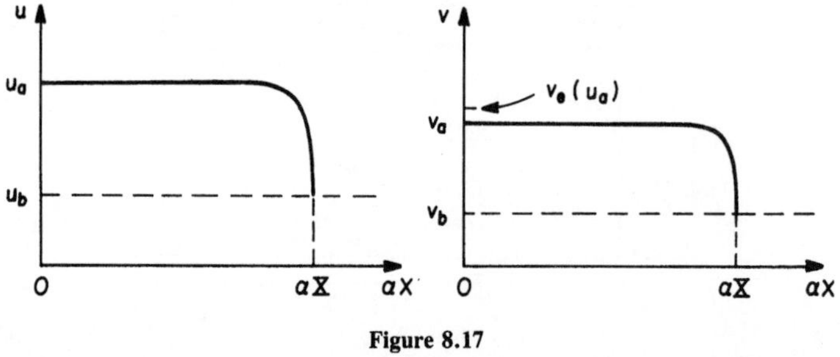

Figure 8.17

so that Eq. (8.4.25) becomes

$$\alpha X = I\left(\left(\frac{u_a - w}{\mu}\right); w, \mu\right) - I(v_b; w, \mu) \tag{8.4.28}$$

A similar argument applies if (u_a, v_b) lies above the isotherm and the integration goes from $(u_a - w)/\mu$ to v_b. Here we will need the integral

$$J(v; w, \mu) = \int_v^1 \frac{\mu dv'}{v' - (w + \mu v')(1 - v')} \tag{8.4.29}$$

for then

$$\alpha X = J\left(\left(\frac{u_a - w}{\mu}\right); w, \mu\right) - J(v_b; w, \mu) \tag{8.4.30}$$

The integrals I and J can be calculated for $w > 0$ and $w > -\mu$, respectively, and serve to calculate the required length of column. For example, if $u_a = 0$, $v_b > 0$, and a 99 percent removal of the solute is specified, $w = u_a - \mu v_a = -\mu v_b/100$ and

$$\alpha X = J\left(\frac{v_b}{100}; -\frac{\mu v_b}{100}, \mu\right) - J\left(v_b; -\frac{\mu v_b}{100}, \mu\right)$$

Let us turn our attention to the case $\mu > 1$ and consider the tangent to the isotherm having slope $1/\mu$. As shown in Fig. 8.18, the intercept of the tangent on the u-axis is $-(\sqrt{\mu} - 1)^2$, and thus if w lies between $-(\sqrt{\mu} - 1)^2$ and zero, there are two intersections between the isotherm and the line $u - \mu v = w$. If $(\bar{u}, \bar{u}/(1 + \bar{u}))$ and $(\tilde{v}/(1 - \tilde{v}), \tilde{v})$ are the upper and lower points of intersection of a chord of slope $1/\mu$, we have

$$\begin{aligned} 2\bar{u} &= (\mu - 1 + w) + \sqrt{(\mu - 1 + w)^2 + 4\mu w} \\ 2\mu\tilde{v} &= (\mu - 1 + w) - \sqrt{(\mu - 1 + w)^2 + 4\mu w} \end{aligned} \tag{8.4.31}$$

It is evident that the point $(\tilde{u}, \tilde{v})$ lies on the line

$$u + \mu v = \mu - 1 \tag{8.4.32}$$

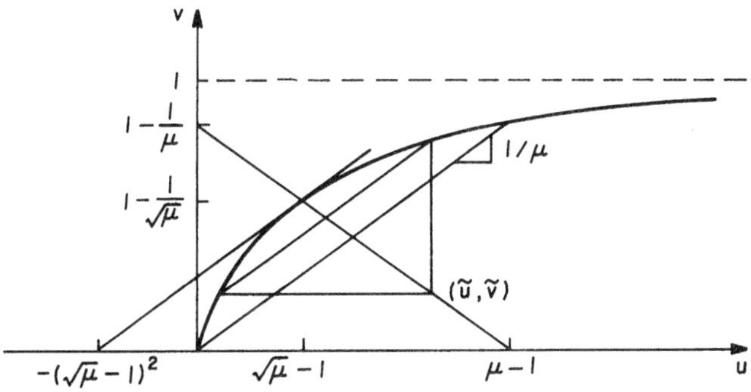

Figure 8.18

We can now describe the steady-state solution completely, both for adsorption equilibrium and for a finite rate of adsorption. There are five cases according to the region of Fig. 8.19 in which the point (u_a, v_b) lies. First, suppose that the point (u_a, v_b) falls in the region A, below the isotherm and to the right of the line given by Eq. (8.4.32). For model III, the trajectory is a segment of a line of slope $1/\mu$ that joins the vertical $u = u_a$ to the horizontal $v = v_b$, and we see from Fig. 8.13 that the direction corresponding to increasing x is downward. Thus the integral (8.4.25) must be evaluated on the line $u - \mu v = w$ for some value of w ranging from $w_f = u_a - \mu v_b$ to $W = u_a \{1 - \mu/(1 + u_a)\}$. As w decreases from w_f to W, αX goes from zero to infinity so that the limit applies either to a very long column ($X \to \infty$) or to the adsorption equilibrium case ($\alpha \to \infty$). In this latter case, there is a boundary discontinuity at $x = X$, and the steady-state solution is given by Eqs. (8.4.23) and (8.4.24). Because the integrand is greater when the point (u, v) is nearer the isotherm, the curves of u and v versus αx must have the form shown in Fig. 8.17. The relation between αX and w is given by Eq. (8.4.28).

Similar arguments obtain when the point (u_a, v_b) lies in each of the regions B, Γ, Δ, and E of Fig. 8.19. Let us look briefly at the region Δ. This is the case of $0 \le u_a < \sqrt{\mu} - 1$, $1 - \sqrt{\mu} < v_b < 1$. Again w ranges from $w_f = u_a - \mu v_b$ to $W = -(\sqrt{\mu} - 1)^2$, and the relation between αX and w is given by Eq. (8.4.30). Since the path is closest to the isotherm at its middle, we expect a plateau in the middle of the adsorber and steeper gradients at the ends. In the limit as $\alpha \to \infty$, this translates into boundary discontinuities both at $x = 0$ and $x = X$ in the steady-state solution:

$$u(x) = \sqrt{\mu} - 1, \qquad v(x) = 1 - \frac{1}{\sqrt{\mu}} \tag{8.4.33}$$

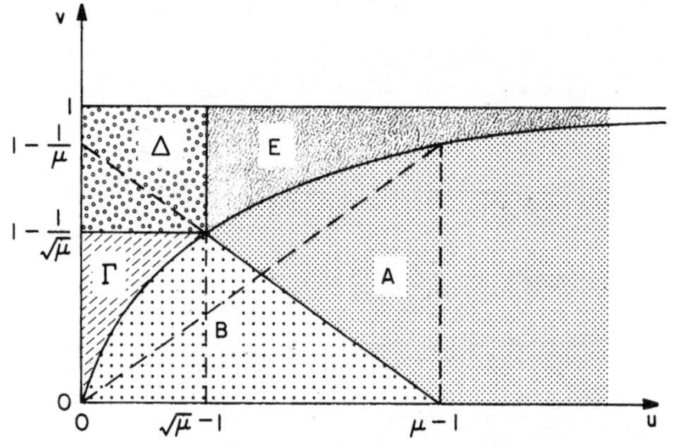

Figure 8.19

TABLE 8.1

(u_a, v_b) in the region:	Range of w for αX to increase from 0 to ∞	αX given by I or J	v_a	$u(x)$	$v(x)$	u_b	Location of B.D.
A	$w_f \searrow u_a - \mu v_e(u_a)$	$I(v; w, \mu)$	$v_e(u_a)$	u_a	$v_e(u_a)$	u_d	$x = X$
B	$w_f \searrow u_e(v_b) - \mu v_b$	$I(v; w, \mu)$	v_d	$u_e(v_b)$	v_b	$u_e(v_b)$	$x = 0$
Γ	$w_f \nearrow u_e(v_b) - \mu v_b$	$J(v; w, \mu)$	v_d	$u_e(v_b)$	v_b	$u_e(v_b)$	$x = 0$
Δ	$w_f \nearrow -(\sqrt{\mu} - 1)^2$	$J(v; w, \mu)$	Eq. (8.4.34)	$\sqrt{\mu} - 1$	$1 - \dfrac{1}{\sqrt{\mu}}$	Eq. (8.4.34)	$x = 0$ and $x = X$
E	$w_f \nearrow u_a - \mu v_e(u_a)$	$J(v; w, \mu)$	$v_e(u_a)$	u_a	$v_e(u_a)$	u_d	$x = X$

Note: $v_e(u) = u/(1 + u)$, $u_e(v) = v/(1 - v\}$, $w_f = u_a - \mu v_b$, $u_d = u_a + \mu\{v_b - v_e(u_a)\}$, and $v_d = v_b + \{u_a - u_e(v_b)\}/\mu$.

$$v_a = \frac{u_a + (\sqrt{\mu} - 1)^2}{\mu}, \qquad u_b = \mu v_b - (\sqrt{\mu} - 1)^2 \tag{8.4.34}$$

The full picture may be displayed in a table as arranged in Table 8.1, in which those columns for concentrations and that for the location of the boundary discontinuity are all for the limiting case ($\alpha \to \infty$), and these may be compared with the result of Sec. 6.5.

Exercise

8.4.1. If the point (u_a, v_b) falls in the region B below the isotherm and to the left of the line $u + \mu v = \mu - 1$ (see Fig. 8.19), show that the steady-state solution is given by the information in the second row of Table 8.1. Repeat the same for the cases Γ and E.

8.5 Countercurrent Adsorber with Reaction

Let us now suppose that the solute not only exchanges between the two streams but that when it is adsorbed on the solid it may react to give a product that desorbs immediately. If the reaction is first order and irreversible, we must modify the second equation of Eq. (8.4.1) by including the reaction rate term $k_r n$. Thus we have

$$\begin{aligned} \varepsilon V_f \frac{\partial c}{\partial z} + \varepsilon \frac{\partial c}{\partial t} &= -k_1 c(N - n) + k_2 n \\ (1 - \varepsilon) V_s \frac{\partial n}{\partial z} - (1 - \varepsilon) \frac{\partial n}{\partial t} &= -k_1 c(N - n) + (k_2 + k_r) n \end{aligned} \tag{8.5.1}$$

Compare this pair of equations with Eq. (1.5.10). We may retain the dimensionless variables and parameters spelled out in Eqs. (8.4.4), (8.4.5), and (8.4.6), but need an additional parameter for the reaction rate constant; that is,

$$\beta = \frac{N k_1 Z}{\varepsilon V_f} \frac{k_r}{k_2} = \alpha \frac{k_r}{k_2} \tag{8.5.2}$$

Thus the pair of equations in Eq. (8.5.1) becomes

$$\frac{\partial u}{\partial x} + \frac{\partial u}{\partial \tau} = -\alpha\{u(1 - u) - v\} \tag{8.5.3}$$

(I)

$$\mu \frac{\partial v}{\partial x} - \nu \frac{\partial v}{\partial \tau} = -\alpha\{u(1 - u) - v\} + \beta v \tag{8.5.4}$$

The boundary conditions, Eqs. (8.4.9) and (8.4.10), and the jump condition (8.4.21) for shocks remain the same, since the boundaries as well as a shock being planar have no volume, and hence reaction can play no role.

Subtracting Eq. (8.5.4) from Eq. (8.5.3) gives

$$\frac{\partial}{\partial x}(u - \mu v) + \frac{\partial}{\partial \tau}(u + \nu v) = -\beta v \tag{8.5.5}$$

If we now consider adsorption and desorption to be very rapid ($\alpha \to \infty$), the adsorption equilibrium relationship

$$v = \frac{u}{1 + u} \tag{8.5.6}$$

must obtain, which is the Langmuir isotherm, and substitution of this in Eq. (8.5.5) gives a single equation:

$$\text{(II)} \quad \left\{1 - \frac{\mu}{(1 + u)^2}\right\}\frac{\partial u}{\partial x} + \left\{1 + \frac{\nu}{(1 + u)^2}\right\}\frac{\partial u}{\partial \tau} = -\frac{\beta u}{1 + u} \tag{8.5.7}$$

which is identical to Eq. (8.3.33).

On the other hand, the assumption of a steady state gives two ordinary differential equations:

$$\text{(III)} \quad \frac{du}{dx} = -\alpha\{u(1 - v) - v\} \tag{8.5.8}$$

$$\mu\frac{dv}{dx} = -\alpha\{u(1 - v) - v\} + \beta v \tag{8.5.9}$$

As in the previous section, these may be combined to give

$$\frac{dv}{du} = \frac{1}{\mu} - \frac{\beta}{\alpha\mu}\frac{v}{u(1 - v) - v} \tag{8.5.10}$$

and the solutions are trajectories that link a point on the vertical $u = u_a$ with the horizontal $v = v_b$. In the limit as $\alpha \to \infty$, the equilibrium relation (8.5.6) must obtain, and if a point on the trajectory lies off this curve, it moves immediately on a line of slope $1/\mu$. Thus the discontinuities are again represented by line segments of slope $1/\mu$ in the (u, v)-plane.

Approaching the equilibrium, steady-state model (model IV) by the other route; that is, discarding the τ-derivative in Eq. (8.5.7), gives

$$\text{(IV)} \quad \frac{du}{dx} = -\frac{\beta u(1 + u)}{(1 + u)^2 - \mu} \tag{8.5.11}$$

Thus, when there is a continuous solution between two values of u, say u_1

$= u(x_1)$ and $u_2 = u(x_2)$,

$$\beta(x_2 - x_1) = -\int_{u_1}^{u_2} \left\{1 + u - \frac{\mu}{1 + u}\right\} \frac{du}{u}$$

$$= (1 - \mu) \ln \frac{u_1}{u_2} + \mu \ln \frac{1 + u_1}{1 + u_2} + u_1 - u_2 \qquad (8.5.12)$$

$$= \chi(u_1; -\mu) - \chi(u_2; \mu)$$

where
$$\chi(u; -\mu) = (1 - \mu) \ln u + \mu \ln(1 + u) + u \qquad (8.5.13)$$

and this function is sketched in Fig. 8.20. We notice that a continuous solution can only exist for $u > \sqrt{\mu} - 1$, for u is the concentration of a component being destroyed by an irreversible process, that is, $du/dx < 0$, and du/dx changes sign at $(1 + u)^2 = \mu$ [cf. Eq. (8.5.11)]. It follows that u can only cross the level $u = \sqrt{\mu} - 1$ by means of a discontinuity that is a shock.

Let us examine the steady-state models first, observing that the isoclines of Eq. (8.5.10) are all hyperbolas of the family

$$v = \frac{qu}{1 + qu} \qquad (8.5.14)$$

where
$$q = \frac{(\alpha/\beta)\{1 - \mu(dv/du)\}}{1 + (\alpha/\beta)\{1 - \mu(dv/du)\}} \qquad (8.5.15)$$

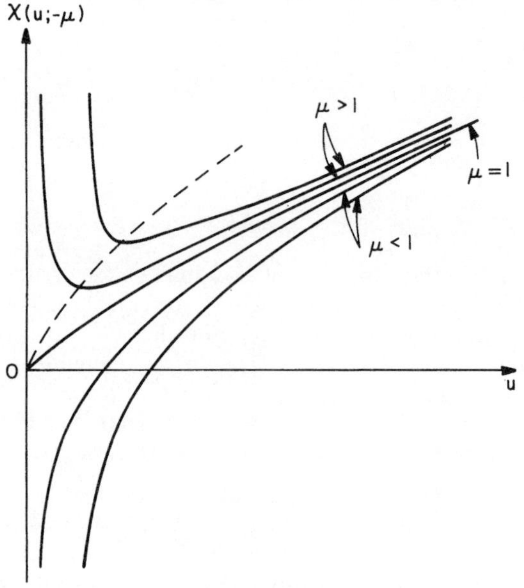

Figure 8.20

As $\alpha \to \infty$, q approaches to 1, and so all the isoclines, except the one corresponding to $dv/du = 1/\mu$, collapse onto the equilibrium isotherm, which is the isocline of $dv/du = \infty$. At all other points away from this curve, the slope approaches $1/\mu$. Again the solution of the equilibrium, steady-state equations is illuminated by the nonequilibrium case, although it is no longer true that the limiting case, $\alpha \to \infty$, corresponds to the limit $X \to \infty$ with finite α. Moreover, in contrast to the nonreactive case, the isotherm is an integral curve (or a trajectory) with a unique point of attraction, $u = \sqrt{\mu} - 1$ and $v = 1 - 1/\sqrt{\mu}$, where the slope of the tangent is $1/\mu$. In other words, once we are on the isotherm, we always move toward this point as x increases because the characteristic equation of Eqs. (8.5.8) and (8.5.9) has two negative roots there. This will become clearer later. This point is only in the positive quadrant if $\mu > 1$, and so, as we have come to expect, the behavior of the system when $\mu < 1$ is essentially that of a fixed bed.

Turning first to this case ($\mu < 1$), we see that the slope of the B.D. lines, $1/\mu$, is greater than the slope of any tangent to the isotherm, and the continuous solution does not turn back on itself. Thus, if the point (u_a, v_b) lies below the isotherm, there is continuity at $x = 0$, where $u_{0+} = u_a$ and $v_a = v_{0+} = u_a/(1 + u_a)$, and the isotherm is followed until

$$\beta X = \chi(u_a; -\mu) - \chi(u_{X-}; -\mu) \tag{8.5.16}$$

gives the value of u_{X-}. By the equilibrium hypothesis, we have $v_{X-} = u_{X-}/(1 + u_{X-})$, and by the boundary discontinuity at $x = X$,

$$u_b = \mu v_b + u_{X-}\left\{1 - \frac{\mu}{1 + u_{X-}}\right\} \tag{8.5.17}$$

Thus the internal state is unaffected by v_b and the system may be regarded as having a pseudo-fixed-bed behavior.

Now that we have seen the kind of solution to expect for the steady state, we can turn to model II, the transient, equilibrium model, to see how it evolves. Although this has been treated in Sec. 8.3, we will redo it in the line of this section with focus on the steady state. The characteristic differential equations are

$$\begin{aligned} \frac{dx}{ds} &= 1 - \frac{\mu}{(1 + u)^2} \\ \frac{d\tau}{ds} &= 1 + \frac{\nu}{(1 + u)^2} \\ \frac{du}{ds} &= -\beta\frac{u}{1 + u} \end{aligned} \tag{8.5.18}$$

Since u is monotone decreasing, it can be used as the parameter along the characteristic. But, then, dividing the first equation by the last gives

$$-\beta\frac{dx}{du} = \frac{1 + u}{u}\left\{1 - \frac{\mu}{(1 + u)^2}\right\} \tag{8.5.19}$$

which is precisely the steady-state equation (8.5.11). Thus

$$\beta(x - x_0) = \chi(u_0; -\mu) - \chi(u; -\mu) \tag{8.5.20}$$

Similarly, dividing the second equation of Eq. (8.5.18) by the third and integrating, we obtain

$$\beta(\tau - \tau_0) = \chi(u_0; \nu) - \chi(u; \nu) \tag{8.5.21}$$

where $\chi(u; \nu)$ is defined by Eq. (8.5.13) with ν in place of $-\mu$. The point (x_0, τ_0) is the point on the characteristic at which $u(x_0, \tau_0) = u_0$, and u, decreasing from u_0 to zero, leads us along the characteristic. If $u_0 = 0$, the concentration cannot decrease, and the characteristic is the straight line

$$\tau - \tau_0 = \frac{1 + \nu}{1 - \mu}(x - x_0) \tag{8.5.22}$$

Not surprisingly, every characteristic is asymptotic to a line of this slope. In particular, the characteristic through the point (x_0, τ_0), where $u = u_0$, is asymptotic to

$$\tau - \tau_0 - \frac{1}{\beta}\chi(u_0; \nu) = \frac{1 + \nu}{1 - \mu}\left\{x - x_0 - \frac{1}{\beta}\chi(u_0; -\mu)\right\} \tag{8.5.23}$$

To illustrate the development of the steady state in the equilibrium case with $\mu < 1$, let us consider first a bed with initial concentration equal to that of the feed (i.e., $u_0 = u_a$). The characteristics emanating from the x-axis and the τ-axis are identical. Along the x-axis, we see decreasing segments of the lower part of OP; above OP the characteristics are images of OP by vertical displacement [see Fig. 8.21(a)]. Thus, after a finite time T, the ordinate of P, the steady state is established; T is obtained by eliminating u_{X-} between

$$\begin{aligned} \beta X &= \chi(u_a; -\mu) - \chi(u_{X-}; -\mu) \\ \beta T &= \chi(u_a; \nu) - \chi(u_{X-}; \nu) \end{aligned} \tag{8.5.24}$$

The steady-state solution is

$$\begin{aligned} \beta x &= \chi(u_a; -\mu) - \chi(u; -\mu) \\ v(x) &= \frac{u(x)}{1 + u(x)} \\ v_a &= \frac{u_a}{1 + u_a} \\ u_b &= \mu v_b + u_{X-}\left\{1 - \frac{\mu}{1 + u_{X-}}\right\} \end{aligned} \tag{8.5.25}$$

If the inlet and initial conditions are

$$u(0, \tau) = u_a, \qquad u(x, 0) = u_0 > u_a$$

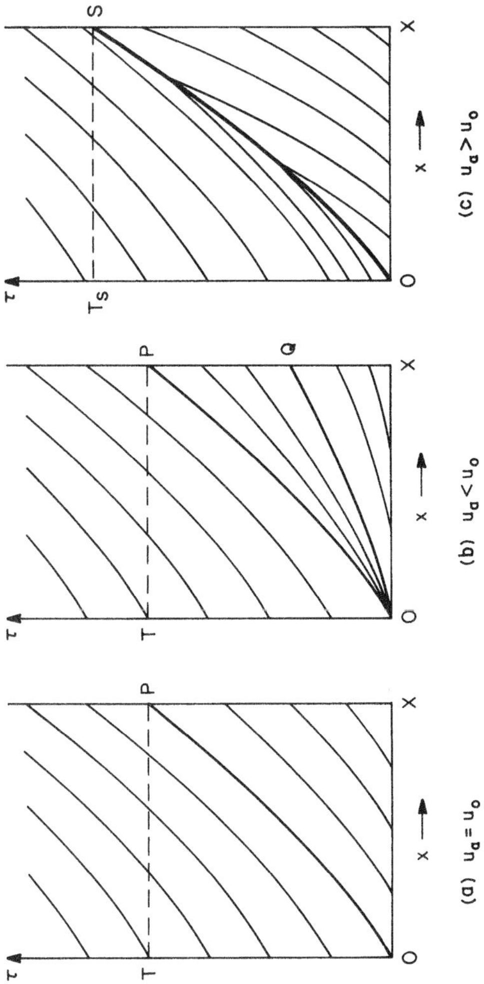

Figure 8.21

then the characteristics do not interfere with one another, as shown in Fig. 8.21(b), but the characteristic OQ through $(0, 0, u_0)$ lies below OP, the characteristic through $(0, 0, u_a)$. Thus the steady state is established at the same time as before although the history of the outlet composition is different. The values of u_{X-} and u_b for times between τ_Q and τ_P can be found by tracing the characteristics through $(0, 0, \bar{u})$ for $\bar{u}$ in (u_a, u_0). The steady state is identical to that of the previous case [i.e., Eq. (8.5.25)].

On the other hand, if

$$u(0, \tau) = u_a, \qquad u(x, 0) = u_0 < u_a$$

the characteristics emanating from the x-axis are steeper than those coming from the τ-axis, and a shock must be introduced to separate them, as shown in Fig. 8.21(c). If (x, τ) is a point on the shock path OS at which u jumps down from u^l to u^r in the x-direction, then

$$\frac{d\tau}{dx} = \frac{(1 + u^l)(1 + u^r) + \nu}{(1 + u^l)(1 + u^r) - \mu} \tag{8.5.26}$$

from Eq. (8.4.21), while

$$\begin{aligned} \beta x &= \chi(u_a; -\mu) - \chi(u^l; -\mu) \\ \beta\tau &= \chi(u_0; \nu) - \chi(u^r; \nu) \end{aligned} \tag{8.5.27}$$

We can eliminate x and τ to give an equation for the relationship between u^l and u^r:

$$\frac{du^l}{du^r} = \frac{u^r(1 + u^r)\{(1 + u^l)^2 - \mu\}\{(1 + u^l)(1 + u^r) + \nu\}}{u^l(1 + u^l)\{(1 + u^r)^2 + \nu\}\{(1 + u^l)(1 + u^r) - \mu\}} \tag{8.5.28}$$

[cf. Eq. (8.3.51)]. This equation can be integrated, since we know the initial values, $u^l = u_a$ and $u^r = u_0$. The shock path comes from Eq. (8.5.27), and the integration stops when $x = X$. The corresponding time T_s is greater than T, but again the steady state is established in a finite time. It is given by the same expressions as Eq. (8.5.25).

To complete the study of the pseudo-fixed bed, let us return to the case of a finite adsorption rate and consider the steady state (model III). From Eq. (8.5.10), we have

$$\mu\frac{dv}{du} = 1 - \frac{\beta}{\alpha}\frac{v}{u(1 - v) - v} \tag{8.5.29}$$

and notice that the origin, $u = v = 0$, is the only critical point. The linearization of the pair of equations (8.5.8) and (8.5.9) near the origin gives

$$\begin{aligned} \frac{du}{dx} &= -\alpha u + \alpha v \\ \frac{dv}{dx} &= -\frac{\alpha}{\mu}u + \frac{\alpha + \beta}{\mu}v \end{aligned} \tag{8.5.30}$$

and, since the matrix of coefficients has eigenvalues of opposite signs, the origin (O) in the (u, v)-plane is a saddle point. Hence there are two trajectories through O, one approaching (BO) and the other receding (OA), as illustrated in Fig. 8.22. The trajectories OA and OB are, respectively, tangent to the lines

$$\frac{v}{u} = 1 + \frac{\beta}{2\alpha} + \sqrt{\left(\frac{\beta}{2\alpha}\right)^2 + \frac{\beta}{\alpha}}$$
$$\frac{v}{u} = 1 + \frac{\beta}{2\alpha} - \sqrt{\left(\frac{\beta}{2\alpha}\right)^2 + \frac{\beta}{\alpha}} \tag{8.5.31}$$

at the origin. Between the arms, OA and OB, of this saddle point lie the isoclines of horizontal ($dv/du = 0$) and vertical ($dv/du = \infty$) directions, OD and OC, the equations of which are

$$v = \frac{u}{\beta/\alpha + 1 + u}$$
$$v = \frac{u}{1 + u} \tag{8.5.32}$$

respectively. Note that the isocline OC is the equilibrium isotherm itself. We can thus discern three regions in which (u_a, v_b) may lie, for as before a solution must be represented by a trajectory that joins the vertical $u = u_a$ to the horizontal $v = v_b$.

First, if the point (u_a, v_b) lies below OD, that is,

$$v_b < \frac{u_a}{1 + u_a + \beta/\alpha}$$

the sequence of paths corresponding to an increasing sequence of X is shown in Fig. 8.23. If (u_a, v_b) is below OB, the value of v_{0+} increases as X increases, and both u and v decrease monotonically with x. This continues through

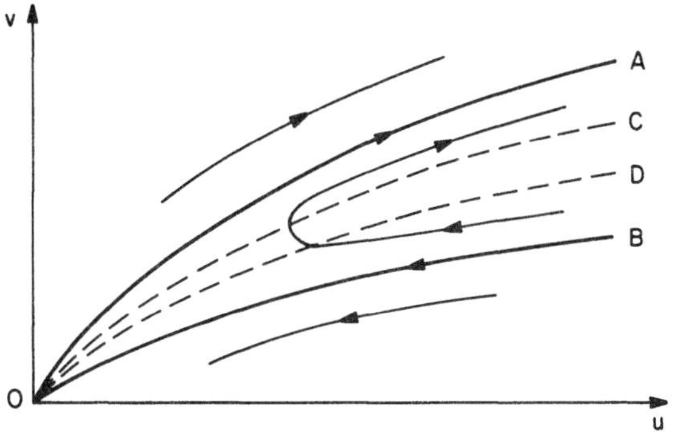

Figure 8.22

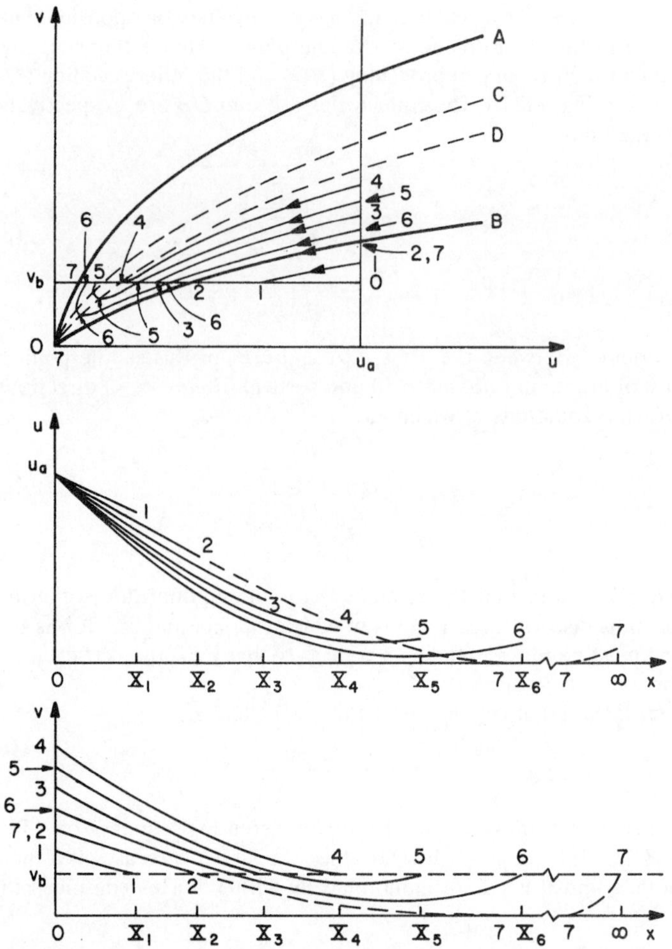

Figure 8.23

the loci 1–1, 2–2 (a segment of OB), 3–3, and 4–4. But the locus 4–4 is just tangent to the horizontal $v = v_b$, and if v_{0+} increased above v_4 there would be no path connecting the vertical $u = u_a$ to the horizontal $v = v_b$. Thus v_{0+} must decrease, and the path no longer gives monotonic u and v, since it drops below the horizontal $v = v_b$ and comes back up again as in the paths 5–5 and 6–6. These paths tend to approach two segments of the separatrices BO and OA (i.e., 7–7–7), and this corresponds to the limit $X \to \infty$ since O is a critical point, so du/dx and dv/dx are very small in its neighborhood. If (u_a, v_b) lies between OB and OD, a similar sequence obtains. If $v_b = 0$, then the paths are all monotone, for the locus 2–2 corresponds to a column of infinite length.

Suppose the point (u_a, v_b) falls between OC and OD, that is,

$$\frac{u_a}{\beta/\alpha + 1 + u_a} < v_b < \frac{u_a}{1 + u_a}$$

then the paths soon develop a single minimum in both u and v. As $X \to \infty$, the paths tend to two segments of BO and OA as depicted in Fig. 8.24. Finally, if the point (u_a, v_b) lies above OC, that is,

$$v_b > \frac{u_a}{1 + u_a}$$

the sequence of trajectories for an increasing sequence of X is as shown in Fig. 8.25. Here again the trajectories tend to approach two segments of BO and OA as $X \to \infty$.

So far we have been concerned with the simplest case of $\mu < 1$ and still have not solved the first model for the evolution of the nonequilibrium steady-state model. Transformation of the independent variables x and τ to the characteristic form

$$\xi = \frac{\nu x + \mu\tau}{\nu + \mu}, \qquad \eta = \frac{x - \tau}{\nu + \mu} \tag{8.5.33}$$

gives

$$\begin{aligned} \frac{\partial u}{\partial \xi} &= -\alpha\{u(1 - v) - v\} \\ \frac{\partial v}{\partial \eta} &= -\alpha\{u(1 - v) - v\} + \beta v \end{aligned} \tag{8.5.34}$$

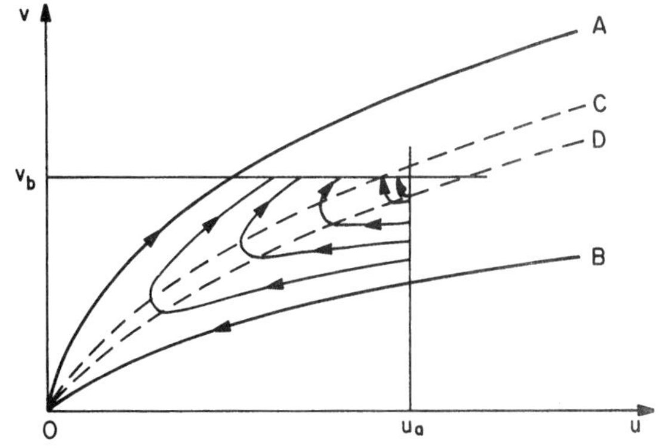

Figure 8.24

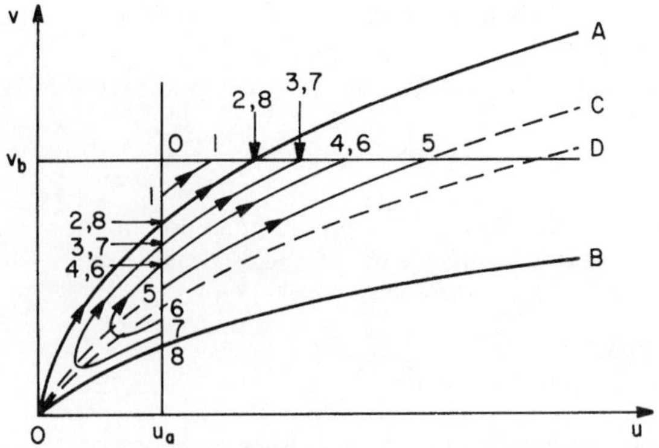

Figure 8.25

from Eqs. (8.5.3) and (8.5.4). The characteristics are straight lines and form a natural finite grid on which to do calculations. These calculations will undoubtedly give a solution that is similar in structure to the equilibrium case and increasingly so as α becomes large. The discontinuities will be smoothed into steep changes both inside the domain and at either boundary.

We turn now to the case $\mu > 1$ and return to the equilibrium steady-state model (model IV). We note here that the equilibrium isotherm is a solution of Eqs. (8.5.8) and (8.5.9), and at any point not on the isotherm the limit $\alpha \to \infty$ in Eq. (8.5.10) gives $dv/du = 1/\mu$; that is, the solution is a straight line. The latter solution corresponds to either a stationary shock or a boundary discontinuity. There is a point P on the isotherm of slope $1/\mu$; in fact, $u = \sqrt{\mu} - 1$ and $v = 1 - 1/\sqrt{\mu}$ at this point. The isotherm as a trajectory is always traversed toward this point because the characteristic equation of Eqs. (8.5.8) and (8.5.9) has two negative roots there; that is,

$$r_{\pm} = -\frac{\beta}{\mu} \pm \sqrt{\left(\frac{\beta}{\mu}\right)^2 - \frac{4\alpha\beta}{\mu}} \tag{8.5.35}$$

Since $d(\mu v - u)/dx = \beta v$ is not zero at P, it can be reached in a finite length. The (u, v)-plane portrait is shown in Fig. 8.26, where the straight line segments all correspond to stationary shocks or boundary discontinuities. It is this feature that gives the case $\mu > 1$ its greater variety and interest. It follows that a chord of slope $1/\mu$ can be drawn to the isotherm, and if (u, v) is one end on the isotherm, the other is at

$$\begin{aligned} u^* &= \frac{\mu}{1 + u} - 1 = \mu - 1 - \mu v \\ v^* &= 1 - \frac{1 + u}{\mu} = 1 - \frac{1}{\mu(1 - v)} \end{aligned} \tag{8.5.36}$$

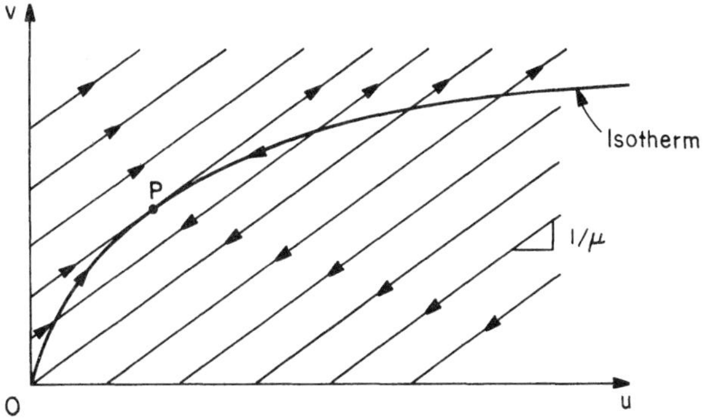

Figure 8.26

It will be convenient to use the asterisk as an abbreviation for "the other end of the chord," and thus u_a^* is $\{\mu/(1 + u_a)\} - 1$, the value of u at the other end of the chord from $(u_a, u_a/(1 + u_a))$.

Figure 8.27 and Table 8.2 give all the categories of steady-state solutions in the equilibrium case. The heading to the table shows all the varieties of solutions that will be encountered. They are D, the trivial solution for a column of zero length; C, solution for a column without discontinuity; O, X, columns with one boundary discontinuity at $x = 0$ or at $x = X$, respectively; S, columns with a stationary shock inside; OX, columns with two boundary discontinuities; SX, columns with one shock and one boundary discontinuity at $x = X$ (OS does not occur); OSX does not occur but is used to show the nomenclature. The table lists the sequences of solutions for X increasing from zero to infinity, and this is done for each region of the (u, v)-plane.

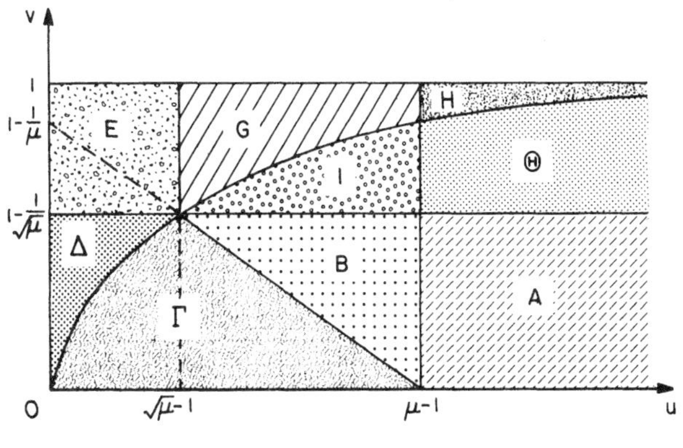

Figure 8.27

TABLE 8.2

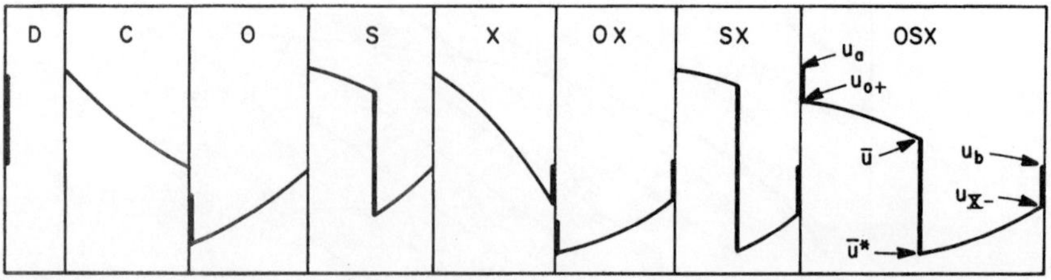

Region	Type	$\beta \mathbf{X}$	u_{o+}	$\bar{u}$	$\bar{u}^*$	$u_{\mathbf{X}-}$	u_b
A	D	0	–	–	–	–	$u_a - \mu[v_e(u_a) - v_b]$
	$\mathbf{X}$	$X(u_a) - X(u_{\mathbf{X}-})$	u_a	–	–	(u_b^*, u_a)	$u_{\mathbf{X}-} - \mu(v_{\mathbf{X}-} - v_b)$
	$\mathbf{X}$	$X(u_a) - X(u_b^*)$	u_a	–	–	u_b^*	$u_e(v_b)$
	S	$X(u_a) - X(\bar{u}) + X(\bar{u}^*) - X(u_b)$	u_a	$(u_b^*, \mu - 1)$	$\bar{u}^*$	$u_e(v_b)$	$u_e(v_b)$
	S	∞	u_a	$\mu - 1$	0	$u_e(v_b)$	$u_e(v_b)$
B	D	0	–	–	–	–	$u_a - \mu[v_e(u_a) - v_b]$
	O	$X(u_{o+}) - X[u_e(v_b)]$	$(0, u_a^*)$	–	–	$u_e(v_b)$	$u_e(v_b)$
	O	∞	0	–	–	$u_e(v_b)$	$u_e(v_b)$
	S	$X(u_a) - X(\bar{u}) + X(\bar{u}^*) - X(u_b)$	u_a	(u_a, u_b^*)	$[u_a^*, u_e(v_b)]$	$u_e(v_b)$	$u_e(v_b)$
	S	$X(u_a) - X[u_e(v_b)^*]$	u_a	$u_e(v_b)^*$	$u_e(v_b)$	$u_e(v_b)$	$u_e(v_b)$
Γ	D	0	–	–	–	–	$u_e(v_b)$
	O	$X(u_{o+}) - X[u_e(v_b)]$	$[0, u_e(v_b)]$	–	–	$u_e(v_b)$	$u_e(v_b)$
	O	∞	0	–	–	$u_e(v_b)$	$u_e(v_b)$
Δ	D	0	–	–	–	–	$u_e(v_b)$
	O	$X(u_{o+}) - X[u_e(v_b)]$	$[u_a, u_e(v_b)]$	–	–	$u_e(v_b)$	$u_e(v_b)$
	C	$X(u_a) - X[u_e(v_b)]$	u_a	–	–	$u_e(v_b)$	$u_e(v_b)$
	O	$X(u_{o+}) - X[u_e(v_b)]$	$(0, u_a)$	–	–	$u_e(v_b)$	$u_e(v_b)$
	O	∞	0			$u_e(v_b)$	$u_e(v_b)$
E	D	0	–	–	–	$\sqrt{\mu} - 1$	$\mu v_b - (\sqrt{\mu} - 1)^2$
	O$\mathbf{X}$	$X(u_{o+}) - X(\sqrt{\mu} - 1)$	$(u_a, \sqrt{\mu} - 1)$	–	–	"	"
	$\mathbf{X}$	$X(u_a) - X(\sqrt{\mu} - 1)$	u_a	–	–	"	"
	O$\mathbf{X}$	$X(u_{o+}) - X(\sqrt{\mu} - 1)$	$(0, u_a)$	–	–	"	"
	O$\mathbf{X}$	∞	0	–	–	"	"

Region	Type	$\beta\mathbf{X}$	u_{o+}	$\bar{u}$	$\bar{u}^*$	$u_{\mathbf{X}-}$	u_b
G	D	0	–	–	–	–	$u_a-\mu[v_e(u_a)-v_b]$
	$\mathbf{X}$	$\chi(u_a)-\chi(u_{\mathbf{X}-})$	u_a	–	–	$(\sqrt{\mu}-1,u_a)$	$u_{\mathbf{X}-}-\mu(v_{\mathbf{X}-}-v_b)$
	$\mathbf{X}$	$\chi(u_a)-\chi(\sqrt{\mu}-1)$	u_a	–	–	$\sqrt{\mu}-1$	$\mu v_b-(\sqrt{\mu}-1)^2$
	S$\mathbf{X}$	$\chi(u_a)-\chi(\bar{u})$ $+\chi(\bar{u}^*)-\chi(\sqrt{\mu}-1)$	u_a	$(\sqrt{\mu}-1,u_a)$	$\bar{u}^*$	"	"
	O$\mathbf{X}$	$\chi(u_{o+})-\chi(\sqrt{\mu}-1)$	$(0,u_a^*)$	–	–	"	"
	O$\mathbf{X}$	∞	0	–	–	"	"
H	D	0	–	–	–	–	$u_a-\mu[v_e(u_a)-v_b]$
	$\mathbf{X}$	$\chi(u_a)-\chi(u_{\mathbf{X}-})$	u_a	–	–	$(\sqrt{\mu}-1,u_a)$	$u_{\mathbf{X}-}-\mu(v_{\mathbf{X}-}-v_b)$
	$\mathbf{X}$	$\chi(u_a)-\chi(\sqrt{\mu}-1)$	u_a	–	–	$\sqrt{\mu}-1$	$\mu v_b-(\sqrt{\mu}-1)^2$
	S$\mathbf{X}$	$\chi(u_a)-\chi(\bar{u})$ $+\chi(\bar{u}^*)-\chi(\sqrt{\mu}-1)$	u_a	$(\sqrt{\mu}-1,\mu-1)$	$\bar{u}^*$	"	"
	S$\mathbf{X}$	∞	u_a	$\mu-1$	0	"	"
Θ	D	0	–	–	–	–	$u_b-\mu[v_e(u_b)-v_b]$
	$\mathbf{X}$	$\chi(u_a)-\chi(u_{\mathbf{X}-})$	u_a	–	–	$[u_e(v_b),u_a]$	$u_{\mathbf{X}-}-\mu(v_{\mathbf{X}-}-v_b)$
	C	$\chi(u_a)-\chi[u_e(v_b)]$	u_a	–	–	$u_e(v_b)$	$u_e(v_b)$
	$\mathbf{X}$	$\chi(u_a)\ \chi(u_{\mathbf{X}-})$	u_a	–	–	$[u_e(v_b),\sqrt{\mu}-1]$	$u_{\mathbf{X}-}-\mu(v_{\mathbf{X}-}-v_b)$
	$\mathbf{X}$	$\chi(u_a)-\chi(\sqrt{\mu}-1)$	u_a	–	–	$\sqrt{\mu}-1$	$\mu v_b-(\sqrt{\mu}-1)^2$
	S$\mathbf{X}$	$\chi(u_a)-\chi(\bar{u})$ $+\chi(\bar{u}^*)-\chi(\sqrt{\mu}-1)$	u_a	$[\sqrt{\mu}-1,u_e(v_b)]$	$\bar{u}^*$	"	"
	SX	∞	u_a	$u_e(v_b)$	0	"	"
I	D	0	–	–	–	–	$u_b-\mu[v_e(u_b)-v_b]$
	$\mathbf{X}$	$\chi(u_a)-\chi(u_{\mathbf{X}-})$	u_a	–	–	$[u_e(v_b),u_a]$	$u_{\mathbf{X}-}-\mu(v_{\mathbf{X}-}-v_b)$
	O$\mathbf{X}$	∞	0	–	–	$\sqrt{\mu}-1$	$\mu v_b-(\sqrt{\mu}-1)^2$
	O$\mathbf{X}$	$\chi(u_{o+})-\chi(\sqrt{\mu}-1)$	$(0,u_a^*)$	–	–	"	"
	S$\mathbf{X}$	$\chi(u_a^*)-\chi(\sqrt{\mu}-1)$	u_a	u_a	u_a^*	"	"
	S$\mathbf{X}$	$\chi(u_a)-\chi(\bar{u})$ $+\chi(\bar{u}^*)-\chi(\sqrt{\mu}-1)$	u_a	$(u_a,\sqrt{\mu}-1)$	$\bar{u}^*$	"	"
	$\mathbf{X}$	$\chi(u_a)-\chi(\sqrt{\mu}-1)$	u_a	–	–	"	"
	$\mathbf{X}$	$\chi(u_a)-\chi(u_{\mathbf{X}-})$	u_a	–	–	$[\sqrt{\mu}-1,u_e(v_b)]$	$u_a-\mu(v_{\mathbf{X}-}-v_b)$
	C	$\chi(u_a)-\chi[u_e(v_b)]$	u_a	–	–	$u_e(v_b)$	$u_e(v_b)$

Note: The parameter in the function χ is always $-\mu$ and is therefore suppressed.

Nine such regions have to be considered, as shown in Fig. 8.27. They are as follows:

A:	$\mu - 1 < u_a$,	$0 < v_b < 1 - 1/\sqrt{\mu}$
B:	$\mu - 1 - v_b < u_a < \mu - 1$,	$0 < v_b < 1 - 1/\sqrt{\mu}$
Γ:	$u_e(v_b) < u_a < \mu - 1 - v_b$,	$0 < v_b < 1 - 1/\sqrt{\mu}$
Δ:	$0 < u_a < u_e(v_b)$,	$0 < v_b < 1 - 1/\sqrt{\mu}$
E:	$0 < u_a < \sqrt{\mu} - 1$,	$1 - 1/\sqrt{\mu} < v_b < 1$
G:	$\sqrt{\mu} - 1 < u_a < \mu - 1$,	$v_e(u_a) < v_b < 1$
H:	$\mu - 1 < u_a$,	$v_e(u_a) < v_b < 1$
Θ:	$\mu - 1 < u_a$,	$1 - 1/\sqrt{\mu} < v_b < v_e(u_a)$
I:	$\sqrt{\mu} - 1 < u_a < \mu - 1$,	$1 - 1/\sqrt{\mu} < v_b < v_e(u_a)$

where the subscript e is used to denote the equilibrium value corresponding to the value inside the parentheses.

It is not necessary to go through the table in full descriptive detail, but the method will become clear if we examine the region Θ with some care. Figure 8.28 shows the solution both in the (u, v)-plane and, below, as curves of $u(x)$. The first (labeled 1 in Fig. 8.28) is the trivial discontinuous solution for a column of zero length. The second is obtained by a continuous solution from $u_a = u_{0+}$ to u_{X-}, followed by a boundary discontinuity to $u_b = u_{X-} - \mu(v_{X-} - v_b)$. It is clear that the value of X will increase as the point u_{X-} decreases from u_a to $\sqrt{\mu} - 1$. Trajectories 2 and 4 have a boundary discontinuity at $x = X$, in the one case downward and in the other upward. Trajectory 3 is the one that just happens to be continuous through having $u_{X-} = u_e(v_b)$. Trajectory 5 is the limiting case of this type with $u_{X-} = \sqrt{\mu} - 1$. The length of the column for these cases is given by

$$\beta X = \chi(u_a; -\mu) - \chi(u_{X-}; -\mu), \qquad \sqrt{\mu} - 1 \leq u_{X-} \leq u_a \tag{8.5.37}$$

Note that in Table 8.2 the parameter in χ is always $-\mu$, and so it is suppressed in the table. To get a solution for a column longer than that given by Eq. (8.5.37) with $u_{X-} = \sqrt{\mu} - 1$, we must introduce a shock, which is stationary, and thus the states on either side are connected by a chord of slope $1/\mu$; that is, if $\overline{u}$ is the state on the left side of the shock, the state on its right side is $\overline{u}^*$. Then, letting $\overline{u}$ vary from $\sqrt{\mu} - 1$ to $\mu - 1$, we have solutions continuous from u_a to $\overline{u}$, with a shock to $\overline{u}^*$ and a continuous segment from $\overline{u}^*$ to $u_{X-} = \sqrt{\mu} - 1$, followed by a jump to $u_b = v_b - (\sqrt{\mu} - 1)^2$ at $x = X$. This is the trajectory labeled 6, and its length is

$$\beta X = \chi(u_a; -\mu) - \chi(\overline{u}; -\mu) + \chi(\overline{u}^*; -\mu) - \chi(\sqrt{\mu} - 1; -\mu), \qquad \sqrt{\mu} - 1 < \overline{u} < \mu - 1 \tag{8.5.38}$$

In the limiting case, the path 7, $\overline{u} = \mu - 1$, and hence $\overline{u}^* = 0$, making the column infinitely long.

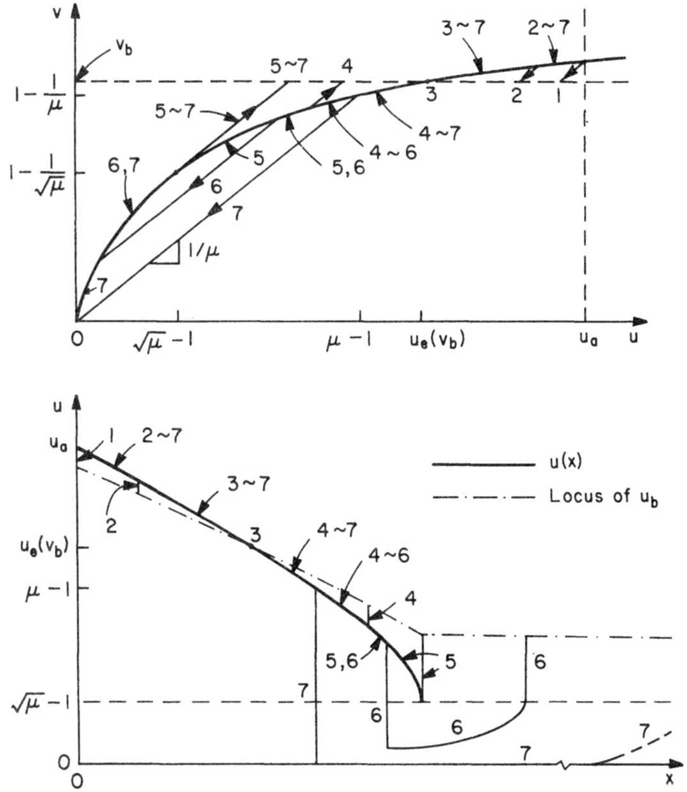

Figure 8.28

Finally, we should note that in the steady state the total reaction rate is proportional to the net flux of the solute into the column from $x = 0$, as well as from $x = X$. This can be expressed in various ways:

$$P = u_a - u_b + \mu(v_b - v_a)$$

$$= (u_{0+} - u_{X-})\left\{1 - \frac{\mu}{(1 + u_{0+})(1 + u_{X-})}\right\} \qquad (8.5.39)$$

For the various categories of solution, Table 8.2 may be used to evaluate the production. Observe that in some cases (e.g., in the region G) the production is independent of v_b. Only with $v_b = 0$ is it possible to have $u_b = 0$, and hence the effluent stream is one of pure product.

We will not attempt to illustrate the development of every single case listed in Table 8.2, for in addition to the 32 distinguishable combinations listed there, which arise from various combinations of u_a, v_b, and X, we would have a further set of variations due to the initial state u_0. Let us take the SX-type solution with (u_a, v_b) in the region Θ as a sufficiently complex solution to illustrate the several phenomena that may arise. In this case,

$$1 - \mu < u_a, \qquad 1 - \frac{1}{\sqrt{\mu}} < v_b < \frac{u_a}{1 + u_a}$$

First, let us consider the case $u_0 > u_a$ for which the solution is shown in Fig. 8.29. In this case, a fan of characteristics corresponding to initial concentrations in the range from u_0 to u_a emanates from the origin in the (x, τ)-plane; this is the fan between OM and OP in the upper part of the figure. The uppermost of these, OM, is the lowest of a family of vertically parallel characteristics emanating from the τ-axis, and these characteristics would overlap with the characteristics of the fan above M if a shock did not begin to develop at the point M. This is the point where the concentration on the characteristic OM drops to $\sqrt{\mu} - 1$; the location of the point M, (x_M, τ_M), is given by

$$\begin{aligned} \beta x_M &= \chi(u_a; -\mu) - \chi(\sqrt{\mu} - 1; -\mu) \\ \beta \tau_M &= \chi(u_a; \nu) - \chi(\sqrt{\mu} - 1; \nu) \end{aligned} \tag{8.5.40}$$

and it is clear that we have been presuming that $X > x_M$. In fact, we will go farther and assume that $X > x_P$, where x_P, given by

$$\beta x_P = \chi(u_0; -\mu) - \chi(\sqrt{\mu} - 1; -\mu)$$

is the position at which the characteristic OP has a vertical tangent. In this case, there will be a set of horizontally parallel characteristics to the right of OPQ. There is a point A with abscissa $X - x_P$ such that B, the point of vertical tangency on the characteristic emanating from A, is just at $x = X$. The characteristics beneath AB all emanate from the x-axis with initial value u_0. Those above BC emanate from the line $x = X$, and all have the value $u = \sqrt{\mu} - 1$ on the line $x = X$ above B. This means that above the point B the concentration u_{X-} is constant in time, and so is u_b. The internal adjustment of the shock position does not affect the concentrations in the outputs of the column as a whole, so the black-box steady state is established at

$$\tau = \tau_B = \chi(u_0; \nu) - \chi(\sqrt{\mu} - 1; \nu) \tag{8.5.41}$$

Two of the cross sections of u as a function of x that are shown in the lower part of Fig. 8.29 are for times less than or equal to τ_M. For example $u(x, \tau_1)$, marked as the curve 1, shows that the steady-state profile has developed up to the point a. Between a and b the fan of characteristics is traversed by the point (x, τ_1), and u accordingly rises to the plateau value beyond the point b. This plateau sinks until at $\tau_2 = \tau_M$ the steady-state profile $u(x, \tau_M)$ to the left of $x = x_M$ has developed a point with a vertical

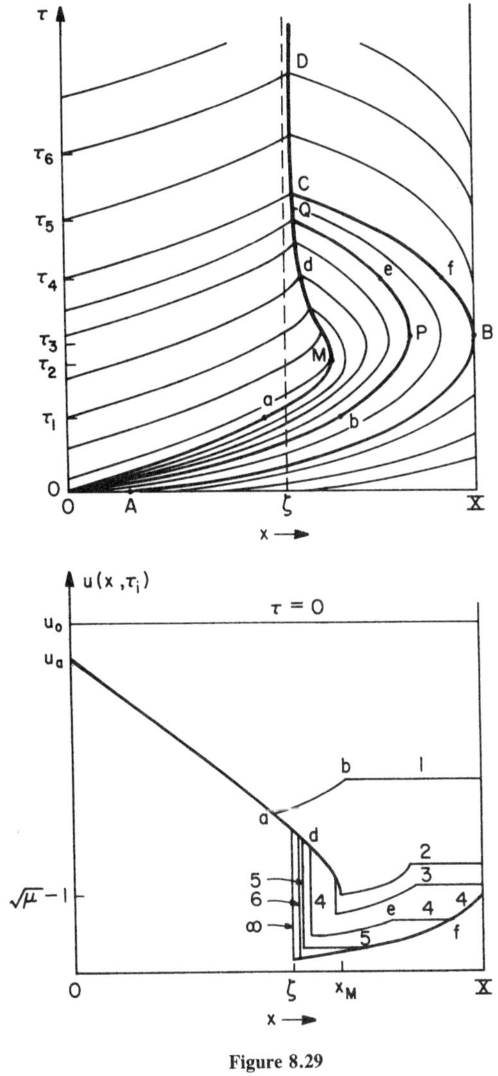

Figure 8.29

tangent at $x = x_M$. This marks the inception of the shock at M, and the path of the shock beyond this point must be found by integrating

$$\frac{d\tau}{dx} = \frac{(1 + u^l)(1 + u^r) + \nu}{(1 + u^l)(1 + u^r) - \mu} \tag{8.5.42}$$

where u^l is given by

$$\beta x = \chi(u_a; -\mu) - \chi(u^l; -\mu) \tag{8.5.43}$$

and u^r is given by

$$\left.\begin{aligned} \beta x &= \chi(u; -\mu) - \chi(u^r; -\mu) \\ \beta\tau &= \chi(u; \nu) - \chi(u^r; \nu) \end{aligned}\right\} \quad \text{for } u_a < u < u_0 \text{ on } MQ \tag{8.5.44}$$

$$\left.\begin{aligned} \beta(x-\xi) &= \chi(u_0; -\mu) - \chi(u^r; -\mu) \\ \beta\tau &= \chi(u_0; \nu) - \chi(u^r; \nu) \end{aligned}\right\} \quad \text{for } 0 < \xi < X - x_p \text{ on } QC \tag{8.5.45}$$

$$\beta(X-x) = \chi(u^r; -\mu) - \chi(\sqrt{\mu}-1; -\mu) \quad \text{beyond } C$$

This path, although not easy to calculate, is asymptotic to the vertical line $x = \zeta$, and it can be shown that it approaches the vertical line monotonically from the right, as shown in Fig. 8.29, by the path $MQCD$. A cross section at the time τ_3 in the interval (τ_M, τ_B) shows that the plateau value of u has not yet dropped to $\sqrt{\mu} - 1$, but that the shock has developed. When $\tau > \tau_B$, the profile near the boundary at $x = X$ is part of the fully developed steady-state profile. Hence, the curve of $u(x, \tau_4)$, shown in the lower part of Fig. 8.29, follows (from left to right) the steady-state profile up to the shock at d, jumps down to u_d^* there, rises to the plateau value at e, and then is constant until the point f on the steady-state profile near the boundary at $x = X$ is reached. At time τ_5, where $\tau_Q < \tau_5 < \tau_C$, the rising section after the shock disappears and the shock is moving back to its final position $x = \zeta$.

Both the cases $\sqrt{\mu} - 1 < u_0 < u_a$ and $0 < u_0 < \sqrt{\mu} - 1 < u_a$ have been treated in detail in the latter part of Sec. 8.3, with $\nu\gamma$, $\mu\gamma$, and $\beta\gamma$ instead of ν, μ, and β and, of course, with the correspondence $Kc = u$. Hence it is not necessary to do it here.

Exercises

8.5.1. Derive Eqs. (8.5.30) and (8.5.31), and show that the point O in Fig. 8.22 is a saddle point with the trajectory BO approaching and the trajectory OA receding.

8.5.2. If the point (u_a, v_b) falls between OC and OD, as shown in Fig. 8.24, sketch the profiles of u and v corresponding to the trajectories given in Fig. 8.24. Repeat the same for the case shown in Fig. 8.25.

8.5.3. Consider the case when the point (u_a, v_b) falls in region I of Fig. 8.27, and sketch the sequence of trajectories in the (u, v)-plane corresponding to an increasing sequence of the column length X, as shown in the upper part of Fig. 8.28. Also, draw the profiles of $u(x)$ corresponding to those trajectories.

8.5.4. Repeat the same procedure as in Exercise 8.5.3 for the region E of Fig. 8.27.

REFERENCES

8.1. For theoretical treatments, see the references given at the end of Chapter 2 and those cited as the basic references at the end of Chapter 7.

The fixed-bed adsorber with reaction is treated in

S. Viswanathan, "Chromatographic Reactors," Ph.D. Thesis, University of Minnesota, Minneapolis, 1973.

J. Loureiro, C. Costa, and A. Rodrigues, "Propagation of Concentration Waves in Fixed-bed Adsorptive Reactors," *Chem. Eng. J.* **27**, 135–148 (1983).

Y. W. Nam, "A Study of Fixed-bed Adsorption-reaction Systems with Freundlich Isotherms," Ph.D. Thesis, Seoul National University, Seoul, Korea, 1984.

8.2. The treatment of volumetric pool boiling follows closely that of

D. W. Condiff and M. Epstein, "Transient Volumetric Pool Boiling—I. Convex Flux Relations," *Chem. Eng. Sci.* **31**, 1139–1148 (1976).

The case of nonconvex flux relations is discussed in

D. W. Condiff and M. Epstein, "Transient Volumetric Pool Boiling—II. Nonconvex Flux Relations," *Chem. Eng. Sci.* **31**, 1149–1161 (1976).

An experimental study of transient volume-heated boiling pools is found in

R. P. Lipkis, C. Liu, and N. Zuber, "Measurement and Prediction of Density Transients in a Volume-heated Boiling System" *Chem. Engng. Progr. Symposium Series* **52** (18), 105–113 (1956).

8.3. The first sample treated here is taken from

R. Aris, "On the Ostensible Steady State of a Dynamical System," *Atti della Academia Nazionale dei Lincei*, Series VIII, Vol. LVII, 1–9 (1974).

For this treatment of the countercurrent adsorber with reaction, see

S. Viswanathan and R. Aris, "On the Concept of the Steady State in Chemical Reactor Analysis," *Chem. Eng. Commun.* **2**, 1–4 (1975).

See also

S. Viswanathan and R. Aris, "Countercurrent Moving Bed Chromatographic Reactors," in *Chemical Reaction Engineering*—II, H. M. Hulburt (ed.), Advances in Chemistry Series 133, American Chemical Society, Washington, D.C., 1974, pp. 191–204.

8.4. For the equilibrium countercurrent adsorber, refer to Sec. 6.5 of this volume and see the references cited for that section at the end of Chapter 6.

The effects of mass-transfer resistance and axial dispersion have been studied extensively by

B. R. Locke, "The Effects of Mass Transfer Resistance and Axial Dispersion on

Multicomponent Moving Bed Adsorption and Ion Exchange," M.S. Thesis, University of Houston, Houston, Tex., 1982.

8.5. The countercurrent adsorber with reaction under nonequilibrium conditions is treated in

S. Viswanathan and R. Aris, "An Analysis of the Countercurrent Moving Bed Reactor," *SIAM-AMS Proc.* **8**, 99–124 (1974).

R. Aris, *Mathematical Modelling Techniques*, Pitman, London, 1978, pp. 86–103.

See also the article by Viswanathan and Aris (1974) in *Chemical Reaction Engineering*—II, cited previously, and

D. Altshuller, "Design Equations and Transient Behavior of the Countercurrent Moving Bed Chromatographic Reactor," *Chem. Eng. Commun.* **19**, 363–375 (1983)

where the Fowler–Guggenheim adsorption isotherm is introduced.

Nonlinear Equations 9

9.1 Nonlinear Equations with Two Independent Variables

Because it is so easy to visualize the geometry of the situation, we will again consider the case of two independent variables first. We will denote these by x and y and the dependent variable by z, and use the traditional abbreviations

$$p \equiv z_x \equiv \frac{\partial z}{\partial x} \quad \text{and} \quad q \equiv z_y \equiv \frac{\partial z}{\partial y}$$

The general nonlinear equation of the first order is

$$F(x, y, z, p, q) = 0 \tag{9.1.1}$$

and as before we will hope to pick out a particular solution by requiring that the surface should pass through an initial curve, which we can specify parametrically:

$$x = X(\xi), \qquad y = Y(\xi), \qquad z = Z(\xi) \tag{9.1.2}$$

We can again use the geometrical fact that the direction cosines of the normal to the solution surface are proportional to $p{:}q{:}\ -1$. Suppose that a particular point (x, y, z) lies on the surface; then the admissible normals are restricted by the relation $F(x, y, z, p, q) = 0$. When the equation was linear or quasi-linear, this relation was particularly simple and implied that the normal should lie in a plane. From this it followed that the normal to this plane must be tangent to the solution surface and we obtained the unique characteristic direction. To put it in another way, we might say that since the normal must lie in a plane, the possible tangent planes to the surface must be the family of planes all containing the characteristic direction. For the nonlinear equation, the relation $F(x, y, z, p, q) = 0$ may be thought of as defining a cone in which the normal to a solution surface must lie. The term *cone* is used for any surface generated by a one-parameter family of straight lines through the fixed point that is its *vertex*. Only in a very special case would it be a circular cone, but it is easier to draw it so, as in Fig. 9.1.

Now if we move the normal direction $p:q: -1$ around in the cone of normals, the tangent plane will envelop another cone. This is known as the *Monge cone*, and it is this that in the quasi-linear case degenerates into the characteristic direction, or, as it is sometimes called, the *Monge axis*. Just as the characteristic direction had to be tangent to the solution surface, so we see that in the fully nonlinear case the Monge cone, or, in particular, one of its generators, must be tangent to the solution surface. If then we can find the generators of the Monge cone at points on the solution surface and know the orientation of the tangent planes along those generators, we will have the skeleton of the solution surface, just as we had it in the characteristics for the quasi-linear case. In fact, a curve with an associated tangent plane is just what we called a ribbon or strip in Sec. 0.12, so what we will try to do is to piece together the solution from characteristic strips.

Before going on to the general case, the Monge cone and cone of normals may be rather easily visualized in the example that we derived geometrically in Sec. 1.16. It will be recalled that the family of unit spheres whose centers lie in the plane $z = 0$ gives rise to the equation $z^2(p^2 + q^2 + 1) = 1$, which we will write in the form

$$F(x, y, z, p, q) \equiv z(p^2 + q^2)^{1/2} - (1 - z^2)^{1/2} = 0$$

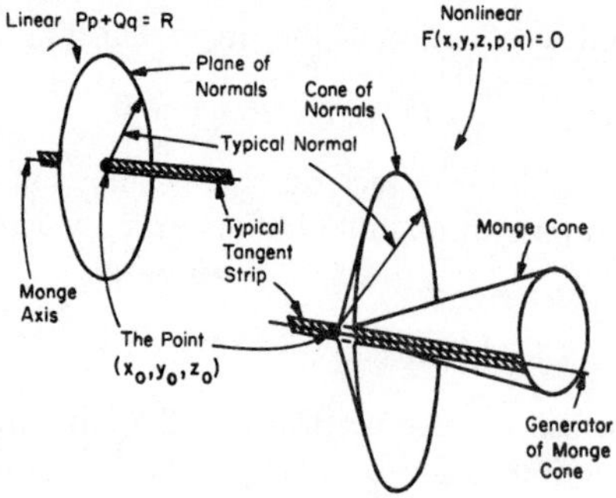

Figure 9.1

Thus the cone of normals through the point (x, y, z) is given by the lines with direction cosines zp, zq, $-z$ such that

$$(p^2 + q^2)^{1/2} = \frac{(1 - z^2)^{1/2}}{z}$$

Since $|z| < 1$, we can conveniently set $z = \cos\theta$, with θ lying between 0 and π. Then p and q will satisfy the equation if

$$p = \tan\theta\cos\mu, \qquad q = \tan\theta\sin\mu$$

and μ becomes the parameter of the one-parameter family of normal directions

$$zp:zq: -z = \sin\theta\cos\mu:\sin\theta\sin\mu: -\cos\theta$$

These clearly all make the angle θ with the vertical direction $0:0: -1$ and so define a right circular cone with vertical axis and semiangle θ. This is no surprise, for the spheres that pass through the point (x, y, z), where $z = \cos\theta$, must have centers on the circle of radius $\sin\theta$ whose center is (x, y), as shown in Fig. 9.2. Their normals through the point of intersection are therefore at an angle θ to the vertical. The surface elements to this family of spheres clearly touch a cone of vertical axis and half-angle $(\pi/2) - \theta$. This is the Monge cone for this point and this equation. In general, the Monge cone is the envelope of the surface elements of the one-parameter family selected out of the two-parameter family of surfaces that make up the complete integral by the requirement of passing through a given point. In this example, the cone degenerates if $z = 0$ or ± 1.

Returning now to the general problem, we wish to find the generators of the Monge cone. To do this, we may imagine that the values of p and q lying in the cone are functions of a generating parameter μ, that is, $p = p(\mu)$, $q = q(\mu)$, and $\mathcal{F}(\mu) \equiv F(x, y, z, p(\mu), q(\mu)) \equiv 0$. If s is a variable of distance in the direction of the generator, then

$$p(\mu)\frac{dx}{ds} + q(\mu)\frac{dy}{ds} - \frac{dz}{ds} = 0 \tag{9.1.3}$$

since the generator, and indeed any direction in the tangent plane, is perpendicular to the direction $p:q: -1$. But a generator may be thought of as the limit of the intersection between tangent planes for μ and $\mu + d\mu$ as

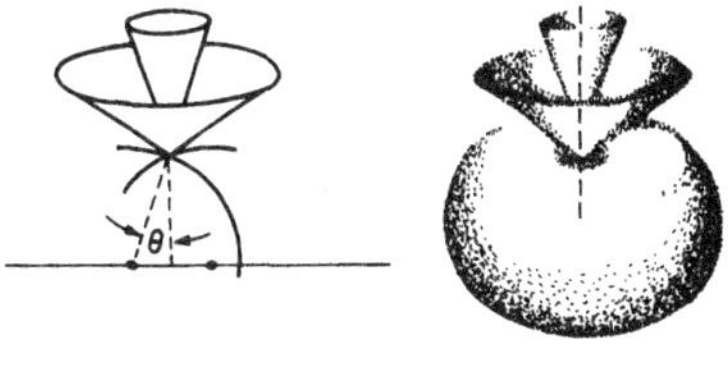

Figure 9.2

$d\mu \to 0$. Hence the generating direction will also satisfy the equation obtained from Eq. (9.1.3) by differentiating with respect to μ,

$$p'(\mu)\frac{dx}{ds} + q'(\mu)\frac{dy}{ds} = 0 \tag{9.1.4}$$

But, since $\mathcal{F}(\mu) \equiv 0$,

$$\frac{d\mathcal{F}}{d\mu} = F_p p'(\mu) + F_q q'(\mu) = 0^{\dagger} \tag{9.1.5}$$

and comparing these equations we see that

$$\frac{(dx/ds)}{F_p} = \frac{(dy/ds)}{F_q}$$

Since we are not concerned with the relation between the variable s and the actual distance along a generator, we may take this common ratio to be 1 and write

$$\frac{dx}{ds} = F_p \tag{9.1.6}$$

$$\frac{dy}{ds} = F_q \tag{9.1.7}$$

From Eq. (9.1.3), we then see that

$$\frac{dz}{ds} = pF_p + qF_q \tag{9.1.8}$$

We have thus found the direction of the generator of the Monge cone; but since F, and so also F_p and F_q, are functions of p and q as well as of x, y, and z, we can no longer integrate these three equations by themselves. The only exception to this would be if F_p, F_q, and $pF_p + qF_q$ were functions only of x, y, z; this is precisely the case for the quasi-linear equation and only for this type. What we need therefore are two more equations for dp/ds and dq/ds. Now the set of five quantities x, y, z, p, q must always satisfy the equation $F(x, y, z, p, q) = 0$. The variables x and y are the independent ones, and the remaining three are functions of them. Thus, differentiating $F(x, y, z(x, y), p(x, y), q(x, y)) = 0$ partially with respect to x, we obtain

$$F_x + F_z p + F_p p_x + F_q q_x = 0 \tag{9.1.9}$$

Similarly, differentiating partially with respect to y gives

$$F_y + F_z q + F_p p_y + F_q q_y = 0 \tag{9.1.10}$$

†F_p denotes the partial derivative of $F(x, y, z, p, q)$ with respect to p, holding x, y, z, and q constant; F_q is the partial derivative of F with respect to q.

Let us also remember that

$$p_y = \frac{\partial}{\partial y}\left(\frac{\partial z}{\partial x}\right) = \frac{\partial^2 z}{\partial y\,\partial x} = \frac{\partial}{\partial x}\left(\frac{\partial z}{\partial y}\right) = q_x \tag{9.1.11}$$

What we want to find is

$$\frac{dp}{ds} = p_x\frac{dx}{ds} + p_y\frac{dy}{ds}$$

But using Eqs. (9.1.6) and (9.1.7) and replacing q_x in Eq. (9.1.9) by p_y gives

$$\frac{dp}{ds} = F_p p_x + F_q p_y = -F_x - pF_z \tag{9.1.12}$$

Similarly,

$$\frac{dq}{ds} = -F_y - qF_z \tag{9.1.13}$$

We now have five ordinary differential equations [(9.1.6), (9.1.7), (9.1.8), (9.1.12), and (9.1.13)] for x, y, z, p, and q as functions of s, and to integrate these is to obtain a characteristic strip, which we have shown must lie in a surface that satisfies the differential equation. All we now have to do is to see how we must start these characteristic strips off from the initial curve in such a way as to allow them to mesh together and form a smooth surface. We are only given an initial curve $x = X(\xi)$, $y = Y(\xi)$, $z = Z(\xi)$, and what we need is another pair of functions, $p = P(\xi)$, $q = Q(\xi)$, to give us an initial strip. But if $p:q: -1$ is to be normal to the tangential direction $dx/d\xi : dy/d\xi : dz/d\xi$, we must have

$$P(\xi)X'(\xi) + Q(\xi)Y'(\xi) = Z'(\xi) \tag{9.1.14}$$

This is one equation for P and Q, and a second may be obtained by requiring that the differential equation itself should be satisfied on the initial curve; that is,

$$F(X(\xi), Y(\xi), Z(\xi), P(\xi), Q(\xi)) = 0 \tag{9.1.15}$$

Thus the solution may be generated by solving the five characteristic differential equations

$$\begin{aligned} \frac{dx}{ds} &= F_p(x, y, z, p, q), & \frac{dy}{ds} &= F_q(x, y, z, p, q), \\ \frac{dz}{ds} &= pF_p + qF_q, & \frac{dp}{ds} &= -F_x - pF_z, \\ \frac{dq}{ds} &= -F_y - qF_z \end{aligned} \tag{9.1.16}$$

for $x(s, \xi)$, $y(s, \xi)$, . . . , $q(s, \xi)$, subject to the initial conditions $x(0, \xi) = X(\xi)$, . . . , $q(0, \xi) = Q(\xi)$.

Obviously, we will be in trouble if F_p and F_q vanish simultaneously, and it is natural to require that

$$F_p^2 + F_q^2 \neq 0 \tag{9.1.17}$$

We will also be unable to carry out the program if the two equations (9.1.14) and (9.1.15), which we have to solve for $P(\xi)$ and $Q(\xi)$, fail to be independent. If we write the first as

$$G(P, Q) \equiv PX' + QY' - Z' = 0$$

and the second as

$$F(P, Q) \equiv F(X, Y, Z, P, Q) = 0$$

the condition for independence is

$$\frac{\partial(F, G)}{\partial(P, Q)} = \begin{vmatrix} F_P & F_Q \\ X' & Y' \end{vmatrix} \neq 0 \tag{9.1.18}$$

But this simply means that the initial curve must not be tangent to a characteristic direction.

As an example, consider the equation that we obtained from the family of unit spheres with centers on the (x, y)-plane. This could be written (see Sec. 1.16)

$$F(x, y, z, p, q) = z(p^2 + q^2)^{1/2} - (1 - z^2)^{1/2} = 0 \tag{9.1.19}$$

Let us suppose that we seek the solution through the line

$$X(\xi) = \xi, \qquad Y(\xi) = \sin\theta, \qquad Z(\xi) = \cos\theta \tag{9.1.20}$$

where θ is an arbitrary constant, $0 < \theta < \pi/2$. Then

$$PX' + QY' - Z' = P(\xi) = 0$$

and

$$F(X, Y, Z, P, Q) = \cos\theta(P^2 + Q^2)^{1/2} - \sin\theta = 0$$

or

$$P(\xi) = 0, \qquad Q(\xi) = \tan\theta \tag{9.1.21}$$

The characteristic differential equations are

$$\frac{dx}{ds} = pz(p^2 + q^2)^{-1/2}, \qquad \frac{dy}{ds} = qz(p^2 + q^2)^{-1/2} \tag{9.1.22}$$

$$\frac{dz}{ds} = z(p^2 + q^2)^{1/2} = (1 - z^2)^{1/2} \tag{9.1.23}$$

$$\frac{dp}{ds} = -p[(p^2 + q^2)^{1/2} + z(1 - z^2)^{-1/2}]$$
$$\frac{dq}{ds} = -q[(p^2 + q^2)^{1/2} + z(1 - z^2)^{-1/2}] \tag{9.1.24}$$

The last pair of equations shows that

$$\frac{1}{p}\frac{dp}{ds} = \frac{1}{q}\frac{dq}{ds} = -[z(1 - z^2)^{-1/2} + (p^2 + q^2)^{1/2}]$$

so p and q are proportional to one another. But since $P(\xi) = 0$, the constant of proportionality must be zero, and so $p(s, \xi) = 0$. It follows from the first equation that $x(s, \xi) = X(\xi) = \xi$. Clearly, the solution of Eq. (9.1.23) is

$$z(s, \xi) = \cos(\theta - s)$$

and so

$$-\frac{dq}{ds} = q^2 + \cot(\theta - s)q$$

whence

$$q(s, \xi) = \tan(\theta - s)$$

and finally

$$y(s, \xi) = -\sin(\theta - s) + 2 \sin \theta$$

The solution is thus the cylinder $(y - 2 \sin \theta)^2 + z^2 = 1$, and it is very easy to see that we have just picked out the envelope of the one-parameter family of spheres that touch the given initial line, as shown in Fig. 9.3. Had we taken the other sign in the square root for Q, we would have had to have been consistent in choosing a negative sign for the square root in Eq. (9.1.22) and so would have obtained the cylinder $y^2 + z^2 = 1$.

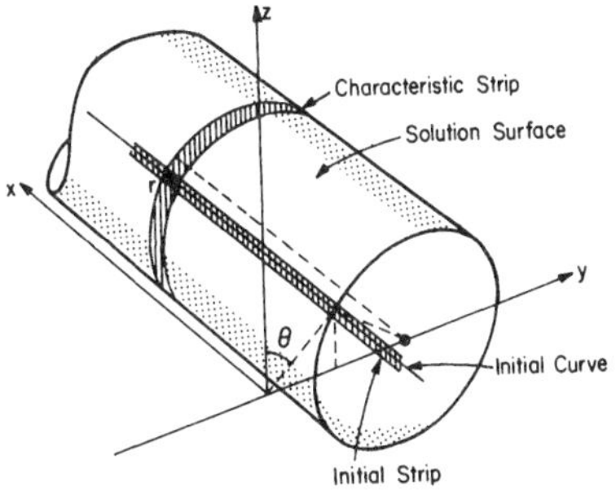

Figure 9.3

Notice that we would run into ambiguity in case $\theta = 0$, for then $Z(\xi) = 1$ and the differential equation

$$\frac{dz}{ds} = -(1 - z^2)^{1/2}$$

would have two solutions, either $z = 1$ or $z = \cos s$. The former would give the whole plane $z = 1$ or the envelope of the complete two-parameter family of spheres; the latter would give the cylinder as before.

Exercises

9.1.1. Obtain the characteristic differential equations for the equation obtained for $v(y/x, z)$ in Exercise 1.14.3 and show that on an optimal path

$$\frac{d\eta}{dk} = r\left(\frac{\rho k^r - k}{k^2}\right)\eta^2 - \left(\frac{r}{k}\right)\eta$$

where $\eta = y/z$. Let η_0 be the initial value of η and k_0 the optimal value of k at η_0. Show that the solution of the above equation is

$$\frac{\eta_0}{\eta} = (1 + K\eta_0)\,\kappa^r + (1 - K)\eta_0\kappa^{r-1} - \eta_0$$

where $$\kappa = \frac{k}{k_0}, \qquad K = 1 - r\rho k_0^{r-1}$$

9.1.2. Use the characteristic differential equations to justify the special methods introduced in Sec. 2.4 for equations of the types

(i) $F(p, q) = 0$, (ii) $F(z, p, q) = 0$, (iii) $F(x, p) = G(y, q)$

9.1.3. Solve Eq. (9.1.19) subject to the conditions (9.1.20) starting from the form $F(x, y, z, p, q) = \frac{1}{2}z^2(p^2 + q^2 + 1) - \frac{1}{2} = 0$.

9.2 Geometry of the Solution Surface

Before going on to the general case of an arbitrary number of independent variables, it will be useful to look a little further at the geometry of the solution surface. Let us recall the definitions. The Monge cone at the point (x, y, z) is the infinitesimal cone whose normals satisfy $F(x, y, z, p, q) = 0$ and to which any solution surface must be tangent. The generators of the Monge cone are given by

$$\frac{dx}{ds} = F_p, \qquad \frac{dy}{ds} = F_q, \qquad \frac{dz}{ds} = pF_p + qF_q \tag{9.2.1}$$

and any curve in space whose direction satisfies these equations is called a *focal curve* or *Monge curve*. If values of p and q satisfying $F(x, y, z, p, q) = 0$ are associated with the space curve, it is called a *focal strip*. These four equations among five quantities form an underdetermined system, any solution of which gives a focal strip. A focal strip that also satisfies the equations

$$\frac{dp}{ds} = -pF_z - F_x, \qquad \frac{dq}{ds} = -qF_z - F_y \tag{9.2.2}$$

is a *characteristic strip*, and we have just seen how these can be pieced together when they emanate from a noncharacteristic *initial curve*. More precisely, the initial curve $X(\xi)$, $Y(\xi)$, $Z(\xi)$ is made into an *initial strip* by the addition of two functions $P(\xi)$ and $Q(\xi)$ satisfying the strip condition $PX' + QY' = Z'$ and the equation $F(X, Y, Z, P, Q) = 0$; this strip can also be called an *integral strip* since it thus lies in the solution surface. It is possible to build the initial curve into a strip in this way provided that

$$\Delta \equiv \begin{vmatrix} F_p & F_q \\ X' & Y' \end{vmatrix} = \frac{dx}{ds}\frac{dY}{d\xi} - \frac{dy}{ds}\frac{dX}{d\xi} \neq 0 \tag{9.2.3}$$

and the derivation of the characteristic equation shows that, as functions of the two parameters s and ξ, the five quantities $x(s, \xi), \ldots, q(s, \xi)$ satisfy the equation $F = 0$. This is because $F = 0$ on the initial strip and $dF/ds = 0$ along a characteristic. That elimination of s and ξ between $x = x(s, \xi)$, $y = y(s, \xi)$, and $z = z(s, \xi)$ gives the solution surface $z = z(x, y)$ will be thoroughly established if it can be shown that $p(x, y)$ [obtained by eliminating s and ξ between $p = p(s, \xi)$, $x = x(s, \xi)$, and $y = y(s, \xi)$] is in fact the same as $z_x(x, y)$ and likewise that $q(x, y) = z_y(x, y)$.

The following existence theorem (although not formally set out) is a consequence of what has just been said: that if $\Delta \neq 0$ in the initial strip, then the solution of the partial differential equation is unique in the neighborhood of the initial strip. However, if $\Delta = 0$ along the initial strip, then the initial-value problem only has a solution if it is a characteristic strip; indeed, the problem may then have infinitely many solutions. The characteristic is thus a branch curve along which two solutions may meet and preserve the continuity of the first derivatives.

This may be illustrated by the geometrical example given in the last section. From the characteristic differential equations (9.1.22) through (9.1.24), we see that

$$\frac{d}{ds}(x - pz) = 0, \qquad \frac{d}{ds}(y - qz) = 0, \qquad \frac{d}{ds}\left(\frac{p}{q}\right) = 0$$

so

$$x - pz = a, \qquad y - qz = b, \qquad p = cq \tag{9.2.4}$$

where a, b, and c are three constants. Substituting from the first two relations into the third shows that the characteristic lies in the plane

$$(x - a) = c(y - b) \tag{9.2.5}$$

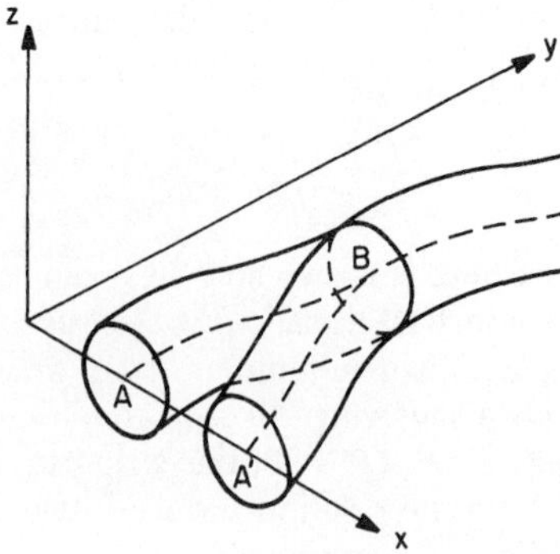

Figure 9.4

Substituting from the first two into the differential equation (9.1.19) gives

$$(x - a)^2 + (y - b)^2 + z^2 = 1 \tag{9.2.6}$$

The characteristics are therefore all great circles of unit spheres with centers on the (x, y)-plane, and this is quite to be expected since the equation was generated from this family of spheres.

The solution that we obtained for the initial curve $y = \sin\theta$, $z = \cos\theta$ in the last section was one of the two cylinders through this line (see Fig. 9.4). It should be fairly clear on intuitive grounds that any tubular surface formed by the envelope of the spheres whose centers lie on a curve in the (x, y)-plane is also a solution. In fact, if this core curve is $b = \phi(a)$, we have a one-parameter family

$$(x - a)^2 + [y - \phi(a)]^2 + z^2 = 1 \tag{9.2.7}$$

and the envelope comes from eliminating a between this equation and

$$2(x - a) + 2\phi'(a)\,[y - \phi(a)] = 0$$

Now we can visualize two curves ABC and $A'BC$ that are tangent at B. These will generate two tubular surfaces with a common part around BC but branching into two distinct parts on the great circle with center B. In fact, we can see that, if the initial curve were the great circle at B, and so wholly characteristic, there would be an infinity of solutions through it, that is, any tubular surface whose core passed through B perpendicularly to the plane of the great circle.

If we follow the characteristics emanating from a particular point, we can generate a surface that satisfies the equation. This is known as the *integral conoid*. For our system, we see that it would generally be an apple-shaped surface [see Fig. 9.5(a)] with a conical node at the fixed point and another at its reflection in the plane $z = 0$. If, however, this point were on either plane $z = \pm 1$, the sphere would be obtained, and if on the plane $z = 0$, the

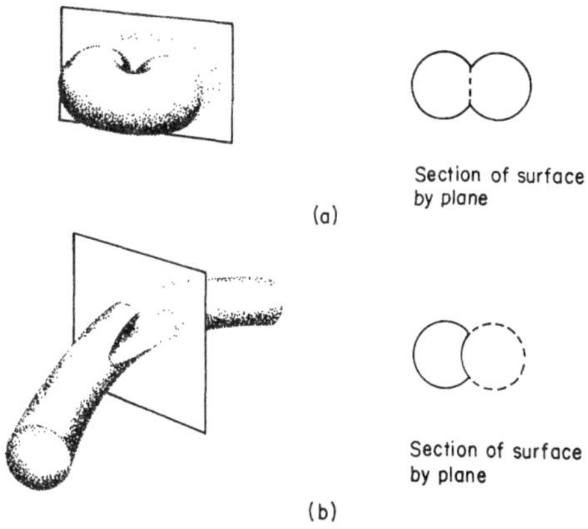

Figure 9.5

two conical nodes would become cusps and meet. Alternatively, we could look at the inner part of the characteristics and obtain a pointed football, which in the two limits would become the sphere or vanish as a point.

What happens when the radius of curvature of the core curve is less than unity? In this case we see that the characteristic great circles must intersect one another, and the surface can only be made continuous by introducing a discontinuity in the first derivative and taking the other great circular arc in the plane of the characteristic [see Fig. 9.5(b)].

Clairaut's equation,

$$z = px + qy + f(p, q) \tag{9.2.8}$$

provides an admirably clear geometrical picture of the various possibilities, for its complete integral is the two-parameter family of planes

$$z = ax + by + f(a, b) \tag{9.2.9}$$

Its characteristic strips are all straight and, having constant $p = a$, $q = b$, are without twist (see Exercise 9.2.4).

For the sake of definiteness, let us consider a particular example,

$$z = px + qy + (1 + p^2 + q^2)^{1/2} \tag{9.2.10}$$

and for brevity denote $(1 + p^2 + q^2)^{1/2}$ by f. Then

$$\begin{aligned} \frac{dx}{ds} &= x + \frac{p}{f}, \qquad \frac{dy}{ds} = y + \frac{q}{f} \\ \frac{dz}{ds} &= px + qy + \frac{p^2 + q^2}{f} = z - \frac{1}{f} \\ \frac{dp}{ds} &= 0, \qquad \frac{dq}{ds} = 0 \end{aligned} \tag{9.2.11}$$

An initial curve for which

$$\frac{x'}{x + p/f} \neq \frac{y'}{y + q/f}$$

will have a unique solution in its neighborhood. For example, if the curve is the circle $x^2 + y^2 = a^2$, we have

$$X(\xi) = a \cos \xi, \qquad Y(\xi) = a \sin \xi, \qquad Z(\xi) = 0 \tag{9.2.12}$$

Then the strip condition gives

$$-P(\xi)a \sin \xi + Q(\xi)a \cos \xi = 0 \tag{9.2.13}$$

and the equation itself gives

$$P(\xi)a \cos \xi + Q(\xi)a \sin \xi + [1 + P^2(\xi) + Q^2(\xi)]^{1/2} = 0 \tag{9.2.14}$$

Equation (9.2.13) will be satisfied if

$$P(\xi) = A(\xi) \cos \xi, \qquad Q(\xi) = A(\xi) \sin \xi$$

and then Eq. (9.2.14) gives

$$aA(\xi)\,(\cos^2 \xi + \sin^2 \xi) + [1 + A^2(\xi)(\cos^2 \xi + \sin^2 \xi)]^{1/2} = 0$$

or

$$(a^2 - 1)A^2(\xi) = 1$$

Hence $A(\xi)$ is a constant, $A(\xi) = \pm(a^2 - 1)^{-1/2}$, which is real provided that $a > 1$. Then, taking the negative sign for A, we have

$$\frac{dx}{ds} = x - \frac{\cos \xi}{a}, \qquad x(0, \xi) = a \cos \xi$$

or

$$x(s, \xi) = \left[\left(a - \frac{1}{a}\right)e^s + \frac{1}{a}\right] \cos \xi = \frac{a^2 - 1}{a}\left(e^s + \frac{1}{a^2 - 1}\right)\cos \xi$$

Similarly,

$$y(s, \xi) = \left[\left(a - \frac{1}{a}\right)e^s + \frac{1}{a}\right]\sin \xi = \frac{a^2 - 1}{a}\left(e^s + \frac{1}{a^2 - 1}\right)\sin \xi$$

and

$$z(s, \xi) = -\frac{(a^2 - 1)^{1/2}}{a}(e^s - 1)$$

Since

$$z - \frac{a}{(a^2 - 1)^{1/2}} = -\frac{(a^2 - 1)^{1/2}}{a}\left(e^s + \frac{1}{a^2 - 1}\right)$$

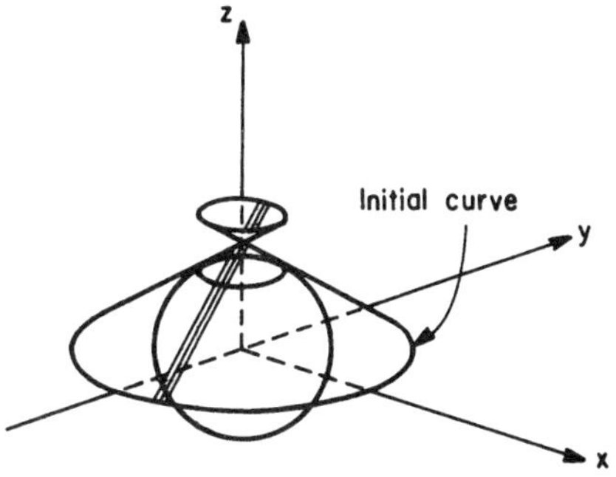

Figure 9.6

we see that the characteristic strips generate the cone (see Fig. 9.6):

$$x^2 + y^2 = (a^2 - 1)\left[z - \frac{a}{(a^2 - 1)^{1/2}}\right]^2 \tag{9.2.15}$$

In general, any solution will be a developable surface (i.e., a ruled surface that can be rolled out onto a plane without distortion). In fact, any solution surface is the envelope of a one-parameter family selected from the complete integral. For example, if in Eq. (9.2.9) $b = \phi(a)$, the envelope of the family

$$z = ax + \phi(a)y + f(a, \phi(a)) \tag{9.2.16}$$

is obtained by eliminating a between this equation and

$$0 = x + \phi'(a)y + f_a + f_b\phi'(a) \tag{9.2.17}$$

But if this envelope is to go through the curve

$$x = X(\xi), \qquad y = Y(\xi), \qquad z = Z(\xi)$$

then

$$Z(\xi) = aX(\xi) + \phi(a)Y(\xi) + f(a, \phi(a))$$
$$0 = X(\xi) + \phi'(a)Y(\xi) + f_a + f_b\phi'(a)$$

and eliminating ξ between these equations gives a first-order ordinary differential equation for $\phi(a)$.

For the particular example we have considered, the family of planes

$$z = ax + by + (1 + a^2 + b^2)^{1/2} \tag{9.2.18}$$

includes those that touch the unit sphere

$$x^2 + y^2 + z^2 = 1 \tag{9.2.19}$$

It is evident then that we will obtain a "conical sort of surface" from the initial curve lying outside the sphere, for it will be generated by the envelope of the planes that are tangent both to the initial curve and to the sphere. In particular, if the initial curve degenerates to a point, the cone with vertex at the point that is tangent to the curve is obviously a solution of the equation.

Exercises

9.2.1. Show that, if $x(s, \xi), \ldots, q(s, \xi)$ are the solutions of the characteristic differential equations of the partial differential equation (9.1.1),

$$U = \frac{\partial z}{\partial s} - p\frac{\partial x}{\partial s} - q\frac{\partial y}{\partial s} \equiv 0$$

on the solution surface. If

$$V = \frac{\partial z}{\partial \xi} - p\frac{\partial x}{\partial \xi} - q\frac{\partial y}{\partial \xi}$$

show that

$$\frac{\partial V}{\partial s} = p_\xi x_s - p_s x_\xi + q_\xi y_s - q_s y_\xi$$

9.2.2. By using the expression for $\partial V/\partial s$ from Exercise 9.2.1 and the partial derivative of F with respect to ξ, show that

$$\frac{\partial V}{\partial s} = -F_z V$$

and deduce that $V \equiv 0$ on the solution surface.

9.2.3. From the fact that U and V are both identically zero on the solution surface, deduce that $p(x, y) = \partial z/\partial x$ and $q(x, y) = \partial z/\partial y$.

9.2.4. Show that the characteristic strips of Clairaut's equation $z = px + qy + f(p, q)$ are all straight and flat and, hence, that the complete integral is the family of planes $z = ax + by + f(a, b)$.

9.3 Nonlinear Equations with n Independent Variables

The general nonlinear equation can be written

$$F(x_1, \ldots, x_n, z, p_1, \ldots, p_n) = 0 \tag{9.3.1}$$

where p_i denotes $\partial z/\partial x_i$. The initial values are given parametrically on an $(n - 1)$-dimensional space

$$x_i = X_i(\xi_1, \ldots, \xi_{n-1}) \tag{9.3.2}$$

and

$$z = Z(\xi_1, \ldots, \xi_{n-1}) \tag{9.3.3}$$

The characteristic differential equations are the immediate generalizations of the case of two independent variables,

$$\frac{dx_i}{ds} = F_{p_i} \tag{9.3.4}$$

$$\frac{dz}{ds} = p_i F_{p_i} \tag{9.3.5}$$

$$\frac{dp_i}{ds} = -F_{x_i} - p_i F_z \tag{9.3.6}$$

[*Note*: Summation of the repeated index i is presumed in Eq. (9.3.5) and elsewhere.] Since

$$\frac{dF}{ds} = F_{x_i}\frac{dx_i}{ds} + F_z\frac{dz}{ds} + F_{p_i}\frac{dp_i}{ds} = 0 \tag{9.3.7}$$

it is clear that if $F = 0$ at one point on a characteristic then it is zero at all.

For the solutions $x_i(s; \xi_1, \ldots, \xi_{n-1})$, $z(s; \xi_1, \ldots, \xi_{n-1})$, $p_i(s; \xi_1, \ldots, \xi_{n-1})$, we have initial conditions

$$\begin{aligned} x_i(0, \xi_1, \ldots, \xi_{n-1}) &= X_i(\xi_1, \ldots, \xi_{n-1}) \\ z(0; \xi_1, \ldots, \xi_{n-1}) &= Z(\xi_1, \ldots, \xi_{n-1}) \end{aligned} \tag{9.3.8}$$

and we must ask how to find the remaining n conditions

$$p_i(0; \xi_1, \ldots, \xi_{n-1}) = P_i(\xi_1, \ldots, \xi_{n-1}) \tag{9.3.9}$$

Certainly, we must expect to satisfy the equation on the initial manifold so that

$$F(X_i, Z, P_i) = 0 \tag{9.3.10}$$

Then the strip relation

$$dZ = P_i dX^i$$

must hold for increments generated by arbitrary increments in the ξ_j. But

$$dZ = Z_{\xi_j}\, d\xi_j \quad \text{and} \quad dX_i = (X_i)_{\xi_j}\, d\xi_j$$

so

$$P_i(X_i)_{\xi_j} = Z_{\xi_j}, \qquad j = 1, \ldots, n-1 \tag{9.3.11}$$

These equations can generally be solved for the P_i provided that

$$\Delta = \frac{\partial(X_1, X_2, \ldots, X_n)}{\partial(s, \xi_1, \ldots, \xi_{n-1})} = \begin{vmatrix} F_{p_1} & \cdots & F_{p_n} \\ (X_1)_{\xi_1} & \cdots & (X_n)_{\xi_1} \\ \vdots & \cdots & \vdots \\ (X_1)_{\xi_{n-1}} & \cdots & (X_n)_{\xi_{n-1}} \end{vmatrix} \neq 0 \tag{9.3.12}$$

If the initial manifold is such that $\Delta \neq 0$, then the problem has a unique solution at least in its neighborhood. If $\Delta = 0$ on the initial manifold, then it must be a characteristic manifold if the solution is to exist, and in fact there are then infinitely many solutions that can branch into one another smoothly across such a manifold. The possibilities for singular surfaces and special integrals of various kinds are naturally much richer in $(n + 1)$-dimensional spaces, but we will not attempt to lay them all out. A proof that the solution surface is indeed so generated is given in various places; see particularly Chapter 2 of Courant and Hilbert (1962).

Exercises

9.3.1. A nonhomogeneous medium has a refractive index $n = n(x_1)$. If θ is the angle between a ray (i.e., a characteristic of $p_1^2 + p_2^2 + p_3^2 = n^2$) and the x_1-axis, show that $n \sin \theta$ is constant along the ray.

9.3.2. The refractive index of a nonhomogeneous medium varies continuously as a function of position. Show that a ray can remain within a surface of constant refractive index.

9.3.3. Show that the characteristic differential equations of Eq. (1.14.20) are

$$\dot{x} = -kx, \qquad \dot{y} = kx - \rho k^r y, \qquad \dot{z} = -1$$

$$\dot{u}_x = -k(1 + u_y - u_x), \qquad \dot{u}_y = \rho k^r(1 + u_y), \qquad \dot{u}_z = 0$$

where k is given by Eq. (1.14.19). Hence show that along a characteristic

$$\dot{k} = -\frac{k^2 x}{ry}$$

9.4 Some Questions of Existence and Continuity

We have used geometrical and intuitive arguments to make the existence of solutions plausible and to visualize their continuity. Although it would be beyond the intention of this book to give too much attention to the details of existence proofs, it is desirable to give some account of the relations between the different formulations of the existence problem and to outline the way in which one of the classical proofs proceeds.

Exercises 9.2.1, 9.2.2, and 9.2.3 provided simple steps to the proof that the solution found by the method of characteristics did indeed provide a function $z(x, y)$ whose derivatives were p and q and that satisfied the equation. If the initial strip is not characteristic, the existence and uniqueness of the solution are assured in its neighborhood. If the initial strip is characteristic, an infinity of solutions can pass through the strip, all touching one another there. However, there may be discontinuities in the higher derivatives. If $\Delta = 0$ and the initial strip is a focal strip but not characteristic, then a solution may be generated; but it is not necessarily unique nor need its derivatives

TABLE 9.1

Problem	Equation	Solution	Conditions
	$F(x,y,z,p,q)=0$ $F \in C$	$z=\phi(x,y), p=\phi_x, q=\phi_y$ $\phi \in C'$	
P			$Z(\xi) \equiv \phi(X(\xi),Y(\xi))$ $X(\xi),Y(\xi),Z(\xi) \in C'$
S			$g(y) \equiv \phi(h(y),y)$ $g(y),h(y) \in C'$
G			$g(y) \equiv \phi(x_0,y)$ $g(y) \in C'$, x_0 fixed
I			$0 \equiv \phi(0,y)$
	$p=f(x,y,z,q)$ $f \in C$	$\phi = f(x,y,z,\phi_y)$ $\phi \in C'$	
GN			$g(y) \equiv \phi(x_0,y)$ $g(y) \in C'$, x_0 fixed
IN			$0 \equiv \phi(0,y)$

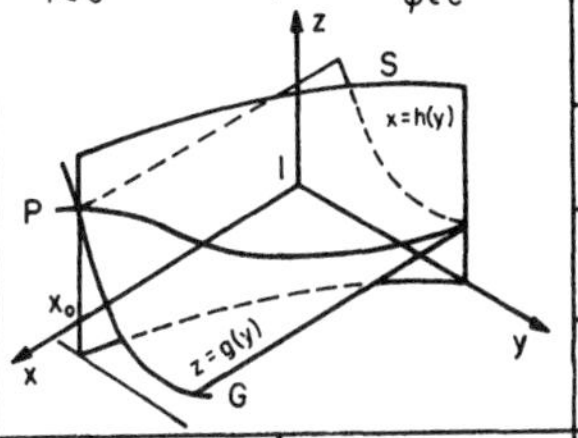

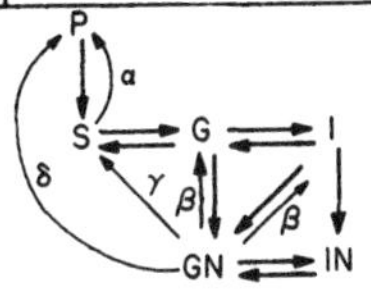

α) $Y'(\xi) \neq 0$

β) $F \in C'$, $F_p \neq 0$

γ) $F \in C'$, $F_p - h' F_q \neq 0$

δ) $F \in C'$, $Y'(\xi) \neq 0$
$F_p Y'(R(y)) - F_q X'(R(y)) \neq 0$
$R(Y(\xi)) \neq \xi$

above the first order be continuous. It may be shown that every solution with continuous first derivatives is made up of characteristic strips.

The various forms that the *Cauchy problem* may take are illustrated in Table 9.1. It is an adaptation of material presented in Bernstein's collection of existence theorems. This collection was made in what Tompkins' introduction describes as "the parlous times just after the war when it was apparent to a large and vociferous set of engineers that the electronic digital calculating machines they were then developing would replace all mathematicians, indeed, all thinkers except the engineers building more machines." Many such engineers have acquired a little more wisdom, as well as building machines of a capability then scarcely imagined, and certainly the tribe of mathematicians has not decreased; but the point is well made in connection with the computational advances of the last few years. It is still necessary to have some understanding of why a solution exists at all, if the danger of computing a chimera is to be avoided.

We will take the nonlinear equation with two independent variables as the typical case and denote the solution surface by $z = \phi(x, y)$. The notation $f \in C^m$ denotes that f is a continuous function with continuous derivatives up to order m, and C^0, the class of continuous functions, is written as C. As we will see later, it is convenient to write the equation in a normal form by solving $F(x, y, z, p, q) = 0$ for $p = f(x, y, z, q)$. The problems P, S, G, and I are posed for the first form and GN and IN for the second. In P the initial curve is given parametrically, but in S it is the intersection of two cylindrical surfaces. In G the initial curve is specified in a plane of constant x, and in I the initial curve is the y-axis itself.

In the diagram of relationship in the lower part of the figure in the table, the heavy arrows denote an immediate specialization. Thus, if we can find a solution to P, we can get one also for S by putting $\xi = y$, $X(\xi) = h(y)$, $Y(\xi) = y$, $Z(\xi) = g(y)$, since $y \in C^1$. Given the solution of S, G is solved by making $h(y) = x_0$ and, given G, we can pass to I by setting $x_0 = 0$, $g(y) = 0$. From G and I, we can go to GN and IN by giving $F(x, y, z, p, q) = 0$ the special form $p - f(x, y, z, q) = 0$. But if we know how to solve S, we can only solve P if $Y'(\xi) \neq 0$; for then $y = Y(\xi)$ can be inverted to give $\xi = R(y)$, so that $h(y) = X(R(y))$, $g(y) = Z(R(y))$ and these are all of class C^1. This is the condition α. If a solution to I is known and problem S is posed, we can use the transformation $x = \hat{x} + h(\hat{y})$, $y = \hat{y}$, $z = \hat{z} + g(\hat{y})$, $p = \hat{p}$, $q = \hat{q} - h'(\hat{y})\hat{p} + g'(\hat{y})$. Thus, if we find a solution to the problem I for the equation

$$\hat{F}(\hat{x}, \hat{y}, \hat{z}, \hat{p}, \hat{q}) \equiv F(\hat{x} + h(\hat{y}), \hat{y}, \hat{z} + g(\hat{y}), \hat{p}, \hat{q} - h'(\hat{y})\hat{p} + g(\hat{y})) = 0 \qquad (9.4.1)$$

we will have a solution to problem S, for the inverse transform is $\hat{x} = x - h(y)$, $\hat{y} = y$, $\hat{z} = z - g(y)$, and when $\hat{x} = 0$, $\hat{y} = 0$, $x = h(y)$, and $z = g(y)$. The reader can check that the Jacobian of the transformation never vanishes and that no further conditions are necessary. Since I leads to S and S to G, I must lead to G and problems S, G, and I are equivalent. Similarly, GN and IN are equivalent.

However, if we know how to solve GN and we are given problem G, we need to be able to solve $F(x, y, z, p, q) = 0$ for p in terms of x, y, z, and q. By the implicit function theorem, this is possible if $F_p \neq 0$. This is the condition β, which also allows the connection $GN \rightarrow I$ to be made. The connection from GN to S is possible if we can make the same substitutions $\hat{x} + h(\hat{y}) = x$, $\hat{y} = y$, and so on, as were used previously and if in addition $\hat{F}_{\hat{p}} \neq 0$. But from Eq. (9.4.1), we see that $\hat{F}_{\hat{p}} = F_p - h'(y)F_q$, so the condition γ is required. Finally, the connection $GN \rightarrow P$ requires conditions α and γ. But $h'(y) = X'(\xi)/Y'(\xi)$, which gives the condition δ.

The same set of formulations can be made requiring the several functions to belong to the class C^m. The relationship between these problems, which might be denoted by P_m, S_m, and so on, is the same as outlined previously. In particular, if we require the functions to be analytic, we have an important classical case, to which we now turn.

The discussion of this existence theorem, initiated by Cauchy and carried through in great generality by Kowalewski, falls into three parts. We will first show how the equation $F(x, y, z, p, q) = 0$ can be reduced to a system of quasi-linear equations in normal form. Second, we will discuss the notion of majorant series and, third, show how this can be used to establish the convergence of a series solution in the neighborhood of the initial curve. For simplicity, we will confine attention to the equation $F(x, y, z, p, q) = 0$, but the methods and ideas are much more widely applicable.

The reduction of $F(x, y, z, p, q) = 0$ to a normalized quasi-linear system in z, p, and q is quite straightforward (see Sec. 2.2). For by definition $z_x = p$ and $q_x = p_y$ and, by differentiating the equation with respect to x, $F_x + F_z p + F_p p_x + F_q q_x = 0$. Thus, if $F_p \neq 0$,

$$\begin{aligned} z_x &= p \\ p_x &= -\left(\frac{F_x}{F_p}\right) - \left(\frac{F_z}{F_p}\right)p - \left(\frac{F_q}{F_p}\right)p_y \\ q_x &= p_y \end{aligned} \tag{9.4.2}$$

and the right sides are linear in the derivatives and otherwise functions of x, y, z, p, and q. If we wish to remove reference to x and y, this can be done by the procedure discussed in Sec. 2.2; that is, we replace x and y by ξ and η, respectively, and add the two equations

$$\begin{aligned} \xi_x &= \eta_y \\ \eta_x &= 0 \end{aligned} \tag{9.4.3}$$

to the system. If we denote the solution of $F = 0$ for p by $p = f(x, y, z, q)$ and make use of the fact that $\eta_y = 1$, the five equations can be written as homogeneous quasi-linear equations:

$$\begin{aligned} \xi_x &= \eta_y \\ \eta_x &= 0 \\ z_x &= p\eta_y \\ p_x &= (f_x + f_z p)\eta_y + f_q p_y \\ q_x &= p_y \end{aligned} \tag{9.4.4}$$

Suppose that the formulation of G of the table is used with $x_0 = 0$; then the initial conditions are

$$\begin{gathered} \xi(0, y) = 0, \qquad \eta(0, y) = y, \qquad z(0, y) = g(y) \\ p(0, y) = f(0, y, g(y), g'(y)), \qquad q(0, y) = g'(y) \end{gathered} \tag{9.4.5}$$

Since we are to deal with systems of quasi-linear equations in normal form, let us introduce the notation

$$u_1 = \xi, \qquad u_2 = \eta, \qquad u_3 = z, \qquad u_4 = p, \qquad u_5 = q$$

so our equations come generally under the form

$$\frac{\partial u_i}{\partial x} = \sum_{j=1}^{m} G_{ij}(u_1, \ldots, u_m)\frac{\partial u_j}{\partial y} \tag{9.4.6}$$

with

$$u_i(0, y) = g_i(y)$$

Without loss of generality, we can require $g_i(0) = 0$. We are going to consider analytical differential equations, and it is therefore natural to replace the functions by their power series, say

$$g_i(y) = \sum_{n=1}^{\infty} a_n^i y^n \tag{9.4.7}$$

and

$$G_{ij}(u_1, \ldots, u_m) = \sum_{n_1, \ldots, n_m = 0}^{\infty \cdots \infty} b_{n_1 n_2 \ldots n_m}^{ij} u_1^{n_1} u_2^{n_2} u_3^{n_3} \ldots u_m^{n_m} \tag{9.4.8}$$

These series converge in certain regions, say $|y| \leq \rho$ and $|u_i| \leq r$, respectively. It is natural to seek a solution of the equations in the form

$$u_i(x, y) = \sum_{\substack{k=0 \\ l=1}}^{\infty}\sum^{\infty} c_{kl}^i x^k y^l \tag{9.4.9}$$

where the condition $u_i(0, y) = g_i(y)$ will be immediately satisfied if we set

$$c_{0l}^i = a_l^i \tag{9.4.10}$$

The other coefficients will be determined by comparing coefficients of powers of x and y on the two sides of the equation. Perhaps the easiest way to see how this process would proceed is to calculate the c_{1l}^i. These will be the coefficients of the y^l in $\partial u^i/\partial x$ evaluated on $x = 0$. But

$$\left(\frac{\partial u_i}{\partial x}\right)_{x=0} = \sum_{j=1}^{m} G_{ij}(g_1, \ldots, g_m) g_j'(y)$$

$$= \sum_{j=1}^{m} \sum_{n_1, \ldots, n_m} b_{n_1, \ldots, n_m}^{ij} \left(\sum a_n^1 y^n\right)^{n_1} \cdots \left(\sum a_n^m y^n\right)^{n_m} \left(\sum_l l c_{0l}^j y^{l-1}\right)$$

Thus

$$c_{11}^i = \sum_{j=1}^{m} b_{0,\ldots,0}^{ij} c_{01}^j = \sum_{j=1}^{m} b_{0,\ldots,0}^{ij} a_1^j$$

$$c_{12}^i = \sum_{j=1}^{m} (2b_{0,\ldots,0}^{ij} c_{02}^j + b_{1,0,\ldots,0}^{ij} a_1^1 c_{01}^j + \cdots + b_{0,\ldots,1}^{ij} a_1^m c_{01}^j)$$

and so forth. To calculate the c_{2l}^i, we differentiate the equation with respect to x and proceed as before. Horrendous as these formulas appear to be in their algebraic complexity, it is quite clear that the c_{kl}^i are always polynomials in the a_n^i and $b_{n_1, \ldots, n_m}^{ij}$ and that the coefficient of any particular product of a's and b's is a positive integer. We will use this fact to show that the series (9.4.9) does converge.

Let us turn aside for the moment and describe the concept of a dominant series. If we can find a series with positive coefficients

$$\sum_{n=1}^{\infty} A_n y^n, \qquad A_n \geq |a_n|$$

then this series is said to dominate the series $\Sigma\, a_n y^n$. Moreover, if the dominating series converges, the dominated series must also converge. It is easy to find a dominating series for a series $\Sigma\, a_n y^n$, which is known to converge if $|y| < \rho$. For this convergence implies that $|a_n \rho^n|$ is bounded, say by M, and so

$$|a_n| \leq A_n = M\rho^{-n}$$

The dominating series is thus

$$M \sum_{n=0}^{\infty} \frac{y^n}{\rho^n} = M\left(1 - \frac{y}{\rho}\right)^{-1} = \frac{M\rho}{\rho - y}$$

If, as is the case where we wish to use it later, the series $\Sigma_{n=1}^{\infty} a_n y^n$ lacks a constant term, we can more efficiently dominate it by

$$M \sum_{n=0}^{\infty} \left(\frac{y}{\rho}\right)^n - M = \frac{My}{\rho - y}$$

Similarly, the power series

$$\sum_{n_1,\ldots,n_m=0}^{\infty} b_{n_1,\ldots,n_m}\, u_1^{n_1} \ldots u_m^{n_m}$$

is said to be dominated by

$$\sum_{n_1,\ldots,n_m=0}^{\infty} B_{n_1,\ldots,n_m}\, u_1^{n_1} \ldots u_m^{n_m}$$

if

$$|b_{n_1,\ldots,n_m}| \leq B_{n_1,\ldots,n_m}$$

If the first series converges for $|u_i| \leq r_i$, then we can take M to be an upper bound of

$$b_{n_1,\ldots,n_m} r_1^{n_1}, r_2^{n_2}, \ldots, r_m^{n_m}$$

and choose

$$B_{n_1,\ldots,n_m} = \frac{(n_1 + \cdots + n_m)!}{n_1! \cdots n_m!} \frac{M}{r_1^{n_1} \cdots r_m^{n_m}} \geq \frac{M}{r_1^{n_1} \cdots r_m^{n_m}} \geq |b_{n_1,\ldots,n_m}|$$

so the dominating series converges to

$$M\left(1 - \frac{u_1}{r_1} - \cdots - \frac{u_m}{r_m}\right)^{-1}$$

One additional fact about dominating series needs to be pointed out: that if $\Sigma\, a_n^i y^n$ and $\Sigma\, b_{n_1,\ldots,n_m}^{ij} u_1^{n_1} \ldots u_m^{n_m}$ are dominated by $\Sigma\, A_n^i y^n$ and $\Sigma\, B_{n_1,\ldots,n_m}^{ij} u_1^{n_1} \ldots u_m^{n_m}$, respectively, and if we have a rule for forming the coefficients in the series $\Sigma\, c_{kl}^i x^k y^l$ from the a's and b's that involves only addition and multiplication by positive numbers, then the series $\Sigma\, C_{kl}^i x^k y^l$, with the C_{kl}^i formed by the same rule from the A's and B's, dominates $\Sigma\, c_{kl}^i x^k y^l$.

Let us return now to our equations

$$\frac{\partial u_i}{\partial x} = \sum_{j=1}^{m} G_{ij}(u_1, \ldots, u_m) \frac{\partial u_j}{\partial y}$$

subject to

$$u_i(0, y) = g_i(y)$$

Since we have claimed that the power series for the G_{ij} and g_i are convergent for $|u_i| < r$ and $|y| < \rho$, respectively, we can take M to be an upper bound for both $a_n^i \rho^n$ and $b_{n_1,\ldots,n_m}^{ij} r^{n_1+\cdots+n_m}$ and set

$$G(u_1, \ldots, u_m) = M\left(1 - \frac{u_1}{r} - \cdots - \frac{u_m}{r}\right)^{-1}$$

and

$$\gamma(y) = My(\rho - y)^{-1}$$

The series for G dominates each of those for the G_{ij} and that for γ dominates all those for the g_i. Consider the next problem:

$$\frac{\partial v_i}{\partial x} = G(v_1, \ldots, v_m) \sum \frac{\partial v_j}{\partial y} \tag{9.4.11}$$

with

$$v_i(0, y) = \gamma(y) \tag{9.4.12}$$

If we can prove that this has a solution with convergent power-series expansions

$$v_i(x, y) = \sum_{k,l} C_{kl}^i x^k y^l$$

then the solution to the original problem given by Eq. (9.4.9) will converge, and we will have constructed a unique, analytic solution to the original equation. It remains only to show that we can find a solution to Eqs. (9.4.11) and (9.4.12) that is convergent in some neighborhood of $x = 0$.

Since the boundary condition is the same for each v_i, we will seek a solution in which all the v_i are the same function $v(x, y)$. This will then have to satisfy

$$\frac{\partial y}{\partial x} = \frac{mMr}{r - mv} \frac{\partial v}{\partial y} \tag{9.4.13}$$

with

$$v(0, y) = \frac{My}{\rho - y} \tag{9.4.14}$$

But we have seen that a solution of such an equation can be written as a function of

$$y + \frac{mMr}{r - mv} x$$

Moreover, the function that reduces to Eq. (9.4.14) when $x = 0$ is clearly

$$v = \frac{M\left(y + \dfrac{mMrx}{r - mv}\right)}{\rho - y - \dfrac{mMrx}{r - mv}} \tag{9.4.15}$$

This is a quadratic in v,

$$v^2 - v\left(\frac{r}{m} + M\frac{y - rx}{\rho - y}\right) + \frac{Mr(y + Mmx)}{m(\rho - y)} = 0$$

which has a solution that tends to zero as x and y tend to zero. Also, if $x = 0$, it can be written

$$\left(v - \frac{My}{\rho - y}\right)\left(v - \frac{r}{m}\right) = 0$$

so this solution tends to the required boundary condition as x tends to zero.

This completes the proof of the existence of an analytic solution to an analytic differential equation. In physical problems we are of course often interested in equations and solutions that have a limited degree of smoothness. These we will build up as we need them. The general situation is extremely complicated; for example, it is known that there are linear equations with smooth coefficients that do not have any solutions at all.

9.5 A Problem in Optimization

In Sec. 1.14 we obtained a highly nonlinear partial differential equation for $u(x, y, z)$, the maximum yield of B from the batch reaction $A \to B \to C$ when the initial concentrations of A and B are x and y, respectively, and the duration of the reaction is z. This equation can be written

$$F \equiv \sigma x^m y^{-n}(1 + u_y - u_x)^m(1 + u_y)^{-n} - u_z = 0 \tag{9.5.1}$$

where $m = r/(r - 1) = n + 1, n = 1/(r - 1) = m - 1$, and $\sigma = (mr^n\rho^n)^{-1}$. The optimal choice of k is given by Eq. (1.14.19):

$$k = \left[\frac{x(1 + u_y - u_x)}{r\rho y(1 + u_y)}\right]^n \tag{9.5.2}$$

In exercise 9.3.3, the question of writing down the charecteristic differential equations was posed, and after some algebraic manipulation these were shown to be

$$\frac{dx}{ds} = -kx$$

$$\frac{dy}{ds} = kx - \rho k^r y \tag{9.5.3}$$

$$\frac{dz}{ds} = -1$$

$$\frac{du}{ds} = -kx + \rho k^r y \tag{9.5.4}$$

$$\frac{du_x}{ds} = -k(1 + u_y - u_x)$$

$$\frac{du_y}{ds} = \rho k^r(1 + u_y) \tag{9.5.5}$$

$$\frac{du_z}{ds} = 0$$

A notable feature of these equations is that the characteristics are solutions of the basic equations of the process. When the translation of symbols $x = a, y = b, z = t$ is made, the only difference between the first three equations, Eqs. (9.5.3), and the equations of the batch reactor, Eqs. (1.14.13) and (1.14.14), is that the sense of time is reversed. This we should expect, since $z = 0$ denotes a remaining process of zero duration (i.e., the *end* of the process). Thus z increases in the opposite direction from that of real time, whose role is taken by s. The boundary condition is Eq. (1.14.21),

$$u(x, y, 0) = 0 \tag{9.5.6}$$

Hence we have an initial manifold $z = 0$ giving initial conditions

$$x = X, \quad y = Y, \quad z = 0, \quad u = 0, \quad u_x = 0, \quad u_y = 0, \quad u_z = \sigma X^m Y^{-n}$$

The last of these is obtained from the equation itself, Eq. (9.5.1).

We can integrate these equations back from any final composition and, by choosing k according to Eq. (9.5.2), we will have an optimal trajectory. However, the optimal problem is usually posed with the composition x, y given at the beginning of the process, whereas here the composition X, Y is given at the end. We are thus faced with a two-point boundary-value problem, for the conditions on the first three equations are more naturally given at the beginning of the process and those for the last three apply at the end. Thus we must either find the complete surface $u(x, y, z)$ and so pick out the trajectory from a particular x, y, z or else search for the X, Y that will lead back to x, y along an optimal trajectory of duration z. Some form of this two-point boundary-value problem is encountered in any formulation of this type of optimization problem.

But we foresee another difficulty in the way we have formulated the problem up to now; for if y is very small, as might be the case at the beginning of the process, Eq. (9.5.2) shows that the optimal k becomes very large. But $k = A \exp(-E/RT)$ and so is bounded above by A for all temperatures, and Eq. (9.5.2) might not have a solution. In practice, the temperature T will certainly be bounded, say $T_* \leq T \leq T^*$, and so k must be between $k_* = k(T_*)$ and $k^* = k(T^*)$,

$$k_* \leq k \leq k^* \tag{9.5.7}$$

To see how we must now proceed, we should return to the equation from which the nonlinear form (9.5.1) was derived. This is

$$\max[kx(1 - u_x + u_y) - \rho k^r(1 + u_y)] - u_z = 0 \tag{9.5.8}$$

The expression in the brackets has a unique maximum given by Eq. (9.5.2) and is monotonic on either side. Let us write Eq. (9.5.8) as

$$\max H(x, y, u_x, u_y; k) - u_z = 0 \tag{9.5.9}$$

where k, the control or decision variable, is distinguished from the others. Equation (9.5.2) was obtained by solving

$$\frac{\partial H}{\partial k} = 0$$

for k in terms of x, y, u_x, u_y. Let us write this as

$$k = G(x, y, u_x, u_y) \tag{9.5.10}$$

Now for any set of values x, y, u_x, and u_y the form of H as a function of k must be one of the three shown in Fig. 9.7. In the first case, if the maximum A' falls to the left of $k = k_*$ (i.e., $G < k_*$), the greatest value of H is obtained when $k = k_*$ at A. In the third case, where the maximum C' lies to the right of $k = k_*$, the greatest value of H is at C, where $k = k^*$. However, when the position of the maximum given by Eq. (9.5.10) is in the admissible

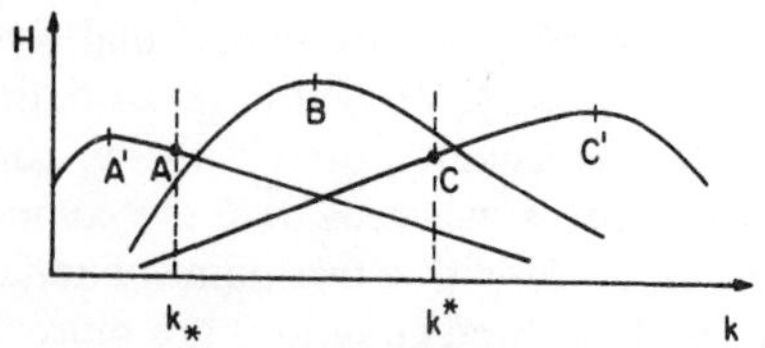

Figure 9.7

range, we can substitute as we did in obtaining Eq. (9.5.1), which could, in fact, be written

$$F \equiv H(x, y, u_x, u_y; G(x, y, u_x, u_y)) - u_z \tag{9.5.11}$$

Thus a complete statement of the partial differential equation would be

$$0 = F \equiv \begin{cases} H(x, y, u_x, u_y; k_*) - u_z, & G < k_* \\ H(x, y, u_x, u_y; G) \quad - u_z, & k_* \le G \le k^* \\ H(x, y, u_x, u_y; k^*) - u_z, & k^* < G \end{cases} \tag{9.5.12}$$

In the two extreme cases, where k is constant, Eq. (9.5.8) is clearly a linear equation and can be solved by integrating the first three characteristic differential equations with $k = k_*$ or $k = k^*$. We have also seen, at the expense of some algebraic labor, that these equations are obtained in the general case, although they must be augmented with the last three equations if the optimal value of k is to be determined. To see why we always obtain the process equations, let us write the characteristic differential equations for

$$F \equiv H(x, y, u_x, u_y; k) - u_z = 0$$

remembering that k may be a function of x, y, u_x, and u_y. Thus

$$\frac{dx}{ds} = \frac{\partial F}{\partial u_x} = \frac{\partial H}{\partial u_x} + \frac{\partial H}{\partial k}\frac{\partial k}{\partial u_x}$$

$$\frac{dy}{ds} = \frac{\partial F}{\partial u_y} = \frac{\partial H}{\partial u_y} + \frac{\partial H}{\partial k}\frac{\partial k}{\partial u_y}$$

$$\frac{dz}{ds} = \frac{\partial F}{\partial u_z} = -1$$

But in the extreme cases, k is constant and $\partial k/\partial u_x = 0$, while in the intermediate case the very basis for choosing $k = G(x, y, u_x, u_y)$ was the equation $\partial H/\partial k = 0$. Similarly, the characteristic differential equations for u_x and u_y will be

$$\frac{du_x}{ds} = -\frac{\partial H}{\partial x} \quad \text{and} \quad \frac{du_y}{ds} = -\frac{\partial H}{\partial y}$$

It can be shown (see Exercise 9.5.1) that the same simplification obtains when there is more than one control variable.

It is sometimes useful to convert the equation for the optimal value of k into a differential equation:

$$\frac{dk}{ds} = G_x \frac{dx}{ds} + G_y \frac{dy}{ds} + G_{u_x} \frac{du_x}{ds} + G_{u_y} \frac{du_y}{ds} \tag{9.5.13}$$

In this problem, the resulting differential equation is surprisingly simple (see Exercise 9.3.3),

$$\frac{dk}{ds} = -\frac{k^2 x}{ry} \tag{9.5.14}$$

in which all reference to u_x and u_y has disappeared. It can be shown that this feature obtains whenever only two dependent process variables and one control variable are involved (see Exercise 9.5.2). More general cases have been treated by Denn and Aris (1965). Now, although Eqs. (9.5.3) and (9.5.14) form a complete set for the three variables x, y, and k, we are again faced with a two-point boundary-value problem, because we can specify k at the end of the process, where it is $(x/\rho ry)^n$, but not at the beginning, where we normally know x and y.

To show the form of the solution of this problem, it is best to make use of the fact that the equations are homogeneous and of the first order in x and y. In Exercise 1.14.3, the substitution $u(x, y, z) = xv(y/x, z)$ was suggested, and if this is made, Eq. (9.5.1) reduces to

$$\sigma\eta^{-n}[1 + v_\eta(1 + \eta) - v]^m(1 + v_\eta)^{-n} - v_z = 0 \tag{9.5.15}$$

where

$$\eta = \frac{y}{x} \tag{9.5.16}$$

Equation (9.5.2) becomes

$$k = \left[\frac{1 + v_\eta(1 + \eta) - v}{r\rho\eta(1 + v_\eta)}\right]^n \tag{9.5.17}$$

The characteristic differential equations for this system are readily found and, after some manipulation (see Exercise 9.1.1), give

$$\begin{aligned}
\frac{d\eta}{ds} &= k + (k - \rho k^r)\eta \\
\frac{dz}{ds} &= -1 \\
\frac{dv}{ds} &= k(v - 1) + \rho k^r \eta \\
\frac{dv_\eta}{ds} &= \rho k^r(1 + v_\eta) \\
\frac{dv_z}{ds} &= k v_z
\end{aligned} \tag{9.5.18}$$

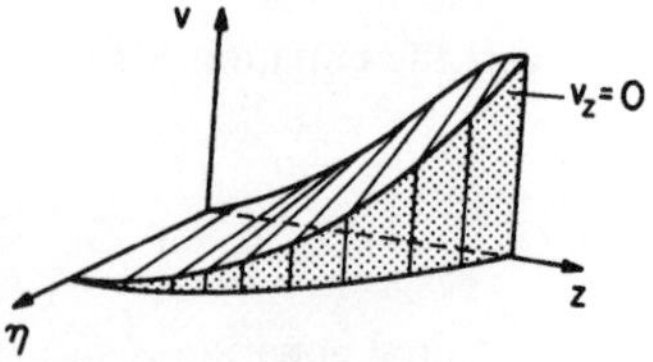

Figure 9.8

The differentiation of the expression for k again gives

$$\frac{dk}{ds} = -\frac{k^2}{r\eta}$$

which may be solved together with the first characteristic differential equation to give η as a function of k. Then z and v can be calculated afterward by quadratures. Figure 9.8 shows the surface $v(\eta, z)$ up to the line on which $v_z = 0$. The quantity v has the physical meaning of the ratio of the yield of B to the remaining A. This is not the quantity we set out to maximize, but since $u_x = xv_z$, the variable u_z is also zero on this locus. But we are only interested in the region for which $u_z \geq 0$, since u_z being negative implies that increase in reaction time would decrease the amount of product B. Hence the region shown is the only part of the surface that is economically attractive.

We will state without proof the partial differential equation that is satisfied by a rather general problem in optimal control. This equation can be obtained by applying the principle of optimality in the style of Exercise 1.14.4. We will then present the characteristic differential equations and use them to state the necessary conditions for optimality associated with the name *Pontryagin's maximum principle*. This approach, although lacking in rigor, will serve to form a link between our subject and the vast domain of modern control theory on which many references may be found. The notation common in control theory will be used with appropriate translations.

The system to be controlled is specified by n state variables $x_i(t)$ and m control variables $u_j(t)$. The point $(x_1, \ldots, x_n)$, or more briefly $\mathbf{x}$, is in $\mathbf{R}_n$, the real space of n dimensions, and the control point $\mathbf{u}$, or $(u_1, \ldots, u_m)$, lies in a subset U of $\mathbf{R}_n$. The system is governed by

$$\dot{x}_i = f_i(x_1, \ldots, x_n; u_1, \ldots, u_m), \qquad i = 1, \ldots, n \tag{9.5.19}$$

or, more briefly,

$$\dot{\mathbf{x}} = \mathbf{f}(\mathbf{x}; \mathbf{u})$$

where the f_i and all their first derivatives are continuous. Let the problem specify the initial values

$$x_i(0) = \xi_i \tag{9.5.20}$$

and ask for the permissible control that minimizes

$$\int_0^\theta f_0(\mathbf{x}, \mathbf{u})\, dt + h(\mathbf{x}(\theta)) \tag{9.5.21}$$

for a given duration of control θ. In this expression, $f_0(\mathbf{x}, \mathbf{u})$ is the cost per unit time when the state is $\mathbf{x}(t)$ and the control $\mathbf{u}(t)$, and $h(\mathbf{x}(\theta))$ is the cost associated with terminating in the state $\mathbf{x}(\theta)$. It is convenient to have this flexibility in the choice of cost functional, although the two terms are to some extent interchangeable. Thus, if h is differentiable and $h(\boldsymbol{\xi}) = 0$, we might replace f_0 by an $f_0 + \nabla h \cdot \mathbf{f}$ and drop the term in h. For symmetry, we denote θ by ξ_{n+1} and take $x_{n+1} = \theta - t$ so that the minimum value of this integral is a function of the $(n + 1)$ variables $\xi_1, \ldots, \xi_{n+1}$, say

$$g(\xi_1, \ldots, \xi_{n+1}) = \min_{\mathbf{u}} \left[\int_0^{\xi_{n+1}} f_0(\mathbf{x}, \mathbf{u})\, dt + h(\mathbf{x}(\xi_{n+1})) \right] \tag{9.5.22}$$

Then the function g satisfies

$$\min_{\mathbf{u}} [H(\boldsymbol{\xi}, \mathbf{p}, \mathbf{u})] - p_{n+1} = 0 \tag{9.5.23}$$

where

$$\boldsymbol{\xi} = (\xi_1, \ldots, \xi_n), \qquad p_i \equiv \frac{\partial g}{\partial \xi_i}$$

and

$$H \equiv f_0(\boldsymbol{\xi}, \mathbf{u}) + f_j(\boldsymbol{\xi}, \mathbf{u}) p_j \tag{9.5.24}$$

Note that a repeated index (the j in this equation) is summed from 1 to n.

The characteristic differential equations are obtained by ignoring the minimization as we have seen above (see also Exercise 9.5.1), and are

$$\frac{d\xi_i}{ds} = \frac{\partial H}{\partial p_i} = f_i(\boldsymbol{\xi}, \mathbf{u}), \qquad i = 1, \ldots, n \tag{9.5.25}$$

$$\frac{d\xi_{n+1}}{ds} = \frac{\partial H}{\partial p_{n+1}} = -1 \tag{9.5.26}$$

$$\frac{dg}{ds} = -f_0(\boldsymbol{\xi}, \mathbf{u}) \tag{9.5.27}$$

$$\frac{dp_i}{ds} = -\frac{\partial H}{\partial \xi_i} = -\frac{\partial f_0}{\partial \xi_i} - \frac{\partial f_j}{\partial \xi_i} p_j \tag{9.5.28}$$

$$\frac{dp_{n+1}}{ds} = 0 \tag{9.5.29}$$

The optimal control is of course obtained by finding the **u** that minimizes H. We observe that the first equations are precisely those that govern the system with x_i replaced by ξ_i. We could, in fact, return to x_i as the designation of the current variable and use ξ_i for its initial value. The p_i are called adjoint variables, and since

$$g(\xi_1, \ldots, \xi_n, 0) = h(\xi_1, \ldots, \xi_n)$$

when the process is of zero duration, we have final values for the p_i:

$$p_i = \frac{\partial h}{\partial x_i} \tag{9.5.30}$$

Along the characteristics, H is constant, and if the duration is unspecified, it may be chosen so as to maximize g with respect to ξ_{n+1} for which a necessary condition is $p_{n+1} = 0$. In this case, Eq. (9.5.24) shows that $H \equiv 0$. In vector notation, the Hamiltonian

$$H \equiv f_0(\mathbf{x}, \mathbf{u}) + \mathbf{f}(\mathbf{x}, \mathbf{u}) \cdot \mathbf{p}$$

$$\dot{\mathbf{x}} = \nabla_p H = \mathbf{f}, \qquad \dot{\mathbf{p}} = \nabla_x H = -\nabla f_0 - (\nabla \mathbf{f})^T \cdot \mathbf{p} \tag{9.5.31}$$

Pontryagin's maximum (or minimum) principle then states that a necessary condition for this to be an optimal control is that H should have an absolute minimum with respect to $\mathbf{u} \in U$ at each point of the path and that the adjoint variables should be continuous and satisfy the final conditions (9.5.30).

We will return to variational problems and their connections with first-order partial differential equations in Chapter 10.

Exercises

9.5.1. Consider the equation

$$F \equiv \max_{\mathbf{u}}[H(x_1, \ldots, x_n, p_1, \ldots, p_n; u_1, \ldots, u_m)] - p_{n+1} = 0$$

where the maximization is by choice of **u** subject to restrictions

$$S_k(u_1, \ldots, u_m) \geq 0, \qquad k = 1, \ldots, l$$

Show that the characteristic differential equations are

$$\frac{dx_i}{ds} = \frac{\partial H}{\partial p_i}, \qquad \frac{dx_{n+1}}{ds} = -1, \qquad \frac{dz}{ds} = p_i\frac{\partial H}{\partial p_i} - H,$$

$$\frac{dp_i}{ds} = -\frac{\partial H}{\partial x_i}, \qquad \frac{dp_{n+1}}{ds} = 0$$

where **u** is chosen to effect the maximization subject to the given restrictions.

9.5.2. The process $\dot{x}_i = f_i(x_1, x_2, u)$ is to be governed so that the quantity $a_i x_i(\theta) - a_i x_i(0)$ is maximized by the optimal choice of u. If there are no restrictions on u, show that it satisfies the equation

$$\dot{u} = \frac{f_2' f_{1j} f_j' - f_1' f_{2j} f_j' + f_2' f_{1j}' f_j - f_1' f_{2j}' f_j}{f_1' f_2'' - f_2' f_1''}$$

where $f_i' = \partial f_i/\partial u$, $f_i'' = \partial^2 f_i/\partial u^2$, $f_{ij} = \partial f_i/\partial x_j$, $f_{ij}' = \partial^2 f_i/\partial u \partial x_j$

9.5.3. Show that by solving the additional four equations

$$\frac{dX_x}{ds} = -k(1 - X_x + X_y), \qquad X_x(0) = 1$$

$$\frac{dX_y}{ds} = \rho k^r(1 + X_y), \qquad X_y(0) = 0$$

$$\frac{dY_x}{ds} = -k(1 - Y_x + Y_y), \qquad Y_x(0) = 0$$

$$\frac{dY_y}{ds} = \rho k^r(1 + Y_y), \qquad Y_y(0) = 1$$

using k determined for the optimal problem by Eq. (9.5.2), the elements of the Jacobian matrix $\partial(X, Y)/\partial(x, y)$ can be calculated. How would you use this matrix in an iterative search for the solution of the optimal problem with $x = x_0$, $y = y_0$, $z = z_0$?

9.5.4. A batch reactor is governed by the equation

$$\frac{dc}{dt} = r(c, T), \qquad \frac{dT}{dt} = Jr(c, T) - Q, \qquad \frac{dQ}{dt} = q$$

where $0 \le Q \le Q^*$ and $q_* \le q \le q^*$. Let γ be a fixed final conversion c and the initial values of c, T, and Q be u, v, and w, respectively. If $f(u, v, w)$ denotes the minimum time to achieve a conversion $c = \gamma$ starting from u, v, w, show that f satisfies

$$\min_q[1 + (f_u + Jf_v)r(u, v) - f_v w - f_w q] = 0$$

and write down the characteristic differential equations.

9.5.5. Show that q must take on one of its extreme values in Exercise 9.5.4 unless Q takes one of its extreme values or the point (u, v, w) lies on the cylinder $r_v(u, v) = 0$. Deduce that f can only be smooth if the point on $r_v(u, v)$ at which $Q = Q^*$ is approached adiabatically.

REFERENCES

As always

R. Courant and D. Hilbert, *Methods of Mathematical Physics: Vol. II. Partial Differential Equations*, Wiley-Interscience, New York, 1962.

provides the deepest insights into all the questions treated in this chapter.

The method of characteristics is, of course, treated in all textbooks, among which

G. F. D. Duff, *Partial Differential Equations*, University of Toronto Press, Toronto, 1956.

is notable for its clarity.

See also

N. Bleistein, *Mathematical Methods for Wave Phenomena*, Academic Press, Orlando, Florida, 1984.

9.4. The questions of existence that are lightly touched on here are dealt with more fully in Courant and Hilbert. The table of relationship among various formulations has been adapted from

D. L. Bernstein, *Existence Theorems in Partial Differential Equations* (Annals of Mathematics Studies No. 23), Princeton University Press, Princeton, N.J., 1950.

The book by Duff, referred to previously, gives a clear presentation of the basic existence theorem. The treatment in Sec. 9.4 draws on all these sources.

9.5. A comprehensive exposition of Pontryagin's principle and an excellent discussion of the precise conditions that must be imposed is given in

E. B. Lee and L. Markus, *Foundations of Optimal Control Theory*, Wiley, New York, 1967.

The problem on the optimal control of a batch reactor is taken from

R. Aris, *Optimal Design of Chemical Reactors*, Academic Press, New York, 1961.

[Unfortunately, the exposition there is marred by a number of minor misprints and an incorrect formula (27, p. 147); see Exercise 9.1.1.]

The simplifying feature of this problem whereby the adjoint variables can be eliminated is generalized in

M. M. Denn and R. Aris, "Generalized Euler Equations," *Z. Angew. Math. Phys.* **16**, 290–295 (1965).

Variational Problems

10

We have already encountered several problems in optimization and control that are of variational nature. Thus the very simplest problem was seen to lead to a partial differential equation in Sec. 1.14. The second example given there, a batch reactor with consecutive reactions, has been used repeatedly. In Sec. 3.6, a special case illustrates linear equations, and in Exercise 9.3.3 the characteristic differential equations of the nonlinear equation are given. The batch reactor problem is again the concrete example used in Sec. 9.5 to motivate the introduction of Pontryagin's maximum principle. Let us now turn to some problems in the classical theory of the calculus of variations to see the connections between this and the study of first-order equations. In particular, it will be shown in Sec. 10.3 how the equations of Pontryagin can be obtained from the formulation of the control problem as a problem in the calculus of variations. Conversely, we will see there how the principle of optimality leads to the Hamilton–Jacobi equations.

In Sec. 10.1, we derive the Euler equations for the basic variational problem, that of minimizing the integral $L[t, x^i(t), \dot{x}^i(t)]$ by the choice of n functions $\dot{x}^i(t)$. The complete variation of this integral is given so that the end and corner conditions can be discussed. These Euler equations are thrown into canonical form in Sec. 10.2 in preparation for discussing the Hamilton–Jacobi equation (Sec. 10.3). This is a first-order partial differential equation, and the conditions under which any given equation can be made the Hamilton–Jacobi equation of a variational problem are discussed in Sec. 10.4. Finally (Sec. 10.5) the principles of Fermat and Huygens are used to reinforce the connection between variational problems and partial differential equations.

In this chapter only we will use tensor notation and a summation convention. Thus x^i, $i = 1, 2, \ldots, n$, is a typical member of a set $\{x^i\}$ or element of a vector $\mathbf{x}$. A suffix repeated once in the upper and once in the lower position is held to be summed over its range. Thus $p_i x^i$ is an abbreviation for $p_1x^1 + p_2x^2 + \cdots + p_nx^n$. The x^i are usually functions of an independent variable t, and $\dot{x}^i$ denotes dx^i/dt.

10.1 Basic Problem of the Calculus of Variations

Let $x^i(t)$, $i = 1, 2, \ldots, n$, be a set of continuous functions with continuous derivatives defined in the interval $\alpha \leq t \leq \beta$. It thus represents a smooth curve in n-dimensional space with end points.

$$x^i(\alpha) = a^i, \qquad x^i(\beta) = b^i \tag{10.1.1}$$

The integral

$$J[\mathbf{x}] = \int_\alpha^\beta L[t, x^i, \dot{x}^i]\, dt \tag{10.1.2}$$

where L has continuous second derivatives in all arguments, is a functional of the x^i, for its value, which is a scalar, depends on the whole course of the function $x^i(t)$ in the interval $\alpha \leq t \leq \beta$. We wish to choose the functions $x^i(t)$ in such a way as to minimize the integral J. For simplicity, we will consider that the end points of the curve are to remain fixed, so $\mathbf{a}$ and $\mathbf{b}$ are specified.

Let $\xi^i(t)$ be a set of continuously differentiable functions with $\xi^i(\alpha) = \xi^i(\beta) = 0$ and consider

$$J[\mathbf{x} + \varepsilon\boldsymbol{\xi}] = \int_\alpha^\beta L[t, x^i + \varepsilon\xi^i, \dot{x}^i + \varepsilon\dot{\xi}^i]\, dt \tag{10.1.3}$$

For fixed $\boldsymbol{\xi}(t)$ and any given $\mathbf{x}(t)$, this is just a function of the parameter ε, continuous and differentiable to the same degree as L. Figure 10.1 shows this. If $\mathbf{x}(t)$ is the set of functions that minimizes J, then it is certainly necessary to have $dJ/d\varepsilon = 0$ when $\varepsilon = 0$. But

$$\delta J = \left(\frac{dJ}{d\varepsilon}\right)_{\varepsilon=0} = \int_\alpha^\beta \left[\frac{\partial L}{\partial x^i}\xi^i + \frac{\partial L}{d\dot{x}^i}\dot{\xi}_i\right] dt \tag{10.1.4}$$

and, since this is the first time we have used it, let us recall that a summation

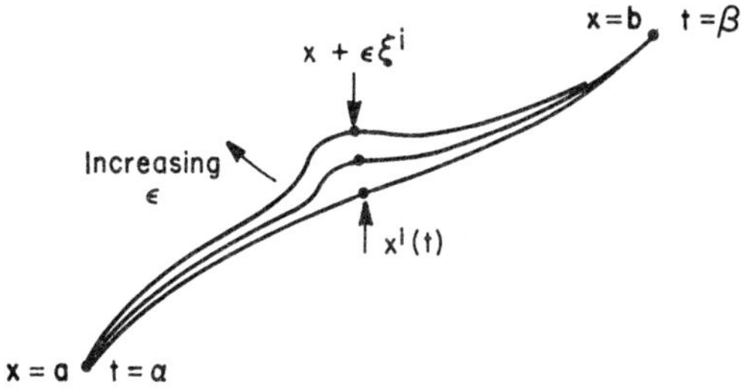

Figure 10.1

over i from 1 to n is implied in each of these terms. But by integrating the second set of terms by parts and using the fact that all the ξ^i vanish at both ends, this gives

$$\int_\alpha^\beta \left[\frac{\partial L}{\partial x^i} - \frac{d}{dt}\frac{\partial L}{\partial \dot{x}^i}\right]\xi^i\, dt = 0 \qquad (10.1.5)$$

If this is to be true for an arbitrary variation $\boldsymbol{\xi}(t)$, the quantity in the brackets must vanish everywhere. For if it did not vanish, it would have to be of one sign in some interval (α', β') since we know it to be continuous. We could then take a smooth $\boldsymbol{\xi}(t)$ that vanished outside the interval (α', β') but was of the same sign as the quantity in the brackets within (α', β'). In this case the integral would definitely be greater than zero. Thus the functions $x^i(t)$ that minimize J must necessarily satisfy the set of Euler equations

$$L_{x^i} - \frac{d}{dt} L_{\dot{x}^i} = 0, \qquad i = 1, \ldots, n \qquad (10.1.6)$$

These form a set of second-order nonlinear equations:

$$L_{x^i} - L_{\dot{x}^i t} - L_{\dot{x}^i x^j}\dot{x}^j - L_{\dot{x}^i \dot{x}^j}\ddot{x}^j = 0 \qquad (10.1.7)$$

Let us remove the restriction that the end points should be fixed and write

$$\xi^i(\alpha) = \delta a^i, \qquad \xi^i(\beta) = \delta b^i \qquad (10.1.8)$$

In this case the integrated parts from Eq. (10.1.4) will not vanish, and we have

$$\delta J = (L_{\dot{x}^i})_{t=\beta}\delta b^i - (L_{\dot{x}^i})_{t=\alpha}\,\delta a^i + \int_\alpha^\beta \left[L_{x^i} - \frac{d}{dt} L_{\dot{x}^i}\right]\xi^i\, dt \qquad (10.1.9)$$

Finally, we may suppose that the interval (α, β) is not fixed, but that we should write $\alpha + \varepsilon\delta\alpha$ and $\beta + \varepsilon\delta\beta$ for the limits of the integral. Then differentiating

$$\int_{\alpha+\varepsilon\delta\alpha}^{\beta+\varepsilon\delta\beta} L[t, x^i + \varepsilon\xi^i, \dot{x}^i + \varepsilon\dot{\xi}^i]\, dt$$

with respect to ε gives additional terms:

$$L[\beta, x^i(\beta), \dot{x}^i(\beta)]\delta\beta - L[\alpha, x^i(\alpha), \dot{x}^i(\alpha)]\delta\alpha \qquad (10.1.10)$$

Moreover, the total variation x^i at $t = \alpha$ is $\delta x^i(\alpha) = \xi^i(\alpha) + \dot{x}^i\delta\alpha$. We should therefore write $\delta a^i = \xi^i(\alpha) = (\delta x^i - \dot{x}^i\delta t)_{t=\alpha}$ and, similarly, $\delta b^i = (\delta x^i - \dot{x}^i\delta t)_{t=\beta}$. Thus, with the obvious notation $\delta\alpha = (\delta t)_{t=\alpha}$ and $\delta\beta = (\delta t)_{t=\beta}$, we can include the terms (10.1.10) in Eq. (10.1.9) and rearrange to give

$$\delta J = \int_\alpha^\beta \left[L_{x^i} - \frac{d}{dt} L_{\dot{x}^i} \right] \xi^i \, dt + [L_{\dot{x}^i} \, \delta x^i]_\alpha^\beta + [(L - \dot{x}^i L_{\dot{x}^i})\delta t]_\alpha^\beta \qquad (10.1.11)$$

This first variation must vanish if $x^i(t)$ is to minimize the functional.

The complete form of the first variation given in Eq. (10.1.11) shows what conditions have to be applied if the end points are not fixed. For example, if the end $x^i(\beta)$ were constrained to lie on the curve $x^i = \phi^i(t)$, as in Fig. 10.2(a), we would have $\delta x^i = \dot{\phi}^i\delta t$ at $t = \beta$, and so $(\dot{\phi}^i - \dot{x}^i) L_{\dot{x}^i} + L = 0$ there. If, as in Fig. 10.2(b), the end point has to lie on a surface in $(n + 1)$-dimensional space,

$$\Phi(x^1, \ldots, x^n, t) = \text{constant}$$

then $\Phi_t \, \delta t + \Phi_{x^i}\delta x^i = 0$, so the last two terms could be combined to give

$$L_{\dot{x}^i} \, \delta x^i - \frac{1}{\Phi_t} \Phi_{x^i} \, \delta x^i (L - \dot{x}^j L_{\dot{x}^j})$$

But this variation would have to vanish for an arbitrary displacement δx^i, and so

$$\Phi_t L_{x^i} - (L - \dot{x}^j L_{\dot{x}^j})\Phi_{x^i} = 0 \qquad (10.1.12)$$

These are known as *transversality conditions*.

The complete form of the variation also shows what conditions have to apply at a corner of an extremal curve. Suppose that the $x^i(t)$ are continuous in $\alpha \leq t \leq \beta$ and have continuous derivatives except at the point $t = \gamma$. This represents a sharp corner in an otherwise smooth curve. Within the intervals $\alpha \leq t \leq \gamma$ and $\gamma \leq t \leq \beta$, the Euler equations must be satisfied, but the position of the corner must be such that variations $\delta x^i(\gamma)$ and $\delta\gamma$ leave $\delta J = 0$. But the corner is the upper end of one interval and the lower end of the other, so the quantities

$$L_{\dot{x}^i} \quad \text{and} \quad L - \dot{x}^i L_{\dot{x}^i} \qquad (10.1.13)$$

must be continuous at $t = \gamma$. The continuity of these $n + 1$ quantities is known as the *Weierstrass–Erdmann condition*.

A compact form of the first variation that anticipates some of the development we are about to embark on can be obtained by putting

$$p_i = L_{\dot{x}^i} \qquad (10.1.14)$$

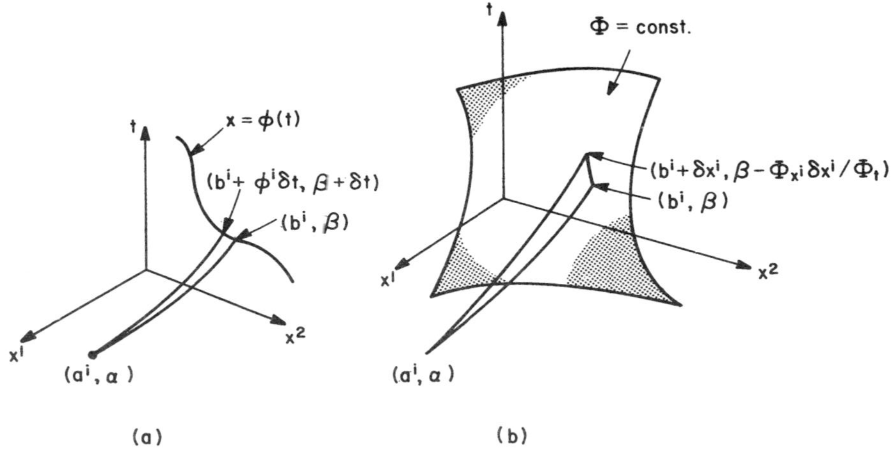
t
x = φ(t)
(bⁱ+ φⁱδt, β + δt)
(bⁱ, β)
x¹
x²
(aⁱ, α)
(a)
t
Φ = const.
(bⁱ+δxⁱ, β − Φ_{xⁱ}δxⁱ/Φ_t)
(bⁱ, β)
x¹
x²
(aⁱ, α)
(b)

Figure 10.2

Then if we suppose that the Jacobian

$$\frac{\partial(p_1, \ldots, p_n)}{\partial(\dot{x}^1, \ldots, \dot{x}^n)} = | L_{\dot{x}^i\dot{x}^j} | \neq 0 \tag{10.1.15}$$

we can solve Eqs. (10.1.14) for the $\dot{x}^i$ as functions of t, $\mathbf{x}$, and $\mathbf{p}$. Now let

$$H(t, x^i, p_i) = -L(t, x^i \dot{x}^i) + \dot{x}^j p_j \tag{10.1.16}$$

where on the right side we assume that the $\dot{x}^i$ have been expressed as functions of t, x^i, and p_i. Then we see that Eq. (10.1.11) can be written

$$\delta J = \int_\alpha^\beta [L_{x^i} - \dot{p}_i]\xi^i(t)\, dt + [p_i\, \delta x^i - H\, \delta t]_\alpha^\beta$$

Exercises

10.1.1. Show that the shortest distance between two nonintersecting, closed, smooth surfaces lies along a common normal. What will be the situation if the surfaces are not closed or only piecewise smooth?

10.1.2. By Fermat's principle, light travels on a path that takes the shortest time between two points. If its speed in the medium is constant, show that this is a straight line. If the end points are in two different media in which the speeds are V_1 and V_2, show that Snell's law of refraction is obeyed at the interfacial surface.

10.2 Canonical Form of the Euler Equations

The function H, known as the *Hamiltonian*, was introduced at the end of the last paragraph as a convenient way of writing the variation compactly, but it has a deeper significance than this. The Euler equations themselves, as Eq. (10.1.7) shows, form a set of n second-order equations. These can be reduced to a set of $2n$ first-order equations in various ways. The obvious way would be to write the $\dot{x}^i$ as new variables, but, although this would reduce the equations, it is far less illuminating than the *canonical reduction* suggested by Eq. (10.1.14). Suppose then that we set

$$p_i = L_{\dot{x}^i} \tag{10.2.1}$$

and use these equations to define $\dot{x}^i$ in terms of t, x^i, and p_i. What will the Euler equations look like in these variables?

From the definition,

$$H = -L + \dot{x}^i p_i \tag{10.2.2}$$

we have

$$dH = -dL + \dot{x}^i\, dp_i + p_i\, d\dot{x}^i \tag{10.2.3}$$

Also, $L = L(t, x^i, \dot{x}^i)$, so

$$dL = L_t\, dt + L_{x^i} dx^i + L_{\dot{x}^i} d\dot{x}^i$$

and when we substitute this in Eq. (10.2.3), the terms in $d\dot{x}^i$ cancel by virtue of the definition (10.2.1). Thus

$$dH = -L_t\, dt - L_{x^i}\, dx^i + \dot{x}^i dp_i \tag{10.2.4}$$

and if H is regarded as a function of t, x^i, and p_i, we see that

$$\frac{\partial H}{\partial t} = -L_t, \qquad \frac{\partial H}{\partial x^i} = -L_{x^i}, \qquad \frac{\partial H}{\partial p_i} = \dot{x}^i \tag{10.2.5}$$

Since the Euler equations are

$$\frac{d}{dt} L_{\dot{x}^i} = L_{x^i}$$

we have, from the definition of p_i and the evaluation of L_{x^i} in Eq. (10.2.5), that

$$\dot{x}^i = \frac{\partial H}{\partial p_i}, \qquad \dot{p}_i = -\frac{\partial H}{\partial x^i} \tag{10.2.6}$$

and this is the canonical form.

Another way of arriving at the canonical equation makes use of the Legendre transformation, which was introduced in Sec. 7.3. Let us compare the two variational problems:

1. Minimize $J_1[\mathbf{x}] = \int_\alpha^\beta L(t, x^i, \dot{x}^i)\, dt$ (10.2.7)

   by the choice of n functions $x^i(t)$, $\alpha \le t \le \beta$.

2. Minimize $J_2\,[\mathbf{x}, \mathbf{p}] = \int_\alpha^\beta \{-H(t, x^i, p_i) + p_j\dot{x}^j\}\, dt$ (10.2.8)

   by the choice of $2n$ functions $x^i(t)$, $p_i(t)$, $\alpha \le t \le \beta$.

The second problem has $2n$ unknown functions, but the derivatives of only n of them appear and these in a particularly simple fashion. Thus the Euler equations of the second problem are

$$\frac{d}{dt}\frac{\partial}{\partial \dot{x}^i}[-H(t, x^i, p_i) + p_j\dot{x}^j] - \frac{\partial}{\partial x^i}[-H(t, x^i, p_i) + p_j\dot{x}^j] = 0$$

and

$$\frac{d}{dt}\frac{\partial}{\partial \dot{p}_i}[-H(t, x^i, p_i) + p_j\dot{x}^j] - \frac{\partial}{\partial p_i}[-H(t, x^i, p_i) + p_j\dot{x}^j] = 0$$

These of course simplify vastly to

$$\dot{p}_i = -H_{x^i}, \qquad \dot{x}^i = H_{p_i}$$

which are the canonical equations (10.2.6). Now the integrands of the two functionals are the same, by virtue of the definition of the Hamiltonian, so that all we have to show is that the extreme of Eqs. (10.2.7) and (10.2.8) occur for the same set of curves.

To this end, we first notice that the transformation from t, x^i, $\dot{x}^i$, L to t, x^i, p_i, H is an involution. For we have the comparisons

$$\begin{aligned} p_i &= L_{\dot{x}^i}, & \dot{x}^i &= H_{p_i} \\ H &= -L + p_j\dot{x}^j, & L &= -H + \dot{x}^j p_j \end{aligned} \tag{10.2.9}$$

or

$$H = -L + L_{\dot{x}^j}\dot{x}j, \qquad L = -H + H_{p_j}p_j$$

Then we observe that we could find the minimum of the second integral $J_2\,[\mathbf{x}, \mathbf{p}]$ in two stages first by finding the functions $p_i(t)$ that minimize J_2 for any given $\mathbf{x}$. These minimizing functions will of course depend on $\mathbf{x}$, say $p_i^0(\mathbf{x})$, and when they are substituted into J_2 to give $J_2[\mathbf{x}, \mathbf{p}^0(\mathbf{x})]$, we can then proceed to minimize with respect to the $\mathbf{x}$. But this means that, if the problems are equivalent,

$$\min_{\mathbf{p}} J_2[\mathbf{x}, \mathbf{p}] = J_2[\mathbf{x}, \mathbf{p}^0(\mathbf{x})] = J_1[\mathbf{x}]$$

Now to minimize J_2 with respect to $\mathbf{p}$ for fixed $\mathbf{x}$ gives the Euler equations

$$\frac{\partial}{\partial p_i}(-H + p_j\dot{x}^j) = 0$$

or

$$\dot{x}^i = H_{p_i}$$

Substituting back into the integrand of J_2 and using the last equation in Eq. (10.2.9) gives

$$J_2[\mathbf{x},\mathbf{p}^0(\mathbf{x})] = \int_\alpha^\beta (-H + p_j H_{p_j})\,dt = \int_\alpha^\beta L\,dt = J_1\,[\mathbf{x}]$$

Exercises

10.2.1. Show that $G(x^1, \ldots, x^n, p_1 \ldots, p_n)$ is constant along trajectories of the

Euler equations if

$$[G, H] = \sum_i \frac{\partial(G, H)}{\partial(x^i, p_i)} = 0$$

G is known as a first integral and $[G, H]$ as the Poisson bracket. Deduce that H is a first integral provided it is not explicitly a function of t.

10.2.2. The equation for diffusion and first-order reaction in a sphere is

$$\frac{1}{r^2}\frac{d}{dr}\left(r^2\frac{dc}{dr}\right) = \phi^2 c; \qquad c(R) = c_0; \qquad \frac{dc}{dr} = 0 \qquad \text{at } r = 0$$

where r is the radial distance $0 \leq r \leq R$ and ϕ is the Thiele modulus. Show that $c(r)$ minimizes the functional

$$J = \int_0^R r^2\left[\left(\frac{dc}{dr}\right)^2 + \phi^2c^2\right] dr$$

and find the canonical form of the equation.

10.2.3. The n particles occupying positions $x^i(t)$, $y^i(t)$, $z^i(t)$ at time t have a total kinetic energy $T = \frac{1}{2}\Sigma_i m_i\,[(\dot{x}^i)^2 + (\dot{y}^i)^2 + (\dot{z}^i)^2]$. If their potential energy is a function $U(x^1, \ldots, z^n)$ of the $3n$ position variables, the "action" of the system in the interval (α, β) is the integral

$$\int_\alpha^\beta (T - U)\, dt$$

Show that minimizing this action leads to the equations of particle motion and obtain the canonical forms. Show also that the total energy $T + U$ is conserved during the motion.

10.3 Hamilton–Jacobi Equation

Consider the problem of minimizing

$$J[\mathbf{x}] = \int_\alpha^\beta L[t, x^i, \dot{x}^i]\, dt \tag{10.3.1}$$

subject to the constraints

$$x^i(\alpha) = a^i, \qquad x^i(\beta) = b^i \tag{10.3.2}$$

where **a** and **b** are given points. The solution of this problem is a curve $x^i(t)$ called the *geodesic* between the points **a** and **b**, the value of J is the *geodesic distance*. Let **a** be fixed, regard **b** as a variable, and denote the geodesic distance as a function

$$S(\mathbf{b}) = S(b^1, \ldots, b^n, \beta) \tag{10.3.3}$$

It is clearly a function only of the position of the end point, for the path is the geodesic from this end point to the fixed **a**. Now the differential

$$dS = S(b^1 + db^1, \ldots, b^n + db^n, \beta + d\beta) - S(b^1, \ldots, b^n, \beta)$$

can be evaluated by Eq. (10.1.17), for since the integral vanishes, δx^i and δt are zero at the fixed end, and

$$\begin{aligned} \delta x^i = db^i, \qquad \delta t &= d\beta \quad \text{at} \quad t = \beta \\ dS = p_i db^i &- H\, d\beta \end{aligned} \tag{10.3.4}$$

whence

$$\frac{\partial S}{\partial b^i} \equiv S_{b^i} = p_i, \qquad \frac{\partial S}{\partial \beta} \equiv S_\beta = -H$$

But $H = H(t, x^i, p_i)$ at $t = \beta$, so the last equation can be written

$$S_\beta + H(\beta, b^i, S_{b^i}) = 0 \tag{10.3.5}$$

This is a first-order partial differential equation known as the Hamilton–Jacobi equation.

The characteristic differential equations of the Hamilton–Jacobi equation are

$$\frac{d\beta}{ds} = 1, \qquad \frac{db^i}{ds} = \frac{\partial H}{\partial S_{b^i}}, \qquad \frac{dS}{ds} = S_\beta + S_{b^i}\frac{\partial H}{\partial S_{b^i}}$$

$$\frac{dS_\beta}{ds} = -\frac{\partial H}{\partial \beta}, \qquad \frac{dS_{b^i}}{ds} = -\frac{\partial H}{\partial b_i}$$

Changing β to t, b^i to $x^i(t)$, S_{b^i} to $p_i(t)$ and dividing all these equations by the first, we have the canonical equations

$$\begin{aligned} \dot{x}^i = H_{p^i}, \qquad \dot{p}_i &= -H_{x^i} \\ \dot{S} = S_t + p_i H_{p^i}, \qquad \dot{S}_i &= -H_t \end{aligned} \tag{10.3.6}$$

as the characteristic differential equations of

$$\frac{\partial S}{\partial t} + H\left(t, x^i, \frac{\partial S}{\partial x^i}\right) = 0 \tag{10.3.7}$$

An entirely similar equation could be obtained by keeping the end **b** fixed and regarding **a** as the variable point.

The family of hypersurfaces $S(t, x^i) =$ constant in $(n + 1)$-dimensional space, together with the family of trajectories $\dot{x}^i = H_{p^i}$ was called "the complete solution of the variational problem" by Caratheodory. It establishes the intimate connection that exists between the calculus of variations and the theory of first-order partial differential equations.

If $L[t, x^i, \dot{x}^i]$ is independent of t and homogeneous of the first degree in $\dot{x}^i$, then H as defined by Eqs. (10.2.1) and (10.2.2) vanishes identically. In this case, the fundamental integral is independent of the choice of t, for if $t = \phi(\tau)$ and $\mathbf{x}' = d\mathbf{x}/d\tau$,

$$\int L[x^i, \dot{x}^i]\, dt = \int L\left[x^i, \frac{x'^i}{\phi'(\tau)}\right] \phi'(\tau)\, d\tau$$

$$= \int L[x^i, x'^i]\, d\tau$$

The relations $S_{b^i} = L_{\dot{x}^i}$ and $S_\beta + H = 0$ still hold, however. In fact, the last is consonant with the independence of the time scale, for it says that S is not a function of the final time. Since the $L_{\dot{x}^i}$ are homogeneous and of degree zero in the $\dot{x}^i$, we can determine the ratios of the $\dot{x}^i$ as functions of the S_{b^i} and substitute them in the homogeneity relation. This may be written in the form

$$\frac{\dot{x}^i}{L} L_{\dot{x}^i} = 1$$

where it is evidently homogeneous of degree zero and so a function only of these ratios.

Let us consider the control problem of minimizing

$$\int_\alpha^\beta L(t, x^i(t), u^k(t))\, dt \tag{10.3.8}$$

where the $x^i(t)$ are governed by differential equations

$$\dot{x}^i = f^i(\mathbf{x}, \mathbf{u}) \tag{10.3.9}$$

With some obvious changes of notation, this is basically the same problem that we enunciated in Sec. 9.5. In this problem, the u^k are the unknown functions and the x^i are constrained to obey Eq. (10.3.9). We use the standard way of changing a constrained one into an unconstrained optimization problem, by introducing Lagrange multipliers. Anticipating later developments, we will denote these by p_i and minimize

$$\int_\alpha^\beta \{L(t, x^i, u^k) - p_j[\dot{x}^j - f^j(x^i, u^k)]\}\, dt$$

$$= \int_\alpha^\beta [H^*(t, x^i, p_i, u^k) - p_j \dot{x}^j]\, dt = \int_\alpha^\beta L^*\, dt \tag{10.3.10}$$

where

$$H^* = L + p_j f^j \tag{10.3.11}$$

Regarding x^i, p_i, and u^k as independent variables, we now have the Euler equations

$$\frac{d}{dt}\frac{\partial L^*}{\partial \dot{x}^i} - \frac{\partial L^*}{\partial x^i} = 0, \qquad \frac{d}{dt}\frac{\partial L^*}{\partial \dot{p}^i} - \frac{\partial L^*}{\partial p^i} = 0, \qquad \frac{d}{dt}\frac{\partial L^*}{\partial \dot{u}^k} - \frac{\partial L^*}{\partial u^k} = 0$$

But these give, respectively,

$$\dot{p}_i = -\frac{\partial H^*}{\partial x^i} = -\frac{\partial L}{\partial x^i} - p_j \frac{\partial f^j}{\partial x^i} \tag{10.3.12}$$

$$\dot{x}^i = f^i(\mathbf{x}, \mathbf{u}) \tag{10.3.13}$$

and (10.3.14)

$$\frac{\partial H^*}{\partial u^k} = 0$$

The last set is the set of necessary conditions for the Hamiltonian H^* to be minimum and so governs the choice of the optimal control. The second simply recovers the governing equations (10.3.9), and the first is the set of adjoint equations for the adjoint variables p_i. Moreover, we see that these variables can be interpreted in two ways. First, they are Lagrange multipliers. Second, if we denote the minimum value of the functional (10.3.8) by $g(\xi^1, \ldots, \xi^{n+1})$, where $\xi^i = x^i(\alpha)$, $\xi^{n+1} = \beta - \alpha$, the Hamilton–Jacobi equation becomes

$$H(\xi^1, \ldots, \xi^{n+1}, g_{\xi^1}, \ldots, g_{\xi^n}) - g_{\xi^{n+1}} = 0$$

or

$$\min_{\mathbf{u}}[L(\xi^1, \ldots, \xi^{n+1}, u^k) + g_{\xi^j} f^j(\xi^i, u^k)] - g_{\xi^{n+1}} = 0 \tag{10.3.15}$$

The adjoint variables are thus the partial derivatives of the minimum loss with respect to the initial state.

Now the method of dynamic programming goes straight to the Hamilton–Jacobi equations by the application of the principle of optimality. For if we write

$$g(\xi^1, \ldots, \xi^{n+1}) = \min \int_\alpha^\beta L(t, x^i, u^k)\, dt$$

we can divide the interval up into two parts and write

$$g(\xi^1, \ldots, \xi^{n+1}) = \min \left[\int_\alpha^{\alpha+\delta\alpha} L\, dt + \int_{\alpha+\delta\alpha}^\beta L\, dt \right]$$

The principle of optimality asserts that the choice of $u^k(t)$ in the second interval, $\alpha + \delta\alpha \le t \le \beta$, must be optimal with respect to the state at time $\alpha + \delta\alpha$, for otherwise the whole integral could not be minimized. At this time, the state is $\xi^j + f^j(\xi^i, u^k)\delta\alpha$, and the remaining duration of the process is $\xi^{n+1} - \delta\alpha$, so

$$g(\xi^1, \ldots, \xi^{n+1}) = \min \left[\int_\alpha^{\alpha+\delta\alpha} L\, dt + g(\xi^1 + f^1\delta\alpha, \ldots, \xi^{n+1} - \delta\alpha) \right]$$

If $\delta\alpha$ is small, we can approximate the integral by $L(\xi^1, \ldots, \xi^{n+1}, u^k)\delta\alpha$ and expand the second term in a Taylor series, giving

$$g(\boldsymbol{\xi}) = \min_{\mathbf{u}} [L(\boldsymbol{\xi}, \mathbf{u})\delta\alpha + g(\boldsymbol{\xi}) + g_{\xi^j} f^j \,\delta\alpha - g_{\xi^{n+1}}\,\delta\alpha + 0(\delta\alpha^2)]$$

Since the minimization only affects the first and third terms in the brackets, we can take the others outside. Then $g(\boldsymbol{\xi})$ cancels and, if we divide by $\delta\alpha$ and take the limit as $\delta\alpha \to 0$, we have the partial differential equation (10.3.15).

Let us summarize the relation between the optimization problems and what we have done in this chapter as follows. Apart from the specific optimization problem, we merely started the Pontryagin formulation in Sec. 9.5. It can be proved with varying degrees of rigor and at various levels of complication by the calculus of variations, dynamic programming, or the use of Green's functions (see the references to this section). In this section we have shown how the ordinary differential equations of Pontryagin can arise as the characteristic differential equations of the Hamilton–Jacobi equation and second how the Hamilton–Jacobi equation can be derived by the methods of dynamic programming.

Exercise

10.3.1. If $L^2 = g_{ij}(\mathbf{x})\dot{x}^i\dot{x}^j$, show that the Hamilton–Jacobi equation is $g^{ij}(\mathbf{x})S_{x^i}S_{x^j} = 1$, where g_{ij} and g^{ij} are elements of a symmetric positive matrix and its inverse.

10.4 Equivalence of First-order Partial Differential Equations and Variational Problems

Having shown that a variational problem is equivalent to a first-order partial differential equation, we may now turn the question around and ask when a given partial differential equation may be regarded as the Hamilton–Jacobi equation of a variational problem. Let us retain the distinction between the variables x^i, $i = 1, \ldots, n$ and t and consider a general first-order equation for $S(x^i, t)$.

$$F(x^i, t, S, S_{x^i}, S_t) = 0 \tag{10.4.1}$$

Comparing this with Eq. (10.3.7),

$$H(x^i, t, S_{x^i}) + S_t = 0 \tag{10.4.2}$$

we see immediately that F should not depend explicitly on S. Moreover, we should be able to solve $F = 0$ for S_t if the equation is to be equivalent to Eq. (10.4.2). By the implicit function theorem, the condition for this is that

$$F_{S_t} \neq 0 \tag{10.4.3}$$

Now it it will be recalled that the Hamiltonian H was given by $-L + p_i\dot{x}^i$, that L was assumed to be twice differentiable, and that in order to change the transformation to canoncial form the Hessian determinant

$$\left|\frac{\partial^2 H}{\partial S_{x^i}\,\partial S_{x^j}}\right| \neq 0 \tag{10.4.4}$$

The twice differentiability of the H obtained by solving for S_t in Eq. (10.4.1) is no problem since this is a consequence of the implicit function theorem, but, unless H can be found explicitly, it is difficult to test the condition (10.4.4).

However, by considering the characteristics, we can show that Eq. (10.4.1) is equivalent to the Hamilton–Jacobi equation (10.4.2) provided that

$$F_{p_{n+1}} \neq 0, \qquad \left|\frac{\partial^2 F}{\partial p_r\,\partial p_s}\right| \neq 0 \tag{10.4.5}$$

where p_i denotes S_{x^i}, p_{n+1} denotes S_t, and the Hessian determinant is an $(n + 1) \times (n + 1)$ determinant involving all the p_i's. (We can keep track of this by letting i, j, k have the range $1, 2, \ldots, n$ and r, s the range $1, 2, \ldots, n + 1$.) Since we must certainly assume that F does not depend explicitly on S, the characteristic differential equations of

$$F(x^i, t, S_{x^i}, S_t) \equiv F(x^r, p_r) = 0 \tag{10.4.6}$$

are

$$\frac{dx^r}{ds} = \frac{\partial F}{\partial p_r}, \qquad \frac{dp_r}{ds} = -\frac{\partial F}{\partial x^r} \tag{10.4.7}$$

If we can find a Lagrangian L such that the Euler equations of the variational problem are the same as those of the characteristics, then their solution will define a complete figure equivalent to that of the variational problem with L.

The parameter s along the characteristics is arbitrary, and since $F_{p_{n+1}} \neq 0$, we can also use $x^{n+1} = t$ as a parameter along the curves. Denoting d/ds by a prime and d/dt by an overdot, we have the forms

$$x'^r = F_{p_r}, \qquad p'_r = -F_{x^r}$$

and

$$\dot{x}^r = \frac{F_{p_r}}{F_{p_{n+1}}} = \frac{F_{p_r}}{x'^{n+1}}, \qquad \dot{p}^r = -\frac{F_{x^r}}{F_{p_{n+1}}} = -\frac{F_{x^r}}{x'^{n+1}}, \tag{10.4.8}$$

where naturally $\dot{x}^{n+1} = 1$. Now if the determinant $|F_{p_r p_s}|$ does not vanish, we can express the p_r as functions of x^r and x'^r, say

$$p_r = \phi_r(\mathbf{x}, \mathbf{x}') \tag{10.4.9}$$

Finally, let us construct the Lagrangian

$$L(x^r, x'^r) = \frac{[-F(x_r, \phi_r(\mathbf{x}, \mathbf{x}')) + x'^s\phi_s(\mathbf{x}, \mathbf{x}')]}{x'^{n+1}} \tag{10.4.10}$$

and see what its Euler equations are.

First, we see that

$$\frac{\partial L}{\partial x^r} = \left[-\frac{\partial F}{\partial x^r} + \frac{\partial \phi_s}{\partial x'^r}\left(x'^s - \frac{\partial F}{\partial p_s}\right)\right]\frac{1}{x'^{n+1}} = \frac{-1}{x'^{n+1}}\frac{\partial F}{\partial x^r} \tag{10.4.11}$$

Also, we have

$$\frac{\partial L}{\partial x'^i} = \left(-\frac{\partial F}{\partial p_s} \frac{\partial \phi_s}{\partial x'^i} + \phi_i + x'^s \frac{\partial \phi_s}{\partial x'^i}\right)\frac{1}{x'^{n+1}} = \frac{p_i}{x'^{n+1}}$$

But since $x'^i = \dot{x}^i x'^{n+1}$ and $\partial x'^i/\partial \dot{x}^j = \delta^i_j x'^{n+1}$

$$\frac{\partial L}{\partial \dot{x}^j} = \frac{\partial L}{\partial x'^i}\frac{\partial x'^i}{\partial \dot{x}^j} = \frac{p_i}{x'^{n+1}}\delta^i_j x'^{n+1} = p_j$$

Thus, by Eqs. (10.4.8) and (10.4.11),

$$\frac{d}{dt}\left(\frac{\partial L}{\partial \dot{x}^j}\right) = \frac{dp_j}{dt} = \frac{p'_j}{x'^{n+1}} = \frac{-1}{x'^{n+1}}\frac{\partial F}{\partial x^j} = \frac{\partial L}{\partial x^j}$$

which is the Euler equation for the variational problem with the integrand L.

10.5 Principles of Fermat and Huygens

In Sec. 10.3 we regarded

$$J[\mathbf{x}] = \int_\alpha^\beta L[t, x^i, \dot{x}^i]\, dt \tag{10.5.1}$$

as a function of its end point $x^i = b^i$, $t = \beta$, with the other end point $x^i = a^i$, $t = \alpha$ fixed. Let us now regard it as a function of both end points; that is, of $2n + 2$ variables,

$$S(a^i, \alpha; b^i, \beta) = \min \int_\alpha^\beta L[t, x^i, \dot{x}^i]\, dt \tag{10.5.2}$$

where the path taken is a geodesic. Since by Eq. (10.1.17)

$$dS = p_i(\beta)\, db^i - p_i(\alpha)\, da^i - H(\beta)\, d\beta + H(\alpha)\, d\alpha$$

we have at $t = \alpha$,

$$S_{a^i} = -p_i(\alpha), \qquad S_\alpha = H(\alpha, a^i, p_i(\alpha))$$

and at $t = \beta$,

$$S_{b^i} = p_i(\beta), \qquad S_\beta = -H(\beta, b^i, p_i(\beta)) \tag{10.5.3}$$

Thus the Hamilton–Jacobi equation (10.3.7) is satisfied with either $x^i = a^i$, $t = \alpha$ or $x^i = b^i$, $t = \beta$.

If we fix (a^i, α) and consider the set of points (x^i, t) for which

$$K_R(a^i, \alpha) = \{S(a^i, \alpha; x^i, t) = R\} \tag{10.5.4}$$

we have a set of points geodesically equidistant from the fixed point (a^i, α). This might be called the geodesic sphere of radius R. Figure 10.3 illustrates this schematically. The set of all geodesic spheres of radius R forms a complete integral of the Hamilton–Jacobi equations, since it is a family of solutions depending on $(n + 1)$ parameters.

Consider now a solution of the Hamilton–Jacobi equation for which $S = 0$ on the surface $\Sigma\ (x^i, t) = 0$. Such a surface can be parameterized by a set of n parameters $u^1, \ldots, u^n$ and its equation written as

$$x^i = x^i(u^1, \ldots, u^n), \qquad t = t(u^1, \ldots, u^n) \tag{10.5.5}$$

The surface on which $S(x^i, t) = R$ might be called parallel to Σ and distant R from it, since the geodesic distance between $S = R$ and Σ is always equal to R. It can be found by solving the characteristic equations of the Hamilton–Jacobi equation starting from any given point on Σ and integrating them until $S = R$. Thus the point corresponding to $(u^1, \ldots, u^n)$ will be given by integrating

$$\frac{dx^i}{ds} = \frac{\partial H}{\partial p_i}, \qquad \frac{dt}{ds} = 1, \qquad \frac{dS}{ds} = -H + p_i \frac{\partial H}{\partial p_i}, \qquad \frac{dp_i}{\partial s} = \frac{\partial H}{\partial x^i} \tag{10.5.6}$$

subject to $x^i = x^i(u^k)$, $t = t(u^k)$, $S = 0$, and $p_i = \Sigma_{x^i}$ when $s = 0$, and continuing the integration until $S = R$. This is shown in Fig. 10.4.

But if we consider the family of spheres, $K_R(x^i, t)$, with centers on the surface $\Sigma = 0$ and radius R, we see that it forms an n-parameter subfamily of the complete integral. Now we know that the envelope of a subfamily of the complete integral gives a particular solution. In this case there is a correspondence between each point on the envelope and the center of the tangent geodesic sphere, a point on Σ. But this means that the envelope is

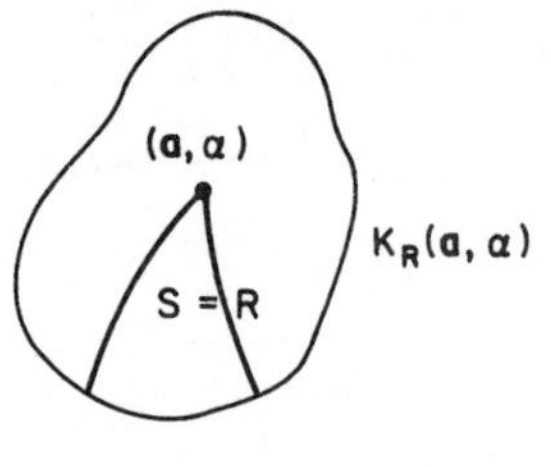

Figure 10.3

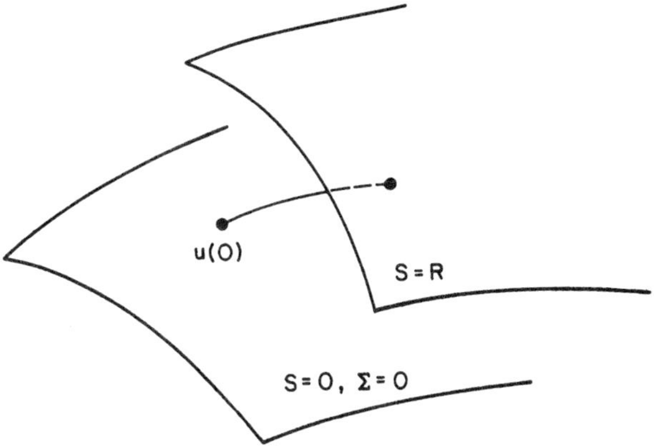

Figure 10.4

at a constant geodesic distance R from Σ and so is the geodesically parallel surface.

These two ways of constructing the surface $S = R$ from the initial surface correspond to tracing the rays and taking the envelope of the wave fronts, respectively. Huygen's principle, that at any instant each point on the surface of a propagating disturbance can be regarded as the center of a new disturbance, is thus seen to be an aspect of the structure of the solution of first-order partial differential equations.

Let us consider the propagation of light in a three-dimensional isotropic medium. If the path of the light ray is $x^1(t)$, $x^2(t)$, $x^3(t)$ and its velocity at any point is $c/n(x^1, x^2, x^3)$, where c is the velocity of light in vacuo and n the refractive index, the time of travel between two points is

$$T = c^{-1} \int_\alpha^\beta n(x^1, x^2, x^3) \, [(\dot{x}^1)^2 + (\dot{x}^2)^2 + (\dot{x}^3)^2]^{1/2} \, dt \tag{10.5.7}$$

Fermat's principle asserts that the path of a light ray is such as to minimize the time between any two points with respect to local variations of the path. Hence a light ray should follow a path minimizing

$$cT = \int_\alpha^\beta L[\mathbf{x}, \dot{\mathbf{x}}] \, dt$$

where

$$L[\mathbf{x}, \dot{\mathbf{x}}] = n(\mathbf{x})(\dot{\mathbf{x}} \cdot \dot{\mathbf{x}})^{1/2} \tag{10.5.8}$$

Instead of merely following the method mentioned in Sec. 10.3, we will use this homogeneity to write

$$cT = \int_{a^3}^{b^3} L' \, dx^3 = \int_{a^3}^{b^3} n(x^1, x^2, x^3) \, [(x'^1)^2 + (x'^2)^2 + 1]^{1/2} \, dx^3 \tag{10.5.9}$$

where $x'^j = dx^j/dx^3$. Then with x^3 playing the role of t, we have

$$p'_1 = L'_{x'^1} = \frac{nx'^1}{L'}, \qquad p'_2 = L'_{x'^2} = \frac{nx'^2}{L'}$$

and

$$\begin{aligned} H' &= L' + x'^1 L_{x'^1} + x'^2 L_{x'^2} \\ &= -n(x^1, x^2, x^3)[(x'^1)^2 + (x'^2)^2 + 1]^{1/2} \\ &= -(n^2 - p_1'^2 - p_2'^2)^{1/2} \end{aligned}$$

Thus the Hamilton–Jacobi equation is

$$\begin{aligned} S_{x^3} - (n^2 - S_{x^1}^2 - S_{x^2}^2)^{1/2} &= 0 \\ S_{x^1}^2 + S_{x^2}^2 + S_{x^3}^2 &= n^2 \end{aligned} \tag{10.5.10}$$

This is the equation that we have met before for the geometric limit of wave optics in Sec. 1.13. It also agrees with the result of Exercise 10.3.1.

The characteristic differential equations of Eq. (10.5.10) are

$$\frac{dx^i}{ds} = 2S_{x^i}, \qquad \frac{dS}{ds} = 2n^2, \qquad \frac{dS_{x^i}}{ds} = 2n\frac{\partial n}{\partial x^i} \tag{10.5.11}$$

If the medium is homogeneous, the S_{x^i} are constant, and thus the light paths are straight lines. We have already obtained Snell's law from Fermat's principle (Exercise 10.1.2) and from Huygens principle (Exercise 2.4.1). The wave fronts are the surfaces $S(x^1, x^2, x^3) = \text{constant}$, and the first of these characteristic differential equations shows that the light rays are always normal to the wave fronts.

Exercises

10.5.1. Show that the geodesic sphere in $\mathbf{R}_3$, when S is the shortest Cartesian distance between two points, is the usual sphere. Generalize to $\mathbf{R}_n$.

10.5.2. Show that the surface that is geodesically parallel to and distant from the ellipse $(\xi/a^2) + (\eta/b)^2 + (\zeta/c)^2 = 1$ is given parametrically by

$$x = \xi\left[1 + \frac{R}{a^2\sigma}\right]$$

$$y = \eta\left[1 + \frac{R}{b^2\sigma}\right]$$

$$z = \zeta\left[1 + \frac{R}{c^2\sigma}\right]$$

where (ξ, η, ζ) is a point on the ellipse and

$$\sigma^2 = \frac{\xi^2}{a^4} + \frac{\eta^2}{b^4} + \frac{\zeta^2}{c^4}$$

10.5.3. A medium has refractive index $n(x^1)$. Show that the wave front from a spark at (a^1, a^2, a^3) will in time T reach a surface given parametrically by

$$\int_0^{\xi} [n^2(a^1 + \xi) - \alpha^2]^{1/2}\, d\xi + \alpha\rho = cT$$

$$\alpha \int_0^{\xi} [n^2(a^1 + \xi) - \alpha^2]^{-1/2}\, d\xi = \rho$$

where $\quad \rho^2 = (x^2 - a^2)^2 + (x^3 - a^3)^2, \qquad \xi = (x^1 - a^1)$

REFERENCES

There are books without number on the calculus of variations and we can only give a few references. One of the most readable is

I. M. Gelfand and S. V. Fomin, *Calculus of Variations* (translated from the Russian by R. A. Silverman), Prentice-Hall, Englewood Cliffs, N.J., 1963.

See also

O. Bolza, *Lectures on the Calculus of Variations*, Dover Publications, New York, 1946.

C. Caratheodory, *Calculus of Variations and Partial Differential Equations of the First Order* (translated by R. B. Dean and J. J. Brandstatter), Holden-Day, San Francisco, 1965.

As usual, there is a succinct and penetrating analysis in

R. Courant and D. Hilbert, *Methods of Mathematical Physics*, Vol. II, Wiley-Interscience, New York, 1962.

10.2. For a general exposition of complementary variational principles and the role of the Legendre transformation, see

B. Noble and M. J. Sewell, "On Dual Extremum Principles in Applied Mathematics," *J. Inst. Math. Appl.* **9**, 123–193 (1972).

10.3. A very thorough presentation is given in

H. Rund, *The Hamilton–Jacobi Theory in the Calculus of Variations*, Van Nostrand Reinhold, New York, 1966.

A sophisticated treatment of the control problem is given by Falb in Chapter 3 of

R. E. Kalman, P. L. Falb, and M. A. Arbib, *Topics in Mathematical System Theory*, McGraw-Hill, New York, 1969.

See also

M. R. Hestenes, *Calculus of Variations and Optimal Control Theory*, Wiley, New York, 1967.

L. C. Young, *Lectures on the Calculus of Variations and Optimal Control Theory*, W. B. Saunders, Philadelphia, 1969.

10.4. The equivalence of an equation to a Hamilton–Jacobi equation has been exploited by Fleming in proving the existence and uniqueness of generalized solutions. See

W. H. Fleming, "The Cauchy Problem for a Non-linear First Order Partial Differential Equation," *J. Diff. Eqns.* **5**, 515–550 (1969).

10.5. A valuable monograph is

B. B. Baker and E. T. Copson, *The Mathematical Theory of Huygen's Principle*, Oxford University Press, New York, 1939.

Author Index

D

E

F

G

H

I

J

K

Subject Index

D

E

F

G

H

I

J

K

L

M

N

O

P

R

S

T

U

V

W

A CATALOG OF SELECTED

DOVER BOOKS

IN SCIENCE AND MATHEMATICS

A CATALOG OF SELECTED

DOVER BOOKS

IN SCIENCE AND MATHEMATICS

Astronomy

BURNHAM'S CELESTIAL HANDBOOK, Robert Burnham, Jr. Thorough guide to the stars beyond our solar system. Exhaustive treatment. Alphabetical by constellation: Andromeda to Cetus in Vol. 1; Chamaeleon to Orion in Vol. 2; and Pavo to Vulpecula in Vol. 3. Hundreds of illustrations. Index in Vol. 3. 2,000pp. 6⅛ x 9¼.
23567-X, 23568-8, 23673-0 Pa., Three-vol. set $46.85

THE EXTRATERRESTRIAL LIFE DEBATE, 1750–1900, Michael J. Crowe. First detailed, scholarly study in English of the many ideas that developed between 1750 and 1900 regarding the existence of intelligent extraterrestrial life. Examines ideas of Kant, Herschel, Voltaire, Percival Lowell, many other scientists and thinkers. 16 illustrations. 704pp. 5⅜ x 8½. 40675-X Pa. $19.95

A HISTORY OF ASTRONOMY, A. Pannekoek. Well-balanced, carefully reasoned study covers such topics as Ptolemaic theory, work of Copernicus, Kepler, Newton, Eddington's work on stars, much more. Illustrated. References. 521pp. 5⅜ x 8½.
65994-1 Pa. $15.95

AMATEUR ASTRONOMER'S HANDBOOK, J. B. Sidgwick. Timeless, comprehensive coverage of telescopes, mirrors, lenses, mountings, telescope drives, micrometers, spectroscopes, more. 189 illustrations. 576pp. 5⅝ x 8¼. (Available in U.S. only.)
24034-7 Pa. $13.95

STARS AND RELATIVITY, Ya. B. Zel'dovich and I. D. Novikov. Vol. 1 of *Relativistic Astrophysics* by famed Russian scientists. General relativity, properties of matter under astrophysical conditions, stars, and stellar systems. Deep physical insights, clear presentation. 1971 edition. References. 544pp. 5⅝ x 8¼.
69424-0 Pa. $14.95

Chemistry

CHEMICAL MAGIC, Leonard A. Ford. Second Edition, Revised by E. Winston Grundmeier. Over 100 unusual stunts demonstrating cold fire, dust explosions, much more. Text explains scientific principles and stresses safety precautions. 128pp. 5⅜ x 8½. 67628-5 Pa. $5.95

THE DEVELOPMENT OF MODERN CHEMISTRY, Aaron J. Ihde. Authoritative history of chemistry from ancient Greek theory to 20th-century innovation. Covers major chemists and their discoveries. 209 illustrations. 14 tables. Bibliographies. Indices. Appendices. 851pp. 5⅜ x 8½. 64235-6 Pa. $24.95

CATALYSIS IN CHEMISTRY AND ENZYMOLOGY, William P. Jencks. Exceptionally clear coverage of mechanisms for catalysis, forces in aqueous solution, carbonyl- and acyl-group reactions, practical kinetics, more. 864pp. 5⅜ x 8½.
65460-5 Pa. $19.95

THE HISTORICAL BACKGROUND OF CHEMISTRY, Henry M. Leicester. Evolution of ideas, not individual biography. Concentrates on formulation of a coherent set of chemical laws. 260pp. 5⅜ x 8½. 61053-5 Pa. $8.95

A SHORT HISTORY OF CHEMISTRY, J. R. Partington. Classic exposition explores origins of chemistry, alchemy, early medical chemistry, nature of atmosphere, theory of valency, laws and structure of atomic theory, much more. 428pp. 5⅜ x 8½. (Available in U.S. only.) 65977-1 Pa. $12.95

GENERAL CHEMISTRY, Linus Pauling. Revised 3rd edition of classic first-year text by Nobel laureate. Atomic and molecular structure, quantum mechanics, statistical mechanics, thermodynamics correlated with descriptive chemistry. Problems. 992pp. 5⅜ x 8½. 65622-5 Pa. $19.95

Engineering

DE RE METALLICA, Georgius Agricola. The famous Hoover translation of greatest treatise on technological chemistry, engineering, geology, mining of early modern times (1556). All 289 original woodcuts. 638pp. 6¾ x 11. 60006-8 Pa. $21.95

FUNDAMENTALS OF ASTRODYNAMICS, Roger Bate et al. Modern approach developed by U.S. Air Force Academy. Designed as a first course. Problems, exercises. Numerous illustrations. 455pp. 5⅜ x 8½. 60061-0 Pa. $12.95

DYNAMICS OF FLUIDS IN POROUS MEDIA, Jacob Bear. For advanced students of ground water hydrology, soil mechanics and physics, drainage and irrigation engineering and more. 335 illustrations. Exercises, with answers. 784pp. 6⅛ x 9¼. 65675-6 Pa. $24.95

ANALYTICAL MECHANICS OF GEARS, Earle Buckingham. Indispensable reference for modern gear manufacture covers conjugate gear-tooth action, gear-tooth profiles of various gears, many other topics. 263 figures. 102 tables. 546pp. 5⅜ x 8½. 65712-4 Pa. $16.95

MECHANICS, J. P. Den Hartog. A classic introductory text or refresher. Hundreds of applications and design problems illuminate fundamentals of trusses, loaded beams and cables, etc. 334 answered problems. 462pp. 5⅜ x 8½. 60754-2 Pa. $13.95

MECHANICAL VIBRATIONS, J. P. Den Hartog. Classic textbook offers lucid explanations and illustrative models, applying theories of vibrations to a variety of practical industrial engineering problems. Numerous figures. 233 problems, solutions. Appendix. Index. Preface. 436pp. 5⅜ x 8½. 64785-4 Pa. $15.95

STRENGTH OF MATERIALS, J. P. Den Hartog. Full, clear treatment of basic material (tension, torsion, bending, etc.) plus advanced material on engineering methods, applications. 350 answered problems. 323pp. 5⅜ x 8½. 60755-0 Pa. $11.95

Mathematics

HANDBOOK OF MATHEMATICAL FUNCTIONS WITH FORMULAS, GRAPHS, AND MATHEMATICAL TABLES, edited by Milton Abramowitz and Irene A. Stegun. Vast compendium: 29 sets of tables, some to as high as 20 places. 1,046pp. 8 x 10½. 61272-4 Pa. $32.95

FUNCTIONAL ANALYSIS (Second Corrected Edition), George Bachman and Lawrence Narici. Excellent treatment of subject geared toward students with background in linear algebra, advanced calculus, physics and engineering. Text covers introduction to inner-product spaces, normed, metric spaces, and topological spaces; complete orthonormal sets, the Hahn-Banach Theorem and its consequences, and many other related subjects. 1966 ed. 544pp. 6⅛ x 9¼. 40251-7 Pa. $18.95

ASYMPTOTIC EXPANSIONS OF INTEGRALS, Norman Bleistein & Richard A. Handelsman. Best introduction to important field with applications in a variety of scientific disciplines. New preface. Problems. Diagrams. Tables. Bibliography. Index. 448pp. 5⅜ x 8½. 65082-0 Pa. $13.95

FAMOUS PROBLEMS OF GEOMETRY AND HOW TO SOLVE THEM, Benjamin Bold. Squaring the circle, trisecting the angle, duplicating the cube: learn their history, why they are impossible to solve, then solve them yourself. 128pp. 5⅜ x 8½. 24297-8 Pa. $6.95

VECTOR AND TENSOR ANALYSIS WITH APPLICATIONS, A. I. Borisenko and I. E. Tarapov. Concise introduction. Worked-out problems, solutions, exercises. 257pp. 5⅝ x 8¼. 63833-2 Pa. $10.95

THE ABSOLUTE DIFFERENTIAL CALCULUS (CALCULUS OF TENSORS), Tullio Levi-Civita. Great 20th-century mathematician's classic work on material necessary for mathematical grasp of theory of relativity. 452pp. 5⅝ x 8¼. 63401-9 Pa. $14.95

AN INTRODUCTION TO ORDINARY DIFFERENTIAL EQUATIONS, Earl A. Coddington. A thorough and systematic first course in elementary differential equations for undergraduates in mathematics and science, with many exercises and problems (with answers). Index. 304pp. 5⅜ x 8½. 65942-9 Pa. $9.95

FOURIER SERIES AND ORTHOGONAL FUNCTIONS, Harry F. Davis. An incisive text combining theory and practical example to introduce Fourier series, orthogonal functions and applications of the Fourier method to boundary-value problems. 570 exercises. Answers and notes. 416pp. 5⅜ x 8½. 65973-9 Pa. $13.95

COMPUTABILITY AND UNSOLVABILITY, Martin Davis. Classic graduate-level introduction to theory of computability, usually referred to as theory of recurrent functions. New preface and appendix. 288pp. 5⅜ x 8½. 61471-9 Pa. $12.95

ASYMPTOTIC METHODS IN ANALYSIS, N. G. de Bruijn. An inexpensive, comprehensive guide to asymptotic methods–the pioneering work that teaches by explaining worked examples in detail. Index. 224pp. 5⅜ x 8½ 64221-6 Pa. $9.95

THEORY OF MATRICES, Sam Perlis. Outstanding text covering rank, nonsingularity and inverses in connection with the development of canonical matrices under the relation of equivalence, and without the intervention of determinants. Includes exercises. 237pp. 5⅜ x 8½. 66810-X Pa. $8.95

INTRODUCTION TO ANALYSIS, Maxwell Rosenlicht. Unusually clear, accessible coverage of set theory, real number system, metric spaces, continuous functions, Riemann integration, multiple integrals, more. Wide range of problems. Undergraduate level. Bibliography. 254pp. 5⅜ x 8½. 65038-3 Pa. $11.95

MODERN NONLINEAR EQUATIONS, Thomas L. Saaty. Emphasizes practical solution of problems; covers seven types of equations. ". . . a welcome contribution to the existing literature...."–*Math Reviews*. 490pp. 5⅜ x 8½. 64232-1 Pa. $13.95

MATRICES AND LINEAR ALGEBRA, Hans Schneider and George Phillip Barker. Basic textbook covers theory of matrices and its applications to systems of linear equations and related topics such as determinants, eigenvalues and differential equations. Numerous exercises. 432pp. 5⅜ x 8½. 66014-1 Pa. $12.95

MATHEMATICS APPLIED TO CONTINUUM MECHANICS, Lee A. Segel. Analyzes models of fluid flow and solid deformation. For upper-level math, science and engineering students. 608pp. 5⅜ x 8½. 65369-2 Pa. $18.95

ELEMENTS OF REAL ANALYSIS, David A. Sprecher. Classic text covers fundamental concepts, real number system, point sets, functions of a real variable, Fourier series, much more. Over 500 exercises. 352pp. 5⅜ x 8½. 65385-4 Pa. $11.95

AN INTRODUCTION TO MATRICES, SETS AND GROUPS FOR SCIENCE STUDENTS, G. Stephenson. Concise, readable text introduces sets, groups, and most importantly, matrices to undergraduate students of physics, chemistry, and engineering. Problems. 164pp. 5⅜ x 8½. 65077-4 Pa. $7.95

SET THEORY AND LOGIC, Robert R. Stoll. Lucid introduction to unified theory of mathematical concepts. Set theory and logic seen as tools for conceptual understanding of real number system. 496pp. 5⅜ x 8¼. 63829-4 Pa. $14.95

TENSOR CALCULUS, J.L. Synge and A. Schild. Widely used introductory text covers spaces and tensors, basic operations in Riemannian space, non-Riemannian spaces, etc. 324pp. 5⅜ x 8¼. 63612-7 Pa. $13.95

ORDINARY DIFFERENTIAL EQUATIONS, Morris Tenenbaum and Harry Pollard. Exhaustive survey of ordinary differential equations for undergraduates in mathematics, engineering, science. Thorough analysis of theorems. Diagrams. Bibliography. Index. 818pp. 5⅜ x 8½. 64940-7 Pa. $19.95

INTEGRAL EQUATIONS, F. G. Tricomi. Authoritative, well-written treatment of extremely useful mathematical tool with wide applications. Volterra Equations, Fredholm Equations, much more. Advanced undergraduate to graduate level. Exercises. Bibliography. 238pp. 5⅜ x 8½. 64828-1 Pa. $8.95

Math–Decision Theory, Statistics, Probability

ELEMENTARY DECISION THEORY, Herman Chernoff and Lincoln E. Moses. Clear introduction to statistics and statistical theory covers data processing, probability and random variables, testing hypotheses, much more. Exercises. 364pp. 5⅜ x 8½. 65218-1 Pa. $12.95

STATISTICS MANUAL, Edwin L. Crow et al. Comprehensive, practical collection of classical and modern methods prepared by U.S. Naval Ordnance Test Station. Stress on use. Basics of statistics assumed. 288pp. 5⅜ x 8½. 60599-X Pa. $8.95

SOME THEORY OF SAMPLING, William Edwards Deming. Analysis of the problems, theory and design of sampling techniques for social scientists, industrial managers and others who find statistics important at work. 61 tables. 90 figures. xvii +602pp. 5⅜ x 8½. 64684-X Pa. $16.95

STATISTICAL ADJUSTMENT OF DATA, W. Edwards Deming. Introduction to basic concepts of statistics, curve fitting, least squares solution, conditions without parameter, conditions containing parameters. 26 exercises worked out. 271pp. 5⅜ x 8½. 64685-8 Pa. $9.95

LINEAR PROGRAMMING AND ECONOMIC ANALYSIS, Robert Dorfman, Paul A. Samuelson and Robert M. Solow. First comprehensive treatment of linear programming in standard economic analysis. Game theory, modern welfare economics, Leontief input-output, more. 525pp. 5⅜ x 8½. 65491-5 Pa. $17.95

DICTIONARY/OUTLINE OF BASIC STATISTICS, John E. Freund and Frank J. Williams. A clear concise dictionary of over 1,000 statistical terms and an outline of statistical formulas covering probability, nonparametric tests, much more. 208pp. 5⅜ x 8½. 66796-0 Pa. $8.95

PROBABILITY: An Introduction, Samuel Goldberg. Excellent basic text covers set theory, probability theory for finite sample spaces, binomial theorem, much more. 360 problems. Bibliographies. 322pp. 5⅜ x 8½. 65252-1 Pa. $11.95

GAMES AND DECISIONS: Introduction and Critical Survey, R. Duncan Luce and Howard Raiffa. Superb nontechnical introduction to game theory, primarily applied to social sciences. Utility theory, zero-sum games, n-person games, decision-making, much more. Bibliography. 509pp. 5⅜ x 8½. 65943-7 Pa. $14.95

FIFTY CHALLENGING PROBLEMS IN PROBABILITY WITH SOLUTIONS, Frederick Mosteller. Remarkable puzzlers, graded in difficulty, illustrate elementary and advanced aspects of probability. Detailed solutions. 88pp. 5⅜ x 8½. 65355-2 Pa. $5.95

PROBABILITY THEORY: A Concise Course, Y. A. Rozanov. Highly readable, self-contained introduction covers combination of events, dependent events, Bernoulli trials, etc. 148pp. 5⅜ x 8¼. 63544-9 Pa. $8.95

CURVATURE AND HOMOLOGY: Enlarged Edition, Samuel I. Goldberg. Revised edition examines topology of differentiable manifolds; curvature, homology of Riemannian manifolds; compact Lie groups; complex manifolds; curvature, homology of Kaehler manifolds. New Preface. Four new appendixes. 416pp. 5⅜ x 8½.
40207-X Pa. $14.95

TOPOLOGY, John G. Hocking and Gail S. Young. Superb one-year course in classical topology. Topological spaces and functions, point-set topology, much more. Examples and problems. Bibliography. Index. 384pp. 5⅝ x 8¼. 65676-4 Pa. $13.95

LECTURES ON CLASSICAL DIFFERENTIAL GEOMETRY, Second Edition, Dirk J. Struik. Excellent brief introduction covers curves, theory of surfaces, fundamental equations, geometry on a surface, conformal mapping, other topics. Problems. 240pp. 5⅜ x 8½. 65609-8 Pa. $9.95

Math–History of

A SHORT ACCOUNT OF THE HISTORY OF MATHEMATICS, W. W. Rouse Ball. One of clearest, most authoritative surveys from the Egyptians and Phoenicians through 19th-century figures such as Grassman, Galois, Riemann. Fourth edition. 522pp. 5⅜ x 8½. 20630-0 Pa. $13.95

THE HISTORICAL ROOTS OF ELEMENTARY MATHEMATICS, Lucas N. H. Bunt, Phillip S. Jones, and Jack D. Bedient. Fundamental underpinnings of modern arithmetic, algebra, geometry and number systems derived from ancient civilizations. 320pp. 5⅜ x 8½. 25563-8 Pa. $9.95

GAMES, GODS & GAMBLING: A History of Probability and Statistical Ideas, F. N. David. Episodes from the lives of Galileo, Fermat, Pascal, and others illustrate this fascinating account of the roots of mathematics. Features thought-provoking references to classics, archaeology, biography, poetry. 1962 edition. 304pp. 5⅜ x 8½. (Available in U.S. only.) 40023-9 Pa. $9.95

HISTORY OF MATHEMATICS, David E. Smith. Nontechnical survey from ancient Greece and Orient to late 19th century; evolution of arithmetic, geometry, trigonometry, calculating devices, algebra, the calculus. 362 illustrations. 1,355pp. 5⅜ x 8½. Two-vol. set. Vol. I: 20429-4 Pa. $13.95
Vol. II: 20430-8 Pa. $14.95

A CONCISE HISTORY OF MATHEMATICS, Dirk J. Struik. The best brief history of mathematics. Stresses origins and covers every major figure from ancient Near East to 19th century. 41 illustrations. 195pp. 5⅜ x 8½. 60255-9 Pa. $8.95

THE HISTORY OF THE CALCULUS AND ITS CONCEPTUAL DEVELOPMENT, Carl B. Boyer. Origins in antiquity, medieval contributions, work of Newton, Leibniz, rigorous formulation. Treatment is verbal. 346pp. 5⅜ x 8½.
60509-4 Pa. $9.95

Physics

OPTICAL RESONANCE AND TWO-LEVEL ATOMS, L. Allen and J. H. Eberly. Clear, comprehensive introduction to basic principles behind all quantum optical resonance phenomena. 53 illustrations. Preface. Index. 256pp. 5⅜ x 8½. 65533-4 Pa. $10.95

ULTRASONIC ABSORPTION: An Introduction to the Theory of Sound Absorption and Dispersion in Gases, Liquids and Solids, A. B. Bhatia. Standard reference in the field provides a clear, systematically organized introductory review of fundamental concepts for advanced graduate students, research workers. Numerous diagrams. Bibliography. 440pp. 5⅜ x 8½. 64917-2 Pa. $11.95

QUANTUM THEORY, David Bohm. This advanced undergraduate-level text presents the quantum theory in terms of qualitative and imaginative concepts, followed by specific applications worked out in mathematical detail. Preface. Index. 655pp. 5⅜ x 8½. 65969-0 Pa. $16.95

ATOMIC PHYSICS (8th edition), Max Born. Nobel laureate's lucid treatment of kinetic theory of gases, elementary particles, nuclear atom, wave-corpuscles, atomic structure and spectral lines, much more. Over 40 appendices, bibliography. 495pp. 5⅜ x 8½. 65984-4 Pa. $14.95

AN INTRODUCTION TO HAMILTONIAN OPTICS, H. A. Buchdahl. Detailed account of the Hamiltonian treatment of aberration theory in geometrical optics. Many classes of optical systems defined in terms of the symmetries they possess. Problems with detailed solutions. 1970 edition. xv + 360pp. 5⅜ x 8½. 67597-1 Pa. $10.95

THIRTY YEARS THAT SHOOK PHYSICS: The Story of Quantum Theory, George Gamow. Lucid, accessible introduction to influential theory of energy and matter. Careful explanations of Dirac's anti-particles, Bohr's model of the atom, much more. 12 plates. Numerous drawings. 240pp. 5⅜ x 8½. 24895-X Pa. $8.95

ELECTRONIC STRUCTURE AND THE PROPERTIES OF SOLIDS: The Physics of the Chemical Bond, Walter A. Harrison. Innovative text offers basic understanding of the electronic structure of covalent and ionic solids, simple metals, transition metals and their compounds. Problems. 1980 edition. 582pp. 6⅛ x 9¼. 66021-4 Pa. $19.95

HYDRODYNAMIC AND HYDROMAGNETIC STABILITY, S. Chandrasekhar. Lucid examination of the Rayleigh-Benard problem; clear coverage of the theory of instabilities causing convection. 704pp. 5⅜ x 8¼. 64071-X Pa. $17.95

INVESTIGATIONS ON THE THEORY OF THE BROWNIAN MOVEMENT, Albert Einstein. Five papers (1905–8) investigating dynamics of Brownian motion and evolving elementary theory. Notes by R. Fürth. 122pp. 5⅜ x 8½. 60304-0 Pa. $7.95